Springer Texts in Statistics

More information about this series at `http://www.springer.com/series/417`

Rabi Bhattacharya • Lizhen Lin
Victor Patrangenaru

A Course in Mathematical
Statistics and Large Sample Theory

 Springer

Rabi Bhattacharya
Department of Mathematics
The University of Arizona
Tucson, AZ, USA

Victor Patrangenaru
Department of Statistics
Florida State University
Tallahssee, FL, USA

Lizhen Lin
Department of Applied and Computational
Mathematics and Statistics
The University of Notre Dame
Notre Dame, IN, USA

ISSN 1431-875X   ISSN 2197-4136   (electronic)
Springer Texts in Statistics
ISBN 978-1-4939-4030-1   ISBN 978-1-4939-4032-5   (eBook)
DOI 10.1007/978-1-4939-4032-5

Library of Congress Control Number: 2016941742

Printed on acid-free paper

This Springer imprint is published by Springer Nature
The registered company is Springer Science+Business Media LLC New York

Preface

This book is primarily aimed at graduate students of statistics, mathematics, science and engineering who have had an undergraduate course in statistics, an upper division course in analysis and some acquaintance with measure theoretic probability. We have often taught courses based on it with very little emphasis on measure theory. Part I is designed as a one-semester course on basic parametric mathematical statistics whose presentation owes a great deal to the classic texts by Lehmann (1959) and Ferguson (1967). Part II deals with the large sample theory of statistics—parametric and nonparametric—and received a somewhat greater emphasis than Part I. But its main contents may be covered in a semester as well. Part III provides brief accounts of a number of topics of current interest. We expect the book to be used also as a reference by practitioners in other disciplines whose work involves the use of statistical procedures.

The Appendices at the end of the book provide a ready access to a number of standard results, with many proofs. Also, solutions are given to a number of selected exercises from Part I. For Part II, instead, exercises with a certain level of difficulty appear with detailed hints.

Statistics is a very big discipline and is growing fast in even new directions. The present book attempts to provide a rigorous presentation of what we consider to be the core of mathematical statistics.

It took us a long time to write this book which began with a set of class notes used over many years at Indiana University for a two-semester course in theoretical statistics. Its present incarnation, however, is quite different—much expanded and with many changes from the original.

We would like to take this opportunity to thank the NSF for its support over the years which enabled us to spend time on this project. In particular, we would like to acknowledge support from NSF grants DMS 1406872, IIS 1546331 and DMS 1106935. In addition, a UT-Austin start-up grant helped defray some of the expenses in the preparation of the manuscript. The Springer editors dealing with this book project deserve our sincere appreciation for their patience and for their counsel. Finally, we are extremely grateful to Virginia Jones, or Ginny, for her meticulous conversion of often badly handwritten material into beautiful LaTeX; her expertise with the alternate formats acceptable to the publisher and her judgement on which of them to adopt have also helped us greatly.

Tucson, AZ, USA
Austin, TX, USA
Tallahassee, FL, USA
October 2015

Rabi Bhattacharya
Lizhen Lin
Victor Patrangenaru

Contents

Part I
Mathematical Statistics: Basic (Nonasymptotic) Theory

Chapter 1
Introduction

Abstract After describing the general inductive nature of statistical inference, this chapter introduces two popular methods of collecting data: simple random sampling—with or without replacement, and stratified random sampling. A comparison of these methods is made for estimating the mean of a population. The last section is devoted to some simple examples illustrating parametric, nonparametric and semiparametric inference problems.

1.1 What is Statistical Inference?

Mathematics, including probability theory, is mostly concerned with *deductive inference:* derivation of consequences of a given mathematical model. The goal of Statistics, on the other hand, is the inverse problem of *inductive inference,* i.e., to figure out the (probability) model, or at least some features of the model, from some of its consequences (or observations). Since the model cannot be completely recovered from a finite set of observations from it, this inverse problem can only be solved approximately. The present course deals with the problem of finding an optimal approximation or, at least, a "good" approximation.

The underlying probability *model,* usually referred to as the *population* or *population distribution,* is said to be *parametric,* if it can be identified by the value of a finite-dimensional parameter. One then engages in *parametric statistical inference.* If the model can not be so identified, or indexed, by a finite-dimensional parameter, the model and the corresponding inference are said to be *nonparametric.* A special class of the nonparametric models are the so-called *semi-parametric* models, whose structure usually involves a finite-dimensional parameter of interest, as well as a non-parametric family of distributions. If a member of the latter family is specified (as "true"), then the model is parametric with a finite-dimensional unknown parameter.

Often the interest lies only in knowing some features of the model, and not the model itself. This is generally the case with non-parametric and semi-parametric models. But even in the case of a parametric model one may only want to know the value of a subset of the parameter vector. The remaining parameters are then called *nuisance parameters.*

© Springer-Verlag New York 2016

R. Bhattacharya et al., *A Course in Mathematical Statistics and Large Sample Theory*, Springer Texts in Statistics, DOI 10.1007/978-1-4939-4032-5_1

A simple illustration of the role of uncertainty in statistical analysis is provided by the problem of estimation of unknown parameters involved in the description of a deterministic physical law. If there are k unknown parameters, it would only require k observations of appropriate characteristics to compute the parameters, provided the observations are without error. In practice, however, there are always random *measuremental errors,* and no matter how many observations are taken, one can only obtain *estimates* of the parameters, and the estimates are subject to random errors. More commonly, the object of interest itself is random. For example, the life length of an electric bulb, the yearly rainfall in Tucson, an opinion poll. One may be interested in estimating the mean or a population proportion in such cases. Statistics analyzes the random variability inherent in the estimates and makes inferences based on them.

1.2 Sampling Schemes

(a) Simple Random Samples As mentioned above, the underlying probability model in a statistical inference problem is often referred to as the *population.* Sometimes this term is used broadly to also denote an actual finite (but generally large) group of individuals or objects whose characteristics are being explored. It is to be distinguished from the term *sample* which usually comprises a relatively small number of units, chosen from the population (distribution) at *random,* i.e., by a suitable probability mechanism, to ensure that the mathematical laws of probability may be applied for a scientific analysis of the sample.

We do not deal here much with the theory of *sample surveys* which is concerned with the design of efficient and cost effective sampling schemes, taking into account the practical difficulties that arise in the implementation of such schemes. For although this is a subject of great importance in many problems such as the early projection of the results of an election, or constructing cost of living indices, or even conducting an opinion poll, for the most part in this course we consider either (1) a simple random sample of observations which may be taken to be *i.i.d.* or independent and identically distributed (each having the population distribution), or (2) a set of observations provided a priori by nature or some agency, and which may be reasonably assumed to be governed by a certain probability model. We will below briefly consider two types of simple random sampling from a finite population, as well as a stratified random sample from it, in order to estimate the mean of a population characteristic.

From a population of size N a *simple random sample* of size n is drawn *with replacement,* if (1) each of the n observations is drawn such as to give each of the N members of the population the same chance (namely, $1/N$) to be selected, and (2) the n observations are statistically independent. One may think of implementing this by having cards bearing numbers $1, \ldots, N$, identifying the N members of the population, (1) mixing these up thoroughly in a hat, and then (2) picking one from the hat blindfolded, and observing or measuring the characteristic of the individual so chosen. Next, (3) return the chosen card back to the hat, and (4) repeat the procedures (1), (2) and (3), until n cards have been picked. For a *simple random sample without replacement,* the step (3) is skipped. That is, the r-th card is picked from the group of $N - (r - 1)$ cards remaining in the hat after the first

$r - 1$ cards are drawn and put away, giving each of these remaining cards the same chance of selection, namely, $1/(N - r + 1)$ $(r = 1, 2, \ldots, n)$. For sampling without replacement, one requires $n < N$. Various statistical softwares are available, which use sophisticated random number generators for obtaining a random sample.

A simple random sample without replacement is more efficient than a simple random sample of the same size taken with replacement. For the latter allows the wasteful possibility of observing the same individual in the population more than once. However, since the population size is in most cases enormously large compared to the sample (i.e., n/N is extremely small), these two methods are virtually the same. Since it is always easier to analyze independent and identically distributed (i.i.d.) observations, we will assume that the sample observations are independent (i.e., sampling is with replacement), unless stated otherwise. When random data are not provided by sampling, e.g., repeated measurements of length of an object using the same instrument, or amount of yearly rainfall in Tucson over the past 50 years, etc., one may consider these as independent observations or simple random samples with replacement.

(b) Stratified Random Samples A perfectly homogeneous population (distribution), namely, one without variability, can be estimated without error by just one observation. For good statistical inference about a population with a great deal of heterogeneity, on the other hand, one needs a sufficiently large random sample. Thus it would make sense to divide a highly heterogeneous population into a small number of relatively homogeneous subpopulations, or *strata,* and draw randomly and independently from each of these. Such a sampling scheme is called *stratified random sampling.* One popular way to draw a stratified random sample of size n from a population divided into k strata $1, 2, \ldots, k$, with given sizes $N_1, N_2 \ldots, N_k$ (whose sum is N) is to draw k independent simple random samples, one from each stratum and with sample of size $n(N_i/N) = nw_i$, say, drawn from the i-th stratum $(i = 1, \ldots, k)$. For simplicity, we assume nw_i to be an integer. This is the type of stratified sampling most commonly used. When good estimates of the variances v_i $(i = 1, \ldots, k)$ of the k strata are available, the mean of the population is more effectively estimated by letting the size n_i of the sample from the i-th stratum be proportional (approximately) to v_i as well as N_i (See Exercise 1.2(d)). This procedure is used instead only when reasonably reliable past information is available on the variances v_i. On the other hand, stratum sizes are often available from census data, e.g., in the case of opinion polls, surveys of cost of living indices, etc. It may be a little surprising from the intuitive point of view that one can have arbitrarily chosen strata, without any reference to their homogeneity, and still have smaller expected squared error in estimating the mean m than one would have using a simple random sample of the same size, unless the means of all the strata are the same (Exercise 1.2(c)). Two different explanations one may advance to explain this seeming puzzle are the following. First, the statistician is using additional information here, namely, the knowledge of the sizes of sub-populations. Consider drawing a random observation X from the population in two steps. In the first step, choose a stratum at random, with the probability N_i/N of choosing the i-th stratum. At the second step, choose an observation at random from the chosen stratum giving equal chance of selection to every member of the stratum. Repeat these two steps independently n times to have a simple random sample of size n. Then note that the conditional expectation $v = E\{(X - m))^2|\text{given the outcome of the first step}\}$, is larger than $v_i \equiv E\{(X - m_i)^2| \text{ given the outcome of the first}$

step}, if the first step outcome is the choice of the i-th stratum. Taking expectations on both sides of the inequality, one gets the desired result (Exercise 1.3). For the second intuitive explanation, note that differences among means of strata may be attributed to having on the whole less within stratum variability than the overall variability in the whole population. [However, it is easy to construct a stratum with a larger variance than in the population as a whole, for example by putting a stratum together by taking extreme observations from the right and left tails of the population. But that will reduce variability in the remaining population.]

We emphasize again that there are many practical issues that arise (1) in appropriately (and efficiently) *designing statistical experiments* for data collection, and (2) in finding corrective methods for *non-response, missing data, dishonest reporting of data,* etc., even in a properly designed experiment. Unfortunately, we would have little time in this course to deal with these important problems. A good classical reference on some of these matters is the book by W.G. Cochran (1977) entitled *Sampling Techniques* (Wiley).

1.3 Some Simple Examples of Inference

The most common inference problems are those of (1) *estimation* of the model or of some features of it, and (2) *testing hypotheses* concerning the model. Among other somewhat different problems, we mention the one of (3) *classification* of some observation as belonging to one of several populations. One may think of these as special cases of the general *statistical decision problem* as described in the next chapter.

Convention Henceforth we use the term *random sample* to mean a simple random sample with replacement, unless stated otherwise.

Example 1.1. The Normal distribution $N(\mu, \sigma^2)$ provides a parametric model for measurements of lengths, etc., of some physical object. One may seek to estimate the mean length (or the "true" length) μ and the standard deviation σ (a measure of imprecision of the measuring device), based on a random sample of size n. If the sole interest is in μ, and not in σ, then the parameter σ may be viewed as a "nuisance parameter". This is somewhat of a misnomer in the present example, since to judge the precision of any reasonable estimator of μ, one must estimate σ. Note that (i) the sample mean $\overline{X}$ is an unbiased estimator of μ, i.e., $E(\overline{X}) = \mu$; (ii) $E(\overline{X} - \mu)^2 = \sigma^2/n$, and (iii) $s^2 \equiv \sum_{1 \leq i \leq n}(X_i - \overline{X})^2/(n-1)$ is an unbiased estimator of σ^2.

Example 1.2. One wishes to compare two brands of automobile tires by observing the life times $X_1, \ldots, X_m$ of m tires of the first brand and those of n tires of the second brand, namely, $Y_1, \ldots, Y_n$. There is a claim that the second brand has a greater longevity than the first one. One may assume a parametric model with the two distributions Q_1, Q_2 on $[0, \infty)$ (of life lengths of tires of brands 1 and 2) being exponential with means μ_1 and μ_2. The claim may be stated as a hypothesis $H : \mu_2 > \mu_1$. One may test H by comparing the means $\overline{X}$ and $\overline{Y}$ of the two sets of observations, and accept the claim if $\overline{Y}$ is much larger than $\overline{X}$, and reject the claim, or H, otherwise.

One may also consider the nonparametric model comprising all pairs (Q_1, Q_2) of distributions with finite means μ_1, μ_2, respectively, and follow a similar procedure for testing H as above, based on the observed means.

Another interesting nonparametric approach to the problem in this example is the following. Instead of comparing the means of the two distributions of life lengths, it may be more appropriate to compare the two distribution functions F_1 and F_2. The claim may be stated as a test of the hypothesis involving random variables X and Y with distribution function F_1 and F_2, respectively: $H : 1 - F_2(t) \equiv \text{Prob}(Y > t) \geq \text{Prob}(X > t) \equiv 1 - F_1(t)$ for all positive t, with a strict inequality for a least some t. Suppose one only assumes that F_1, F_2 are continuous. Then the underlying model is the set of all pairs of continuous distributions on $[0, \infty)$. This is a nonparametric model. An appropriate and popular test of this hypothesis H is based on first ranking all the $m + n$ observations from 1 through $m + n$ in increasing order of their magnitudes, and then computing the mean rank of the Y observations among these. If this mean is sufficiently large compared to the mean of all the ranks, namely $(m + n + 1)/2$, then one would accept the claim, and otherwise reject it. This test is known as the *Wilcoxon rank test*.

Example 1.3. Consider the linear regression equation

$$Y = \alpha + \beta X + \varepsilon \tag{1.1}$$

where X is a non-stochastic "predictor variable", Y is the "response variable", and ε is Normal $N(0, \sigma_e^2)$. If n independent observations (X_i, Y_i), $i = 1, \ldots, n$, follow this relation, then the underlying model is parametric as it is specified by the parameter $\theta = (\alpha, \beta, \sigma_e^2)$. One is generally interested in estimating the values of α and β, as these allow one to make a prediction of the value of Y based on (a value of) X by the formula: $Y = \hat{\alpha} + \hat{\beta} X$, where $\hat{\alpha}$ and $\hat{\beta}$ are *least squares estimates* of the corresponding parameters derived from the observations. If X is stochastic one assumes (X_i, ε_i), $1 \leq i \leq n$, i.i.d. two-dimensional Normal, with $E(X_i) = \mu_x$, $E\varepsilon_i = 0$, $\text{var}(X_i) = \sigma_x^2$, $\text{var}(\varepsilon_i) = \sigma_\varepsilon^2$, $\text{cov}(X_i, \varepsilon_i) = 0$, and carries out the same analysis.

If one drops the assumption of Normality of ε and, instead, simply assumes that it has finite second moments and $E(\varepsilon) = 0$, then the model is semi-parametric.

1.4 Notes and References

The present book is not concerned with the methodology for obtaining sample data; it simply assumes that the data conform to the hypothesis of randomness. There are many instances where nature provides data automatically (monthly rainfall, daily temperature, number of traffic accidents per week at a city cross section, etc.). In contrast, proper designing of sampling schemes are very important in the sampling of items for testing for defects or other features, public opinion polls, etc. Cochran (1977) is a classic text on the pitfalls involved in taking samples and sample surveys (bias arising from non-response and other forms of missing data, deliberate falsification of data, etc.), and on how to deal with them. It also describes optimal sampling schemes taking both accuracy and costs into consideration. In view of the advent of the computer and the internet, newer methods of taking samples are being developed. Still, some of the basic issues of sampling remain the same.

Exercises for Chap. 1

Ex. 1.1. From a population of size N a simple random sample of size n is drawn *without replacement,* and a real-valued characteristic X measured to yield observations X_j $(j = 1, 2, \ldots, n)$. Show that

(a) the sample mean $\overline{X}$ is an unbiased estimator of the population mean m (i.e., $E(\overline{X}) = m$).
(b) the expected squared error of $\overline{X}$ as an estimator of m, i.e., the variance of $\overline{X}$, is smaller than that of the mean of a simple random sample of the same size n drawn *with replacement,* and
(c) the difference between the expected squared errors of the two estimators is $O(n/N)$, as n/N goes to zero.

Ex. 1.2. Suppose a population of size N is divided into k strata of sizes N_i $(i = 1, 2, \ldots, k)$. Let m_i be the mean of the i-th stratum and v_i its variance $(i = 1, 2, \ldots, k)$. For each i, a simple random sample $\{X_{ij} : j = 1, \ldots, n_i\}$ of size n_i is drawn with replacement from the i-th stratum, and let $\overline{X}_i$ denote the corresponding sample mean. Let $\{x_{ij} : j = 1, \ldots, N_i\}$ be the values of the N_i units in the ith population (stratum), and $\{x_i : i = 1, \ldots, N\}$ be an enumeration of the values of the N units in the population overall. For the following statements, assume $v_i > 0$ and $n_i > 0$ for all i, to avoid trivialities.

(a) Show that (1) $\overline{Y} := \sum w_i \overline{X}_i$ is an unbiased estimator of the population mean m (i.e., $E(\overline{Y}) = m$), where $w_i = N_i/N$, and (2) $E(\overline{Y} - m)^2 = \sum w_i^2 (v_i/n_i)$.
(b) Let v denote the population variance: $v = (\frac{1}{N}) \sum_{1 \leq i \leq N} (x_i - m)^2 = E(X - m)^2$, where x_i is the characteristic of the i-th member of the population, and X is a random observation from the population, i.e., $P(X = x_i) = 1/N$ for all i. Show that $v = \sum_{1 \leq i \leq k} w_i v_i + \sum_{1 \leq i \leq k} w_i (m_i - m)^2$.
(c) Let $\overline{X} = (\frac{1}{n}) \sum_{1 \leq j \leq n} X_j$ be the mean of a simple random sample with replacement. Show that if $n_i = nw_i$ for all i, then $E(\overline{Y} - m)^2 < E(\overline{X} - m)^2$, unless m_i's are all equal, in which case equality holds.
(d) Suppose you know v_i for all i. Show that the optimal choices of n_i (for minimizing the expected squared error of estimation of m by $\overline{Y}$) are

$$n_i = \frac{nw_i \sqrt{v_i}}{\sum_{1 \leq r \leq k} w_r \sqrt{v_r}} \qquad (1 \leq i \leq k), \tag{1.2}$$

assuming the right sides are integers. [Hint: Use calculus and Lagrange multipliers for the constraint $\sum n_i = n$.]

Ex. 1.3. (a) Justify the inequality $V \equiv E((X - m)^2 \mid I) \geq v_I \equiv E((X - m_I)^2 \mid I)$ in Sect. 1.2(b) on stratified random samples, where I is the index of the stratum picked at random, and $m_I = E(X \mid I)$.
(b) Use this to show that $v \equiv E(X - m)^2 \geq \sum_{i=1}^{k} \frac{N_i}{N} v_i$, with equality if and only if m_i's are all the same.

Ex. 1.4 (A Problem of Non-response). In a large random sample of size n from a very big population of size N, there were a sizeable number $n\tilde{}$ of non-responses. To avoid the possibility of a systematic bias in the estimation of the population mean m of a certain variable X, a random sub-sample of size s is drawn from the $n\tilde{}$ non-respondents and their X-values obtained, with additional efforts and costs. Let X_j, $1 \le j \le n - n\tilde{}$ denote the (observed) X-values of the original respondents, and X_j, $n - n\tilde{} + 1 \le j \le n$ those of the (unobserved) respondents. Let Y_j, $1 \le j \le s$, be the sub-sample observations.

Assume that the population comprises two groups—one (of size $N - N\tilde{}$, say,) from which the responses came and the other from which non-responses occurred. The two groups have possibly different means m_R and $m\tilde{}$, respectively, and variances σ_R^2 and $\sigma^{2}\tilde{}$. Let $\overline{X}_R = \sum_{j=1}^{n-n\tilde{}} X_j/(n-n\tilde{})$ denote the mean of the responses, $\overline{Y} = \sum_{j=1}^{s} Y_j/s$, $\overline{X}\tilde{} = \sum_{j=n-n\tilde{}+1}^{n} X_j/n\tilde{}$.

(a) Show that $\overline{Z} = [(n-n\tilde{})\overline{X}_R + n\tilde{}\overline{Y}]/n$ is an unbiased estimate of the population mean m. [Hint: $m = [(N - N\tilde{})m_R + N\tilde{}m\tilde{}]/N$, $E\overline{Z} = E(E[\overline{Z} \mid \xi n\tilde{}; X_j, n - n\tilde{} + 1 \le j \le n]) = E((n - n\tilde{})m_R + n\tilde{}\overline{X}\tilde{})/n.]$

(b) Compute the variance of $\overline{Z}$. [Hint: $E(\overline{Z} - m)^2 = E[\overline{Z} - E(\overline{Z} \mid \mathcal{F}) + E(\overline{Z} \mid \mathcal{F}) - m]^2 = E(E((\overline{Z} - E)\overline{Z} \mid \mathcal{F})^2 \mid \mathcal{F}) + E(E(\overline{Z} \mid \mathcal{F}) - m)^2$, where $\mathcal{F} = \sigma\{n\tilde{}; X_j, n - n\tilde{} + 1 \le j \le n\}.]$

Reference

Cochran, G. W. (1977). *Sampling techniques* (3rd ed.). New York: Wiley.

Chapter 2
Decision Theory

Abstract Statistical inference problems such as estimation and testing come under the purview of decision theory, in which one is given a *parameter space* Θ indexing a family of distributions P_θ of an *observation* (vector) $\mathbf{X}$ ($\theta \in \Theta$), an *action space* $\mathscr{A}$, and a *loss function* $L(\theta, a)$ signifying the loss incurred when θ is the true parameter value and the action taken by the statistician is $\boldsymbol{a}$. The statistician's strategy for action based on the observation $\mathbf{X}$ is a *decision rule* $d(\mathbf{X})$, a function on the space $\mathscr{X}$ of observations into $\mathscr{A}$. A decision rule d is *admissible* if there does not exist any rule d_1 such that $R(\theta, d_1) \equiv E_\theta L(\theta, d_1(\mathbf{X})) \leq R(\theta, d) \equiv E_\theta L(\theta, d(\mathbf{X})) \; \forall \, \theta$, with strict inequality for some θ.

2.1 Decision Rules and Risk Functions

For a substantial part of this course our main interest will be in parametric models. The theory here is well developed and it helps one understand the issues that arise in the analysis of more complex models.

In the following the *observation* $\mathbf{X}$ comprises all that is observed. For example, it may be given as a vector of i.i.d. random variables constituting the sample from a population: $\mathbf{X} = (X_1, \ldots, X_n)$, as in the case of Example 1.1. In Example 1.2, $\mathbf{X} = (X_1, \ldots, X_m, Y_1, \ldots, Y_n)$ where the X_i's are i.i.d. observations from one population while the Y_j's are i.i.d. observations from a second population, the two sets being independent of each other. In Example 1.3, $\mathbf{X} = ((X_1, Y_1), \ldots, (X_n, Y_n))$, obeying (1.1) as specified. Sometimes, with a slight abuse of this terminology, we will also describe the *individual components of* $\mathbf{X}$ *as observations,* when the context is clear.

We begin informally. Let Θ be the parameter space, $\mathbf{X}$ the observation (vector), and $\mathscr{A}$ the set of all possible decisions or actions the statistician can take. A decision rule is a function $d(\cdot)$ of the observation, taking values in $\mathscr{A}$. A loss function $L(\theta, a)$ is prescribed, measuring the loss incurred when an action a is taken while θ is the true parameter value. The risk function $R(\theta, d)$ associated with a decision rule d is defined by

$$R(\theta, d) = E_\theta L(\theta, d(\mathbf{X})), \tag{2.1}$$

where E_θ denotes expectation under θ (i.e., when the true parameter value is θ).

© Springer-Verlag New York 2016

R. Bhattacharya et al., *A Course in Mathematical Statistics and Large Sample Theory*, Springer Texts in Statistics, DOI 10.1007/978-1-4939-4032-5_2

Suppose $(S_1, \mathscr{B}(S_1))$, $(S_2, \mathscr{B}(S_2))$ are two measurable spaces, i.e., $\mathscr{B}(S_i)$ is a sigma-field on S_i ($i = 1, 2$). A function $f : S_1 \to S_2$ is *measurable* if $f^{-1}(B_2) \in \mathscr{B}(S_1)$ $\forall$ $B_2 \in \mathscr{B}(S_2)$. We will often express this by saying f is measurable on $(S_1, \mathscr{B}(S_1))$ into $(S_2, \mathscr{B}(S_2))$.

Definition 2.1. The *parameter space, action space,* and *observation space* are three measurable spaces $(\Theta, \mathscr{B}(\Theta))$, $(\mathscr{A}, \mathscr{B}(\mathscr{A}))$, $(\mathscr{X}, \mathscr{B}(\mathscr{X}))$, respectively. In the case of a metric space S, $\mathscr{B}(S)$ denotes the Borel sigma-field on S. A *loss function* is a real-valued measurable function on $(\Theta \times \mathscr{A}, \mathscr{B}(\Theta) \otimes \mathscr{B}(\mathscr{A}))$ (into $\mathbb{R}, \mathscr{B}(\mathbb{R})$). Here $\otimes$ is used to denote the product sigma-field of its factors. For each parameter value θ, there exists a (specified) probability measure P_θ on a measurable space $(\Omega, \mathscr{F})$, with the corresponding probability space $(\Omega, \mathscr{F}, P_\theta)$. An *observation* $\mathbf{X}$ is a measurable map on $(\Omega, \mathscr{F})$ into the observation space $(\mathscr{X}, \mathscr{B}(\mathscr{X}))$. A (non-randomized) *decision rule* d is a measurable map on the observation space $(\mathscr{X}, \mathscr{B}(\mathscr{X}))$ into the action space $(\mathscr{A}, \mathscr{B}(\mathscr{A}))$. The *risk function* R of a decision rule d is given by (2.1), where E_θ denotes expectation w.r.t. P_θ.

In most problems that we deal with, the spaces Θ, $\mathscr{A}$, and $\mathscr{X}$ are either countable or separable metric spaces. In the case a space is countable one uses the discrete topology on it, so that the sigma-field on it comprises all its subsets.

We consider four examples, the first two concern the problem of *estimation*, while the other two deal with the problem of *testing hypotheses*.

Example 2.1. For the estimation of the mean of a Normal distribution based on a random sample $\mathbf{X} = (X_1, X_2, \ldots, X_n)$ of size n, considered in Example 1.1 in Chap. 1, the parameter space is $\Theta = (-\infty, \infty) \times (0, \infty) = \mathbb{R} \times \mathbb{R}_{++}$, and the action space is $\mathscr{A} = \mathbb{R}$. Here the observation space is $\mathscr{X} = \mathbb{R}^n$. The most commonly used loss function is *squared error loss*

$$L(\boldsymbol{\theta}, \boldsymbol{a}) = |\mu - a|^2, \qquad (\boldsymbol{\theta} = (\mu, \sigma^2)). \tag{2.2}$$

One may take $\Omega = \mathbb{R}^\infty$, the space of all infinite sequences of reals $(x_1, x_2, \ldots)$. Then the sigma-field $\mathscr{F}$ is the Kolmogorov product sigma-field $\mathscr{B}(\mathbb{R}^\infty)$, and P_θ is the product probability measure with all factors being the Normal distribution $N(\mu, \sigma^2)$. That is, P_θ is the distribution of a sequence of i.i.d. $N(\mu, \sigma^2)$ random variables. The observation $\mathbf{X}$ is the projection map on Ω into its first n coordinates. Alternatively, one may take $\Omega = \mathscr{X} = \mathbb{R}^n$, $\mathbf{X}$ as the *identity map*: $\mathbf{X}(\omega) = \omega = (x_1, \ldots, x_n)$ and P_θ as the product probability measure with all n factors the same, namely, $N(\mu, \sigma^2)$.

For the decision rule $d(\mathbf{x}) = \overline{x} \equiv (x_1 + x_2 + \cdots + x_n)/n$ ($\mathbf{x} = (x_1, x_2, \ldots, x_n) \in \mathscr{X} = \mathbb{R}^n$), $d(\mathbf{X}) = \overline{X}$, and the risk function is given by

$$R(\boldsymbol{\theta}, d) = E_\theta(\mu - \overline{X})^2 = \frac{\sigma^2}{n}, \qquad \theta = (\mu, \sigma^2) \in \Theta. \tag{2.3}$$

If, instead of μ, the parameter of interest is σ^2, then the action space is $\mathbb{R}_{++}$. One may use the decision rule $d(\mathbf{x}) = \sum_{1 \leq i \leq n}(x_i - \overline{x})^2/(n-1)$. Then $d(\mathbf{X}) = s^2$, as defined in Example 1.1 in Chap. 1. One may let $\mathscr{A} = [0, \infty)$, and show (Exercise 2.1) that with squared error loss function $L(\boldsymbol{\theta}, a) = (\sigma^2 - a)^2$, the risk function of this decision rule d is given by

$$R(\boldsymbol{\theta}, d) = E_\theta(\sigma^2 - s^2)^2 = 2\sigma^4/(n-1). \tag{2.4}$$

Finally, suppose one wishes to estimate the vector parameter $\boldsymbol{\theta}$, then the action space is $\mathbb{R} \times [0, \infty) = \mathbb{R} \times \mathbb{R}_+$. If the loss function is squared error (in Euclidean distance),

$$L(\boldsymbol{\theta}, \mathbf{a}) = |(\mu, \sigma^2) - \mathbf{a}|^2 \qquad \boldsymbol{\theta} = (\mu, \sigma^2), \quad \mathbf{a} = (a_1, a_2) \in \mathbb{R} \times \mathbb{R}_+, \tag{2.5}$$

and one uses the *estimator* $d(\mathbf{X}) = (\overline{X}, s^2)$, then the risk function is given by the sum of the right sides of (2.3) and (2.4),

$$R(\boldsymbol{\theta}, d) = \frac{\sigma^2}{n} + \frac{2\sigma^4}{n-1}, \qquad \boldsymbol{\theta} = (\mu, \sigma^2). \tag{2.6}$$

Since the units of μ and σ^2 are different, it may be more appropriate to define the loss function as $(\mu - a_1)^2 + (\sigma - \sqrt{a_2})^2$. Then the risk function of d above is $c(n)\sigma^2$, where $c(n)$ only depends on n and is of the order $O(n^{-1})$, as $n \to \infty$ (Exercise 2.1).

For computations related to this example, the following proposition is useful. It is also an important property of samples from a Normal distribution. We will write $U \overset{d}{\sim} G$ to indicate that the random variable U has distribution G. Also $U \overset{d}{=} V$ means U has the same distribution as V.

Proposition 2.1. *Let* $X_1, X_2, \ldots, X_n$ *be i.i.d.* $N(\mu, \sigma^2)$. *Then* $\overline{X}$ *and* s^2 *are independent, with* $\overline{X} \overset{d}{\sim} N(\mu, \frac{\sigma^2}{n})$ *and* $(n-1)s^2/\sigma^2 \overset{d}{=} \sum_{i=2}^{n} Y_i^2$, *where* $Y_2, \ldots, Y_n$ *are i.i.d. standard Normal* $N(0, 1)$.

Proof. First let $\mu = 0$, $\sigma^2 = 1$. Write $\mathbf{X}$ as the column vector $\mathbf{X} = (X_1, \ldots, X_n)'$. Let O be an $n \times n$ orthogonal matrix whose first row is $\left(\frac{1}{\sqrt{n}}, \frac{1}{\sqrt{n}}, \ldots, \frac{1}{\sqrt{n}}\right)$ [For example, take the $(k+1)$-th row of O as $\underbrace{(a, a, \ldots, a}_{k \text{ elements}}, -ka, 0, 0, \ldots, 0)$, with $a = (k(k+1))^{-\frac{1}{2}}$ $(k = 1, 2, \ldots, n-1)$]. Define $\mathbf{Y} = O\mathbf{X}$. Then the probability density function of $\mathbf{Y} = (Y_1, Y_2, \ldots, Y_n)'$ is the same as that of $\mathbf{X}$, i.e., $Y_1, Y_2, \ldots, Y_n$ are i.i.d. $N(0, 1)$ [The Jacobian of the transformation has determinant $|O'| = \pm 1$; $|\mathbf{X}|^2 = |\mathbf{Y}|^2$]. But $Y_1 = \sqrt{n}\,\overline{X}$, and $Y_2^2 + Y_3^2 + \cdots + Y_n^2 \equiv \sum_{i=1}^{n} Y_i^2 - Y_1^2 = \sum_{i=1}^{n} X_i^2 - n\overline{X}^2 \equiv (n-1)s^2$ are independent.

To complete the proof in the general case $X_i \overset{d}{\sim} N(\mu, \sigma^2)$, write $Z_i = (X_i - \mu)/\sigma$, and apply the above argument to $\mathbf{Z} = (Z_1, Z_2, \ldots, Z_n)'$. $\qquad\square$

Example 2.2. One of the most common statistical problems is that of estimation of the proportion θ of members of a population possessing some characteristic. For example, θ may be the proportion of adult Americans who are supportive of a pending gun control legislation, or it may be the proportion of defective items among all items of some kind manufactured by a company. A random sample $\mathbf{X} = (X_1, \ldots, X_n)$ is observed, where X_i is 1 or 0 according as the i-th observation possesses the characteristic ("support gun control", "defective") or it does not. The most common estimate is the *sample proportion* $d_1(\mathbf{X}) = (X_1 + \cdots + X_n)/n$ of those in the sample possessing the characteristic. Its risk function is (Exercise 2.2)

$$R(\theta, d_1(\mathbf{X})) = \theta(1 - \theta)/n, \qquad \theta \in \Theta = [0, 1]. \tag{2.7}$$

Here the observation space is $\mathscr{X} = \{0,1\}^n$, and one may take $\Omega = \mathscr{X}$, and P_θ as the product probability: $P_\theta(\{\mathbf{x}\}) = \theta^{\sum x_i}(1 - \theta)^{\sum(1-x_i)}$, for all $\mathbf{x} = (x_1, x_2, \ldots, x_n) \in \{0,1\}^n$.

If, instead of the sample proportion, one uses the (seemingly bad) estimator $d_2(\mathbf{X}) \equiv 1/3$, then the risk function is

$$R(\theta, d_2(\mathbf{X})) = (\theta - 1/3)^2, \quad \text{which is 0 for } \theta = 1/3. \tag{2.8}$$

No other estimator has a risk as low as that of $d_2(\cdot)$ at the value $\theta = 1/3$. Although for values of θ a bit away from $1/3$, this estimator has a large risk, one can not get an estimator as good as d_2 everywhere on the parameter space! Thus one can not hope for a decision rule which is better (or, at least, as good as) every other estimator uniformly over the entire parameter space. We will introduce later less stringent notions of optimality.

We next consider the problem of testing hypotheses about the underlying model, which may be expressed in terms of a *null hypothesis* $H_0 : \theta \in \Theta_0$ where Θ_0 is a subset of Θ, and an *alternative hypothesis* $H_1 : \theta \in \Theta_1 = \Theta \backslash \Theta_0$. Here the action space is $\mathscr{A} = \{a_0, a_1\}$, where a_i accepts the hypothesis H_i $(i = 0, 1)$. The most commonly used loss function in this case is the $0 - 1$ *loss function*,

$$L(\theta, a_i) = 1 - \delta_{ij} \quad \text{if } \theta \in \Theta_j, \qquad (i, j \in \{0,1\}), \tag{2.9}$$

where δ_{ij} is Kronecker's delta, which equals 1 or 0 according as $i = j$ or $i \neq j$. Thus the loss incurred is 0 if a_i is the correct decision (to accept H_i), and 1 if it is the wrong decision.

Definition 2.2. The error of accepting the alternative hypothesis H_1 when the null hypothesis H_0 is correct, is called a *Type I Error*. Its probability is $P_\theta(d(\mathbf{X}) = a_1) = P_\theta(\text{Accept the alternative hypothesis})$, when $\theta \in \Theta_0$. The error of accepting the null hypothesis H_0 when the alternative hypothesis H_1 is correct is called a *Type II Error*. Its probability is $P_\theta(d(\mathbf{X}) = a_0) = P_\theta(\text{Accept the null hypothesis})$, when $\theta \in \Theta_1$.

Example 2.3. In quality control experiments, an inspector is often confronted with the problem of deciding whether to "pass" or "fail" a product. For example, a product or a large batch of it may be considered good only if no more than 5% of the items are defective. The inspector's decision is to be based on a random sample $\mathbf{X} = (X_1, X_2, \ldots, X_n)$, where $X_i = 0$ or 1 according as the i-th item picked is defective or not. The parameter space Θ here is the set of all possible proportions θ of defective items, conveniently taken as the unit interval $[0, 1]$. The spaces $\mathscr{X}$, Ω, Θ, and the probability distribution P_θ are as in Example 2.2. Let $\Theta_0 = [0, 0.05]$, while $\Theta_1 = (0.05, 1.0]$. The action space is $\{a_0, a_1\}$, where the action a_0 means the product is passed and a_1 means it is failed. Suppose one uses the decision rule: $d(\mathbf{x}) = a_0$ if the sample proportion $\hat{p}$ of defectives is 0.03 or less, and $d(\mathbf{x}) = a_1$ otherwise. Under the loss function (2.9), the risk function of the decision rule $d(\cdot)$ is (Exercise 2.3)

$$R(\theta,d)=E_\theta L(\theta,d(\mathbf{X})) = \begin{cases} P_\theta(\hat{p}>0.03) = \sum\limits_{r>0.03n}^{n} C_r\theta^r(1-\theta)^{n-r} \text{ for } 0 < \theta \le 0.05, \\[2em] P_\theta(\hat{p} \le 0.03) = \sum\limits_{r\le 0.03n}^{n} C_r\theta^r(1-\theta)^{n-r} \text{ for } 0.05 < \theta \le 1. \end{cases}$$

$$(2.10)$$

The top probability in (2.10) is the probability of a Type I Error, while the bottom probability is that of a Type II Error.

Example 2.4. Consider Example 1.2 of Chap. 1, assuming that the lifetime distributions are exponential with means μ_1 and μ_2. We let H_1 denote the claim: $\mu_2 > \mu_1$, so that $\Theta_1 = \{(\mu_1,\mu_2) \in (0,\infty)^2 : \mu_2 > \mu_1\}$, while $\Theta_0 = \{(\mu_1,\mu_2) \in (0,\infty)^2 : \mu_2 \le \mu_1\}$. The observation space is $\mathscr{X} = \{(\mathbf{x},\mathbf{y}) : \mathbf{x} = (x_1,\ldots,x_m) \in [0,\infty)^m, \mathbf{y} = (y_1,\ldots,y_n) \in [0,\infty)^n\}$. Let us denote the observation vector by $\mathbf{X} = (\{X_j : j = 1,\ldots,m\}, \{Y_j : j = 1,\ldots,n\})$. Write $\bar{x} = \sum x_j/m$, $\bar{y} = \sum y_j/n$, and $\overline{X}, \overline{Y}$ for the means of the corresponding sample observations.

Suppose one uses the decision rule $d(\mathbf{x},\mathbf{y}) = a_1$ if $\bar{y} > (1 + c)\bar{x}$ for some constant $c > 0$, and $d(\mathbf{x},\mathbf{y}) = a_0$ if $\bar{y} \le (1 + c)\bar{x}$. Then the risk function of d is $E_\theta L(\theta,d(\mathbf{X}))$, which equals $E_\theta(\mathbf{1}_{[\overline{Y}\le(1+c)\overline{X})]})$ for θ in Θ_1, and $E_\theta(\mathbf{1}_{[\overline{Y}>(1+c)\overline{X}]})$ for θ in Θ_0. That is,

$$R(\theta,d) = \begin{cases} P_\theta(\overline{Y} \le (1+c)\overline{X}) \text{ if } \theta \in \Theta_1 \text{ (i.e., if } \mu_2 > \mu_1), \text{ [Type II Error Probability]} \\ P_\theta(\overline{Y} > (1+c)\overline{X}) \text{ if } \theta \in \Theta_0 \text{ (i.e., if } \mu_2 \le \mu_1), \text{ [Type I Error Probability]}. \end{cases}$$

$$(2.11)$$

2.2 Randomized Decision Rules, Admissibility

Sometimes for a given observation one may wish to assign probabilities among different (sets of) actions, instead of choosing a single action. We will later see that such allowance for randomization among several actions is particularly important in the case of hypothesis testing. Thus a general decision rule δ (allowing randomization in the action space) may be defined as a measurable map from the observation space $(\mathscr{X},\mathscr{B}(\mathscr{X}))$ into the space $\mathscr{P}(S)$ of all probability measures on the action space $(\mathscr{A},\mathscr{B}(\mathscr{A}))$. The sigma-field on $\mathscr{P}(S)$ is generally taken to be the Borel sigma-field under the topology of weak convergence (assuming $\mathscr{A}$ is a metric space).

To avoid confusing notation, we will continue to denote a decision rule by the symbol d, with or without subscripts or superscripts, whether the rule is a randomized one or not.

We have seen in Example 2.2 that no estimator of the population proportion θ, based on a random sample of size n, has the smallest risk function on the whole parameter space. Indeed, the smallest risk at every point θ is zero. This is true not just for the squared error loss function, but for any loss function L such that $L(\theta,a) > 0$ for all $\theta \ne a$, and $= 0$ for $\theta = a$ (Exercise 2.2). We must then relax our requirement for optimality. One reasonable requirement would seem to be the following. From now on we will assume that the loss function L is given, so that all comparisons among decision rules are based on the risk function $R(\theta,d) = E_\theta L(\theta,d(\mathbf{X}))$ $(\theta \in \Theta)$.

Definition 2.3. A decision rule d is said to be *inadmissible* if there exists a decision rule d_1 such that

$$R(\theta, d_1) \leq R(\theta, d) \qquad \text{for all } \theta \in \Theta, \tag{2.12}$$

with strict inequality for a least one θ. A decision rule which is not inadmissible is said to be *admissible*.

It turns out, unfortunately, that although it is easy to establish the admissibility of such estimators as $d_2\,(\mathbf{X}) \equiv 1/3$ in Example 2.2 (Exercise 2.4), it is not so easy to prove the admissibility of time honored estimators such as the sample proportion d_1! We will later introduce other notions of optimality, or restrictions which will rule out frivolous estimators such as d_2.

2.3 Notes and References

For basic notions of loss and risk functions in testing (probabilities of Type 1 and Type 2 errors) and estimation (expected squared error) and admissibility one may refer to Bickel and Doksum (2001), Sect. 1.3, or Ferguson (1967), Sects. 1.3, 2.1.

Exercises for Chap. 2

Ex. 2.1. In Example 2.1, prove (2.4) and (2.6). Also with the loss function as $L(\theta,\mathbf{a}) = (\mu - a_1)^2 + (\sigma - \sqrt{a_2}\,)^2$, instead of (2.5), compute the risk function of the estimator $d(\mathbf{X}) = (\overline{X}, s^2)$.

Ex. 2.2. (a) Prove (2.7).
(b) In Example 2.2, with the loss function as specified, show that $\min_d\{R(\theta, d(\mathbf{X}))\} = 0 \;\forall\, \theta$, where the minimum is taken over all decision rules.
(c) Justify the corresponding statement for any loss function for which $L(\theta, a) > 0$ if $\theta \neq a$, $L(\theta, \theta) = 0$.

Ex. 2.3. Prove (2.10) in Example 2.3.

Ex. 2.4. Show that (a) the estimator $d_2(\mathbf{X}) \equiv \frac{1}{3}$ in Example 2.2 is admissible, and

(b) the estimator $d_3(\mathbf{X}) = \frac{X_1 + \cdots + X_{n-1}}{n-1}$ is inadmissible ($n > 1$).
(c) Find the range of values of θ over which $R(\theta, d_2) \leq R(\theta, d_3)$, and show that as n increases this interval converges to the singleton $\{\frac{1}{3}\}$.

Ex. 2.5. Graph the two error probabilities in (2.10) in the case $n = 20$.

Ex. 2.6. Express the two error probabilities in (2.11) in terms of the parameters (and the sample sizes m, n).

References

Bickel, P. J., & Doksum, K. (2001). *Mathematical statistics* (2nd ed.). Englewood
Cliffs, NJ: Prentice Hall.

Ferguson, T. (1967). *Mathematical statistics: A decision theoretic approach.*
Boston: Academic.

Chapter 3
Introduction to General Methods of Estimation

Abstract In this chapter we will introduce some important methods for finding reasonable estimators of parameters of a population. Of these, *Bayes estimators* will be treated in some detail, especially to illustrate various admissible estimators. The *maximum likelihood estimators* (MLE) and the *method of moments* are discussed briefly. Although the MLE is generally regarded as the most important method of estimation, its asymptotic optimality properties are best described in Part II on large sample theory.

3.1 The Maximum Likelihood Estimator

Perhaps the most important estimator is statistics is the maximum likelihood estimator, originally used by Gauss for estimating the parameters of a Normal distribution $N(\mu, \sigma^2)$ in connection with his astronomical observations. The method in its general form is due to the British statistician R.A. Fisher who introduced it in the early part of the twentieth century. Fisher is widely regarded as the father of modern statistical theory.

Definition 3.1. Let $f(\mathbf{x}|\theta)$ denote the density of the distribution of the observation (vector) $\mathbf{X}$, with respect to some sigma-finite measure ν. For example, ν may be Lebesgue measure on $\mathscr{X} = R^n$, in which case f is the *classical density* of $\mathbf{X}$, or ν may be the counting measure on a countable observation space $\mathscr{X}$ such as $\{0, 1\}^n$, in which case f is called a *probability mass function (pmf)*. The function $\ell(\theta) = f(\mathbf{X}|\theta)$, $\theta \in \Theta$, is called the *likelihood function*. The *maximum likelihood estimator* (**MLE**) of θ is the value of θ where the likelihood function attains its maximum, assuming the existence of a unique such point in Θ.

One may argue (and we will make this precise in connection with the notion of *sufficiency* later) that all the statistical information about the true parameter value in the (sample) observation $\mathbf{X}$ is contained in the likelihood function $\ell(\theta)$. The magnificent intuition of Gauss and Fisher is that the proper estimator of the true parameter value is the one that maximizes the likelihood—it is the value of the parameter which makes the given observation $\mathbf{X}$ as the most likely to occur [You may think of this as a rather vain point of view!]

© Springer-Verlag New York 2016

R. Bhattacharya et al., *A Course in Mathematical Statistics and Large Sample Theory*, Springer Texts in Statistics, DOI 10.1007/978-1-4939-4032-5_3

Example 3.1 ($N(\mu, \sigma^2)$). In the Normal example, Example 1.1, the density is with respect to Lebesgue measure ν on R^n. The *likelihood function* is given by (with $\theta = (\mu, \sigma^2) \in \Theta = R \times (0, \infty)$)

$$\ell(\theta) = (2\pi\sigma^2)^{-n/2} \exp\left\{-\frac{1}{2}\sum(X_i - \mu)^2/\sigma^2\right\}$$

$$= (2\pi\sigma^2)^{-n/2} \exp\left\{-\left(\frac{1}{2\sigma^2}\right)\left[\sum(X_i - \overline{X})^2 + n(\overline{X} - \mu)^2\right]\right\}. \quad (3.1)$$

The maximizer of ℓ maximizes the strictly increasing function of ℓ given by the *log-likelihood function*

$$\ln \ell(\mu, \sigma^2) = -\frac{n}{2}\ln(2\pi) - \frac{n}{2}\ln(\sigma^2) - \left(\frac{1}{2\sigma^2}\right)\left[\sum(X_i - \overline{X})^2 + n(\overline{X} - \mu)^2\right]. \quad (3.2)$$

Setting the derivative of this with respect to μ to zero yields the MLE $\widehat{\mu}$ of μ:

$$0 = \frac{\partial}{\partial\mu}\ln\ell = \left(\frac{n}{\sigma^2}\right)(\overline{X} - \mu), \qquad \text{or } \widehat{\mu} = \overline{X}. \quad (3.3)$$

Also, differentiation with respect to σ^2 (using the solution for μ) yields

$$0 = \left(-\frac{n}{2\sigma^2}\right) + \left(\frac{1}{2\sigma^4}\right)\sum(X_i - \overline{X})^2, \qquad \text{or } \widehat{\sigma}^2 = \sum\frac{(X_i - \overline{X})^2}{n}. \quad (3.4)$$

Thus the MLE of $\theta = (\mu, \sigma^2)$ is given by $(\overline{X}, \sum(X_i - \overline{X})^2/n)$. One may easily check that the matrix of second derivatives of $\ell(\theta)$ is negative-definite (Exercise 3.1), so that the above solution is the unique maximizer of the likelihood function. Equations (3.3), (3.4) yielding this solution are called *likelihood equations*.

Example 3.2 (Bernoulli (θ)). Consider Example 2.2 of Chap. 2.1, on estimating the proportion θ, based on a random sample of n i.i.d. Bernoulli random variables with probabilities θ and $1 - \theta$ for values 1 and 0, respectively. Here we take $\Theta = (0, 1)$ and consider the density $f(\mathbf{x}|\theta)$ with respect to the counting measure ν on $\{0, 1\}^n$. The likelihood and the log-likelihood functions are

$$\ell(\theta) = \theta^{\sum X_i}(1 - \theta)^{\sum(1 - X_i)}, \quad \ln\ell(\theta) = \left(\sum X_i\right)\ln\theta + \left(n - \sum X_i\right)\ln(1 - \theta). \quad (3.5)$$

The likelihood equation is then

$$0 = \frac{\partial\ln\ell(\theta)}{\partial\theta} = \frac{1}{\theta}\sum X_i - \frac{n - \sum X_i}{1 - \theta}, \quad \text{or } (1 - \theta)\sum X_i - \theta(n - \sum X_i) = 0,$$

$$\text{or } \sum X_i - n\theta = 0, \quad (3.6)$$

which has the solution $\widehat{\theta} = \sum X_i/n$. Hence the MLE of the population proportion θ is the sample proportion of 1's. You may check that the second derivative of the log-likelihood function is negative (Exercise 3.1).

Example 3.3 ($U(0,\theta)$). Let $\mathbf{X} = (X_1, \ldots, X_n)$ be a random sample from the uniform distribution on the interval $(0, \theta]$ where $\theta \in \Theta = (0, \infty)$. Here the density $f(\mathbf{x}|\theta) = \prod_{1 \leq i \leq n}(1/\theta)\mathbf{1}\{0 < x_i \leq \theta\}$, $\mathbf{x} = (x_1, \ldots, x_n)$, is with respect to Lebesgue measure ν on $(0, \infty)^n$. The likelihood function is

$$\ell(\theta) = \frac{1}{\theta^n}\mathbf{1}_{\{X_i \leq \theta,\, 1 \leq i \leq n\}}, \tag{3.7}$$

$$\text{or } \ell(\theta) = \theta^{-n}\mathbf{1}\{\theta \geq M_n \equiv \max(X_1, \ldots, X_n)\}, \quad \theta \in (0, \infty).$$

Here $\mathbf{1}\{\ldots\}$ denotes the indicator function of the set $\{\ldots\}$. Since the likelihood function has the value zero for $\theta < M_n$, and decreases monotonically as θ increases from M_n to infinity, its maximum is attained at $\theta = M_n$. Thus the MLE of θ is $M_n = \max(X_i : i = 1, \ldots, n)$. Note that the maximum here occurs at a point where $\ell(\theta)$ is discontinuous (Exercise 3.2).

3.2 Method of Moments

Classically, in order to estimate an r-dimensional parameter $\theta = (\theta_1, \ldots, \theta_r)$ by the method of moments, one equates the first r population moments with the corresponding r sample moments. More generally, one may equate r functionally independent population averages, say, $E_\theta g_j(X_1)$, $1 \leq j \leq r$, with the corresponding sample averages $(1/n)\sum_{1 \leq i \leq n} g_j(X_i)$, $1 \leq j \leq r$, to solve for θ_j, $j = 1, \ldots, r$. Although the MLE can be generally shown to be superior to these estimates, the latter are sometimes a lot easier to compute and, may be used as initial solutions in an iterative process to numerically compute the MLE (Exercise 3.11). In Examples 3.1, 3.2, however, the classical method of moments, of equating as many sample moments with population moments as there are parameters, yield the same estimates as the MLE.

Example 3.4 (Gamma $\mathscr{G}(\alpha, \beta)$). Suppose the observation vector $\mathbf{X}$ comprises n i.i.d. random variables X_i $(i = 1, \ldots, n)$ each with the *gamma density* (with respect to Lebesgue measure on $(0, \infty)$) given by

$$\gamma(x; \alpha, \beta) = (\alpha^\beta \Gamma(\beta))^{-1} x^{\beta-1} e^{-x/\alpha}, \qquad x > 0, \ (\alpha > 0, \beta > 0). \tag{3.8}$$

The first two moments of the gamma distribution are easily computed as (Exercise 3.3)

$$E(X_1) = \alpha\beta, \qquad E(X_1^2) = \alpha^2\beta(\beta + 1), \tag{3.9}$$

which, when equated with the corresponding sample moments $m_1 = \sum X_i/n$ and $m_2 = \sum X_i^2/n$, yield the solutions

$$\widetilde{\alpha} = \frac{(m_2 - m_1^2)}{m_1}, \qquad \widetilde{\beta} = \frac{m_1}{\alpha} = \frac{m_1^2}{(m_2 - m_1^2)}, \tag{3.10}$$

that is, the right sides of the two equations in (3.10) are the method-of-moment estimates of α and β.

3.3 Bayes Rules and Bayes Estimators

Sometimes the statistician has prior information about the true parameter value. This information does not generally specify a particular value, but, instead, is given as a probability distribution τ on Θ, called *the prior distribution,* or simply the *prior.* This distribution gives more weight to those values which, according to the statistician, are more likely to be true compared to other values. Following the same line of thought, if the statistician does not have any particular preference for any value, then a prior to use may be the uniform distribution (provided the parameter space admits one). We will not here get into the philosophical viewpoint of a *Bayesian.* Indeed, in this course our approach may appear closer to that of the so-called *frequentist,* although that is not entirely intentional. Hopefully this apparent bias will be corrected in a follow up course. We remind the reader once again that we fix an arbitrarily chosen loss function $L(\theta, a)$, which may satisfy certain general properties to be specified as needed.

Given a prior τ, the statistician computes the *Bayes risk* $r(\tau, d)$ of a decision rule d given by

$$r(\tau, d) = \int_{\Theta} R(\theta, d) d\tau(\theta). \tag{3.11}$$

Definition 3.2. A *Bayes rule* d_0 is a decision rule, if one exists, which has the smallest Bayes risk among all decision rules:

$$r(\tau, d_0) = \inf r(\tau, d), \tag{3.12}$$

where the infimum is over all decision rules d. In the case of an estimation problem, a Bayes rule is called a *Bayes estimator.*

Observe that the Bayes risk of a decision rule d (with respect to a prior τ) may be expressed as

$$r(\tau, d) = \int_{\Theta} R(\theta, d) d\tau(\theta) = \int_{\Theta} \left\{ \int_{\mathcal{X}} L(\theta, d(\mathbf{x})) dP_\theta(\mathbf{x}) \right\} d\tau(\theta). \tag{3.13}$$

Here P_θ is defined canonically as the distribution of the observation $\mathbf{X}$ on $\mathcal{X}$ when θ is the true parameter value. Suppose we define on (a possibly enlarged probability space $(\Omega, \mathcal{F}, P)$) a random variable ϑ whose distribution is τ, and a random variable X (with values in $\mathcal{X}$) *whose conditional distribution, given* $\vartheta = \theta$, *is* P_θ. Indeed, one can define this space canonically as $\Omega = \mathcal{X} \times \Theta$ with the product sigma-field, and with the probability measure P specified by

$$P(C \times D) = \int_{D} P_\theta(C) d\tau(\theta) \quad \text{for all } C \in \mathcal{B}(\mathcal{X}) \text{ and } D \in \mathcal{B}(\Theta). \tag{3.14}$$

Denoting expectation with respect to P as E, (3.13) then may be expressed as

$$r(\tau, d_0) = EL(\vartheta, d(\mathbf{X})). \tag{3.15}$$

Note that (3.13) computes this expectation by (1) first taking the conditional expectation given $\vartheta = \theta$ (the inner integral in (3.13)) and (2) then integrating this conditional expectation with respect to the (marginal) distribution of ϑ (i.e., with respect to the prior τ). That is, (3.13) says

$$r(\tau, d) = EL(\vartheta, d(\mathbf{X})) = E[E(L(\vartheta, d(\mathbf{X})) \mid \vartheta)]. \tag{3.16}$$

One may reverse the order of integration, by first taking the conditional expectation, given $\mathbf{X}$, and then integrating this conditional expectation over the (marginal) distribution of $\mathbf{X}$:

$$r(\tau, d) = EL(\vartheta, d(\mathbf{X})) = E[E(L(\vartheta, d(\mathbf{X})) \mid \mathbf{X})]. \tag{3.17}$$

The conditional expectation in (3.17) is obtained by integrating $L(\theta, d(\mathbf{x}))$ with respect to the conditional distribution of ϑ, given $\mathbf{X} = \mathbf{x}$. This is the so-called *posterior distribution of ϑ*, denoted $d\tau(\theta|\mathbf{x})$. Next this conditional expectation is integrated with respect to the *marginal distribution G* of $\mathbf{X}$, which has the density (with respect to ν) given by

$$g(\mathbf{x}) = \int_{\Theta} f(\mathbf{x}|\theta)d\tau(\theta), \qquad [dG(\mathbf{x}) = g(\mathbf{x})d\nu(\mathbf{x})]. \tag{3.18}$$

Hence

$$r(\tau, d) = \int \left[\int L(\theta, d(\mathbf{x}))d\tau(\theta \mid x) \right] g(\mathbf{x})d\nu(\mathbf{x}) \tag{3.19}$$

If the prior τ has a density $t(\theta)$ with respect to some sigma-finite measure λ (for example, λ may be Lebesgue measure on Θ), then the *conditional density* of ϑ, given $\mathbf{X} = \mathbf{x}$, that is, the posterior density, is given by

$$t(\theta \mid \mathbf{x}) = \frac{t(\theta)f(\mathbf{x} \mid \theta)}{g(\mathbf{x})}, \qquad [d\tau(\theta \mid \mathbf{x}) = t(\theta \mid \mathbf{x})d\lambda(\mathbf{x})]. \tag{3.20}$$

One may think of the posterior distribution as the updating of the prior given the observed data $\mathbf{X}$. In the Bayesian paradigm, all inference is to be based on the posterior.

Theorem 3.1. *Let τ be a given prior with a finite second moment. In the problem of estimating the parameter $\boldsymbol{\theta}$ belonging to a (measurable) set $\Theta \subset R^k$, with the action space $\mathscr{A}$ a (measurable) convex subset of $\mathbb{R}$ containing Θ, and under squared error loss, the posterior mean of ϑ is a Bayes estimator of $\boldsymbol{\theta}$.*

To prove this we first need a simple lemma.

Lemma 3.1. *Let $\mathbf{Z}$ be a random vector with mean $\boldsymbol{\mu}$ and a finite second moment. Then*

$$E|\mathbf{Z} - \boldsymbol{\mu}|^2 < E|\mathbf{Z} - \mathbf{c}|^2 \qquad \text{for every } \mathbf{c} \neq \boldsymbol{\mu}. \tag{3.21}$$

Proof. Let $\mathbf{Z} = (Z_1, \ldots, Z_k)$, $\boldsymbol{\mu} = (\mu_1, \ldots, u_k)$. Then (3.21) follows from the relations

$$E(Z_i - c_i)^2 \equiv E(Z_i - \mu_i + \mu_i - c_i)^2 = E(Z_i - \mu_i)^2 + (\mu_i - c_i)^2 + 2(\mu_i - c_i)E(Z_i - \mu_i)$$
$$= E(Z_i - \mu_i)^2 + (\mu_i - c_i)^2 \qquad (i = 1, \ldots, k).$$

$$\square$$

Proof of Theorem 3.1. The posterior mean of ϑ is $E(\vartheta \mid \mathbf{X}) = d_0(\mathbf{X})$, say. If d is any other decision rule (estimator), then one has, by applying the Lemma to the conditional distribution of ϑ, given $\mathbf{X}$,

$$E(L(\vartheta, d(\mathbf{X})) \mid \mathbf{X}) \equiv E(|\vartheta - d(\mathbf{X})|^2 \mid \mathbf{X}) \geq E(|\vartheta - d_0(\mathbf{X})|^2 \mid \mathbf{X})$$
$$\equiv E(L(\vartheta, d_0(\mathbf{X})) \mid \mathbf{X}). \tag{3.22}$$

Hence

$$r(\tau, d) = E(L(\vartheta, d(\mathbf{X})) = E[E(L(\vartheta, d(\mathbf{X})) \mid \mathbf{X}] \geq E[E(L(\vartheta, d_0(\mathbf{X})) \mid \mathbf{X})]$$
$$= E(L(\vartheta, d_0(\mathbf{X})) = r(\tau, d_0). \tag{3.23}$$

$\square$

Remark 3.1. The convexity of $\mathscr{A}$ ensures that $d_0(X) \in \mathscr{A}$ a.s. The conclusion of Theorem 3.1 and its proof apply to any (measurable) real or vector valued function $g(\theta)$ of θ having a finite second moment under the given prior τ:

Proposition 3.1. *If the action space $\mathscr{A}$ is a (measurable) convex set C, containing the range of g, then under squared error loss $L(\theta, \mathbf{a}) = |g(\theta) - \mathbf{a}|^2$, $E(g(\vartheta) \mid \mathbf{X})$ is a Bayes estimator of $g(\theta)$.*

Theorem 3.2. *Let $g(\theta)$ be a real-valued measurable function on Θ having a finite absolute first moment under the prior τ. Let the action space $\mathscr{A}$ be an interval containing the range of g. Under absolute error loss $L(\theta, a) = |g(\theta) - a|$, a (the) median of the posterior distribution of ϑ (i.e., the conditional distribution of ϑ, given $\mathbf{X}$) is a Bayes estimator of θ.*

Remark 3.2. Recall that a *median* of a probability measure Q on R is a number M such that

$$Q((-\infty, M]) \equiv Q(\{x \leq M\}) \geq \frac{1}{2}, \quad \text{and} \quad Q([M, \infty)) \equiv Q(\{x \geq M\}) \geq \frac{1}{2}. \tag{3.24}$$

Unlike the mean, a median of a distribution need not be unique. There are two cases in which the median is unique: (1) There is a *unique M* such that $F_Q(M) = \frac{1}{2}$, where F_Q is the (right-continuous) distribution function of Q, and (2) there is a (necessarily unique) point M such that $F_Q(M) > \frac{1}{2}$, and $F_Q(M-) < \frac{1}{2}$. In the remaining case, F_Q has a flat stretch of x-values where it has the value $\frac{1}{2}$. This may be either of the form (iii) $[M_0, M_1)$ (with a jump discontinuity at M_1), or of the form (iv) $[M_0, M_1]$, $M_0 < M_1$. Such an interval comprises the set of medians of Q, illustrating the case of *non-uniqueness*.

We will need the following Lemma to prove Theorem 3.2.

Lemma 3.2. *Let Z be a real-valued random variable with a finite mean. If M is a median of the distribution Q of Z, then*

$$\inf\{E|Z - c| : c \in \mathbb{R}\} = E|Z - M|. \tag{3.25}$$

Proof. Let M be a median of Q, and $a < M$. Then

$$
\begin{aligned}
E|Z - a| &= E\left[(Z - a)\mathbf{1}_{\{Z \geq a\}}\right] + E\left[(a - Z)\mathbf{1}_{\{Z < a\}}\right] \\
&= E\left[(Z-M)\mathbf{1}_{\{Z \geq a\}} + (M - a)\mathbf{1}_{\{Z \geq a\}} + (a - M)\mathbf{1}_{\{Z < a\}} + (M - Z)\mathbf{1}_{\{Z < a\}}\right] \\
&= E\left[(Z - M)(\mathbf{1}_{\{Z \geq M\}} + \mathbf{1}_{\{a \leq Z < M\}}) + (M - Z)(\mathbf{1}_{\{Z < M\}} - \mathbf{1}_{\{a \leq Z < M\}})\right] \\
&\qquad + (M - a)[P(Z \geq a) - P(Z < a)] \\
&= E|Z - M| - 2E[(M - Z)\mathbf{1}_{\{a \leq Z < M\}}] + (M - a)[2P(Z \geq a) - 1] \\
&\geq E|Z - M| - 2(M - a)P(a \leq Z < M) + (M - a)[2P(Z \geq a) - 1] \\
&= E|Z - M| + 2(M - a)[P(Z \geq a) - P(a \leq Z < M)] - (M - a) \\
&= E|Z - M| + 2(M - a)P(Z \geq M) - (M - a) \geq E|Z - M|
\end{aligned}
$$

using $P(Z \geq M) \geq \frac{1}{2}$ in the last step. Similarly, one can show that for $a > M$ one has $E|Z - a| \geq E|Z - M|$. Else one may use the fact that $-M$ is a median of the distribution of $-Z$. Hence if $a > M$, $-a < -M$, so that the above argument shows that $E|-Z - (-a)| \geq E|-Z - (-M)|$. $\qquad\square$

Proof of Theorem 3.2. This follows by first taking conditional expectation, given $\mathbf{X}$: $E[L(\vartheta, d(\mathbf{X})) \mid \mathbf{X}] \equiv E[(|g(\vartheta) - d(\mathbf{X})|)|\mathbf{X}] \geq E[(|g(\vartheta) - d_0(\mathbf{X})|) \mid \mathbf{X}] \equiv E[L(\vartheta, d_0(\mathbf{X})) \mid \mathbf{X}]$, where $d_0(\mathbf{X})$ is a (the) median of the conditional distribution of $g(\vartheta)$, given $\mathbf{X}$. Integrating this over the marginal distribution of $\mathbf{X}$, the proof is completed. $\qquad\square$

Remark 3.3. In the case of a vector parameter $\boldsymbol{\theta} = (\theta_1, \ldots, \theta_k)$, if one uses the additive loss function, $L(\boldsymbol{\theta}, \mathbf{a}) = |\theta_1 - a_1| + \cdots + |\theta_k - a_k|$, then it follows by Lemma 3.1 that the vector of medians of the coordinates of ϑ, given $\mathbf{X}$, is a Bayes estimator of $\boldsymbol{\theta}$.

Example 3.5. Let $\Theta = [0, 1] = \mathscr{A}$, $L(\theta, a) = c(\theta - a)^2$ (where $c > 0$ does not depend on θ), $\mathscr{X} = \{0, 1\}^n \equiv$ set of all n-tuples of 0's and 1's (the observation space for a random sample of size n from a Bernoulli distribution $\mathscr{B}(\theta)$), and

$$
f(\mathbf{x} \mid \theta) = P_\theta(\{\mathbf{X} = \mathbf{x}\}) = \theta^r(1 - \theta)^{n-r}, \quad \mathbf{x} = (x_1, \ldots, x_n) \in \mathscr{X},
$$
$$
r := \sum_{i=1}^{n} x_i \quad \text{is the number of 1's in the sample.} \tag{3.26}
$$

We wish to compute the Bayes estimator of θ for the prior τ with density t (with respect to Lebesgue measure on $\Theta = [0, 1]$) given by

$$
d\tau(\theta) = t(\theta)d\theta, \qquad t(\theta) = \frac{\Gamma(\alpha + \beta)}{\Gamma(\alpha)\Gamma(\beta)} \theta^{\alpha-1}(1 - \theta)^{\beta-1} \tag{3.27}
$$

where $\alpha > 0$, $\beta > 0$ are parameters of this *beta distribution* $\mathscr{B}_e(\alpha, \beta)$. Note that (See the *Appendix on Univariate Distributions*)

$$
\int_0^1 \theta^{\alpha-1}(1 - \theta)^{\beta-1}d\theta = \frac{\Gamma(\alpha)\Gamma(\beta)}{\Gamma(\alpha + \beta)}. \tag{3.28}
$$

The mean of the $\mathscr{B}_e(\alpha, \beta)$ distribution is

$$
\int_0^1 \theta t(\theta)d\theta = \frac{\Gamma(\alpha + \beta)}{\Gamma(\alpha)\Gamma(\beta)} \int_0^1 \theta^{\alpha}(1 - \theta)^{\beta-1}d\theta
$$

$$= \frac{\Gamma(\alpha+\beta)}{\Gamma(\alpha)\Gamma(\beta)} \cdot \frac{\Gamma(\alpha+1)\Gamma(\beta)}{\Gamma(\alpha+\beta+1)} = \frac{\alpha}{\alpha+\beta}, \tag{3.29}$$

using (3.28) and the *identity:* $\Gamma(\alpha+1) = \alpha\Gamma(\alpha)$ for all $\alpha > 0$. The *posterior density* or ϑ (i.e., *the conditional density of ϑ, given* $(\mathbf{X} = \mathbf{x})$ is computed as in (3.20),

$$t(\theta \mid \mathbf{x}) = \frac{t(\theta)f(\mathbf{x} \mid \theta)}{g(\mathbf{x})} = \frac{c(\alpha,\beta)}{g(\mathbf{x})} \theta^{\alpha+r-1}(1-\theta)^{\beta+n-r-1}$$

$$\left[c(\alpha,\beta) = \frac{\Gamma(\alpha+\beta)}{\Gamma(\alpha)\Gamma(\beta)} \right] \qquad (r = \sum_i^n x_i), \tag{3.30}$$

where the marginal density (w.r.t. counting measure on $\mathscr{X}$) $g(\mathbf{x})$ is given by (see (3.18))

$$g(\mathbf{x}) = \int_\Theta f(\mathbf{x} \mid \theta)t(\theta)d\theta. \tag{3.31}$$

Although we can surely compute $g(\mathbf{x})$ from (3.31), note that a simpler way is to recognize that (3.30) gives a probability density function (in θ) for every given $\mathbf{x}$ and, therefore, must integrate out to 1. Then

$$g(\mathbf{x}) = c(\alpha,\beta) \int_0^1 \theta^{\alpha+r-1}(1-\theta)^{\beta+n-r-1}d\theta = \frac{\Gamma(\alpha+\beta)}{\Gamma(\alpha)\Gamma(\beta)} \frac{\Gamma(\alpha+\beta+n)}{\Gamma(\alpha+r)\Gamma(\beta+n-r)}. \tag{3.32}$$

But even this computation is unnecessary! For the functional form (3.30), as a density in θ, shows that the *posterior distribution of ϑ is* $\mathscr{B}_e(\alpha+r, \beta+n-r)$. Hence the *Bayes estimator of θ*, namely, the mean of its (or ϑ's) posterior distribution is, by (3.29),

$$d_0(\mathbf{x}) = E(\mathscr{O} \mid \mathbf{X} = \mathbf{x}) = \frac{r+\alpha}{n+\alpha+\beta} \qquad \left(r = \sum_{i=1}^n x_i \right). \tag{3.33}$$

If n is large (and α, β are relatively small), then this is not significantly different from the traditional (maximum likelihood) estimator $\hat{\theta} = \bar{x} \equiv \sum_{i=1}^n x_i/n$ (the *sample proportion* of 1's).

To compute the Bayes risks of d_0 and $\hat{\theta}$ (Exercise 3.4), first compute the second moment of a $\mathscr{B}_e(\alpha,\beta)$ distribution as

$$\frac{\Gamma(\alpha+\beta)}{\Gamma(\alpha)\Gamma(\beta)} \int_0^1 \theta^2 \theta^{\alpha-1}(1-\theta)^{\beta-1}d\theta = \frac{\Gamma(\alpha+\beta)}{\Gamma(\alpha)\Gamma(\beta)} \cdot \frac{\Gamma(\alpha+2)\Gamma(\beta)}{\Gamma(\alpha+\beta+2)}$$

$$= \frac{\alpha(\alpha+1)}{(\alpha+\beta)(\alpha+\beta+1)}. \tag{3.34}$$

Example 3.6. Let $\Theta = \mathbb{R} = \mathscr{A}$, $\mathscr{X} = \mathbb{R}^n = \Omega$, P_θ has density $f(\mathbf{x} \mid \theta)$ (w.r.t. Lebesgue measure on $\mathbb{R}^n$), given by

$$f(\mathbf{x} \mid \theta) = (2\pi\sigma^2)^{-n/2} \exp\left\{ -\sum_{i=1}^n (x_i - \theta)^2/2\sigma^2 \right\},$$

$$\mathbf{x} = (x_1, \ldots, x_n) \in \mathbb{R}^n. \tag{3.35}$$

Here $\sigma^2 > 0$ *is assumed to be known.* Finally, let $L(\theta, a) = c(\theta - a)^2$, with $c > 0$ not dependent on θ. We will construct the Bayes estimator of θ for the Normal prior τ with density (w.r.t. Lebesgue measure on $\Theta = \mathbb{R}$) $t(\cdot)$ given by

$$t(\theta) = (2\pi\beta^2)^{-\frac{1}{2}} \exp\left\{-\frac{\theta^2}{2\beta^2}\right\}, \qquad \theta \in \mathbb{R}, \tag{3.36}$$

where $\beta > 0$ is a given constant. The *posterior density* (w.r.t. Lebesgue measure) of ϑ given $\mathbf{X}$ is

$$t(\theta \mid \mathbf{x}) = \frac{t(\theta)f(\mathbf{x} \mid \theta)}{g(\mathbf{x})} = \frac{(2\pi\beta^2)^{-\frac{1}{2}}(2\pi\sigma^2)^{-\frac{n}{2}} \exp\left\{-\frac{\theta^2}{2\beta^2} - \frac{1}{2\sigma^2}\sum(x_i - \theta)^2\right\}}{g(\mathbf{x})}$$

$$= c_1(\beta, \sigma^2, \mathbf{x})\exp\left\{-\frac{\theta^2}{2\beta^2} - \frac{n(\theta - \overline{x})^2}{2\sigma^2}\right\}$$

$$= c_2(\beta, \sigma^2, \mathbf{x})\exp\left\{-\left(\frac{1}{2\beta^2} + \frac{n}{2\sigma^2}\right)\theta^2 + \frac{n\overline{x}}{\sigma^2}\theta\right\} \tag{3.37}$$

$$= c_3(\beta, \sigma^2, \mathbf{x})\exp\left\{-\frac{1}{2}\left(\frac{n\beta^2 + \sigma^2}{\beta^2\sigma^2}\right)\left(\theta - \frac{n\beta^2}{n\beta^2 + \sigma^2}\overline{x}\right)^2\right\} \quad \text{on } \mathbf{X} = \mathbf{x},$$

where $c_i(\beta, \sigma^2, \mathbf{x})$ $(i = 1, 2, 3)$ do not involve θ. Hence the posterior distribution $d\tau(\theta \mid \mathbf{x})$ is Normal $N\left(\frac{n\beta^2\overline{x}}{n\beta^2 + \sigma^2}, \frac{\beta^2\sigma^2}{n\beta^2 + \sigma^2}\right)$, with mean $[n\beta^2/(n\beta^2 + \sigma^2)]\overline{x}$. Therefore, the Bayes estimator of θ is

$$d_0(\mathbf{x}) = \frac{n\beta^2}{n\beta^2 + \sigma^2}\,\overline{x}. \tag{3.38}$$

In this example $d_0(\mathbf{X})$ is also the *median* of the posterior, in view of the symmetry of the Normal distribution. Hence d_0 in (3.38) is also the Bayes estimator under the loss function $L(\theta, a) = |\theta - a|$.

This example is easily extended to k-dimensional i.i.d. $N(\boldsymbol{\theta}, I_k)$ observations, where $\boldsymbol{\theta} \in \mathbb{R}^k$, and I_k is the $k \times k$ identity matrix. The prior distribution of $\boldsymbol{\theta}$ is taken to be that of independent $N(0, \beta_i^2)$ random variables $(i = 1, \ldots, k)$ (Exercise 3.10).

A parametric family of priors is said to be *conjugate* if, for every prior τ in the family, the posterior distribution belongs to the same family $(\forall \mathbf{x} \in \mathscr{X})$. Examples 3.5, 3.6 and Exercise 3.5 above provide examples of such priors.

Bayes estimators are, under mild restrictions, admissible. Our next couple of results make this precise.

Definition 3.3. A Bayes rule d_0 (w.r.t. a prior τ, and a given loss function) is said to be *unique up to equivalence* if for any other Bayes rule d_1 (w.r.t. τ) one has $R(\theta, d_0) = R(\theta, d_1) \ \forall \theta \in \Theta$.

Theorem 3.3. *If, for a given prior τ, a Bayes rule is unique up to equivalence, then it is admissible.*

Proof. Let a Bayes rule d_0 be unique up to equivalence, and suppose, if possible, d_0 is inadmissible. Then there exists a decision rule d_1 satisfying (i) $R(\theta, d_1) \leq R(\theta, d_0) \ \forall \theta \in \Theta$, and (ii) $R(\theta_1, d_1) < R(\theta_1, d_0)$ for some $\theta_1 \in \Theta$. By integrating the first inequality (i) w.r.t. τ, one gets $r(\tau, d_1) \leq r(\tau, d_0)$. But d_0 has the smallest possible Bayes risk among all decision rules. Hence $r(\tau, d_1) = r(\tau, d_0)$, implying d_1 is a Bayes rule. By the hypothesis of uniqueness, one then has $R(\theta, d_1) = R(\theta, d_0) \; \forall \, \theta \in \Theta$, contradicting inequality (ii). $\square$

Often the following stronger uniqueness holds for Bayes rules.

Definition 3.4. A Bayes rule d_0 is said to be *unique* if for every Bayes rule d_1 (w.r.t. the same prior τ) one has

$$P_\theta(d_1(\mathbf{X}) = d_0(\mathbf{X})) = 1 \qquad \forall \, \theta \in \Theta. \tag{3.39}$$

It is clear that (3.39) implies $R(\theta, d_1) = R(\theta, d_0) \; \forall \, \theta \in \Theta$.

Example 3.7. Let $\Theta = [0,1] = \mathscr{A}$, $L(\theta, a) = c(\theta - a)^2$ ($c > 0$ does not depend on θ), $\mathscr{X} = \{0,1\}^n$,

$$f(\mathbf{x} \mid \theta) \equiv P_\theta(\{\mathbf{x}\}) = \theta^r(1 - \theta)^{n-r} \qquad (r = \sum_1^n x_i, \; \mathbf{x} = (x_1, \ldots, x_n)).$$

In this example, let τ be a prior assigning all its mass to $\{0,1\}$ ($\tau(\{0,1\}) = 1$), say, $\tau(\{0\}) = p$, $\tau(\{1\}) = q = 1 - p$. Assume $n \geq 2$. Consider the estimators

$$d_0(\mathbf{X}) \equiv \overline{X}, \qquad d_1(\mathbf{X}) \equiv X_1. \tag{3.40}$$

Then

$$R(\theta, d_0) = \frac{c\theta(1 - \theta)}{n}, \quad R(\theta, d_1) = c\theta(1 - \theta),$$

$$r(\tau, d_0) = \int_\Theta R(\theta, d_0)d\tau(\theta) = R(0, d_0)p + R(1, d_0)q = 0,$$

$$r(\tau, d_1) = R(0, d_1)p + R(1, d_1)q = 0.$$

For $P_0(X_i = 0) = 1$, $P_1(X_i = 1) = 1 \; \forall \, i = 1, 2, \ldots, n$, and $P_0(\overline{X}_i = 0) = 1$, $P_1(\overline{X} = 1) = 1$. Thus d_0 and d_1 are both Bayes estimators. But d_1 is clearly inadmissible, since for all $0 < \theta < 1$, $R(\theta, d_0) < R(\theta, d_1)$, while $R(\theta, d_0) = R(\theta, d_1)$ for $\theta = 0, 1$.

To rule out situations like this, one may require a property such as P_1 below.

Property P_1 There exists a subset $\widetilde{\Theta}$ of Θ such that (1) $\tau(\widetilde{\Theta}) > 0$ and (2) if $P_{\theta_0}(A) = 0$ for some $\theta_0 \in \widetilde{\Theta}$ (and for some event $A \subset \mathscr{X}$), then $P_\theta(A) = 0$ for all $\theta \in \Theta$.

Note that the prior τ in Example 3.7 does not satisfy P_1.

Theorem 3.4. *In a decision problem with loss $c(\theta - a)^2$ ($c > 0$ independent of θ), suppose a prior τ has the property P_1. Then a Bayes estimator w.r.t. τ is unique, assuming $r(\tau, d_0) < \infty$.*

Proof. A Bayes rule w.r.t. to τ is the posterior mean $d_0(\mathbf{X}) = E(\vartheta \mid \mathbf{X})$. Write

$$r(\tau, d_0) = c\int_\mathscr{X} v(\mathbf{x})dG(\mathbf{x}), \qquad v(\mathbf{x}) := E[(\vartheta - d_0(\mathbf{x}))^2 \mid \mathbf{X} = \mathbf{x}],$$

where G is the (marginal) distribution of $\mathbf{X}$. Now suppose d_1 is another Bayes estimator w.r.t. τ, and

$$r(\tau, d_1) = c \int_{\mathscr{X}} v_1(\mathbf{x}) dG(\mathbf{x}), \qquad v_1(\mathbf{x}) := E\left[(\vartheta - d_1(\mathbf{x}))^2 \mid \mathbf{X} = \mathbf{x} \right].$$

Let $A = \{\mathbf{x} \in \mathscr{X} : d_0(\mathbf{x}) \neq d_1(\mathbf{x})\}$. Then

$$r(\tau, d_1) = c \int_A v_1(\mathbf{x}) dG(\mathbf{x}) + c \int_{A^c} v_1(\mathbf{x}) dG(\mathbf{x}),$$

$$r(\tau, d_0) = c \int_A v(\mathbf{x}) dG(\mathbf{x}) + c \int_{A^c} v(\mathbf{x}) dG(\mathbf{x}).$$

Since $r(\tau, d_1) = r(\tau, d_0)$, it follows that $\int_A (v_1(\mathbf{x}) - v(\mathbf{x})) dG(\mathbf{x}) = 0$. But on A, $v_1(\mathbf{x}) > v(\mathbf{x})$. Therefore, $G(A) = 0$. This implies

$$0 = G(A) = \int_\Theta P_\theta(A) d\tau(\theta) \geq \int_{\widetilde{\Theta}} P_\theta(A) d\tau(\theta),$$

so that the last integral must be zero. Hence $P_\theta(A) = 0$ a.e. (w.r.t. τ) on $\widetilde{\Theta}$. But $\tau(\widetilde{\Theta}) > 0$. Therefore, there exists $\theta_0 \in \widetilde{\Theta}$ such that $P_{\theta_0}(A) = 0$. By property P_1, one now gets $P_\theta(A) = 0 \ \forall \theta \in \Theta$. Thus uniqueness holds. $\qquad\square$

Remark 3.4. The above argument extends to the absolute error loss $|\theta - a|$, provided the posterior distribution has a unique median a.s. for all x. This is the case, e.g., when the prior has a strictly positive density on Θ.

Corollary 3.1. *Under the hypothesis of Theorem 3.4, a Bayes estimator is admissible.*

We next turn to the (uneasy) relationship between Bayes estimators and unbiased estimators. Recall $d(\mathbf{X})$ is an *unbiased estimator of a parametric function* $g(\theta)$ if

$$E_\theta d(\mathbf{X}) = g(\theta) \qquad \forall \ \theta \in \Theta.$$

Theorem 3.5. *Let loss be proportional to squared error. Then the Bayes estimator* $d(\mathbf{X}) = E(\theta \mid \mathbf{X})$ *(with a finite Bayes risk) is not unbiased, if* $r(\tau, d) > 0$.

Proof. Suppose, if possible, that $d(\mathbf{X})$ is unbiased. Then

$$E\vartheta d(\mathbf{X}) = E[\vartheta E(d(\mathbf{X}) \mid \vartheta)] = E\vartheta\vartheta = E\vartheta^2, \quad \text{(by unbiasedness of } d\text{)},$$

$$E\vartheta d(\mathbf{X}) = E[d(\mathbf{X}) E(\vartheta \mid \mathbf{X})] = Ed(\mathbf{X}) d(\mathbf{X}) = Ed^2(\mathbf{X}).$$

On the other hand,

$$0 < r(\tau, d) = E[\vartheta - d(\mathbf{X})]^2 = E\vartheta^2 + Ed^2(\mathbf{X}) - 2E\vartheta d(\mathbf{X}) = 0,$$

a contradiction. $\qquad\square$

Remark 3.5. In the proof of Theorem 3.4 the nature of the loss function is only used to require that $v_1(\mathbf{x}) > v(\mathbf{x}) \ \forall \ \mathbf{x} \in \{d_0(\mathbf{x}) \neq d_1(\mathbf{x})\}$. Thus Theorem 3.4 and Corollary 3.1 hold under every loss function such that $E[L(\mathscr{O}, \mathbf{X}) \mid \mathbf{X} = \mathbf{x}]$ has a unique minimizer for every $\mathbf{x}$.

3.4 Minimax Decision Rules

A conservative statistician may try to avoid maximum penalty by choosing a decision rule d^* whose risk function $R(\theta, d^*)$ has the *smallest maximum value*.

Definition 3.5. A decision rule d^* is *minimax* for a decision problem (specified by Θ, $\mathscr{A}$, $L(\theta, a)$, $\mathscr{X}$, and $P_\theta \ \forall \theta \in \Theta$), if for every decision rule d one has

$$\sup_{\theta \in \Theta} R(\theta, d^*) \leq \sup_{\theta \in \Theta} R(\theta, d) \tag{3.41}$$

or, equivalently, if

$$\sup_{\theta \in \Theta} R(\theta, d^*) = \inf_d \sup_{\theta \in \Theta} R(\theta, d), \tag{3.42}$$

where the infimum on the right side is over the class of all decision rules d.

Theorem 3.6. *Suppose τ_N ($N = 1, 2, \ldots$) is a sequence of priors with corresponding Bayes rules d_N such that $r(\tau_N, d_N) \to C < \infty$, as $N \to \infty$. If there exists a decision rule d^* such that*

$$R(\theta, d^*) \leq C \qquad \forall \, \theta \in \Theta, \tag{3.43}$$

then d^* is minimax.

Proof. If d^* is not minimax, there exists d such that $\sup_{\theta \in \Theta} R(\theta, d) < \sup_{\theta \in \Theta} R(\theta, d^*) \leq C$. Let $\epsilon > 0$ be such that $R(\theta, d) \leq C - \epsilon \ \forall \theta \in \Theta$. Then $r(\tau_N, d) \equiv \int_\Theta R(\theta, d) d\tau_N(\theta) \leq C - \epsilon$ for all N, and $r(\tau_N, d) < r(\tau_N, d_N)$ for all sufficiently large N (since $r(\tau_N, d_N) \to C$ as $N \to \infty$). This contradicts the fact that d_N is Bayes (w.r.t. τ_N) for every N. $\qquad\qquad\square$

Example 3.8. Let $\Theta = \mathbb{R} = \mathscr{A}$, $\mathscr{X} = \mathbb{R}^n$, P_θ (on $\Omega = \mathscr{X}$) has the density

$$f(\mathbf{x} \mid \theta) = (2\pi\sigma^2)^{-n/2} \exp\left\{ -\frac{1}{2\sigma^2} \sum_{i=1}^n (x_i - \theta)^2 \right\}.$$

Let the loss function be $L(\theta, a) = c(\theta - a)^2$ ($c > 0$). Consider the rule $d^*(\mathbf{x}) = \bar{x} = \frac{1}{n}\sum_{i=1}^n x_i$. For each $N = 1, 2, \ldots$, consider the prior τ_N which is $N(0, N)$ (Normal with mean 0 and variance N). Then the Bayes estimator d_N for the prior τ_N is (See Example 3.6) is $d_N(\mathbf{x}) = \frac{nN\bar{x}}{nN+\sigma^2}$ with the corresponding Bayes risk

$$r(\tau_N, d_N) = cE(\vartheta - d_N(\mathbf{X}))^2 = cE[E(\vartheta - d_N(\mathbf{X}))^2 \mid \mathbf{X}]$$

$$= cE\left(\frac{N\sigma^2}{Nn + \sigma^2}\right) = c\frac{N\sigma^2}{Nn + \sigma^2} \longrightarrow \frac{c\sigma^2}{n} \qquad \text{as } N \to \infty.$$

Since $R(\theta, d^*) = cE_\theta(\overline{X} - \theta)^2 = c\frac{\sigma^2}{n}$, Theorem 3.6 applies, and d^* is minimax.

Theorem 3.7. *Suppose d^* is a decision rule whose risk function is a constant c',*

$$R(\theta, d^*) = c' \qquad \forall \, \theta \in \Theta. \tag{3.44}$$

If, in addition, (i) there exists some prior τ such that d^* is Bayes w.r.t. to τ, or (ii) d^* is admissible, then d^* is minimax.

Proof. Let d be any other decision rule. Then, if (i) holds,

$$\sup_{\theta \in \Theta} R(\theta, d) \geq \int_{\Theta} R(\theta, d) d\tau(\theta) \equiv r(\tau, d) \geq r(\tau, d^*)$$

$$= \int_{\Theta} R(\theta, d^*) d\tau(\theta) = c' = \sup_{\theta \in \Theta} R(\theta, d^*).$$

Suppose now that (ii) holds, and d^* is not minimax. Then there exists d_1 such that

$$\sup_{\theta \in \Theta} R(\theta, d_1) < R(\theta, d^*) = c' \;\forall\; \theta.$$

This means d^* is not admissible, a contradiction. $\square$

Note that condition (1) in Theorem 3.7 does not necessarily imply condition (2) without some condition such as P_1 ensuring the uniqueness of the Bayes rule for τ.

Example 3.9 (Admissibility of the Sample Proportion). Consider Example 3.7 (or Exercise 3.4), but with $\Theta = (0,1)$, $\mathscr{A} = [0,1]$, and $L(\theta, a) = (\theta - a)^2/\theta(1 - \theta)$. The decision rule $d^* = \overline{x}$ has constant risk

$$R(\theta, d^*) = E_\theta(\theta - \overline{X})^2/\theta(1 - \theta) = \frac{1}{n} \qquad \forall\; \theta \in \Theta. \tag{3.45}$$

We will show that d^* is also Bayes with respect to the beta prior $\tau = \mathscr{B}_e(1,1)$. For any decision rule d, the Bayes risk w.r.t. the uniform prior τ is

$$r(\tau, d) = \int_0^1 \frac{E_\theta(\theta - d(\mathbf{X}))^2}{\theta(1 - \theta)}\, d\theta = \int_0^1 \sum_{\mathbf{x} \in \{0,1\}^n} \frac{(\theta - d(\mathbf{x}))^2}{\theta(1 - \theta)}\, \theta^r (1 - \theta)^{n-r} d\theta$$

$$= \sum_{\mathbf{x} \in \{0,1\}^n} \int_0^1 (\theta - d(\mathbf{x}))^2 \theta^{r-1}(1 - \theta)^{n-r-1} d\theta. \tag{3.46}$$

If $r = 0$, the integral is infinite, unless $d(\mathbf{x}) = 0$, and if $r = n$ the integral is infinite unless $d(\mathbf{x}) = 1$. Hence a Bayes rule d_0 must have the values

$$d_0(\mathbf{x}) = \begin{cases} 0 \text{ if } \mathbf{x} = (0, 0, \ldots, 0), \\ 1 \text{ if } \mathbf{x} = (1, 1, \ldots, 1). \end{cases} \tag{3.47}$$

For every other $\mathbf{x}$, the summand in (3.46) may be expressed as

$$\frac{\Gamma(r)\Gamma(n - r)}{\Gamma(n)} \int_0^1 (\theta - d(\mathbf{x}))^2 b_{r,n-r}(\theta) d\theta, \tag{3.48}$$

where $b_{r,n-r}$ is the density of the beta distribution $\mathscr{B}_e(r, n - r)$. The integral in (3.48) is then $E(\vartheta - d(\mathbf{X}))^2$, where ϑ has the $\mathscr{B}_e(r, n - r)$ distribution, and is therefore minimum when $d(\mathbf{x}) = E\vartheta = \frac{r}{n} = \overline{x}$. Thus the (unique) Bayes estimator is $d^* = \overline{x}$. In particular, d^* is *minimax* (w.r.t. the loss $\frac{(\theta - a)^2}{\theta(1 - \theta)}$), by Theorem 3.7 and admissible (by Theorem 3.3). Admissibility w.r.t. the loss function $L(\theta, a) = (\theta - a)^2/[\theta(1 - \theta)]$ means that there does not exist any decision rule d such that

$$R(\theta, d) \equiv \frac{E_\theta(\theta - d(\mathbf{X}))^2}{\theta(1 - \theta)} \leq \frac{E_\theta(\theta - \overline{X})^2}{\theta(1 - \theta)} \equiv R(\theta, d^*) \quad \forall\; \theta \in (0,1), \tag{3.49}$$

with strict inequality for some $\theta \in (0,1)$. Canceling out $\theta(1-\theta)$ from both sides, this implies that $d^*(\mathbf{x}) \equiv \overline{x}$ *is admissible w.r.t. squared error loss* $L(\theta, a) = (\theta - a)^2$, as well as w.r.t. the loss $[\theta(1-\theta)]^{-1}(\theta - a)^2$.

3.5 Generalized Bayes Rules and the James-Stein Estimator

1. *Improper Priors and Generalized Bayes Rules.* If the (prior) weight measure τ is allowed to be an arbitrary non-zero sigma-finite measure on Θ, it is called an *improper prior* in case $\tau(\Theta) = \infty$. Whether τ is finite or not, a decision rule $d_0(\mathbf{x})$ which minimizes

$$a \longrightarrow \int_{\Theta} L(\theta, a) f(\mathbf{x} \mid \theta) d\tau(\theta) \tag{3.50}$$

for every $\mathbf{x}$ (over all $a \in \mathscr{A}$) is called a *generalized Bayes rule* for the possibly improper prior τ. If τ is finite, then this minimizer is the same as that w.r.t. the (normalized) prior $\tau/\tau(\Theta)$.

Example 3.10 ($\mathbf{N}(\theta, \sigma^2)$*).* Consider the problem of estimating the Normal mean θ, with σ^2 known, discussed in Examples 3.6, 3.8. If τ is the Lebesgue measure on $\Theta = \mathbb{R}$, then the integral (3.50) equals

$$(2\pi\sigma^2)^{-n/2} \int_{-\infty}^{\infty} (\theta - a)^2 e^{-\frac{1}{2\sigma^2} \sum_1^n (x_i - \theta)^2} d\theta$$

$$= \frac{(2\pi\sigma^2)^{-(n-1)/2}}{\sqrt{n}} e^{-\frac{1}{2\sigma^2} \sum_1^n (x_i - \overline{x})^2} \int_{-\infty}^{\infty} (\theta - a)^2 \frac{\sqrt{n}}{\sqrt{2\pi\sigma^2}} e^{-\frac{n}{2\sigma^2}(\overline{x} - \theta)^2} d\theta \tag{3.51}$$

The last integral is the expected value of $(\vartheta - a)^2$ where ϑ is $N(\overline{x}, \frac{\sigma^2}{n})$. Hence the minimum is attained at $a = \overline{x}$, and $d_0(\mathbf{x}) = \overline{x}$ is the generalized Bayes estimator for τ.

We will apply the following theorem to prove that $d_0(\mathbf{x}) = \overline{x}$ is admissible as an estimator of θ in this example.

We continue to use the notation $r(\tau, d)$ as in (3.11), for improper priors as well as proper priors.

Theorem 3.8 (Blyth's Method). *Suppose that* $\theta \to R(\theta, d)$ *is finite and continuous for every decision rule d for which $R(\theta', d) < \infty$ for some $\theta' \in \Theta$. Let d_0 be a decision rule with a finite risk function. Assume there exists a sequence of proper or improper priors τ_N ($N = 1, 2, \ldots$) with the following properties: (i) for every nonempty open subset Θ_0 of Θ there exist N_0 and $b_0 > 0$ such that $\tau_N(\Theta_0) \geq b_0 \; \forall \, N \geq N_0$, and (ii) $r(\tau_N, d_0) - r(\tau_N, d_N) \to 0$ as $N \to \infty$, where d_N is a generalized Bayes rule for τ_N.*

Then d_0 is admissible.

Proof. Suppose d_0 is inadmissible. Then there exists a decision rule d, $\theta_0 \in \Theta$ and $\varepsilon > 0$ such that $R(\theta, d) \leq R(\theta, d_0) \; \forall \, \theta$, $R(\theta_0, d) < R(\theta_0, d_0) - \varepsilon$. By continuity of risk functions, there exists an open subset Θ_0 of Θ, with $\theta_0 \in \Theta_0$, for which $R(\theta, d) < R(\theta, d_0) - \varepsilon \; \forall \theta \in \Theta_0$. By (i) there exist N_0 and $b_0 > 0$ so that $\tau_N(\Theta_0) \geq b_0 \; \forall N \geq N_0$. Hence

$$r(\tau_N, d_0) - r(\tau_N, d) > \varepsilon \tau_N(\Theta_0) \geq \varepsilon b_0 > 0 \qquad \forall \, N \geq N_0.$$

This contradicts (ii), since $r(\tau_N, d) \geq r(\tau_N, d_N)$. $\qquad\qquad\square$

Example 3.11 (Admissibility of the Mean of a Sample from $N(\theta, \sigma^2)$). In context of the Normal example above, let $\tau_N = \sqrt{N} \, \mathbf{N}(0, N)$. To check condition (i) in Theorem 3.8, let Θ_0 be a nonempty open subset of $\Theta = \mathbb{R}$. There exist $\theta_0 < \theta_1$ such that $(\theta_0, \theta_1) \subset \Theta_0$. Now

$$\tau_N(\Theta_0) \geq \tau_N((\theta_0, \theta_1)) = \sqrt{N} \int_{\theta_0}^{\theta_1} \frac{1}{\sqrt{2\pi N}} e^{-\theta^2/2N} d\theta$$

$$= \frac{1}{\sqrt{2\pi}} \int_{\theta_0}^{\theta_1} e^{-\theta^2/2N} d\theta \longrightarrow \frac{1}{\sqrt{2\pi}} (\theta_1 - \theta_0) > 0 \quad \text{as } N \to \infty, \qquad (3.52)$$

from which (i) follows. Also, $r(\tau_N, \overline{x}) = \int R(\theta, \overline{x}) d\tau_N(\theta) = \sqrt{N} \frac{\sigma^2}{n}$, $r(\tau_N, d_N) = \sqrt{N} \left(\frac{N\sigma^2}{nN + \sigma^2} \right)$ (See Example 3.6, with $\beta^2 = N$), so that

$$r(\tau_N, \overline{x}) - r(\tau_N, d_N) = \sqrt{N} \left(\frac{nN + \sigma^2 - Nn}{n(nN + \sigma^2)} \right) \sigma^2 = \frac{\sqrt{N}}{n(nN + \sigma^2)} \sigma^4 \longrightarrow 0,$$

$$(3.53)$$

as $N \to \infty$, proving condition (ii) of Theorem 3.8, and establishing the admissibility of $\overline{x}$.

Remark 3.6. If one tries to extend the proof to the k-dimensional Normal distribution $\mathbf{N}(\boldsymbol{\theta}, I)$ (where $\boldsymbol{\theta} = (\theta_1, \ldots, \theta_k) \in \Theta = \mathbb{R}^k$, and I the $k \times k$ identity matrix), by letting τ_N be the product measure $\sqrt{n} \, \mathbf{N}(0, N) \times \cdots \times \sqrt{n} \, \mathbf{N}(0, N)$, then condition (i) of Theorem 3.8 holds as in (3.52) (with $(\sqrt{N})^k$ canceling out from the numerator and denominator). However, condition (ii) breaks down, since $r(\tau_N, \overline{\mathbf{x}}) - r(\tau_N, d_N) = (\sqrt{N})^{k/2} k / \{n(nN + 1)\}$, which does not go to zero as $N \to \infty$, if $k \geq 2$. It may be shown that $\overline{\mathbf{x}}$ is admissible for the case $k = 2$. *However, $\overline{\mathbf{x}}$ is inadmissible for $k \geq 3$,* a fact first discovered by Charles Stein in 1956 and which came as a big shock to most statisticians. We give below a proof of this inadmissibility due to James and Stein (1961) by showing that the so-called *James–Stein estimator*

$$d^{JS}(\mathbf{x}) := \left(1 - \frac{(k-2)\sigma^2}{n|\overline{\mathbf{x}}|^2} \right) \overline{\mathbf{x}} \qquad (3.54)$$

is uniformly better than $\overline{\mathbf{x}}$ for $k \geq 3$, when the underlying distribution $P_{\boldsymbol{\theta}}$ is $\mathbf{N}(\boldsymbol{\theta}, \sigma^2 I)$ for some known $\sigma^2 > 0$, and the loss function is squared error $L(\boldsymbol{\theta}, \mathbf{a}) = |\boldsymbol{\theta} - \mathbf{a}|^2 = \sum_{j=1}^{k} |\theta_j - a_j|^2$ ($\boldsymbol{\theta} = (\theta_1, \ldots, \theta_k)$, $\mathbf{a} = (a_1, \ldots, a_k) \in \mathbb{R}^k$).

Theorem 3.9. *One has*

$$R(\theta, d^{JS}) = k \frac{\sigma^2}{n} - (k-2)^2 \frac{\sigma^2}{n} E_{\boldsymbol{\theta}} \left(\frac{1}{|Y|^2} \right), \qquad (k \geq 3), \qquad (3.55)$$

where Y is $N(\sqrt{n} \, \boldsymbol{\theta}/\sigma, I)$.

We need two auxiliary properties of the Normal distribution.

Lemma 3.3. *Let g be a real-valued differentiable function on $\mathbb{R}$ such that $E|g'(X)| < \infty$ where X is $N(\theta, 1)$, and assume $g(x)\varphi(x - \theta) \to 0$ as $|x| \to \infty$. Then $Eg'(X) = \mathrm{cov}(X, g(X))$.*

Proof. Integrating by parts, and denoting the standard normal density by φ,

$$Eg'(X) = \int_{-\infty}^{\infty} g'(x)\varphi(x - \theta)dx = \int_{-\infty}^{\infty} g(x)(x - \theta)\varphi(x - \theta)dx$$

$\square$

Lemma 3.4. *Let $\mathbf{g} = (g_1, g_2, \ldots, g_k)$ be differentiable on $\mathbb{R}^k$ into $\mathbb{R}^k$. Let $\mathbf{X} = (X_1, \ldots, X_k)$ have the distribution $N(\boldsymbol{\theta}, I)$, $\boldsymbol{\theta} = (\theta_1, \ldots, \theta_k) \in \mathbb{R}^k$, I $k \times k$ identity matrix. Assume that $E|\mathbf{g}(\mathbf{X})|^2 < \infty$ and define $h_j(y) = E(g_j(\mathbf{X})|X_j)_{X_j = y} = Eg_j(X_1, \ldots, X_{j-1}, y, X_{j+1}, \ldots, X_k)$. Assume that h_j satisfies the hypothesis of Lemma 3.3 (in place of g there), $1 \leq j \leq k$. Then*

$$E|\mathbf{X} + \mathbf{g}(\mathbf{X}) - \boldsymbol{\theta}|^2 = k + E\left(|\mathbf{g}(\mathbf{X})|^2 + 2\sum_{j=1}^{k} \frac{\partial}{\partial x_j} g_j(\mathbf{x})\,|_{\mathbf{x}=\mathbf{X}}\right). \qquad (3.56)$$

Proof. The left side equals

$$E|\mathbf{X} - \boldsymbol{\theta}|^2 + E|\mathbf{g}(\mathbf{X})|^2 + 2E(\mathbf{X} - \boldsymbol{\theta}) \cdot \mathbf{g}(\mathbf{X}) = k + E|\mathbf{g}(\mathbf{X})|^2 + 2\sum_{j=1}^{k} E(X_j - \theta_j)g_j(\mathbf{X}).$$

Now $E(X_j - \theta_j)g_j(\mathbf{X}) = E[(X_j - \theta_j) \cdot E(g_j(\mathbf{X})|X_j)] = E(X_j - \theta_j)h_j(X_j)$. Apply Lemma 3.3 (with $g = h_j$) to get $E(X_j - \theta_j)h_j(X_j) = Eh_j'(X_j) = E\left[\left(\frac{\partial}{\partial x_j} g_j(\mathbf{x})\right)_{\mathbf{x}=\mathbf{X}}\right].$ $\square$

Proof of Theorem 3.9. Since $\overline{\mathbf{X}}$ is distributed as $N\left(\boldsymbol{\theta}, \frac{\sigma^2}{n} I\right)$, by rescaling it as $\frac{\sqrt{n}}{\sigma} \overline{\mathbf{X}}$ (which is distributed as $N(\boldsymbol{\gamma}, I)$, with $\boldsymbol{\gamma} = \frac{\sqrt{n}}{\sigma} \boldsymbol{\theta}$), one may take $n = 1$, $\sigma^2 = 1$ in the Theorem and write $\mathbf{X}$ for $\overline{\mathbf{X}}$. In this case write $\mathbf{g}(\mathbf{x}) = -\frac{(k-2)}{|\mathbf{x}|^2} \mathbf{x}$ to have (by (3.56))

$$E_\theta \left|d^{JS}(\mathbf{X}) - \boldsymbol{\theta}\right|^2 = E|\mathbf{X} + \mathbf{g}(\mathbf{X}) - \boldsymbol{\theta}|^2 = k + (k-2)^2 E\left(\frac{1}{|\mathbf{X}|^2}\right) + 2E\left[\sum_{j=1}^{k} \frac{\partial}{\partial x_j} g_j(\mathbf{x})|_{\mathbf{x}=\mathbf{X}}\right].$$

Now $\sum(\partial/\partial x_j)g_j(\mathbf{x}) = -(k-2)^2/|\mathbf{x}|^2$, so that

$$E_\theta \left|d^{JS}(\mathbf{X}) - \boldsymbol{\theta}\right|^2 = k - (k-2)^2 E\left(\frac{1}{|\mathbf{X}|^2}\right), \qquad (k \geq 3).$$

$\square$

Remark 3.7. It has been shown that even d^{JS} is not admissible (James and Stein, 1961).

We conclude this chapter with a result which implies, in particular, that the sample mean $\overline{X}$ is an admissible estimator of the population mean θ of a Normal distribution $N(\theta, \sigma^2)$ when both θ and $\sigma^2 > 0$ are unknown parameters.

Theorem 3.10. *Let $\Theta = \Theta_1 \times \Theta_2$ and suppose a decision rule d is admissible for the parameter space $\{(\theta_1, \theta_2) : \theta_1 \in \Theta_1\}$ for every given value of $\theta_2 \in \Theta_2$. Then d is admissible when the parameter space is Θ.*

Proof. Suppose d is inadmissible when the parameter space is $\Theta = \Theta_1 \times \Theta_2$. Then there exists a decision rule d_1 and a point $\theta^0 = (\theta_1^0, \theta_2^0)$ such that $R(\theta, d_1) \leq R(\theta, d) \ \forall \ \theta \in \Theta$ and $R(\theta^0, d_1) < R(\theta^0, d)$. But this implies $R((\theta_1, \theta_2^0), d_1) \leq R((\theta_1, \theta_2^0), d) \ \forall \ \theta_1 \in \Theta_1$, $R((\theta_1^0, \theta_2^0), d_1) < R((\theta_1^0, \theta_2^0), d)$, contradicting the fact that d is admissible when the parameter space is $\Theta_1 \times \{\theta_2^0\}$. $\qquad\square$

3.6 Notes and References

For Bayes estimation we refer to Ferguson (1967, Sects. 1.8, 2.1–2.3), and Lehmann and Casella (1998, Chaps. 4 and 5).

Exercises for Chap. 3

Ex. 3.1. (a) In Example 3.1, show that the solution $(\widehat{\mu}, \widehat{\sigma}^2)$ of the likelihood equations (3.3), (3.4) is the unique value of the parameter $\theta = (\mu, \sigma^2)$ which maximizes the likelihood function $\ell(\theta)$.
(b) In Example 3.2, show that $\hat{\theta} = \sum X_i / n$ is the unique maximizer of $\ell(\theta)$.

Ex. 3.2. In Example 3.3,

(a) plot $\ell(\theta)$ for $n = 4$, $M_n = 2$, and
(b) schematically draw the graph of $\ell(\theta)$ for a general $n \geq 2$.

Ex. 3.3. (a) Write down the likelihood function and the likelihood equation in Example 3.4. Note that no explicit solution is available for the MLE.
(b) Verify (3.9).

Ex. 3.4. Calculate the Bayes risk $r(\tau, d_0)$ in Example 3.5, and compare this with the Bayes risk of $\hat{\theta}$, namely,

$$r(\tau, \hat{\theta}) = \int_{\Theta} R(\theta, \hat{\theta}) t(\theta) d\theta = \int_0^1 \frac{\theta(1 - \theta)}{n} t(\theta) d\theta. \tag{3.57}$$

Ex. 3.5. Let the observation space be $\mathscr{X} = \{0, 1, \ldots\}^n \equiv \mathbb{Z}_+^n$,

$$f(\mathbf{x} \mid \theta) = \prod_{i=1}^{n} e^{-\theta} \frac{\theta^{x_i}}{x_i!} \equiv e^{-n\theta} \frac{\theta^{\sum_1^n x_i}}{\prod_{i=1}^n x_i!} \qquad (\mathbf{x} \in \mathscr{X}) \tag{3.58}$$

where $\theta \in \Theta = (0, \infty)$, $\mathscr{A} = [0, \infty)$. Find the Bayes estimator of θ (with squared error loss) for the prior $\mathscr{G}(\alpha, \beta)$ (See definition in Example 3.4).

Ex. 3.6. Let $\Theta = [0,1] = \mathscr{A}$, $\mathscr{X} = \{0,1\}^n$, $P_\theta(\{\mathbf{x}\}) = \theta^r(1-\theta)^{n-r}$ $(r = \sum_1^n x_i)$, $L(\theta,a) = c(\theta - a)^2$ $(c > 0)$. By Example 3.5, the Bayes rule for the beta prior $\mathscr{B}_e(\alpha,\beta)$ is $(r+\alpha)/(n+\alpha+\beta) \equiv d_0(\mathbf{X})$. Show that for $\alpha = \beta = \sqrt{n}/2$, the risk function of the Bayes rule of $d_0(\mathbf{X})$ is a constant $(R(\theta, d_0) = c/[4(\sqrt{n}+1)^2] \ \forall \ \theta \in \Theta)$, and conclude that

$$d^*(\mathbf{X}) := \frac{\sum_i^n X_i + \frac{\sqrt{n}}{2}}{n + \sqrt{n}} \equiv \frac{\overline{X} + \frac{1}{2\sqrt{n}}}{1 + \frac{1}{\sqrt{n}}} \tag{3.59}$$

is minimax, as well as admissible.

Ex. 3.7. *Admissibility of the Sample Mean from a Poisson Distribution.* Let $\Theta = (0,\infty)$, $\mathscr{A} = [0,\infty)$, $\mathscr{X} = \mathbb{Z}_+^n \equiv \{0,1,2,\dots\}^n = \Omega$,

$$P_\theta(\{\mathbf{x}\}) = \prod_{i=1}^n e^{-\theta} \frac{\theta^{x_i}}{x_i!} = e^{-n\theta} \frac{\theta^{\sum_1^n x_i}}{\prod_1^n x_i!} \equiv f(\mathbf{x} \mid \theta).$$

Let $L(\theta, a) = \frac{e^\theta}{\theta}(\theta - a)^2$ for (a), (b) below.

(a) Find the Bayes estimator w.r.t. the prior $\mathscr{G}(\alpha,\beta)$(Gamma).
(b) Show that $\overline{X} = \frac{1}{n}\sum_{i=1}^n X_i$ is Bayes w.r.t. some prior τ, and admissible.
(c) Show that $\overline{X}$ is admissible *under squared error loss:* $L(\theta, a) = (\theta - a)^2$.
(d) Show that $\overline{X}$ is minimax, w.r.t. loss function $\frac{(\theta - a)^2}{\theta}$.

Ex. 3.8. Show that, under squared error loss, (a) $\overline{\mathbf{X}}$ is an admissible estimator of $\boldsymbol{\mu} \in \Theta_1 = \mathbb{R}^k$ when the sample is from $\mathbf{N}(\boldsymbol{\mu}, \sigma^2 \mathbf{I})$ with $\boldsymbol{\mu}$, σ^2 both unknown and $k = 1, 2$, and that (b) $\overline{\mathbf{X}}$ is inadmissible if $k \geq 3$ $(\Theta = \mathbb{R}^k \times (0,\infty))$. [Hint: Assume the admissibility of the sample mean for $k = 1, 2$, when σ^2 is known, and use Theorem 3.10.]

Ex. 3.9. Let $\overline{\mathbf{X}}$ be the mean of a random sample from $N(\boldsymbol{\mu}, \Sigma)$ when $\boldsymbol{\mu} \in \mathbb{R}^k \equiv \Theta_1$, $\Sigma \in \Theta_2 \equiv$ set of all symmetric positive definite $k \times k$ matrices. Let $\Theta = \Theta_1 \times \Theta_2$, $\mathscr{A} = \Theta_1$, and let the loss function be squared error $L(\theta, a) = |\boldsymbol{\mu} - \mathbf{a}|^2$.

(a) Show that $\overline{\mathbf{X}}$ is an admissible estimator of $\boldsymbol{\mu}$ when the parameter space is restricted to $\Theta_1 \times \{\Sigma\}$ for any given $\Sigma \in \Theta_2$, and $k = 1$ or 2.
(b) Show that $\overline{\mathbf{X}}$ is an admissible estimator of $\boldsymbol{\mu}$ if $k = 1$ or 2 and when $\boldsymbol{\mu}$ and Σ are both unknown, i.e., the parameter space is $\mathbb{R}^k \times \Theta_2$.
*(c) *(Optional.)* In the cases (a), (b) show that $\overline{\mathbf{X}}$ is inadmissible if $k \geq 3$. [Hint: Brownian motion with zero drift and arbitrary non-singular diffusion matrix Γ is recurrent iff $k \leq 2$.]

Ex. 3.10. Extend Example 3.6 to obtain

(a) the posterior distribution of the mean $\boldsymbol{\theta}$ of the Normal distribution $N(\boldsymbol{\theta}, I_k)$, based on a random sample $\mathbf{X}_1, \dots, \mathbf{X}_n$, when the prior distribution of $\boldsymbol{\theta}$ is taken to be that of k independent Normal random variables, $\theta_i \overset{d}{\sim} N(0, \beta_i^2)$ $(i = 1, \dots, k)$. Also,
(b) compute the Bayes estimator of $\boldsymbol{\theta}$ under squared error loss.

(c) Compute the Bayes estimator of $\boldsymbol{\theta}$ when the prior distribution is that of k independent Normal random variables $\theta_i \overset{d}{\sim} N(C_i, \beta_i^2)$ for some $c_i \in \mathbb{R}$ and $\beta_i^2 > 0 \ \forall \ i$.

Ex. 3.11. Consider a random sample $(X_1, \ldots, X_n)$ of size $n = 50$ from a gamma distribution $\mathscr{G}(\alpha, \beta)$, with $\overline{X} = 4.5$, $\frac{1}{n}\sum_{j=1}^{n} X_j^2 = 41$, and $\frac{1}{n}\sum_{j=1}^{n} \log X_j = 0.95$.

(a) Find the method-of-moments estimates of α, β.
(b) Use the estimates in (a) as the initial trial solution of the likelihood equations, and apply the Newton–Raphson, or the gradient method, to compute the MLEs $\widehat{\alpha}, \widehat{\beta}$, by iteration.

Ex. 3.12. Consider $\mathbf{X} = (X_1, \ldots, X_n)$ where X_i's are i.i.d. $N(\mu, \sigma^2)$ with μ known and $\theta = \sigma^2 > 0$ is the unknown parameter. Let the prior τ for σ^2 be the *inverse gamma* $\mathscr{IG}(\alpha, \beta)$, i.e., $1/\sigma^2$ has the gamma distribution $\mathscr{G}(\alpha, \beta)$.

(a) Compute the posterior distribution of σ^2. [Hint: First compute the posterior distribution of $1/\sigma^2$.]
(b) Find the Bayes estimator of σ^2 under squared error loss $L(\sigma^2, a) = (\sigma^2 - a)^2$.

References

Ferguson, T. (1967). *Mathematical statistics: A decision theoretic approach.* Boston: Academic.

James, W., & Stein, C. (1961). Estimation with quadratic loss. In *Proceedings of the Fourth Berkeley Symposium on Mathematical Statistics and Probability, Berkeley Symposium on Mathematical Statistics and Probability* (Vol. 1, pp. 361–379). University of California Press

Lehmann, E. L., & Casella, G. (1998). *Theory of point estimation* (2nd ed.). New York: Springer.

Chapter 4
Sufficient Statistics, Exponential Families, and Estimation

Abstract A *sufficient statistic* is a function of the observed data $\mathbf{X}$ containing all the information that $\mathbf{X}$ holds about the model. A *complete sufficient statistic* is one that reduces the data the most, without losing any information. More importantly, according to Rao–Blackwell-, Lehmann–Scheffé-theorems, statistical inference procedures must be based on such statistics for purposes of efficiency or optimality.

4.1 Sufficient Statistics and Unbiased Estimation

For simplicity, we assume $\Omega = \mathscr{X}$, $\mathbf{X} : \mathscr{X} \to \mathscr{X}$ the *identity map*. Let $(\mathscr{T}, \mathscr{B}(\mathscr{T}))$ be a measurable space and $T : \Omega \to \mathscr{T} \equiv \mathscr{R}_T$ (range space of T) a measurable map. Then T is said to be a *statistic*. Generally one requires that T does not depend on unknown population parameters (i.e., it can be computed entirely based on the observation X). We will often write $\mathscr{B}_T$ for $\mathscr{B}(\mathscr{T})$ and $\sigma(T) = T^{-1}(\mathscr{B}_T) \equiv \{T^{-1}(B) : B \in \mathscr{B}_T\}$.

Definition 4.1. A statistic T is said to be *sufficient* for a family $\mathscr{P}$ of probability measures P on $\mathscr{X}$ if the conditional distribution of the observation (vector) $\mathbf{X}$, given T, is the same for all $P \in \mathscr{P}$. If we index $\mathscr{P}$ as $\{P_\theta : \theta \in \Theta\}$ where Θ is an index (or parameter-) set, one says T is *sufficient for θ* if the conditional distribution of $\mathbf{X}$, given T, does not depend on θ.

Example 4.1. $\Omega = \mathscr{X} = \{0,1\}^n$, $P_\theta(\{\mathbf{x}\}) \equiv f(\mathbf{x} \mid \theta) = \theta^{\sum x_i}(1-\theta)^{n-\sum x_i}$, $\theta \in \Theta = (0,1)$. Then $T \equiv \sum_1^n X_i$ (i.e., $T(\mathbf{x}) = \sum_{i=1}^n x_i \ \forall \ \mathbf{x}$) is a sufficient statistic for θ. To see this note that, for any *given* $\mathbf{x} = \mathscr{X}$ and $t \in \mathscr{R}_T = \{0,1,\dots,n\}$,

$$P_\theta(\mathbf{X} = (x_1,\dots x_n) \equiv \mathbf{x} \mid T = t) = \frac{P_\theta(X_1 = x_1,\dots,X_n = x_n \ \& \ T = t)}{P_\theta(T = t)}$$

$$= \begin{cases} 0 & \text{if } T(\mathbf{x}) \neq t, \\ \frac{\theta^t(1-\theta)^{n-t}}{\binom{n}{t}\theta^t(1-\theta)^{n-t}} \equiv \frac{1}{\binom{n}{t}} & \text{if } T(\mathbf{x}) = t. \end{cases} \tag{4.1}$$

R. Bhattacharya et al., *A Course in Mathematical Statistics and Large Sample Theory*, Springer Texts in Statistics, DOI 10.1007/978-1-4939-4032-5_4

For, T has the binomial distribution $B(n, \theta)$, and (i) if $T(\mathbf{x}) \neq t$, the set $\{X_1 = x_1, \ldots, X_n = x_n, \, \& \, T(\mathbf{X}) = t\} = \emptyset$, (ii) if $T(\mathbf{x}) = t$, then the set $\{T = t\} \equiv \{y \in \mathcal{X} : T(y) = t\} \supset \{\mathbf{X} = \mathbf{x}\} \equiv \{\mathbf{x}\}$, so that $\{\mathbf{X} = \mathbf{x}, T = t\} = \{\mathbf{X} = \mathbf{x}\}$.

Note that, equivalently, $\overline{X}$ *is sufficient for* θ.

Example 4.2. Let $\mathcal{X} = \mathbb{R}^n = \Omega$, $\theta \in \Theta = \mathbb{R}$, and P_θ has density (with respect to Lebesgue measure on $\mathbb{R}^n$)

$$f(\mathbf{x} \mid \theta) = (2\pi)^{-n/2} e^{-\frac{1}{2}\sum_{i=1}^{n}(x_i - \theta)^2}, \qquad (\mathbf{x} \in \mathbb{R}^n). \qquad (4.2)$$

In this case $T = \overline{X} = \frac{1}{n}\sum_{i=1}^{n} X_i$ is sufficient for θ. To see this consider the orthogonal transformation $U : \mathbf{x} \to \mathbf{y}$ given by

$$y_1 = \sum_{i=1}^{n} \frac{1}{\sqrt{n}} x_i, \quad y_j = \sum_{i=1}^{n} c_{ij} x_i \qquad (2 \leq j \leq n), \qquad (4.3)$$

where the vectors $(c_{1j}, c_{2j}, \ldots, c_{nj})$, $2 \leq j \leq n$, are of unit length, orthogonal to $(1/\sqrt{n}, 1/\sqrt{n}, \ldots, 1/\sqrt{n})$, and orthogonal to each other. Then $\mathbf{Y} = U(\mathbf{X}) \equiv (Y_1, \ldots, Y_n)$ has the distribution of n independent Normal random variables, with Y_1 having the distribution $N(\sqrt{n}\,\theta, 1)$, while the distribution of each Y_j ($2 \leq j \leq n$) is $N(0, 1)$. Therefore, the conditional distribution of $\mathbf{Y}$, given $Y_1 = y_1$ is the distribution of $(y_1, Y_2, \ldots, Y_n)$ with Y_j's i.i.d. $N(0, 1)$, $2 \leq j \leq n$. Now note that $\overline{X} = \frac{1}{\sqrt{n}} Y_1$. Hence the conditional distribution of $\mathbf{Y}$, given $\overline{X} = z$ is the distribution of $(\sqrt{n}\,z, Y_2, \ldots, Y_n)$. But $\mathbf{X} = U^{-1}(\mathbf{Y})$ ($= U'(\mathbf{Y})$ if U is identified with the matrix of orthogonal rows in (4.3)). Hence the conditional distribution of $\mathbf{X}$ given $\overline{X} = z$ is the distribution of $U^{-1}(\sqrt{n}\,z, Y_2, \ldots, Y_n)$ which does not depend on θ. Hence $\overline{X}$ is sufficient for θ.

Remark 4.1. For statistical inference about θ (or, P_θ), $\theta \in \Theta$, it is enough to know the value of a sufficient statistic T for θ. For, given $T = t$, one can simulate the random variable $\mathbf{X}^0 = (X_1^0, X_2^0, \ldots, X_n^0)$, say, whose distribution is the same as the conditional distribution, given $T = t$. The (unconditional, or) marginal distribution of $\mathbf{X}^0$ is then the same as the distribution P_θ of $\mathbf{X}$. In other words, given T the rest of the data contain no additional information about θ that can not be gleaned from T. (Exercise 4.1)

Since the conditional distribution of $\mathbf{X}$, given a statistic T, is not generally very easy to calculate the following criterion is very useful.

Theorem 4.1 (The Factorization Theorem). *Let $\mathcal{P} = \{P_\theta : \theta \in \Theta\}$ be such that each P_θ is absolutely continuous with respect to a sigma-finite measure μ on $(\mathcal{X}, \mathcal{B}(\mathcal{X}))$ with density $f(\mathbf{x} \mid \theta) = (dP_\theta/d\mu)(\mathbf{x})$. Then T is sufficient for $\mathcal{P}$ if and only if one has a factorization of f in the form*

$$f(\mathbf{x} \mid \theta) = g(T(\mathbf{x}), \theta) h(\mathbf{x}) \qquad a.e. \ (\mu), \qquad (4.4)$$

where, for each $\theta \in \Theta$, $t \to g(t, \theta)$ is a measurable nonnegative function on $\mathcal{T}$, and h is a nonnegative measurable function on $\mathcal{X}$ which does not depend on θ.

Proof. For a complete proof, see *Testing Statistical Hypothesis* (2005), by E. Lehmann and J. Romano, pp. 43–44. We will give a proof for the discrete case (i.e., $\mathcal{X}$ countable). Let $f(\mathbf{x} \mid \theta)$ denote $P_\theta(\{\mathbf{x}\})$. Suppose (4.4) holds. Without loss of generality, assume $h(\mathbf{x}) > 0 \ \forall \ \mathbf{x} \in \mathscr{X}$. For, if $h(\mathbf{x}) = 0$, then (4.4) implies $f(\mathbf{x} \mid \theta) = 0 \ \forall \ \theta$. Hence the set $\{\mathbf{x} \in \mathscr{X} : h(\mathbf{x}) = 0\}$ may be removed from $\mathscr{X}$. Then the distribution of T (under P_θ) is given by

$$P_\theta(T = t) = \sum_{\{\mathbf{x} \in \mathscr{X}, T(\mathbf{x}) = t\}} f(\mathbf{x} \mid \theta) = g(t, \theta) h_1(t), \tag{4.5}$$

where $h_1(t) = \sum_{\{\mathbf{x} \in \mathscr{X} : T(\mathbf{x}) = t\}} h(\mathbf{x})$. Then

$$P_\theta(\mathbf{X} = \mathbf{x}, T = t) = \begin{cases} 0 & \text{if } T(\mathbf{x}) \neq t \\ P_\theta(\mathbf{X} = \mathbf{x}) & \text{if } T(\mathbf{x}) = t, \end{cases}$$

i.e.,

$$P_\theta(\mathbf{X} = \mathbf{x}, T = t) = g(t, \theta) h(\mathbf{x}) \mathbf{1}_{\{T(\mathbf{x}) = t\}}. \tag{4.6}$$

If $P_\theta(T = t) > 0$, then dividing (4.6) by (4.5) one gets

$$P_\theta(\mathbf{X} = \mathbf{x} \mid T = t) = \frac{h(\mathbf{x})}{h_1(t)} \mathbf{1}_{\{T(\mathbf{x}) = t\}}. \tag{4.7}$$

Now note that $P_{\theta'} (T = t) = 0$, for some θ if and only if $\{\mathbf{x} : T(\mathbf{x}) = t\} = \emptyset$. Hence (4.7) holds for all θ, $\mathbf{x}$ and dT, and T is sufficient.

Conversely, suppose T is sufficient for θ. Then, writing $P_\theta(T = t) = g(t, \theta)$ and $P_\theta(\mathbf{X} = \mathbf{x} \mid T = t) = h_2(\mathbf{x}, t)$, one obtains, on the set $\{T(\mathbf{x}) = t\} \subset \mathscr{X}$, for a given t,

$$\begin{aligned} f(\mathbf{x} \mid \theta) \equiv P_\theta(\mathbf{X} = \mathbf{x}) = P_\theta(\mathbf{X} = \mathbf{x}, T = t) = P_\theta(T = t) h_2(\mathbf{x}, t) \\ = g(t, \theta) h_2(\mathbf{x}, t) = g(T(\mathbf{x}), \theta) h_2(\mathbf{x}, T(\mathbf{x})) = g(T(\mathbf{x}), \theta) h(\mathbf{x}), \end{aligned}$$

say. $\square$

Example 4.3. Let $\mathbf{X} = (X_1, \ldots, X_n)$ where X_i's are independent uniform random variables on the interval $[\alpha, \beta]$ ($\alpha < \beta$). That is, $\mathscr{X} = \mathbb{R}^n$, $\theta = (\alpha, \beta) \in \Theta = \{(\alpha, \beta) : \alpha < \beta \text{ real}\}$, P_θ has density (w.r.t. Lebesgue measure on $\mathbb{R}^n$)

$$\begin{aligned} f(\mathbf{x} \mid \theta) &= \frac{1}{(\beta - \alpha)^n} \prod_{i=1}^n \mathbf{1}_{[\alpha, \beta]}(x_i) \\ &= \frac{1}{(\beta - \alpha)^n} \mathbf{1}_{\{\alpha \leq \min(x_1, \ldots, x_n), \beta \geq \max(x_1, \ldots, x_n)\}} \\ &= g(T(\mathbf{x}), \theta), \text{ say,} \end{aligned}$$

where $T(\mathbf{x}) = (\min(x_1, \ldots, x_n), \max(x_1, \ldots, x_n))$ (a measurable map on $\mathscr{X}$ into $\mathscr{T} = \{(m, M) : m, M \text{ real numbers } m \leq M\}$), and $g((m, M), (\alpha, \beta)) = (\beta - \alpha)^{-n} \mathbf{1}_{\{m \geq \alpha\}} \mathbf{1}_{\{M \leq \beta\}}$. Here one may take $h(\mathbf{x}) \equiv 1$ in (4.4). Hence T is sufficient for $\theta = (\alpha, \beta)$.

Example 4.4. Let $\mathbf{X} = (X_1, \ldots, X_n)$ be i.i.d. $N(\mu, \sigma^2)$, $\mu \in \mathbb{R}$ and $\sigma^2 > 0$ being both unknown parameters. That is, $\mathscr{X} = \mathbb{R}^n$, $\Theta = \mathbb{R} \times (0, \infty)$, P_θ has density (w.r.t. Lebesgue measure on $\mathbb{R}^n$), given by

$$f(\mathbf{x} \mid (\mu, \sigma^2)) = (2\pi\sigma^2)^{-n/2} \exp\left\{ -\frac{1}{2\sigma^2} \sum_{i=1}^n (x_i - \mu)^2 \right\}$$

which may be expressed as

$$f(\mathbf{x} \mid (\mu, \sigma^2)) = (2\pi\sigma^2)^{-n/2} \exp\left\{ -\sum_{i=1}^{n} (x_i - \overline{x} + \overline{x} - \mu)^2 / 2\sigma^2 \right\}$$

$$= (2\pi\sigma^2)^{-n/2} \exp\left\{ -\left[\sum_{i=1}^{n} (x_i - \overline{x})^2 + n(\overline{x} - \mu)^2 \right] \Big/ 2\sigma^2 \right\}$$

$$= g(T(\mathbf{x}), \theta), \quad \text{say}, \ (\theta = (\mu, \sigma^2)),$$

with $T(\mathbf{x}) = (\sum_{i=1}^{n} (x_i - \overline{x})^2, \overline{x})$. Hence taking $h(\mathbf{x}) \equiv 1$ in (4.4), it follows that T is sufficient for (μ, σ^2).

The next theorem describes the important role sufficient statistics play in improving decision rules. Before we state and prove it, let us prove

Lemma 4.1 (Jensen's Inequality for Convex Functions in Multi-dimension). *Let C be a (measurable) convex subset of $\mathbb{R}^k$ and f a real-valued convex function on C. If Y is a random variable with values in C such that EY is finite, then $EY \in C$ and*

$$f(EY) \leq Ef(Y).$$

If f is strictly convex, then this inequality is strict unless $P(Y = EY) = 1$.

Proof. Convexity of f means that, for every $z \in C$, there exists $m \in \mathbb{R}^k$ such that

$$f(y) \geq f(z) + m \cdot (y - z) \qquad y \in C, \tag{4.8}$$

with a strict inequality for $y \neq z$ if f is strictly convex. [Note: In the case C is an open convex set and f is twice continuously differentiable with the Hessian matrix $((D_i D_j f(\mathbf{x})))$ positive definite for every $\mathbf{x}$, (4.8) follows by a Taylor expansion of f around z, with $m(z) = (\operatorname{grad} f)(z)]$.

To prove the desired inequality, let $z = EY$, $y = Y$ in (4.8) to get

$$f(Y) \geq f(EY) + m(EY) \cdot (Y - EY). \tag{$*$}$$

Taking expectations on both sides, the inequality in the Lemma is obtained. In case f is strictly convex, the inequality $(*)$ is strict for every $Y(\omega) \neq EY$, and hence the inequality in the Lemma is strict unless Y is a constant $(= EY)$ a.s.
$\qquad\square$

Theorem 4.2 (Rao–Blackwell Theorem). *Let $\mathscr{A}$ be a (measurable) convex subset of $\mathbb{R}^k$, $a \to L(\theta, a)$ a convex function for each $\theta \in \Theta$, and T a sufficient statistic for θ. If d is a (non-randomized) decision rule, then the decision rule*

$$\hat{d}(T) = E_\theta(d(\mathbf{X}) \mid T) = E(d(\mathbf{X}) \mid T) \tag{4.9}$$

is at least as good as d: $R(\theta, \hat{d}) \leq R(\theta, d)$ for all $\theta \in \Theta$.

Proof. Fix $\theta \in \Theta$, and use (4.8) with $f = L(\theta, \cdot)$, $z = \hat{d}(T)$, $y = d(\mathbf{X})$ to get

$$L(\theta, d(\mathbf{X})) \geq L(\theta, \hat{d}(T)) + (d(\mathbf{X}) - \hat{d}(T)) \cdot m(\hat{d}(T)). \tag{4.10}$$

Now take conditional expectation, given T, on both sides to get

$$E(L(\theta, d(\mathbf{X})) \mid T) \geq L(\theta, \hat{d}(T)),$$

and then taking expectation complete the proof. $\qquad\square$

Remark 4.2. It is important to note that $\hat{d}(T)$ as defined by the first equation in (4.9) does not depend on θ, in view of the sufficiency of T. In other words, $\hat{d}(T)$ *is an estimator.* The inequality (4.10) holds whether T is sufficient or not.

Corollary 4.1. *Let $c(\theta)$ be a real-valued parametric function of θ, and $d(\mathbf{X})$ an estimator of $c(\theta)$, with $E_\theta d^2(\mathbf{X})$ finite for all θ. Let T be a sufficient statistic for θ. (a) Then, for every $\theta \in \Theta$,*

$$E_\theta(\hat{d}(T) - c(\theta))^2 \leq E_\theta(d(\mathbf{X}) - c(\theta))^2, \tag{4.11}$$

with a strict inequality unless $P_\theta(d(\mathbf{X}) = \hat{d}(T)) = 1$.
    (b) If $d(\mathbf{X})$ is an unbiased estimator of $c(\theta)$ (i.e., $E_\theta d(\mathbf{X}) = c(\theta) \ \forall \ \theta$), then so is $\hat{d}(T)$.

Proof. (a) One can derive this from Theorem 4.2, by letting $\mathscr{A} = \mathbb{R}$ and noting that $a \to L(\theta, a) \equiv (c(\theta) - a)^2$ is strictly convex. *Alternatively,* one has

$$\begin{aligned}
E_\theta(d(\mathbf{X}) - c(\theta))^2 &= E_\theta(d(\mathbf{X}) - \hat{d}(T) + \hat{d}(T) - c(\theta))^2 \\
&= E_\theta(\hat{d}(T) - c(\theta))^2 + E_\theta(d(\mathbf{X}) - \hat{d}(T))^2,
\end{aligned}$$

since

$$E_\theta[(d(\mathbf{X}) - \hat{d}(T))(\hat{d}(T) - c(\theta))] = E_\theta[(\hat{d}(T) - c(\theta)) \cdot E_\theta(d(\mathbf{X}) - \hat{d}(T) \mid T)] = 0.$$

(b)

$$E_\theta(\hat{d}(T)) = E_\theta[E_\theta(d(\mathbf{X}) \mid T)] = E_\theta d(\mathbf{X}) = c(\theta) \quad \forall \ \theta.$$

$\qquad\square$

Corollary 4.2. *In the hypothesis of Corollary 4.1 assume $E_\theta|d(\mathbf{X})| < \infty$ for all θ. Then (a) one has*

$$E_\theta \mid \hat{d}(T) - c(\theta)| \leq E_\theta|d(\mathbf{X}) - c(\theta)|.$$

Also, (b) $\hat{d}(T)$ is an unbiased estimator of $c(\theta)$ if $d(\mathbf{X})$ is.

Proof. (a) One may apply Theorem 4.2 here, noting that the function $a \to |a - c(\theta)|$ is convex. But, more simply, $|\hat{d}(T) - c(\theta)| = E(|d(\mathbf{X}) - c(\theta) \mid T)| \leq E(|d(\mathbf{X}) - c(\theta)| \mid T)$, and taking expectation with respect to P_θ one obtains the desired result.
(b) The proof of part (b) is the same as that of Corollary 4.1(b). $\qquad\square$

Remark 4.3. As the statement of Theorem 4.2 indicates, Corollaries 4.1, 4.2 extend to the case of estimation of vector valued parametric functions $\mathbf{c}(\theta) = (c_1(\theta), \ldots, c_k(\theta))$ for additive loss functions such as $L(\theta, \mathbf{a}) = \sum_{1 \leq i \leq k}(a_i - c_i(\theta))^2$ and $L(\theta, \mathbf{a}) = \sum_{1 \leq i \leq k}|a_i - c_i(\theta)|$, respectively.

Remark 4.4. In general there are many sufficient statistics for θ. Which one should you use to improve on a given decision rule or estimator in the manner of Theorem 4.2 or its Corollary above? To answer this, consider two sufficient statistics

T_1, T_2 (for θ) such that T_1 is a function of T_2, i.e., $T_1 = f(T_2)$, where $f(T_2)$ is a measurable function of T_2. Then given an estimator $d_2(T_2)$ of a parametric function $c(\theta)$, the estimator $\hat{d}_1(T_1) \equiv E_\theta(d_2(T_2) \mid T_1)$ is at least as good as $d_2(T_2)$ (say, under squared error loss $L(\theta, a) = (c(\theta) - a)^2$). One of course can reverse the argument and begin with an estimator $d_1(T_1)$ and have $\hat{d}_2(T_2) \equiv E_\theta(d_1(T_1) \mid T_2)$, which would be at least as good as $d_1(T_1)$. But, in this case, $\hat{d}_2(T_2) = E_\theta(d_1(f(T_2)) \mid T_2) = d_1(f(T_2)) = d_1(T_1)$. That is, $d_1(T_1)$ being already a function of T_2, the *Rao–Blackwellization* (of taking conditional expectation of a decision rule, given a sufficient statistic) does not alter the estimator! On the other hand, $\hat{d}_1(T_1)$ is a *strict improvement* over $d_2(T_2)$ unless $P_\theta(\hat{d}_1(T_1) = d_2(T_2)) = 1 \ \forall \ \theta \in \Theta$ (i.e., unless $d_2(T_2)$ is essentially a function of T_1 already). *Thus the "smaller" the sufficient statistic T the better.*

Remark 4.5. A statistic T may be identified with the sigma-field

$$\sigma(T) \equiv T^{-1}(\mathscr{B}_T) \equiv \{T^{-1}(C) : C \in \mathscr{B}_T\}.$$

If T_1 and T_2 are statistics (with possibly different range spaces $\mathscr{R}_{T_i}$ and corresponding σ-fields $\mathscr{B}_{T_i}$ $(i = 1, 2)$) are such that there is a bi-measurable one-to-one map g on $\mathscr{R}_{T_1}$ onto $\mathscr{R}_{T_2}$ with $T_2 = g(T_1)$ (so that $T_1 = g^{-1}(T_2)$), then $\sigma(T_1) = \sigma(T_2)$ and the statistics may be viewed as the *same, or to contain the same information about* θ, since knowing one means knowing the other.

It is technically more convenient to say that T_1 *is a smaller statistic than* T_2 if $\sigma(T_1) \subset \sigma(T_2)$, and that *they are equivalent* if $\sigma(T_1) = \sigma(T_2)$. In a statistical decision problem (with a family of distributions P_θ, $\theta \in \Theta$), we say $\sigma(T_1) \subset \sigma(T_2)$ *with* P_θ-*probability one* $\forall \ \theta \in \Theta$, if for every $B_1 \in \sigma(T_1)$ there exists $B_2 \in \sigma(T_2)$ such that $P_\theta(B_1 \Delta B_2) = 0 \ \forall \ \theta \in \Theta$. Here Δ denotes *symmetric difference* between sets: $B_1 \Delta B_2 = (B_1 \cap B_2^c) \cup (B_1^c \cap B_2)$.

Definition 4.2. In a statistical decision problem a sufficient statistic T^* is said to be *minimal sufficient* (for $\theta \in \Theta$) if given any other sufficient statistic T, $\sigma(T^*) \subset \sigma(T)$ with P_θ-probability one $\forall \ \theta \in \Theta$.

It can be shown that minimal sufficient statistics exist under fairly weak assumptions. But one can find examples when a minimal sufficient statistic does not exist.

For the purpose of unbiased estimation the following generally stronger property than minimality is very useful.

Definition 4.3. A sufficient statistic T is said to be *complete* if for any real-valued function $g(T)$ of T, integrable w.r.t. $P_\theta \ \forall \ \theta \in \Theta$,

$$E_\theta g(T) = 0 \ \ \forall \ \theta \in \Theta \implies P_\theta(g(T) = 0) = 1 \ \ \forall \ \theta \in \Theta. \tag{4.12}$$

A sufficient statistic T is said to be *boundedly complete* if (4.12) holds for all bounded measurable functions $g(T)$ of T. A complete sufficient statistic is obviously boundedly complete.

Proposition 4.1. *Suppose a minimal sufficient statistic S exists. Then if a sufficient statistic T exists which is boundedly complete then T is minimal.*

Proof. Suppose S is minimal sufficient and T is boundedly complete sufficient. Let $f(T)$ be any real-valued bounded $\sigma(T)$-measurable function, and consider $g(S) \equiv E_\theta(f(T)|\sigma(S))$. By minimality of S, there exists a $\sigma(T)$-measurable $h(T)$ such that $P_\theta(g(S) = h(T)) = 1 \ \forall \ \theta \in \Theta$. But $E_\theta(f(T) - h(T)) = 0 \ \forall \ \theta \in \Theta$, which, by bounded completeness of T, implies $P_\theta(f(T) = h(T)) = 1 \ \forall \ \theta \in \Theta$. Therefore, $P_\theta(f(T) = g(S)) = 1 \ \forall \ \theta \in \Theta$. Hence, $\sigma(T) \subset \sigma(S)$ with P_θ-probability one $\forall \ \theta \in \Theta$ (See Exercise 4.2). $\qquad\square$

Definition 4.4. A parametric function $c(\theta)$ is said to be *estimable* if there exists an estimator $d(\mathbf{X})$ such that $E_\theta d(\mathbf{X}) = c(\theta) \ \forall \ \theta \in \Theta$. In this case $d(\mathbf{X})$ is said to be an *unbiased estimator of* $c(\theta)$. An unbiased estimator $d^*(\mathbf{X})$ of $c(\theta)$ is said to be an *UMVU (uniformly minimum variance unbiased)* estimator if

$$E_\theta(d^*(\mathbf{X}) - c(\theta))^2 \leq E_\theta(d(\mathbf{X}) - c(\theta))^2 \qquad \forall \theta \in \Theta \qquad (4.13)$$

for all unbiased estimators $d(\mathbf{X})$ of $c(\theta)$.

Theorem 4.3 (Lehmann–Scheffé Theorem). *Let T be a complete sufficient statistic for $\theta \in \Theta$. Then for every estimable parametric function $c(\theta)$ for which there exists an unbiased estimator $d(\mathbf{X})$ satisfying $E_\theta d^2(\mathbf{X}) < \infty$, there exists a unique UMVU estimator given by $\hat{d}(T) = E_\theta(d(\mathbf{X}) \mid T)$.*

Proof. Clearly, $\hat{d}(T)$ is an unbiased estimator of $c(\theta)$. Also, by the Corollary to the Rao–Blackwell Theorem, $E_\theta(\hat{d}(T) - c(\theta))^2 \leq E_\theta(d(\mathbf{X}) - c(\theta))^2 \ \forall \ \theta \in \Theta$. To show that $\hat{d}(T)$ is UMVU, suppose, if possible, there exists an unbiased estimator $d_1(\mathbf{X})$ such that for some $\theta = \theta_0$, say, $E_{\theta_0}(d_1(\mathbf{X}) - c(\theta_0))^2 < E_{\theta_0}(\hat{d}(T) - c(\theta_0))^2$. Let $\hat{d}_1(T) = E(d_1(\mathbf{X}) \mid T)$. By the Rao–Blackwell Theorem, $E_{\theta_0}(\hat{d}_1(T) - c(\theta_0))^2 \leq E_{\theta_0}(d_1(\mathbf{X}) - c(\theta_0))^2$. But this would imply $E_{\theta_0}(\hat{d}_1(T) - c(\theta_0))^2 < E_{\theta_0}(\hat{d}(T) - c(\theta_0))^2$, and, in particular, $P_{\theta_0}(\hat{d}_1(T) - \hat{d}(T) \neq 0) > 0$. But $g(T) \equiv \hat{d}_1(T) - \hat{d}(T)$ satisfies $E_\theta g(T) = 0 \ \forall \ \theta \in \Theta$, and, therefore, by the completeness of T, $P_\theta(g(T) \neq 0) = 0 \ \forall \theta \in \Theta$, contradicting $P_{\theta_0}(g(T) \neq 0) > 0$. $\qquad\square$

Example 4.5. Let $\mathscr{X} = \{0,1\}^n$, $P_\theta(\{x\}) \equiv f(\mathbf{x} \mid \theta) = \theta^{\sum_1^n x_i}(1 - \theta)^{n - \sum_i^n x_i} \ \forall \mathbf{x} \in \mathscr{X}$, $\Theta = (0,1)$. We will show that $T = \sum_1^n X_i$ is a complete sufficient statistic. For this let g be a real-valued function on $\mathscr{R}_T = \{0, 1, \ldots, n\}$ such that $E_\theta g(T) = 0 \ \forall \ \theta \in \Theta$. Now the distribution of T, under P_θ, is the *binomial distribution* $\mathscr{B}(n, \theta)$,

$$P_\theta(T = t) = \binom{n}{t}\theta^t(1 - \theta)^{n-t} \qquad (t = 0, 1, \ldots, n), \qquad (4.14)$$

so that $E_{\theta_0} g(T) = 0 \ \forall \theta \in (0,1)$ may be expressed as

$$\sum_{t=0}^n g(t)\binom{n}{t}\theta^t(1 - \theta)^{n-t} = 0 \qquad \forall \theta \in (0,1). \qquad (4.15)$$

The left side is a polynomial of degree n, and can not have more than n zeroes in $(0,1)$ unless it is identically zero. Thus $g(t) = 0 \ \forall t \in \{0, 1, \ldots, n\}$, proving $T = \sum X_i$ is a complete sufficient statistic for $\theta \in \Theta$. [Note that we could have taken $\Theta = [0, 1]$ also.] To apply Theorem 4.3, let us first identify the set of all estimable parametric functions $c(\theta)$. If $c(\theta)$ is estimable, then there exists $d(\mathbf{X})$ such that $E_\theta d(\mathbf{X}) = c(\theta) \ \forall \theta \in \Theta$. Then $\hat{d}(T) \equiv E_\theta[d(\mathbf{X}) \mid T]$ is an unbiased estimator of $c(\theta)$. From the expression of the expectation on the left in (4.15), it follows that $c(\theta) = \sum_{t=0}^n \hat{d}(t)\binom{n}{t}\theta^t(1 - \theta)^{n-t}$ is a polynomial of degree n (or less). Hence the set C_n of all estimable functions is a subset of the set Γ_n of all polynomials (in θ) of degree n or less. To show that Γ_n is precisely C_n, note that θ^k is estimable for $k = 0, 1, \ldots, n$, with an estimator $d(\mathbf{X}) = X_1 X_2 \cdots X_k$ for $k = 1, \ldots, n$, and $d(\mathbf{X}) \equiv 1$ for $k = 0$. Hence all polynomials of degree n or less are estimable. The UMVU estimator of θ^k is given by

$$\hat{d}_k(T) = E(X_1 \ldots X_k \mid T), \qquad (4.16)$$

(Exercise 4.3).

Example 4.6. Let $X_1, \ldots, X_n$ be a random sample from the *uniform distribution* $\mathscr{U}(0, \theta)$, $\theta > 0$, with p.d.f. $f_1(x \mid \theta) = (1/\theta)\mathbf{1}_{(0,\theta]}(x)$, so that the (joint) density of $\mathbf{X} = (X_1, \ldots, X_n)$ is

$$f(\mathbf{x} \mid \theta) = \frac{1}{\theta^n} \prod_{j=1}^{n} \mathbf{1}_{[0 < x_j \le \theta]} = \frac{1}{\theta^n}\, \mathbf{1}_{[0 < x_j \le \theta \; \forall j = 1, \ldots, n]}$$

$$= \frac{1}{\theta^n}\, \mathbf{1}_{[0 < M(\mathbf{x}) \le \theta]}, \qquad \mathbf{x} \in \mathbf{X} = (0, \infty)^n$$

where $M(\mathbf{x}) = \max\{x_j : 1 \le j \le n\}$. By the Factorization Theorem, M is a sufficient statistic for θ. We will show that M is a complete sufficient statistic for θ. For this note that the distribution function of M is

$$F_M(t) \equiv P(M \le t) = P(X_j \le t \; \forall \; j = 1, \ldots, n) = \begin{cases} 0 & \text{for } t \le 0, \\ \left(\frac{t}{\theta}\right)^n & \text{for } 0 < t \le \theta, \\ 1 & \text{for } t > \theta, \end{cases}$$

so that its p.d.f. is $f_M(t \mid \theta) = \frac{1}{\theta^n} n t^{n-1} \mathbf{1}_{[0 < t \le \theta]}$. Now let $g(t)$ be such that $E_\theta g(M) = 0 \; \forall \theta \in \Theta = (0, \infty)$. This says

$$n\theta^n \int_0^\theta g(t) t^{n-1} dt = 0 \quad \forall \; \theta > 0,$$

implying $g(t) = 0$ a.e. (with respect to Lebesgue measure on $\mathbb{R}$). (Exercise 4.4). From the expression for $E_\theta g(M)$ given by the last integral, it follows that every function (of θ) of this form (i.e., with $g(t) t^{n-1}$ integrable on every interval $[0, a]$, $a > 0$) is estimable. In particular, with $g(t) = t$, one gets

$$E_\theta M = \frac{n}{\theta^n} \cdot \frac{\theta^{n+1}}{n+1} = \frac{n}{n+1}\, \theta,$$

so that $\frac{n+1}{n} M$ is an unbiased estimator of θ (and M is an unbiased estimator of $\frac{n}{n+1}\theta$). By the Lehmann–Scheffé Theorem, $\left(\frac{n+1}{n}\right) M$ is the uniformly minimum variance unbiased estimator (UMVU) of θ. Note that $E_\theta M^2 = \frac{n}{n+2}\theta^2$, so that

$$E_\theta\left(M - \frac{n}{n+1}\theta\right)^2 = \left(\frac{n}{n+2}\theta^2\right) - \left(\frac{n}{n+1}\theta\right)^2 = \theta^2\, n\left[\frac{1}{n+2} - \frac{n}{(n+1)^2}\right]$$

$$= \frac{n}{(n+2)(n+1)^2}\, \theta^2 \sim \frac{\theta^2}{n^2};$$

$$E_\theta\left(\frac{n+1}{n} M - \theta\right)^2 = \left(\frac{n+1}{n}\right)^2 \text{var}_\theta(M) = \frac{1}{n(n+2)}\, \theta^2.$$

We now turn to a large class of parametric families where complete sufficient statistics exist, so that every estimable parametric function has an UMVU estimator (unique, in case its risk function, or variance, is finite). Examples 4.2, 4.4 and 4.5 are such families, although Example 4.6 is not.

4.2 Exponential Families

We begin with the one-parameter case.

Definition 4.5. A *one-parameter exponential family of distributions* $\{P_\theta : \theta \in \Theta\}$, Θ an interval, is such that the probability measure P_θ on a state space S has a density $p(x \mid \theta)$, with respect to a reference measure v on S, of the form

$$p(x \mid \theta) = c(\theta)h(x)e^{\pi(\theta)t(x)}, \ x \in S, \quad \left[c(\theta) = 1/\int_s h(x)e^{\pi(\theta)t(x)}v(dx)\right], \quad (4.17)$$

where $h(x) > 0$ for all x in S, and $\theta \to \pi(\theta)$ is one-to-one, and $t(x)$ is a real-valued (measurable) function on S. Reparametrizing $\theta \to \pi(\theta) = \pi$, and writing $\tilde{p}(x \mid \pi) = p(x \mid \theta)$ for θ such that $\pi(\theta) = \pi$, one has

$$\tilde{p}(x \mid \pi) = \tilde{c}(\pi)h(x)e^{\pi t(x)}, \quad \left(\tilde{c}(\pi) = \left[\int_s h(x)e^{\pi t(x)}v(dx)\right]^{-1}\right). \quad (4.18)$$

The new parameter is called the *natural parameter* and the *natural parameter space* is

$$\Pi = \left\{\pi \in R : \int h(x)e^{\pi t(x)}v(dx) < \infty\right\}. \quad (4.19)$$

In the case S is an interval, finite or infinite, and $v(dx)$ is Lebesgue measure dx, p or $\tilde{p}$ is the usual density on S, while on a countable state space S the measure v is the *counting measure* with $v(\{x\}) = 1$ for every $x \in S$. The distribution with density $\tilde{p}(x \mid \pi)$ (with respect to v) will be written as P_π, or even P_θ.

Consider now a random sample $\mathbf{X} = (X_1, \ldots, X_n)$ from P_θ, with $X_1, \ldots, X_n$ independent having the common distribution P_θ (or P_π). The (joint) density of $\mathbf{X}$ (with respect to the product measure $\mu = v \times v \times \cdots \times v$) on the observation space $\mathscr{X} = S^n$ is written as

$$f(\mathbf{x} \mid \theta) = c^n(\theta)\prod h(x_j)e^{\pi(\theta)\sum t(x_j)}, \ \text{or} \ \tilde{f}(\mathbf{x} \mid \pi) = \tilde{c}^n(\pi)\prod h(x_j)e^{\pi \sum t(x_j)},$$
$$(4.20)$$

where the product $\prod$ and the sum $\sum$ are both over the indices $j = 1, \ldots, n$. Note that if $v(dx) = dx$ is the Lebesgue measure then $\mu(d\mathbf{x}) = dx_1 \ldots dx_n$ is the usual Lebesgue measure in n-dimension. In the case v is the counting measure on S, μ is the counting measure on $\mathscr{X} = S^n : \mu(\{\mathbf{x}\}) = 1$ for every $\mathbf{x} \in \mathscr{X}$. Note that, by the Factorization Theorem, $T(\mathbf{x}) = \sum_{1 \leq j \leq n} t(x_j)$ is a sufficient statistic.

As an example, let $X_1, \ldots, X_n$ be independent normal $N(\theta, \sigma^2)$ random variables (each with mean θ and variance σ^2). Assume σ^2 is known. Then with $S = R$, $v(dx) = dx$, the common pdf of the X_i's is

$$p(x \mid \theta) = (2\pi\sigma^2)^{-\frac{1}{2}} \exp\left\{ -\frac{(x-\theta)^2}{2\sigma^2} \right\} = c(\theta)h(x)\exp\left\{ \left(\frac{\theta}{\sigma^2}\right)x \right\} \qquad (\theta \in \Theta = R),$$

$$\left[c(\theta) = (2\pi\sigma^2)^{-1/2}\exp\left\{ \frac{-\theta^2}{2\sigma^2} \right\}, \quad h(x) = \exp\left\{ \frac{-x^2}{2\sigma^2} \right\} \right]. \tag{4.21}$$

Hence the natural parameter is $\pi = \theta/\sigma^2$ and $t(x) = x$. The natural parameter space is $\prod = R$. Also, the distribution of the sufficient statistic $T = \sum_{1 \leq j \leq n} X_j$ is $N(n\theta, n\sigma^2)$ is a one-parameter exponential family, with density (with respect to Lebesgue measure on R) given by

$$f_T(t \mid \theta) = (2\pi n\sigma^2)^{-1/2}\exp\left\{ -\frac{(t-n\theta)^2}{2n\sigma^2} \right\}$$

$$= c(\theta)h(t)\exp\left\{ \left(\frac{\theta}{\sigma^2}\right)t \right\} = \tilde{c}_1(\pi)h(t)\exp\{\pi t\}, \tag{4.22}$$

where $c(\theta) = (2\pi n\sigma^2)^{-1/2}\exp\{-n\theta^2/2\sigma^2\}$, $h(t) = \exp\{-t^2/2n\sigma^2\}$, and $\pi = \theta/\sigma^2$.

An example of a one-parameter family of discrete distributions is the Bernoulli family considered in Example 4.8.

We next consider the general case of k-parameter exponential families.

Let $\{G_{\boldsymbol{\theta}} : \boldsymbol{\theta} \in \Theta\}$ be a family of probability measures on a measurable space $(S, \mathscr{S})$ which are absolutely continuous with respect to a sigma-finite measure ν. If the density $p(x \mid \boldsymbol{\theta})$ of $G_{\boldsymbol{\theta}}$ (w.r.t. ν) is of the form

$$p(x \mid \boldsymbol{\theta}) = C(\boldsymbol{\theta})h(x)\exp\left\{ \sum_{i=1}^{k} \pi_i(\boldsymbol{\theta})T_i(x) \right\} \qquad (x \in S, \boldsymbol{\theta} \in \Theta), \tag{4.23}$$

where h is a nonnegative measurable function on S, T_i, $1 \leq i \leq k$, are real-valued measurable functions on S, then $\{G_{\boldsymbol{\theta}} : \boldsymbol{\theta} \in \Theta\}$ is said to be a *k-parameter exponential family*. Here π_i are real-valued functions on Θ, and $C(\boldsymbol{\theta})$ is a normalizing constant,

$$C(\boldsymbol{\theta}) = \left(\int_S h(x)\exp\left\{ \sum_{i=1}^{k}\pi_i(\boldsymbol{\theta})T_i(x) \right\}\nu(dx) \right)^{-1}. \tag{4.24}$$

Let ν_T denote the image of the measure $h(x)d\nu(x)$ on $(\mathbb{R}^k, \mathscr{B}(\mathbb{R}^k))$ under the map $x \to T(x) \equiv (T_1(x), \ldots, T_k(x))$. That is,

$$\nu_T(B) = \int_{T^{-1}(B)} h(x)d\nu(x), \quad B \in \mathscr{B}(\mathbb{R}^k) \tag{4.25}$$

Then the *distributions* $G_{\boldsymbol{\theta}}^T$, say, *of T under $G_{\boldsymbol{\theta}}$* have densities with respect to $\nu_{\mathbf{T}}$ given by

$$p_T(\mathbf{t} \mid \boldsymbol{\theta}) = C(\boldsymbol{\theta})\exp\left\{ \sum_{i=1}^{k}\pi_i(\boldsymbol{\theta})t_i \right\} \qquad (\mathbf{t} = (t_1, \ldots, t_k) \in \mathbb{R}^k, \ \theta \in \Theta), \tag{4.26}$$

so that $\{G_{\boldsymbol{\theta}}^T : \boldsymbol{\theta} \in \Theta\}$ is a k-parameter exponential family. Note that [see (4.23)], by the Factorization Theorem, T is a *sufficient statistic for* $\{G_{\boldsymbol{\theta}} : \boldsymbol{\theta} \in \Theta\}$. One may reparametrize $\{G_{\boldsymbol{\theta}} : \boldsymbol{\theta} \in \Theta\}$ with the new parameter $\boldsymbol{\pi} = (\pi_1, \pi_2, \ldots, \pi_k) \in \mathbb{R}^k$, noting that if $\pi(\boldsymbol{\theta}) = \pi(\boldsymbol{\theta}')$ then $G_{\boldsymbol{\theta}} = G_{\boldsymbol{\theta}'}$. One then writes the density (4.23) in the form

$$\tilde{p}(x \mid \boldsymbol{\pi}) = \widetilde{C}(\boldsymbol{\pi})h(x) \exp\left\{ \sum_{i=1}^{k} \pi_i T_i(x) \right\} \qquad (\widetilde{C}(\boldsymbol{\pi})) = (\textstyle\int h(x) e^{\sum_1^k \pi_i T_i(x)} d\nu(x))^{-1}.$$

$$(4.27)$$

$\boldsymbol{\pi}$ is called a *natural parameter,* and the density (4.26) (w.r.t. ν_T) becomes

$$\tilde{p}_T(\mathbf{t} \mid \boldsymbol{\pi}) = \widetilde{C}(\boldsymbol{\pi}) \exp\left\{ \sum_{i=1}^{k} \pi_i t_i \right\} \qquad (\mathbf{t} \in \mathbb{R}^k). \tag{4.28}$$

The *natural parameter space* Π is taken to be the set of all $\boldsymbol{\pi} \in \mathbb{R}^k$ for which the integral within parentheses in (4.27) is finite:

$$\Pi = \left\{ \boldsymbol{\pi} \in \mathbb{R}^k : \int_S h(x) e^{\sum_1^k \pi_i T_i(x)} d\nu(x) < \infty \right\} \subset \mathbb{R}^k. \tag{4.29}$$

Suppose $X_1, X_2, \ldots, X_n$ are i.i.d. random variables having a common density (w.r.t. ν) of the form (4.23) [or, of the form (4.27)]. Then the (joint) distribution $P_{\boldsymbol{\theta}}$ of $\mathbf{X} = (X_1, \ldots, X_n)$ has a density w.r.t. the product measure $d\mu(\mathbf{x}) = d\nu(x_1) \times d\nu(x_2) \times \cdots \times d\nu(x_n)$ (on $(S^n, \mathscr{S}^{\otimes n})$), given by

$$f(\mathbf{x} \mid \boldsymbol{\theta}) = C^n(\boldsymbol{\theta}) \left(\prod_{j=1}^{n} h(x_j) \right) \exp\left\{ \sum_{i=1}^{k} \pi_i(\boldsymbol{\theta}) \left(\sum_{j=1}^{n} T_i(x_j) \right) \right\}, \quad (\mathbf{x} = (x_1, x_2, \ldots, x_n)),$$

$$(4.30)$$

or, in terms of the natural parameter,

$$\tilde{f}(\mathbf{x} \mid \boldsymbol{\pi}) = \widetilde{C}^n(\boldsymbol{\pi}) \left(\prod_{j=1}^{n} h(x_j) \right) \exp\left\{ \sum_{i=1}^{k} \pi_i \left(\sum_{j=1}^{n} T_i(x_j) \right) \right\}. \tag{4.31}$$

Thus the family of distributions $\{P_{\boldsymbol{\theta}} : \boldsymbol{\theta} \in \Theta\}$ on $(\mathscr{X} = S^n, \mathscr{S}^{\otimes n})$ is a k-parameter exponential family, and $\mathbf{T}(\mathbf{x}) = (\sum_{j=1}^{n} T_1(x_j), \ldots, \sum_{j=1}^{n} T_k(x_j))$ is a *sufficient statistic* for $\{P_{\boldsymbol{\theta}} : \boldsymbol{\theta} \in \Theta\}$ (or, $\{\tilde{P}_{\boldsymbol{\pi}} : \boldsymbol{\pi} \in \Pi\}$). The same argument as above shows that the distributions $P_{\boldsymbol{\theta}}^{\mathbf{T}}$ $(\boldsymbol{\theta} \in \Theta)$ (or $\{\tilde{P}_{\boldsymbol{\pi}}^{\mathbf{T}} : \boldsymbol{\pi} : \Pi\}$) of $\mathbf{T}$ form a k-parameter exponential family with density

$$f_{\mathbf{T}}(\mathbf{t} \mid \boldsymbol{\theta}) = C^n(\boldsymbol{\theta}) \exp\left\{ \sum_{i=1}^{k} \pi_i(\boldsymbol{\theta}) t_i \right\} \qquad (\boldsymbol{\theta} \in \Theta, \mathbf{t} \in \mathbb{R}^k)$$

or,

$$\tilde{f}_{\mathbf{T}}(\mathbf{t} \mid \boldsymbol{\pi}) = \widetilde{C}^n(\boldsymbol{\pi}) \exp\left\{ \sum_{i=1}^{k} \pi_i t_i \right\} \qquad (\boldsymbol{\pi} \in \Pi, \mathbf{t} \in \mathbb{R}^k)$$

w.r.t. the measure (on $(\mathbb{R}^k, \mathscr{B}(\mathbb{R}^k))$) given by

$$\mu_{\mathbf{T}}(B) = \int_{\mathbf{T}^{-1}(B)} \left(\prod_{j=1}^{n} h(x_j) \right) d\mu(\mathbf{x}) \qquad (B \in \mathscr{B}(\mathbb{R}^k)). \tag{4.32}$$

Example 4.7 (Normal). Let $X_1, \ldots, X_n$ be i.i.d. with common distribution $G_{\boldsymbol{\theta}} = N(\mu, \sigma^2)$, $(\boldsymbol{\theta} = (\mu, \sigma^2) \in \mathbb{R} \times (0, \infty) \equiv \Theta)$, whose density w.r.t. Lebesgue measure ν on $\mathbb{R}$ is

$$p(x \mid \boldsymbol{\theta}) = \frac{1}{\sqrt{2\pi\sigma^2}} \exp\left\{-\frac{(x-\mu)^2}{2\sigma^2}\right\}$$

$$= \frac{1}{\sqrt{2\pi\sigma^2}} \exp\left\{-\frac{\mu^2}{2\sigma^2}\right\} \exp\left\{\frac{\mu}{\sigma^2}x - \frac{1}{2\sigma^2}x^2\right\}.$$

Thus $\{G_{\boldsymbol{\theta}} : \boldsymbol{\theta} \in \Theta\}$ is a *2-parameter exponential family*, with natural parameters $\pi_1 = \frac{\mu}{\sigma^2}$, $\pi_2 = -\frac{1}{2\sigma^2}$, and (sufficient statistic) $T(x) = (x, x^2)$. The natural parameter space is $\Pi = \mathbb{R} \times (-\infty, 0)$. The (joint) distribution $P_{\boldsymbol{\theta}}$ of $\mathbf{X} = (X_1, \ldots, X_n)$ (on $\mathcal{X} = \mathbb{R}^n$) has the density (w.r.t. $\mu \equiv$ Lebesgue measure on $\mathbb{R}^n$)

$$f(\mathbf{x} \mid \boldsymbol{\theta}) = (2\pi\sigma^2)^{-\frac{n}{2}} \exp\left\{-\frac{n\mu^2}{2\sigma^2}\right\} \exp\left\{\frac{\mu}{\sigma^2}\sum_{j=1}^{n}x_j - \frac{1}{2\sigma^2}\sum_{j=1}^{n}x_j^2\right\},$$

or

$$\tilde{f}(\mathbf{x} \mid \boldsymbol{\pi}) = \tilde{C}^n(\boldsymbol{\pi}) \exp\left\{\pi_1 \sum_{j=1}^{n}x_j + \pi_2 \sum_{j=1}^{n}x_j^2\right\},$$

and $\mathbf{T} = (\sum_{j=1}^{n} X_j, \sum_{j=1}^{n} X_j^2)$ is a sufficient statistic for $\{P_{\boldsymbol{\theta}} : \boldsymbol{\theta} \in \Theta\}$.

Example 4.8 (Bernoulli). Let $X_1, \ldots, X_n$ be i.i.d. Bernoulli $\mathscr{B}(\theta)$, i.e., $\mathrm{Prob}_\theta(X_j = 1) = \theta$, $\mathrm{Prob}_\theta(X_j = 0) = 1 - \theta$, $\theta \in \Theta = (0, 1)$. Then (with $\nu(\{0\}) = \nu(\{1\}) = 1$, as the counting measure on $S = \{0, 1\}$),

$$p(x \mid \theta) = \theta^x (1-\theta)^{1-x} \qquad [x \in \{0,1\}, \theta \in (0,1) = \Theta]$$

$$= (1-\theta)\left(\frac{\theta}{1-\theta}\right)^x = (1-\theta)e^{x\log\left(\frac{\theta}{1-\theta}\right)},$$

so that $\{G_\theta : \theta \in (0,1)\}$ is a *one-parameter exponential family*, with $\pi(\theta) \equiv \pi_1(\theta) = \log(\frac{\theta}{1-\theta})$, $T(x) = x$. Also, the distribution P_θ of $\mathbf{X} = (X_1, \ldots, \mathbf{X}_n)$ has the density (w.r.t. counting measure μ on $S^n = \{0,1\}^n$) given by

$$f(\mathbf{x} \mid \theta) \equiv P_\theta(\{\mathbf{x}\}) = \theta^{\sum_1^n x_j}(1-\theta)^{n-\sum_1^n x_j}$$

$$= (1-\theta)^n e^{(\sum_1^n x_j)\log\left(\frac{\theta}{1-\theta}\right)} = \tilde{C}^n(\pi)e^{\pi \sum_{j=1}^n x_j} \qquad (\mathbf{x} \in \{0,1\}^n).$$

The natural parameter space is $\Pi = (-\infty, \infty) = \mathbb{R}$. The distribution of the sufficient statistic $\mathbf{T} = \sum_{j=1}^{n} X_j$ is Binomial $\mathscr{B}(n, \theta)$:

$$f_{\mathbf{T}}(t \mid \theta) = \binom{n}{t}\theta^t(1-\theta)^{n-t} = (1-\theta)^n \binom{n}{t}e^{\pi t}$$

$$\left(t \in \{0, 1, \ldots, n\}, \quad \pi = \log\left(\frac{\theta}{1-\theta}\right) \in \mathbb{R}\right).$$

Example 4.9 (Poisson). Let $X_1, X_2, \ldots, X_n$ be i.i.d. Poisson $\mathscr{P}(\theta)$, $\theta \in \Theta = (0, \infty)$. That is,

$$G_\theta(\{x\}) = p(x \mid \theta) = e^{-\theta}\frac{\theta^x}{x!} \quad (x \in S = \{0, 1, 2, \ldots\})$$

$$= e^{-\theta}\frac{1}{x!}\,e^{x \log \theta}.$$

The (joint) distribution of $\mathbf{X} = (X_1, \ldots, X_n)$ is given by the density (w.r.t. the counting measure on S^n)

$$f(\mathbf{x} \mid \theta) = e^{-n\theta}\frac{1}{\prod_{j=1}^n x_j!}\,e^{(\log \theta)\sum_1^n x_j} \qquad \mathbf{x} = (x_1, \ldots, x_n) \in S^n.$$

The natural parameter is $\pi = \log \theta \in \prod = (-\infty, \infty)$, and $\mathbf{T} = X_1 + \cdots + X_n$ is a sufficient statistic for $\{P_\theta : \theta \in (0, \infty)\}$ (P_θ being the distribution of $\mathbf{X}$). This is a *one-parameter exponential family.*

Example 4.10 (Gamma). $X_1, \ldots, X_n$ are i.i.d. gamma $\mathscr{G}(\alpha, \beta)$ with common pdf

$$p(x \mid \boldsymbol{\theta}) = \frac{1}{\alpha^\beta \Gamma(\beta)}\,e^{-\frac{x}{\alpha}}x^{\beta-1}\mathbf{1}_{(0,\infty)}(x), \quad (\boldsymbol{\theta} = (\alpha, \beta) \in (0, \infty) \times (0, \infty) = \Theta),$$

$$= \frac{1}{\alpha^\beta \Gamma(\beta)}\frac{1}{x}\,e^{-\frac{1}{\alpha}x+\beta \log x}.$$

This is a *two-parameter exponential family,* with natural parameters $\pi_1 = -\frac{1}{\alpha}$, $\pi_2 = \beta$, $\Pi = (-\infty, 0) \times (0, \infty)$. The (joint) distribution $P_{\boldsymbol{\theta}}$ of $\mathbf{X} = (X_1, \ldots, X_n)$ has pdf (w.r.t Lebesgue measure on $\mathbb{R}^n$)

$$f(\mathbf{x} \mid \boldsymbol{\theta}) = \left(\frac{1}{\alpha^\beta \Gamma(\beta)}\right)^n \frac{1}{\prod_{j=1}^n x_j}\,e^{-\frac{1}{\alpha}\sum_1^n x_j + \beta \sum_{j=1}^n \log x_j}.$$

A sufficient statistic for $\{P_{\boldsymbol{\theta}} : \boldsymbol{\theta} \in \Theta\}$ is $\mathbf{T} = (\sum_1^n X_j, \sum_1^n \log X_j)$.

Example 4.11 (Beta). $X_1, \ldots, X_n$ are i.i.d. (Beta $\mathscr{B}_e(\alpha, \beta)$) with common pdf

$$p(x \mid \boldsymbol{\theta}) = \frac{\Gamma(\alpha + \beta)}{\Gamma(\alpha)\Gamma(\beta)}\,x^{\alpha-1}(1 - x)^{\beta-1}, \quad 0 < x < 1 \ (\boldsymbol{\theta} = (\alpha, \beta) \in \Theta = (0, \infty)^2)$$

$$= \frac{C(\alpha, \beta)}{x(1 - x)}\,e^{\alpha \log x + \beta \log(1-x)},$$

with natural parameters $\pi_1 = \alpha$, $\pi_2 = \beta$. This is a two-parameter exponential family. The (joint) pdf of $\mathbf{X} = (X_1, \ldots, X_n)$ is

$$f(\mathbf{x} \mid \boldsymbol{\theta}) = \tilde{C}^n(\boldsymbol{\pi})\frac{1}{\prod_{j=1}^n x_j(1 - x_j)}\exp\left\{\pi_1 \sum_1^n \log x_j + \pi_2 \sum_1^n \log(1 - x_j)\right\},$$

$$(\mathbf{x} = (x_1, \ldots, x_j) \in (0, 1)^n),$$

with $(\pi_1, \pi_2) \in (0, \infty)^2 = \Pi$. $\mathbf{T}(\mathbf{x}) = (\sum_1^n \log x_j, \sum_1^n \log(1 - x_j))$ is a sufficient statistic for this family of distributions $\tilde{P}_{\boldsymbol{\pi}}$ (with pdf $\tilde{f}(\mathbf{x} \mid \boldsymbol{\pi})$).

Example 4.12 (Multivariate Normal). Let $X_1, \ldots, X_n$ be i.i.d. with common distribution $N(\boldsymbol{\mu}, \Sigma)$ on $\mathbb{R}^m$, with common density (with respect to Lebesgue measure)

$$p(\tilde{x} \mid \boldsymbol{\theta}) = (2\pi)^{-\frac{k}{2}} \, (\text{Det } \Sigma)^{-\frac{1}{2}} \exp\left\{ -\frac{1}{2}(\tilde{x} - \boldsymbol{\mu})' \, \Sigma^{-1}(\tilde{x} - \boldsymbol{\mu}) \right\} \quad (\tilde{x} \in \mathbb{R}^m),$$

$\boldsymbol{\theta} = (\boldsymbol{\mu}, \Sigma) \in \mathbb{R}^m \times M_m = \Theta$ [M_m is the set of all symmetric positive-definite $m \times m$ matrices] which may be expressed as

$$p(\tilde{x} \mid \boldsymbol{\theta}) = C(\theta) \exp\left\{ \sum_{i=1}^{m} \left(\sum_{i'=1}^{m} \sigma^{ii'} \mu_{i'} \right) x_i - \frac{1}{2} \sum_{i=1}^{m} \sigma^{ii} x_i^2 - \sum_{1 \leq i < i' \leq m} \sigma^{ii'} x_i x_{i'} \right\}, \tag{4.33}$$

where $\sigma^{ii'}$ is the (i, i') element of the matrix Σ^{-1}, and $\tilde{x} = (x_1, \ldots, x_m)'$. The natural parameters are $\pi_i = \sum_{i'=1}^{m} \sigma^{ii'} \mu_{i'}$ $(1 \leq i \leq m)$, $\pi_{ii} = -\frac{1}{2}\sigma^{ii}$ $(1 \leq i \leq m)$, $\pi_{ii'} = -\sigma^{ii'}$ $(1 \leq i < i' \leq m)$, so that this is a k-parameter exponential family with $k = m + m + \binom{m}{2} = \frac{m(m+3)}{2}$. If $\widetilde{P}_{\boldsymbol{\pi}}$ denotes the distribution of $\mathbf{X} = (X_1, \ldots, X_n)$, then $\mathbf{T}(\mathbf{x}) = \left((\sum_{j=1}^{n} x_{ij})_{1 \leq i \leq m}, (\sum_{j=1}^{n} x_{ij}^2)_{1 \leq i \leq m}, (\sum_{j=1}^{n} x_{ij} x_{i'j})_{1 \leq i < i' \leq m} \right)$ is a sufficient statistic for $\{\widetilde{P}_{\boldsymbol{\pi}} : \boldsymbol{\pi} \in \Pi\}$. Note that Π is an open subset of $\mathbb{R}^k$ ($k = m(m+3)/2$), since Θ is; and $\boldsymbol{\pi}$ is a relabeling of $\boldsymbol{\theta} \in \Theta$. This is the *multivariate Normal model.*

Example 4.13 (Multinomial). Suppose a population comprises $k + 1$ different classes (sub-populations, or strata), with the i-th class having probability θ_i being picked in random sampling $(i = 1, \ldots, k + 1)$. The probability (mass) function of a randomly selected observation $\tilde{x} = (x_1, \ldots, x_{k+1}) \in S = \{(1, 0, \ldots, 0), (0, 1, 0, \ldots, 0), \ldots, (0, 0, \ldots, 0, 1)\}$, is

$$p(\tilde{x} \mid \boldsymbol{\theta}) = \theta_1^{x_1} \theta_2^{x_2} \ldots \theta_{k+1}^{x_{k+1}} \qquad \left(\theta_{k+1} = 1 - \sum_{i=1}^{k} \theta_i, \; \boldsymbol{\theta} = (\theta_1, \ldots, \theta_k) \right).$$

Here the i-th vector $(0, 0, \ldots, 0, 1, 0, \ldots, 0)$ (with 1 in the i-th position and 0's elsewhere) represents the selection to belong to the i-th class. The parameter space is $\Theta = \{(0_1, \ldots, \theta_k) \in \mathbb{R}^k : \theta_i > 0 \; \forall \; i = 1, \ldots, k, \; \sum_{i=1}^{k} \theta_i < 1\}$, which is an open simplex (an open subset of $\mathbb{R}^k$). One may express $p(\tilde{x} \mid \boldsymbol{\theta})$ as

$$p(\tilde{x} \mid \boldsymbol{\theta}) = \theta_1^{x_1} \theta_2^{x_2} \cdots \theta_k^{x_k} (1 - \theta_1 - \theta_2 - \cdots - \theta_k)^{1 - x_1 - x_2 - \cdots - x_k}$$

$$= (1 - \theta_1 - \cdots - \theta_k) \prod_{i=1}^{k} \left[\frac{\theta_i}{(1 - \theta_1 - \cdots - \theta_k)} \right]^{x_i}$$

$$= C(\boldsymbol{\theta}) \exp\left\{ \sum_{i=1}^{k} \log\left(\frac{\theta_i}{1 - \theta_1 - \cdots - \theta_k} \right) x_i \right\},$$

so that the natural parameter is $\boldsymbol{\pi} = (\pi_1, \ldots, \pi_k) \in \Pi = \mathbb{R}^k$, with $\pi_i = \log(\frac{\theta_i}{1-\theta_1-\cdots-\theta_k})$. [Given $(\pi_1, \ldots, \pi_k) \in \mathbb{R}^k$, $\theta_r = \frac{e^{\pi_r}}{1+\sum_{i=1}^{k} e^{\pi_i}}$ $(1 \leq r \leq k)$.] For a random sample of size n, the observation space is $\mathscr{X} = S^n$, with probability (mass) function

$$f(\mathbf{x} \mid \boldsymbol{\theta}) = \theta_1^{\sum_1^n x_{1j}} \theta_2^{\sum_1^n x_{2j}} \cdots \theta_{k+1}^{\sum_1^n x_{k+1,j}} = C^n(\boldsymbol{\theta}) \exp\left\{ \sum_{i=1}^k \log\left(\frac{\theta_i}{1 - \theta_1 - \cdots - \theta_k}\right) \sum_{j=1}^n x_{ij} \right\}$$

$$= \tilde{f}(\mathbf{x} \mid \boldsymbol{\pi}) = \widetilde{C}^n(\boldsymbol{\pi}) \exp\left\{ \sum_{i=1}^k \pi_i T_i(\mathbf{x}) \right\} \qquad [T_i(\mathbf{x}) = (\textstyle\sum_{j=1}^n x_{ij})_{1 \le i \le k}]$$

where $(x_{1j}, x_{2j}, \ldots, x_{kj},\ x_{k+1,j}) \in S,\ 1 \le j \le n$. $\mathbf{T}(\mathbf{x}) = (T_1(\mathbf{x}), \ldots, T_k(\mathbf{x}))$ is a sufficient statistic for $\{\widetilde{P}_{\boldsymbol{\pi}} : \boldsymbol{\pi} \in \Pi\}$. This is the so-called *multinomial model*.

The following result shows that the sufficient statistic $\mathbf{T}$ in Examples 4.7–4.13 is *complete*. Before we state it, it is worthwhile to note that one may consider (4.27) to be the general form of a k-parameter exponential family with respect to a natural parameter. The joint distribution of $\mathbf{X} = (X_1, \ldots, X_n)$ based on a random sample from such a family happens to be of the same form. One may have similar joint distributions of random variables which are not independent, and the joint distribution may still belong to the exponential family (as is the case, for example, for Gaussian time series or Gaussian random fields). Thus we may regard (4.26), (4.27) to represent a general exponential family (Exercise 4.5).

Theorem 4.4. *Let Π denote the natural parameter space of a k-parameter exponential family. Then the following hold.*

(a) Π is convex.

(b) If Π has a nonempty interior, then the sufficient statistic T is complete for $\{\widetilde{P}_{\pi} : \pi \in \Pi\}$ where $\widetilde{P}_{\pi}$ has the density (4.27) with respect to a sigma-finite measure ν on an observation space $\mathscr{X} = S$, say.

Proof. (a) Let $\boldsymbol{\pi}, \boldsymbol{\pi}' \in \Pi$, and $0 < \alpha < 1$. Then, writing $d\tilde{\nu} = h d\nu$,

$$\int_S h(x) e^{(\alpha\pi + (1-\alpha)\pi') \cdot T(x)} d\nu(x) = \int_S e^{\alpha\pi \cdot T(x)} \cdot e^{(1-\alpha)\pi' \cdot T(x)} d\tilde{\nu}(x)$$

$$\le \left[\int_S \left(e^{\alpha\pi \cdot T(x)} \right)^{\frac{1}{\alpha}} d\tilde{\nu}(x) \right]^{\alpha} \left[\int_S \left(e^{(1-\alpha)\pi' \cdot T(x)} \right)^{\frac{1}{1-\alpha}} d\tilde{\nu}(x) \right]^{1-\alpha} < \infty,$$

by Hölder's inequality.

(b) Let $\boldsymbol{\pi}_0$ be an interior point of Π. There exists $\delta > 0$ such that the open ball $B(\boldsymbol{\pi}_0, \delta) = \{\boldsymbol{\pi} : |\boldsymbol{\pi} - \boldsymbol{\pi}_0| < \delta\}$ is contained in Π. Let g be a $\widetilde{P}_{\boldsymbol{\pi}}$-integrable function $(\forall\, \boldsymbol{\pi} \in \Pi)$ on $\mathbb{R}^k$ such that $\widetilde{E}_{\boldsymbol{\pi}} g(T) = 0\ \forall\, \boldsymbol{\pi} \in \Pi$. Writing $g = g^+ - g^-$ $(g^+(\mathbf{t}) = g(\mathbf{t})\mathbf{1}_{[0,\infty)} g(\mathbf{t})),\ g^-(\mathbf{t}) = -g(\mathbf{t})\mathbf{1}_{(-\infty,0)}(g(\mathbf{t}))$ one then has

$$\widetilde{E}_{\boldsymbol{\pi}} g^+(T) = \widetilde{E}_{\boldsymbol{\pi}} g^-(T) \qquad \forall\, \boldsymbol{\pi} \in \Pi. \tag{4.34}$$

In particular,

$$\widetilde{E}_{\boldsymbol{\pi}_0} g^+(T) = \widetilde{E}_{\boldsymbol{\pi}_0} g^-(T). \tag{4.35}$$

Suppose, if possible, $\widetilde{P}_{\boldsymbol{\pi}_0}(g(T) = 0) < 1$. This means that the expectations in (4.35) are positive. Write $\tilde{\nu}_T^+$ and $\tilde{\nu}_T^-$ for the probability measures

$$d\tilde{\nu}_T^+(\mathbf{t}) = \frac{g^+(\mathbf{t}) d\widetilde{P}_{\boldsymbol{\pi}_0}(\mathbf{t})}{\widetilde{E}_{\boldsymbol{\pi}_0} g^+(T)}, \quad d\tilde{\nu}_T^-(\mathbf{t}) = \frac{g^-(\mathbf{t}) d\widetilde{P}_{\boldsymbol{\pi}_0}(\mathbf{t})}{\widetilde{E}_{\boldsymbol{\pi}_0} g^-(T)}. \tag{4.36}$$

The moment generating function *(mgf)* of $\tilde{\nu}_T^+$ in $B(0, \delta)$ is given by

$$\frac{\widetilde{C}(\pi_0)}{\widetilde{E}_{\pi_0} g^+(T)} \int e^{\boldsymbol{\xi} \cdot \mathbf{t}} g^+(\mathbf{t}) e^{\pi_0 \cdot \mathbf{t}} d\nu_T(\mathbf{t}) = \frac{\widetilde{C}(\pi_0)}{\widetilde{E}_{\pi_0} g^+(T)} \int e^{(\pi_0 + \boldsymbol{\xi}) \cdot \mathbf{t}} g^+(\mathbf{t}) d\nu_T(\mathbf{t})$$
$$< \infty \ \forall |\boldsymbol{\xi}| < \delta.$$

But, by (4.34), the last integral equals $\int e^{(\pi_0 + \boldsymbol{\xi}) \cdot \mathbf{t}} g^-(\mathbf{t}) d\nu_T(\mathbf{t}) \ \forall \, |\boldsymbol{\xi}| < \delta$. Also, $\widetilde{E}_{\pi_0} g^+(T) = \widetilde{E}_{\pi_0} g^-(T)$. Hence, the mgf's of $\tilde{\nu}_T^+$ and $\tilde{\nu}_T^-$ are equal in a neighborhood of the origin, namely, in $B(0, \delta)$. By the proposition below it follows that the two probability measures $\tilde{\nu}_T^+$ and $\tilde{\nu}_T^-$ are identical. This means $\tilde{P}_{\pi_0}(g^+(T) = g^-(T)) = 1$, or, $\tilde{P}_{\pi_0}(g(T) = 0) = 1$, a contradiction. $\quad\square$

Remark 4.6 (General Structure of Exponential Families). On a probability space $(S, \mathscr{S}, Q)$ let $\mathbf{T} = (T_1, \ldots, T_k)$ be a random vector having a finite mgf φ on a set $\Theta \ (\subset \mathbb{R}^k)$ with a non-empty interior. Then the family of probability measures $G_{\boldsymbol{\theta}} \ (\boldsymbol{\theta} = (\theta_1, \ldots, \theta_k) \in \mathbb{R}^k)$ with density $f(\mathbf{t} \mid \boldsymbol{\theta}) = \frac{\exp\{\sum_{i=1}^k \theta_i t_i\}}{\varphi(\boldsymbol{\theta})}$ with respect to the distribution v of $\mathbf{T}$ in a k-parameter exponential family with natural parameter $\boldsymbol{\theta}$. Every exponential family has this form if the parameter space has a non-empty interior.

Proposition 4.2. *If two probability measures Q_1 and Q_2 on $\mathbb{R}^k$ have the same mgf in a neighborhood of the origin, then $Q_1 = Q_2$.*

Proof. First consider the case $k = 1$. Suppose there exists $u_0 > 0$ such that

$$\varphi_1(u) \equiv \int_{\mathbb{R}} e^{ux} dQ_1(x) = \varphi_2(u) \equiv \int_{\mathbb{R}} e^{ux} dQ_2(x) \quad \text{for all } u \text{ in } (-u_0, u_0). \quad (4.37)$$

Since

$$e^{|ux|} \leq e^{ux} + e^{-ux},$$

one has, on integrating both sides with respect to Q_j,

$$\sum_{n=0}^{\infty} \beta_{n,j} \frac{|u|^n}{n!} \leq \varphi_j(u) + \varphi_j(-u) < \infty \quad \text{for } -u_0 < u < u_0, \quad (4.38)$$

where $\beta_{n,j} = \int |x|^n dQ_j(x)$. Since $e^{ux} \leq e^{|ux|}$, it follows that

$$\sum_{n=0}^{\infty} m_{n,j} \frac{u^n}{n!} \quad \text{converges absolutely in } -u_0 < u < u_0, \ (i = 1, 2)$$

and

$$\varphi_j(u) = \sum_{n=0}^{\infty} \frac{m_{n,j}}{n!} u^n \quad (j = 1, 2), \ -u_0 < u < u_0.$$

Here $m_{n,j} = \int x^n dQ_j(x) \ (j = 1, 2)$. Since a power series with a positive radius of convergence is infinitely differentiable within its radius of convergence, and can be differentiated term by term there, it follows in particular, $m_{n,1} = m_{n,2}, \ \forall n$. To prove the proposition, we will show that the characteristic functions

$$f_j(v) = \int_{\mathbb{R}} e^{ivx} dQ_j(x) \qquad (v \in \mathbb{R}), \ (j = 1, 2),$$

are identical. Now f_j is infinitely differentiable on $\mathbb{R}$, and the n-th derivative is

$$(D^n f_j)(\nu) = \int_{\mathbb{R}} (ix)^n e^{i\nu x} dQ_j(x) \quad (\nu \in \mathbb{R}),\, j = 1, 2. \tag{4.39}$$

By a Taylor expansion, using

$$\left| e^{i(\nu+h)x} - \left(\sum_{n=0}^{N} \frac{(ihx)^n}{n!} \right) e^{i\nu x} \right| \le \frac{|hx|^{N+1}}{(N+1)!} \tag{4.40}$$

one obtains, using the convergence of the series in (4.38),

$$\left| f_j(\nu + h) - \sum_{n=0}^{N} \frac{h^n}{n!} (D^n f_j)(\nu) \right| \le \frac{\beta_{N+1,j}}{(N+1)!} |h|^n \quad (j = 1, 2)$$
$$\longrightarrow 0 \quad \text{as } N \to \infty \quad \forall\, |h| < u_0.$$

Therefore,

$$f_j(\nu + h) = \sum_{n=0}^{\infty} \frac{h^n}{n!} D^n f_j(\nu), \quad (j = 1, 2) \ \forall\, |h| < u_0. \tag{4.41}$$

Now letting $\nu = 0$, and using $D^n f_j(0) = i^n m_{n,j}\ (j = 1, 2)$, one gets

$$f_1(h) = f_2(h) \qquad \forall\, |h| < u_0, \tag{4.42}$$

which also implies that $D^n f_1(\nu) = D^n f_2(\nu)\ \forall\, n$, if $|\nu| < u_0$. Thus $f_1(\nu+h) = f_2(\nu+h)\ \forall\, |\nu| < u_0$ and $\forall\, |h| < u_0$. In other words, $f_1(\nu) = f_2(\nu)\ \forall\, \nu \in (-2u_0, 2u_0)\ \forall\, \varepsilon > 0$. Thus $f_1(\nu) = f_2(\nu)\ \forall\, |\nu| < 2u_0$. Continuing in this manner, it follows that $f_1(\nu) = f_2(\nu)\ \forall\, \nu \in \mathbb{R}$.

Now let $k > 1$. Let X, Y have distribution Q_1, Q_2, respectively. *Fix $\nu \in \mathbb{R}^k \backslash \{0\}$.* The mgf's of the distributions of the random variables $\nu.X$ and $\nu.Y$ have finite and equal mgf's in a neighborhood of the origin (namely, $Ee^{\tau \nu.X} = Ee^{\tau \nu.Y}$ for $-\frac{u_0}{|\nu|} < \tau < \frac{u_0}{|\nu|}$, if the mgf's of Q_1, Q_2 are finite and equal in $B(0, u_0)$). Hence $\nu.X$ and $\nu.Y$ have the same distribution. This being true for all $\nu \in \mathbb{R}^k$, $f_1(\nu) \equiv Ee^{i\nu.X} = Ee^{i\nu.Y} = f_2(\nu)\ \forall\, \nu \in \mathbb{R}^k$. Hence $Q_1 = Q_2$. $\quad\square$

Remark 4.7. Theorem 4.4, together with the Lehmann–Scheffé Theorem (Theorem 4.3), provides UMVU estimators of all estimable parametric functions.

4.3 The Cramér–Rao Inequality

We have seen that uniformly minimum variance unbiased (UMVU) estimators exist for estimable parametric functions if a complete sufficient statistic exists. In particular, this is the case with k-parameter exponential families if the natural parameter space has a non-empty interior in $\mathbb{R}^k$. We now derive a lower bound for the variance (expected squared error) of unbiased estimators of parametric functions under a set of regularity conditions which are satisfied by exponential families (if Π is an open subset of $\mathbb{R}^k$) and many other families. Although this lower bound is rarely attained, it is approached in the large sample limit by maximum likelihood estimators under the regularity conditions.

Theorem 4.5 (Cramér–Rao Information Inequality). *Suppose $\mathbf{X}$ has a density $f(\mathbf{x} \mid \theta)$ (with respect to a sigma-finite measure μ on $(\mathscr{X}, \mathscr{B}(\mathscr{X}))$) satisfying the following conditions:*

(i) Θ is an open interval;
(ii) $\exists$ a μ-null set N such that $f(\mathbf{x} \mid \theta) > 0 \ \forall \ \mathbf{x} \in \mathscr{X} \backslash N, \ \forall \ \theta \in \Theta$;
(iii) $\int \frac{d}{d\theta} f(\mathbf{x} \mid \theta) d\mu(\mathbf{x}) = \frac{d}{d\theta} \int_{\mathscr{X} \backslash N} f(\mathbf{x} \mid \theta) d\mu(\mathbf{x}) \equiv 0, \ \forall \ \theta \in \Theta$;
(iv) if $T(\mathbf{X})$ is a real-valued statistic such that $E_\theta T^2 < \infty \ \forall \ \theta \in \Theta$, then writing $a(\theta) = E_\theta T$, one has

$$\frac{d}{d\theta} a(\theta) \equiv \frac{d}{d\theta} \int_{\mathscr{X}} T(\mathbf{x}) f(\mathbf{x} \mid \theta) d\mu(\mathbf{x}) = \int_{\mathscr{X} \backslash N} T(\mathbf{x}) \frac{d}{d\theta} f(\mathbf{x} \mid \theta) d\mu(x).$$

Then

$$\mathrm{var}_\theta T \equiv E_\theta (T - a(\theta))^2 \geq \frac{(a'(\theta))^2}{E_\theta \left(\frac{d \log f(\mathbf{X}|\theta)}{d\theta} \right)^2}. \tag{4.43}$$

Proof. Condition (iii) may be restated as

$$0 = E_\theta \frac{d \log f(\mathbf{X} \mid \theta)}{d\theta} \qquad \left(= \int_{\mathscr{X} \backslash N} \frac{\frac{d}{d\theta} f(\mathbf{x} \mid \theta)}{f(\mathbf{x} \mid \theta)} f(\mathbf{x} \mid \theta) d\mu(\mathbf{x}) \right). \tag{4.44}$$

Similarly, condition (iv) says (writing cov_θ for *covariance* under P_θ)

$$\mathrm{cov}_\theta(T, \frac{d}{d\theta} \log f(\mathbf{X} \mid \theta)) = a'(\theta), \tag{4.45}$$

since $\mathrm{cov}_\theta(T, \frac{d \log f(\mathbf{X}|\theta)}{d\theta}) = E_\theta T \frac{d \log f(\mathbf{X}|\theta)}{d\theta}$, as $E_\theta(\frac{d \log f(\mathbf{X}|\theta)}{d\theta}) = 0$. The inequality (4.43) now follows from the Cauchy–Schwarz inequality applied to (4.45). $\square$

Remark 4.8. The most common case covered in this course is $\mathbf{X} = (X_1, \ldots, X_n)$, X_j's being i.i.d. with a (common) pdf $f_1(x \mid \theta)$ (w.r.t. a sigma-finite measure ν on a measurable space $(S, \mathscr{B}(S))$). In this case, assumption (ii) may be replaced by $(\mathrm{ii})_1 : f_1(x_1 \mid \theta) > 0 \ \forall \ x_1 \in S \backslash N_1$, *where* $\nu(N_1) = 0$. Note that, in this case

$$f(\mathbf{x} \mid \theta) = \prod_{j=1}^{n} f_1(x_j \mid \theta), \ \log f(\mathbf{x} \mid \theta) = \sum_{j=1}^{n} \log f_1(x_j \mid \theta) \ (\mathbf{x}=(x_1, \ldots, x_n) \in S^n = \mathscr{X}),$$

so that (iii) [or, (4.44)] is equivalent to

$$(\mathrm{iii})_1 : \quad E_\theta \frac{d}{d\theta} \log f_1(X_1 \mid \theta) = 0. \tag{4.46}$$

Also,

$$E_\theta \left(\frac{d \log f(\mathbf{X} \mid \theta)}{d\theta} \right)^2 = \mathrm{var}_\theta \left(\frac{d \log f(\mathbf{X} \mid \theta)}{d\theta} \right) = n E_\theta \left(\frac{d \log f_1(X_1 \mid \theta)}{d\theta} \right)^2$$

$$= n \, \mathrm{var}_\theta \left(\frac{d \log f_1(X_1 \mid \theta)}{d\theta} \right), \tag{4.47}$$

under (iii)$_1$. Then the Cramér–Rao inequality (4.43) takes the form

$$\mathrm{var}_\theta T \geq \frac{(a'(\theta))^2}{nE_\theta\left(\frac{d\log f_1(X_1|\theta)}{d\theta}\right)^2}. \tag{4.48}$$

Remark 4.9. A set of sufficient conditions for (iii) and (iv) in Theorem 4.5, in presence of (i), (ii), are the following: For each $\theta_0 \in \Theta$ there exist $h = h(\theta_0) > 0$, and functions g_1, g_2 on $\mathscr{X}$ such that $\int g_1(\mathbf{x})d\mu(\mathbf{x}) < \infty$, $E_{\theta_0}g_2^2(\mathbf{X}) < \infty$, and

(R$_1$): $\left|\dfrac{df(\mathbf{x}\mid\theta)}{d\theta}\right| \leq g_1(\mathbf{x})\ \forall\,\theta$ satisfying $|\theta - \theta_0| \leq h$, $\forall\,\mathbf{x} \in \mathscr{X}$,

(R$_2$): $\left|\dfrac{df(\mathbf{x}\mid\theta)}{d\theta}\right| \leq g_2(\mathbf{x})\ \forall\,\theta$ satisfying $|\theta - \theta_0| \leq h$, $\forall\,\mathbf{x} \in \mathscr{X}$.

Remark 4.10. The assumptions of Theorem 4.5 hold for one-parameter exponential families where the natural parameter space is an open interval (Exercise 4.6).

Remark 4.11. For $\mathbf{X} = (X_1, \ldots, X_n)$ with X_j's i.i.d., as in Remark 4.8, let $\ell_n(\theta)$ be the *log-likelihood function*

$$\ell_n(\theta) = \sum_{j=1}^{n} \log f_1(X_j \mid \theta). \tag{4.49}$$

Assume that the likelihood equation

$$D_n(\theta) \equiv \frac{d}{d\theta}\,\ell_n(\theta) = 0 \tag{4.50}$$

has a solution $\hat\theta_n$ which is *consistent*, i.e., $P_\theta(\hat\theta_n \to \theta) = 1\ \forall\,\theta$. Assuming $\theta \to f_1(x \mid \theta)$ is twice continuously differentiable, one may use the Taylor expansion

$$0 = \frac{d}{d\theta}\,\ell_n(\theta)\Big|_{\hat\theta_n} \equiv D_n(\hat\theta_n) = D_n(\theta_0) + (\hat\theta_n - \theta_0)\frac{d}{d\theta}\,D_n(\theta)\Big|_{\theta=\theta_n^*}$$

where θ_n^* lies between $\hat\theta_n$ and θ_0. Multiplying both sides by $\sqrt{n}$, one has, by the law of large numbers and the central limit theorem,

$$\sqrt{n}(\hat\theta_n - \theta_0) =$$
$$= -\frac{1}{\sqrt{n}}\sum_{j=1}^{n}\left(\frac{d}{d\theta}\log f_1(X_j \mid \theta)\right)_{\theta=\theta_0}\Bigg/\frac{1}{n}\sum_{j=1}^{n}\left\{\left(\frac{d^2}{d\theta^2}\log f_1(X_j \mid \theta)\right)_{\theta=\theta_n^*}\right\}$$
$$\xrightarrow{\mathscr{L}} N(0, \sigma^2(\theta_0)). \tag{4.51}$$

under P_{θ_0}. Here

$$\sigma^2(\theta_0) = \left[E_{\theta_0}\left(\frac{d\log f_1(X_1 \mid \theta)}{d\theta}\right)^2_{\theta=\theta_0}\Bigg/\left(E_{\theta_0}\left\{\frac{d^2}{d\theta^2}\log f_1(X_1 \mid \theta)\right\}_{\theta=\theta_0}\right)^2\right]. \tag{4.52}$$

Now, under regularity conditions (allowing the interchange of the order of differentiation and integration below),

$$
\begin{aligned}
E_\theta \frac{d^2 \log f_1(X_1 \mid \theta)}{d\theta^2} &= \int_{\mathscr{X}} \frac{d}{d\theta} \left(\frac{\frac{d}{d\theta} f_1(x \mid \theta)}{f_1(x \mid \theta)} \right) f_1(x \mid \theta) d\nu(x) \\
&= \int_{\mathscr{X}} \frac{f_1''(x \mid \theta) f_1(x \mid \theta) - (f_1'(x \mid \theta))^2}{f_1^2(x \mid \theta)} f_1(x \mid \theta) dv(x) \\
&= \int_{\mathscr{X}} f_1''(x \mid \theta) dv(x) - \int_{\mathscr{X}} \left(\frac{d \log f_1(x \mid \theta)}{d\theta} \right)^2 f_1(x \mid \theta) dv(x) \\
&= \frac{d^2}{d\theta^2} \int_{\mathscr{X}} f_1(x \mid \theta) dv(x) - E_\theta \left(\frac{d \log f_1(X_1 \mid \theta)}{d\theta} \right)^2 \\
&= -E_\theta \left(\frac{d \log f_1(X_1 \mid \theta)}{d\theta} \right)^2 .
\end{aligned}
\tag{4.53}
$$

Substituting this in (4.52), we get

$$
\sigma^2(\theta_0) = \left[\frac{1}{E_{\theta_0}} \left(\frac{d \log f_1(X_1 \mid \theta)}{d\theta} \right)^2_{\theta=\theta_0} \right]^{-1} = \frac{1}{I(\theta_0)},
\tag{4.54}
$$

say. The quantity $I(\theta)$ is referred to as the *Fisher information*. Thus $\hat{\theta}_n$ is *asymptotically Normal with mean* θ_0 *and variance* $\frac{1}{nI(\theta_0)}$, the latter being the lower bound in Theorem 4.5 for the variance of unbiased estimators of the parametric function $a(\theta) = \theta$ [see (4.48)].

Consider next a k-parameter exponential family (4.17) or (4.21), $k > 1$, but with q restrictions on the parameters given by $q < k$ smooth functionally independent relations among the parameters. The reduced model is expressed in terms of $d = k - q$ independent parameters and is called a *curved exponential family* since the new parameter space is a d-dimensional surface in the original k-dimensional parameter space. Generally, this term is restricted to those cases where the new model is not a d-dimensional exponential family. The following examples are taken from Bickel and Doksum (2001, pp. 126, 405).

Example 4.14 (Estimation of Mean of a Gaussian with a Fixed Signal-to-Noise Ratio). Here $X_1, \ldots, X_n$ $(n \geq 2)$ are i.i.d. $N(\mu, \sigma^2)$, $\mu > 0$ with the ratio $\frac{\mu}{\sigma} = \lambda > 0$ known. The common density may then be expressed as

$$
\begin{aligned}
f(x; \mu) &= \frac{1}{\sqrt{2\pi\mu^2/\lambda^2}} \exp\left\{ -\frac{\lambda^2}{2\mu^2} x^2 + \frac{\lambda^2}{\mu} x - \frac{\lambda^2}{2} \right\} \\
&= \frac{\lambda/u}{\sqrt{2\pi}} e^{-x^2/2} \exp\left\{ \pi_1(\mu)x + \pi_2(\mu)x^2 \right\},
\end{aligned}
\tag{4.55}
$$

where $\pi_1(\mu) = \lambda^2/\mu$, $\pi_2(\mu) = -\lambda^2/2\mu^2$. If λ was not known, this would be a two-parameter exponential family in natural parameters $\pi_1 = -1/2\sigma^2 \in (-\infty, 0)$, $\pi_2 = \mu/\sigma^2 \in (0, \infty)$. But with the given relation $\mu/\sigma = \lambda$ known, it is a curved exponential family with $k - q = 2 - 1 = 1 = d$. The log-likelihood function is

$$
\ell = \log f_n(\mathbf{X}; \mu) = -\frac{n}{2} \log 2\pi + n \log \lambda - n \log \mu - \frac{\lambda^2}{2\mu^2} T_2 + \frac{\lambda^2}{\mu} T_1 - \frac{n\lambda^2}{2}
$$

$$
\left(T_2 = \sum_1^n X_j^2, \ T_1 = \sum_1^n X_j \right),
$$

and the likelihood equation is

$$0 = -\frac{n}{\mu} + \frac{\lambda^2 T_2}{\mu^3} - \frac{\lambda^2}{\mu^2} T_1, \text{ or, } \mu^2 = \lambda^2 m_2 - \lambda^2 \overline{X} \mu,$$

where $m_2 = \sum_1^n X_j^2/n$. The solutions of this quadratic equation are

$$\mu = -\frac{1}{2}\lambda^2 \overline{X} \pm \sqrt{\frac{1}{4}\lambda^4 \overline{X}^2 + \lambda^2 m_2}.$$

Since $\mu > 0$, the MLE of μ is therefore given by

$$\hat{\mu} = -\frac{1}{2}\lambda^2 \overline{X} + \frac{1}{2}\sqrt{\lambda^4 \overline{X}^2 + 4\lambda^2 m_2}. \tag{4.56}$$

Example 4.15 (The Fisher Linkage Model). The following genetic model was considered by Fisher (1958, p. 301). Self-crossing of maize heterozygous on two alleles yield four types of offspring: sugary-white, sugary-green, starchy-white, starchy-green. If θ_1, θ_2, θ_3, θ_4 are the probabilities of having an offspring of these four types and if N_1, N_2, N_3, N_4 are the numbers of offspring of these types among n offspring, then (N_1, N_2, N_3, N_4) has a multinomial distribution (See Example 4.13)

$$P_{\boldsymbol{\theta}}(N_i = n_i; i = 1, \ldots, 4) = \frac{n!}{n_1! n_2! n_3! n_4!} \theta_1^{n_1} \theta_2^{n_2} \theta_3^{n_3} (1 - \theta_1 - \theta_2 - \theta_3)^{n_4},$$

a $k = 3$-parameter exponential family. According to a linkage model, $\theta_1 = \frac{1}{4}(2+\eta)$, $\theta_2 = \theta_3 = \frac{1}{4}(1 - \eta)$ and, consequently, $\theta_4 = \frac{1}{4}\eta$, where the unknown parameter η lies in $[0, 1]$. One then arrives at a curved exponential family with $d = 1$. The likelihood equation is

$$\frac{n_1}{2 + \eta} - \frac{n_2 + n_3}{1 - \eta} + \frac{n_4}{\eta} = 0, \text{ or,}$$
$$-n\eta^2 + [n_1 - 2(n_2 + n_3) - n_4]\eta + 2n_4 = 0,$$

whose solutions are

$$\eta = -\frac{1}{2}\left[1 - 2\hat{p}_1 + \hat{p}_2 + \hat{p}_3 \pm \sqrt{(1 - 2\hat{p}_1 + \hat{p}_2 + \hat{p}_3)^2 + 4\hat{p}_4}\right],$$

where $\hat{p}_i = n_i/n$ $(i = 1, 2, 3, 4)$. The positive solution is the MLE:

$$\hat{\eta} = -\frac{1}{2}(1 - 2\hat{p}_1 + \hat{p}_2 + \hat{p}_3) + \frac{1}{2}\sqrt{(1 - 2\hat{p}_1 + \hat{p}_2 + \hat{p}_3)^2 + 4\hat{p}_4}. \tag{4.57}$$

4.4 Notes and References

Our presentation is influenced by Ferguson (1967, Chap. 3), Lehmann (1959, Chap. 2), and Bickel and Doksum (2001, Sect. 3.4).

The notion of sufficiency is due to R.A. Fisher (1922), who also stated the factorization criterion. A rigorous derivation of the criterion for general dominated families is due to Halmos and Savage (1949), and further generalized by Bahadur (1954). The Rao–Blackwell theorem is due to Rao (1945) and Blackwell (1947).

The Lehmann–Scheffé theorem is due to Lehmann and Scheffé (1947, 1950, 1955), where the notion of minimal sufficiency is also introduced. Exponential families were introduced by R.A. Fisher (1934) in one dimension and extended to higher dimensions by G. Darmois, B.O. Koopman and E.J.G. Pitman. Barndorff-Nielsen (1978) and Brown (1986) provide rigorous accounts of general exponential families.

Diaconis and Ylvisaker (1979) explore the existence of conjugate priors for Bayes estimation in exponential families. Generally, the computation of the posterior distribution poses numerical challenges if the prior is not conjugate. See chap. 14 in the context.

Exercises for Chap. 4

Ex. 4.1. Show that, irrespective of the hypothesis of convexity of the loss function in Theorem 4.2, given any decision rule $d(\mathbf{X})$ one can construct $\mathbf{X}^0$ based entirely on the value of the sufficient statistic T such that $\mathbf{X}$ and $\mathbf{X}^0$ have the same distribution P_θ, $\forall\ \theta \in \Theta$. Hence $d(\mathbf{X}_0)$ has the same risk function as $d(\mathbf{X})$. In this sense, no information is lost by recording only the value of T (and nothing else) from the observation $\mathbf{X}$,

Ex. 4.2. Consider a family of distributions P_θ, $\theta \in \Theta$, of the observation $\mathbf{X}$, and let T_1, T_2 be two statistics. Show that $\sigma(T_1) \subset \sigma(T_2)$ with P_θ-probability one $\forall\ \theta \in \Theta$, if and only if for every bounded $\sigma(T_1)$-measurable real-valued function $f(T_1)$ there exists a $\sigma(T_2)$-measurable bounded $g(T_2)$ such that $P_\theta(f(T_1) = g(T_2)) = 1\ \forall\ \theta \in \Theta$.

Ex. 4.3. (a) In Example 4.5, find the UMVU estimator of θ^k $(k = 1, \ldots, n)$. (b) In Example 4.6, find the UMVU estimators of (i) $\sin\theta$, (ii) e^θ.

Ex. 4.4. In Example 4.6, show that '$\int_0^\theta g(t)t^{n-1}dt = 0\ \forall\ \theta > 0$' implies $g(t) = 0$ a.e. (w.r.t. Lebesgue measure on $(0,\infty)$). [Hint: Write $g(t) = g^+(t) - g^-(t)$, where $g^+(t) = \max\{0, g(t)\}$, $g^-(t) = -\min\{0, g(t)\}$. Then $F^+(\theta) \equiv \int_0^\theta g^+(t)t^{n-1}dt$ is the distribution function of the Lebesgue–Stieltjes (L–S) measure μ^+ on $(0,\infty)$, which equals $F^-(\theta) \equiv \int_0^\theta g^-(t)t^{n-1}dt$—the distribution function of an L–S measure μ^-, say. Hence $\mu^+ = \mu^-$. In particular, $g^+(t) = g^-(t)$ a.e. on $(0,\infty)$].

Ex. 4.5 (Ornstein–Uhlenbeck Process). Let $\{V_t : 0 \leq t < \infty\}$ be a stationary Gaussian process with V_t having distribution $N(0, \sigma^2/2\gamma)$, and $\mathrm{cov}(V_s, V_{s+t}) = \frac{\sigma^2}{2\gamma} e^{-\gamma t}$ for some $s \geq 0$, $t \geq 0$ $(\sigma^2 > 0, \gamma > 0)$. Consider observations $X_j = V_{t_j}$ $(j = 1, \ldots, n)$, where $0 = t_1 < t_2 < \cdots < t_n$. Find UMVU estimators of σ^2 and γ.

Ex. 4.6. Show that the assumptions of Theorem 4.5 hold for one-parameter exponential families where the natural parameter space is an open interval.

Ex. 4.7. In Example 4.12, find the UMVU estimators of (a) $\boldsymbol{\mu}$ and (b) Σ. That is, find $\hat{\boldsymbol{\mu}}$ and $\widehat{\Sigma}$ such that (ia) $E\hat{\boldsymbol{\mu}} = \boldsymbol{\mu}$, (ib) $E\,\widehat{\Sigma} = \Sigma$ and (iia) $\hat{\boldsymbol{\mu}} = \arg\min E|d_1(\mathbf{X}) - \boldsymbol{\mu}|^2$ and (iib) $\widehat{\Sigma} = \arg\min E|d_2(\mathbf{X}) - \Sigma|^2$, over the class of all unbiased estimators $d_1(\mathbf{X})$ of $\boldsymbol{\mu}$, and the class of all unbiased estimators $d_2(\mathbf{X})$ of Σ. Here the distances in (iia,b) are Euclidean distances between vectors.

Ex. 4.8. Let $X_1, \ldots, X_n$ be i.i.d. $N(\mu, \sigma^2)$, $\theta = (\mu, \sigma^2) \in \mathbb{R} \times (0, \infty)$. Find the UMVU estimator of μ/σ. [Hint: Use the fact that $\overline{X}$, and s^2 are independent, unbiased estimators of μ, and of σ^2. Also, it is known that $U = \frac{(n-1)s^2}{\sigma^2}$ has a χ^2_{n-1} distribution. Therefore, $s^{-2} = \frac{n-1}{\sigma^2} U^{-1}$, where $U \sim \chi^2_{n-1}$, and $s^{-1} = \frac{\sqrt{n-1}}{\sigma} U^{-\frac{1}{2}}$, where $U \sim \chi^2_{n-1}$. We get $E(s^{-1}) = \frac{\sqrt{n-1}}{\sigma} E(U^{-\frac{1}{2}})$, where $U \sim \chi^2_{n-1}$. Note that for $n \geq 3$, since $U \sim \chi^2_{n-1}$, we have

$$c_n = E(U^{-\frac{1}{2}}) = \int_0^\infty u^{\frac{n}{2}-2} \exp\left(-\frac{u}{2}\right),$$

a finite integral that can be expressed in terms of the Gamma function. Thus $\frac{1}{\sigma} = E(\frac{1}{c_n \sqrt{n-1} s_n})$, and $d(X) = \frac{\overline{X}}{c_n \sqrt{n-1} s_n}$ is an unbiased estimator of $\frac{\mu}{\sigma}$, and one may apply Lehmann–Scheffé's theorem.]

Ex. 4.9. Let X_j $(j = 1, 2, \ldots, n)$ be i.i.d. real-valued observations from an unknown distribution P.

(a) Let Θ be the (infinite dimensional) set of all *continuous distributions* on $(\mathbb{R}, \mathscr{B}(\mathbb{R}))$ (i.e., $P(\{x\}) = 0 \ \forall \ x \in \mathbb{R}$). Show that the *order statistic* $T = (X_{(1)}, X_{(2)}, \ldots, X_{(n)})$ is sufficient for $\{P : P \in \Theta\}$. [Hint: Compute the conditional distribution of $\mathbf{X}$ given T.]
(b) Let Θ be the set of all *discrete distributions* on $\{0, 1, 2, \ldots\}$ (i.e., $\sum_j P(\{j\}) = 1$). Let $n_j = \#\{i : X_i = j\}$ $(j = 0, 1, \ldots)$. Show that $T = \{n_j : j = 0, 1, \ldots\}$ is a sufficient statistic for $\{P : P \in \Theta\}$.

Ex. 4.10 (Negative Binomial Distribution). Let X_j $(j = 1, 2, \ldots, n)$ be i.i.d. observations, with $P_\theta(X_1 = x) = \binom{r+x-1}{x}(1 - \theta)^r \theta^x$ $(x = 0, 1, \ldots)$. Here r is a positive integer (known) and $\theta \in \Theta = (0, 1)$. In a coin tossing experiment, $r + X_1$ may be the first time the r-th tail shows up.

(a) Calculate $E_\theta X_1$ [Hint: Think of X_1 as the sum of r i.i.d. random variables each of which has the P_θ-distribution above, but with $r = 1$].
(b) Find the UMVU estimator of $\frac{\theta}{1-\theta}$, and calculate its variance.
(c) Find the MLE of θ and compute its asymptotic distribution as $n \to \infty$.

Ex. 4.11. Let $X_1, \ldots, X_n$ be i.i.d. observations with the common density (w.r.t. Lebesgue measure on $(0, 1)$) $f(x \mid \theta) = \theta x^{\theta-1}$ $(0 < x < 1)$, $\theta \in \Theta = (0, \infty)$.

(a) Find the UMVU estimator of θ.
(b) Find the UMVU estimator of $1/\theta$.
(c) Find the MLE of θ. [Hint: Look at $Y_j = -\ln X_j$.]

Ex. 4.12. (a) Derive the asymptotic distribution of the MLE $\hat{\mu}$ in (4.56).
(b) Derive the asymptotic distribution of the MLE $\hat{\eta}$ in (4.57).

Ex. 4.13. Assume $\theta \in (0, 1)$, and let $X_1, \ldots, X_n$ be i.i.d. from a geometric distribution $P_\theta(X = x) = \theta(1 - \theta)^{x-1}$, $x = 1, 2, \ldots$. Show that $T = \sum_{i=1}^n X_i$ is sufficient for θ. Is T a complete sufficient statistic?

A Project for Students

Project: Space Shuttle Disaster In 1986, the space shuttle Challenger exploded during take off, killing the seven astronauts aboard. It was determined that the explosion was the result of an O-ring failure, a splitting of a ring of rubber that seals different parts of the ship together. The flight accident was believed to be caused by the unusually cold weather ($31\,^\circ$F) at the time of the launch. The past O-ring failure data along with temperature at launch time are given below (in increasing order of temperature) for 23 prior flights. The flight numbers denote the (unimportant) time order of launch. The numbers 0 and 1 indicate "no O-ring failure" and "O-ring failure", respectively.

Flight#	14	9	23	10	1	5	13	15	4	3	8	17	2
Failure	1	1	1	1	0	0	0	0	0	0	0	0	1
Temp. in Degrees F	53	57	58	63	66	67	67	67	68	69	70	70	70

Flight#	11	6	7	16	21	19	22	12	20	18
Failure	1	0	0	0	1	0	0	0	0	0
Temp. in Degrees F	70	72	73	75	75	76	76	78	79	81

Project Objective Estimate the probability of O-ring failure at temperature $31\,^\circ$F and at $65\,^\circ$F.

Suggested Model Let Y denote the failure status (response variable), and X the temperature in degrees F at launch time (explanatory variable). Use the *logistic regression model,*

$$P(Y = 1 \mid X = x) = \exp \frac{\{\alpha + \beta x\}}{[1 + \exp\{\alpha + \beta x\}]} = P(X), \text{ say, and}$$

$$P(Y = 0 \mid X = x) = 1 - p(x). \tag{4.58}$$

Note that one may express the model as

$$\log\left[\frac{p(x)}{(1 - p(x))}\right] = \alpha + \beta x. \tag{4.59}$$

Hence the name logistic regression.

Assume that the regressor x is *stochastic* and (X_i, Y_i) are i.i.d. random vectors.

For 23 independent Y observations $(y_1, \ldots, y_{23})$ the conditional likelihood function (i.e., the conditional p.d.f. of Y_i, given $X_i = x_i$ ($i = 1, \ldots, 23$)), is

$$\ell(\mathbf{y} \mid \mathbf{x}; \alpha, \beta) = \prod_{i=1,\ldots,23} \left[p(x_i)^{y_i} (1 - p(x_i))^{1-y_i}\right], \tag{4.60}$$

and the (conditional) log likelihood is

$$\log \ell = \sum_i [y_i(\alpha + \beta x_i)] - \sum_i \log[1 + \exp\{\alpha + \beta x_i\}]. \tag{4.61}$$

Assume that the distribution of X_i does not involve α, β.

(a) (i) Find the maximum likelihood estimates $\hat{\alpha}, \hat{\beta}$ of α and β, and (ii) use these to estimate the desired failure probabilities at $x = 31^0\mathrm{F}$, and at $x = 65\,^\circ\mathrm{F}$, using (4.58). [Hint: You will need to compute the estimates numerically e.g., by the gradient method, as solutions of the likelihood equations $\partial \log \ell / \partial \alpha = 0$, $\partial \log \ell / \partial \beta = 0$, or using direct maximizing algorithms (e.g., as is available on Matlab). There is also a fast algorithm for this called a Re-weighted Least Squares Algorithm (See Wasserman 2003, pp. 223–224).]

(b) By *bootstrapping* from the i.i.d. observations $\{(X_i, Y_i); 1 \leq i \leq 23\}$, (i) find a lower 90 % confidence bound for the probability of failure at launch temperature $x = 31\,^\circ\mathrm{F}$. In other words, you are to find a number U such that Prob $(p(31) \geq U) = 0.90$. [Hint: Find the lower tenth percentile $q_{0.10}$ of the bootstrap values $\hat{\alpha}^* + \hat{\beta}^*(31)$, using estimates $\hat{\alpha}, \hat{\beta}$ of α, β as in (a), but from each bootstrap resample from $\{(X_i, Y_i) : i = 1, \ldots, 23\}$ instead of the original sample. Now use $\exp\{q_{0.10}\}/(1+\exp\{q_{0.10}\})$ as the desired lower bound for the probability of failure (noting that $e^q/(1 + e^q)$ is a strictly increasing function of q). Observe that this says that you are (approximately) 90 % sure that the probability of failure at $31\,^\circ\mathrm{F}$ is at least $\exp\{q_{0.10}\}/(1 + \exp\{q_{0.10}\})$).]

(ii) Also find an upper 90 % bound for the probability of O-ring failure at the temperature $x = 65\,^\circ\mathrm{F}$. That is, find a value q such that the probability of O-ring failure is less than $e^q/(1 + e^q)$, with a probability 0.90.

[Note: **Bootstrapping** means taking repeated samples of size 23 (with replacement) from the observations $\{(y_i, x_i) : i = 1, \ldots, 23\}$. Between 500 and 1000 such re-samples from the observed data should be enough. Each bootstrap sample (of size 23) is used to compute $(\hat{\alpha}^*, \hat{\beta}^*)$ as in (a)].

Appendix for Project: The Nonparametric Percentile Bootstrap of Efron

Let $\hat{\sigma}_n$ be the standard error of $\hat{\theta}_n$ (That is, $\hat{\sigma}_n$ is an estimate of the standard deviation of $\hat{\theta}_n$). An asymptotic confidence interval of confidence level $1 - \alpha$ for θ would follow from the relation $P(z_{\alpha/2}\hat{\sigma}_n \leq \hat{\theta}_n - \theta \leq z_{1-\alpha/2}\hat{\sigma}_n) \approx 1-\alpha$, namely, it is the interval $[\hat{\theta}_n - z_{1-\alpha/2}\hat{\sigma}_n, \hat{\theta}_n - z_{\alpha/2}\hat{\sigma}_n] = [\hat{\theta}_n - z_{\alpha/2}\hat{\sigma}_n, \hat{\theta}_n + z_{1-\alpha/2}\hat{\sigma}_n] = [l, u]$, say. Now the bootstrap version $\hat{\theta}_n^*$ of $\hat{\theta}_n$ is, under the empirical $\mathbf{P}^* = \widehat{P}_n$, asymptotically Normal $N(\hat{\theta}_n, \hat{\sigma}_n^2)$, so that the $\alpha/2$-th and $(1 - \alpha/2)$-th quantiles of $\hat{\theta}_n^*$, $q_{\alpha/2}^*$ and $q_{1-\alpha/2}^*$ say, are asymptotically equal to $\hat{\theta}_n + z_{\alpha/2}\hat{\sigma}_n = l$ and $\hat{\theta}_n + z_{1-\alpha/2}\hat{\sigma}_n = u$, respectively.

Hence the *percentile bootstrap* based confidence interval for θ is given by

$$\left[q_{\alpha/2}^*, q_{1-\alpha/2}^* \right]. \tag{4.62}$$

Note that the construction of this interval only involves resampling from the data repeatedly to construct bootstrap versions $\hat{\theta}_n^*$ of $\hat{\theta}_n$; it does not involve the computation of the standard error $\hat{\theta}_n$.

Although (4.62) does not involve computing the standard error $\hat{\sigma}_n$, the latter is an important object in statistical analysis. It follows from the above that the variance $\hat{\sigma}_n^{*2}$ of the $\hat{\theta}_n^*$ values from the repeated resamplings provide an estimate of $\hat{\sigma}_n^2$ [A rough estimate of $\hat{\sigma}_n$ is also provided by $([q_{1-\alpha/2}^* - q_{\alpha/2}^*]/2z_{1-\alpha/2})^{\frac{1}{2}}]$.

When the standard error $\hat{\sigma}_n$ of $\hat{\theta}_n$ is known in closed form, one may use the studentized or pivoted statistic $T_n = (\hat{\theta}_n - \theta)/\hat{\sigma}_n$ which is asymptotically standard Normal $N(0,1)$. The usual CLT-based symmetric confidence interval for θ is given by

$$\left[\hat{\theta}_n + z_{\alpha/2}\hat{\sigma}_n, \hat{\theta}_n + z_{1-\alpha/2}\hat{\sigma}_n\right] = \left[\hat{\theta}_n - z_{1-\alpha/2}\hat{\sigma}_n, \hat{\theta}_n - z_{\alpha/2}\hat{\sigma}_n\right], \qquad (4.63)$$

using $P(|T_n| \le z_{1-\alpha/2}) = 1 - \alpha$. The corresponding pivotal bootstrap confidence interval is based on the resampled values of $T_n^* = (\hat{\theta}_n^* - \hat{\theta}_n)/\hat{\sigma}_n^*$, where $\hat{\sigma}_n^*$ is the bootstrap estimate of the standard error as described in the preceding paragraph. Let $c_{\alpha/2}^*$ be such that $P^*(|T_n^*| \le c_{\alpha/2}^*) = 1 - \alpha$. The bootstrap pivotal confidence interval for θ is then

$$\left[\hat{\theta}_n - c_{\alpha/2}^*\hat{\sigma}_n^*, \hat{\theta}_n + c_{\alpha/2}^*\hat{\sigma}_n^*\right]. \qquad (4.64)$$

Suppose $\hat{\theta}_n$ is based on i.i.d. observations $X_1, \ldots, X_n$, whose common distribution has a density (or a nonzero density component), and that it is a smooth function of sample means of a finite number of characteristics of X, or has a stochastic expansion (Taylor expansion) in terms of these sample means (such as the MLE in regular cases). It may then be shown that the *coverage error* of the CLT-based interval (4.63) is $O(n^{-1})$, while that based on (4.64) is $O(n^{-3/2})$, a major advantage of the bootstrap procedure. The coverage error of the percentile interval (4.62) is $O(n^{-1/2})$, irrespective of whether the distribution of X is continuous or discrete.

Definition 4.6. The *coverage error* of a confidence interval for a parameter θ is the (absolute) difference between the actual probability that the true parameter value belongs to the interval and the target level $1 - \alpha$.

References

Bahadur, R. R. (1954). Sufficiency and statistical decision functions. *The Annals of Mathematical Statistics, 25*, 423–462.

Barndorff-Nielsen, O. (1978). *Information and exponential families in statistical theory*. New York: Wiley.

Bickel, P. J., & Doksum, K. (2001). *Mathematical statistics* (2nd ed.). Englewood Cliffs, NJ: Prentice Hall.

Blackwell, D. (1947). Conditional expectation and unbiased sequential estimation. *The Annals of Mathematical Statistics, 18*(1), 105–110.

Brown, L. (1986). *Fundamentals of statistical exponential families: With applications in statistical decision theory* (Vol. 9). Hayward: Institute of Mathematical Statistics.

Diaconis, P. and Ylvisaker, D. (1979). Conjugate priors for exponential families. *Ann. Statist. 7*, 269–281.

Fisher, R. A. (1922). On the mathematical foundations of theoretical statistics. *Philosophical Transactions of the Royal Society of London, 222*, 310–366.

Fisher, R. A. (1934). Two new properties of mathematical likelihood *Proceedings of the Royal Society A, 144*, 285–307.

Fisher, R. A. (1958). *Statistical methods for research workers* (13th ed.). New York: Hafner

Halmos, P., & Savage, J. (1949). Application of the Radon-Nikodym theorem to the theory of sufficient statistics. *The Annals of Mathematical Statistics, 20,* 225–241.

Lehmann, E. (1959). *Testing statistical hypothesis.* New York: Wiley.

Lehmann, E. L., & Scheffé, H. (1947). On the problem of similar regions. *Proceedings of the National Academy of Sciences of the United States of America, 33,* 382–386.

Lehmann, E. L., & Scheffé, H. (1950). Completeness, similar regions, and unbiased estimation. I. *Sankhya, 10,* 305–340.

Lehmann, E. L., & Scheffé, H. (1955). Completeness, similar regions, and unbiased estimation. II. *Sankhya, 15,* 219–236.

Rao, C. R. (1945). Information and accuracy attainable in the estimation of statistical parameters. *Bulletin of the Calcutta Mathematical Society, 37*(3), 81–91.

Wasserman, L. (2003). *All of statistics: A concise course in statistical inference.* New York: Springer.

Chapter 5
Testing Hypotheses

Abstract This chapter develops the theory of optimal parametric tests. The Neyman–Pearson Lemma provides the most powerful test of any given size for a simple null hypothesis H_0 against a simple alternative hypothesis H_1. For one parameter exponential models such tests are uniformly most powerful (UMP) against one-sided alternatives. For two-sided alternatives here one obtains a UMP test among all unbiased tests of a given size. In multiparameter exponential models one may similarly obtain UMP unbiased tests in the presence of nuisance parameters. For statistical models which are invariant under a group of transformations all reasonable tests should be invariant under the group. The theory of UMP tests among all invariant tests is developed for linear models.

5.1 Introduction

An observation (vector) $\mathbf{X}$ is distributed according to P_θ on the observation space $\mathscr{X}$. Here θ is an unknown parameter lying in a set Θ. Suppose there are two competing hypotheses for θ:

$$\begin{aligned}
&\textit{Null Hypothesis } H_0: &&\theta \in \Theta_0,\\
&\textit{Alternative Hypothesis } H_1: &&\theta \in \Theta_1
\end{aligned} \qquad (5.1)$$

where Θ_0 and Θ_1 are nonempty and $\Theta_0 \cup \Theta_1 = \Theta$. On the basis of the observation $\mathbf{X}$, the statistician must decide whether to *accept* H_0 or to *accept* H_1 *(reject H_0)*. Generally, H_0 is such that one can not afford to reject it unless the evidence against it is very compelling. This creates an asymmetry in the problem.

As discussed in Chap. 2, on page 14, one may take the action space here as $\mathscr{A} = \{a_0, a_1\}$, where $a_0 = $ "accept H_0", $a_1 = $ "accept H_1". The usual loss function is

$$L(\theta, a_i) = \begin{cases} 0 \text{ if } \theta \in \Theta_i, \\ 1 \text{ if } \theta \notin \Theta_i \ (i = 0, 1). \end{cases} \qquad (5.2)$$

Definition 5.1. A *nonrandomized test* d is of the form

$$d(\mathbf{x}) = \begin{cases} a_0 \text{ if } \mathbf{x} \in A, \\ a_1 \text{ if } \mathbf{x} \in C = \mathscr{X} \backslash A, \end{cases} \qquad (5.3)$$

© Springer-Verlag New York 2016
R. Bhattacharya et al., *A Course in Mathematical Statistics and Large Sample Theory*, Springer Texts in Statistics, DOI 10.1007/978-1-4939-4032-5_5

where A (the *acceptance region* for H_0), or C (the *rejection region for H_0, or critical* region), is a measurable subset of $\mathscr{X}$. The *risk function* is given by

$$R(\theta, d) = E_\theta L(\theta, d(\mathbf{X})) = \begin{cases} \alpha_d(\theta) = P_\theta(\mathbf{X} \in C) \text{ if } \theta \in \Theta_0, \\ \beta_d(\theta) = P_\theta(\mathbf{X} \in A) \text{ if } \theta \in \Theta_1. \end{cases} \qquad (5.4)$$

The quantity $\alpha_d(\theta)$ is called the probability of a *Type I Error*, or the *level of significance* of the test, while $\beta_d(\theta)$ is the probability of a *Type II Error*, and $\gamma_d(\theta) = 1 - \beta_d(\theta) \equiv P_\theta(\mathbf{X} \in C)$, $\theta \in \Theta_1$, is called the *power of the test*.

Definition 5.2. More generally, *a (randomized) test δ* is a measurable assignment of probabilities $\mathbf{x} \to (1 - \varphi(\mathbf{x}), \varphi(\mathbf{x}))$, $0 \leq \varphi(\mathbf{x}) \leq 1$, so that, given $\mathbf{X} = \mathbf{x}$, one takes the action a_1 (accept H_1 or, equivalently, reject H_0) with probability $\varphi(\mathbf{x})$ and takes the action a_0 (accept H_0) with probability $1 - \varphi(\mathbf{x})$. We will generally refer to φ as the *test,* since δ is determined by it. Then

$$R(\theta, \delta) = \begin{cases} \alpha_\delta(\theta) = E_\theta \varphi(\mathbf{X}) \quad \text{ if } \theta \in \Theta_0, \\ \beta_\delta(\theta) = 1 - E_\theta \varphi(\mathbf{X}) \text{ if } \theta \in \Theta_1. \end{cases} \qquad (5.5)$$

The power of the test is $\gamma_\delta(\theta) = E_\theta \varphi(\mathbf{X})$ (for $\theta \in \Theta_1$).

In the case of a non-randomized test, $\varphi(\mathbf{x}) = 1_C(\mathbf{x})$.

The maximum value (or, the supremum) of $\alpha_\delta(\theta)$ over Θ_0 is referred to as the *size of the test* and is also sometimes called the *level of significance of the test:*

$$\alpha_\delta \equiv \text{size of the test } \delta \; = \sup_{\theta \in \Theta_0} \alpha_\delta(\theta). \qquad (5.6)$$

In view of the asymmetry of the nature of the hypotheses mentioned above, the classical testing procedure aims at keeping the size of the test small (say $\alpha_\delta = 0.05$ or 0.01) while trying to minimize $\beta_\delta(\theta)$, or maximize the power $\gamma_\delta(\theta)$, $\theta \in \Theta_1$, as far as possible.

Definition 5.3. A test δ^* is said to be *uniformly most powerful (UMP) of size α* if

$$\alpha_{\delta^*} \equiv \sup_{\theta \in \Theta_0} \alpha_\delta(\theta) = \alpha, \qquad (5.7)$$

and

$$\gamma_{\delta^*}(\theta) \geq \gamma_\delta(\theta) \qquad \forall \, \theta \in \Theta_1, \qquad (5.8)$$

for all tests δ of size α or less.

As we will see UMP tests exist only under special circumstances.

Choice of H_0 and H_1 in One-Sided Tests To appreciate the importance of the choice of H_0 and H_1 in practical situations, consider the problem of a retailer deciding whether to buy a large consignment of a manufactured item. He would be happy if the proportion p of defectives did not exceed $5\,\%$. One may then consider $H_0 : p \leq 0.05$, $H_1 : p > 0.05$. Suppose a random sample of size n yields a proportion of defectives $4\,\%$. At any reasonable level of significance, say $\alpha = 0.05$, the optimal test will "accept" H_0. But accepting H_0 is not in general a strong endorsement of H_0, since it is given so much protection, and the statistician declares that H_0 is "not rejected". On the other hand, let $H_0 : p \geq 0.05$, $H_1 : p < 0.05$. It may very well happen that the corresponding optimal test will reject H_0. This would be a rather strong indictment against H_0, since it got rejected in spite of such a strong protection (namely, $\alpha = 0.05$). The retailer may now feel confident that the quality of the product meets his criterion.

5.2 Simple Hypotheses and the Neyman–Pearson Lemma

Consider the case where $\Theta_0 = \{\theta_0\}$ is a singleton. Then H_0 is called a *simple null hypothesis* (Else it is a *composite null hypothesis*). Similarly, if $\Theta_1 = \{\theta_1\}$, then H_1 is called a *simple alternative hypothesis* (Else it is a *composite alternative hypothesis*). We first consider the case of a simple null hypothesis $H_0 : \theta = \theta_0$ and a simple alternative hypothesis $H_1 : \theta = \theta_1$ (so that $\Theta = \{\theta_0, \theta_1\}$). We will show that there exists a most powerful test δ^* for a given size α $(0 \leq \alpha \leq 1)$. Let P_θ have density $f(\mathbf{x} \mid \theta)$ (w.r.t. a σ-finite measure μ).

Theorem 5.1 (The Neyman–Pearson Lemma). *For $H_0 : \theta = \theta_0$, $H_1 : \theta = \theta_1$, consider the test $\varphi^*(\mathbf{x})$ of the form*

$$\varphi^*(\mathbf{x}) = \begin{cases} 1 & \text{if } f(\mathbf{x} \mid \theta_1) > kf(\mathbf{x} \mid \theta_0), \\ \gamma & \text{if } f(\mathbf{x} \mid \theta_1) = kf(\mathbf{x} \mid \theta_0), \\ 0 & \text{if } f(\mathbf{x} \mid \theta_1) < kf(\mathbf{x} \mid \theta_0), \end{cases} \tag{5.9}$$

where $0 \leq k < \infty$ and $0 \leq \gamma \leq 1$ are constants.

(a) Then φ^* is a most powerful test of its size.
(b)   The test

$$\varphi^*(\mathbf{x}) = \begin{cases} 1 & \text{if } f(\mathbf{x} \mid \theta_0) = 0, \\ 0 & \text{if } f(\mathbf{x} \mid \theta_0) > 0, \end{cases} \tag{5.10}$$

  is most powerful of size 0.
(c)   For every α, $0 \leq \alpha \leq 1$, there exists a test of the above form.

Proof. (a) Let φ^* be as in (5.9), and consider a test φ of size no more than that of φ^*. Then, writing $\beta^* = \beta_{\delta^*}(\theta_1)$, $\beta = \beta_\delta(\theta_1)$,

$$\beta - \beta^* = E_{\theta_1}\left[(1 - \varphi(\mathbf{X})) - (1 - \varphi^*(\mathbf{X}))\right] = E_{\theta_1}(\varphi^*(\mathbf{X}) - \varphi(\mathbf{X}))$$

$$= \int_{A_1 = \{\mathbf{x}: f(\mathbf{x}|\theta_1) > kf(\mathbf{x}|\theta_0)\}} (\varphi^*(\mathbf{x}) - \varphi(\mathbf{x}))f(\mathbf{x} \mid \theta_1)d\mu(\mathbf{x})$$

$$+ \int_{A_2 = \{\mathbf{x}: f(\mathbf{x}|\theta_1) = kf(\mathbf{x}|\theta_0)\}} (\varphi^*(\mathbf{x}) - \varphi(\mathbf{x}))f(\mathbf{x} \mid \theta_1)d\mu(\mathbf{x})$$

$$+ \int_{A_3 = \{\mathbf{x}: f(\mathbf{x}|\theta_1) < kf(\mathbf{x}|\theta_0)\}} (\varphi^*(\mathbf{x}) - \varphi(\mathbf{x}))f(\mathbf{x} \mid \theta_1)d\mu(\mathbf{x})$$

$$\geq \int_{A_1} (\varphi^*(\mathbf{x}) - \varphi(\mathbf{x}))kf(\mathbf{x} \mid \theta_0)d\mu(\mathbf{x})$$

$$+ \int_{A_2} (\varphi^*(\mathbf{x}) - \varphi(\mathbf{x}))kf(\mathbf{x} \mid \theta_0)d\mu(\mathbf{x}) + \int_{A_3} (\varphi^*(\mathbf{x}) - \varphi(\mathbf{x}))kf(\mathbf{x} \mid \theta_0)d\mu(\mathbf{x}),$$

$$\tag{5.11}$$

since (1) on A_1, $\varphi^*(\mathbf{x}) - \varphi(\mathbf{x}) \geq 0$, (2) on A_2, $f(\mathbf{x} \mid \theta_1) = kf(\mathbf{x} \mid \theta_0)$, and (3) on A_3, $\varphi^*(\mathbf{x}) - \varphi(\mathbf{x}) \leq 0$ (and the factor $f(\mathbf{x} \mid \theta_1)$ is replaced by a smaller quantity on A_3). But the (extreme) right side of (5.11) equals $k(E_{\theta_0}\varphi^*(\mathbf{X}) - E_{\theta_0}\varphi(\mathbf{X})) = k(\alpha_{\delta^*} - \alpha_\delta) \geq 0$, since by hypothesis $\alpha_\delta \leq \alpha_{\delta^*}$.

(b) The size of φ^*, given by (5.10), is $\alpha_{\delta^*} = E_{\theta_0}\varphi^*(\mathbf{X}) = \int_{\{\mathbf{x}:f(\mathbf{x}|\theta_0)=0\}} f(\mathbf{x} \mid \theta_0)d\mu(\mathbf{x}) + \int_{\{\mathbf{x}:f(\mathbf{x}|\theta_0)>0\}} 0 \cdot f(\mathbf{x} \mid \theta_0)d\mu(\mathbf{x}) = 0$. If φ is a test of size 0, then $E_{\theta_0}\varphi(\mathbf{X}) = 0$, so that $\varphi(\mathbf{x}) = 0$ a.e. μ on $\{\mathbf{x} : f(\mathbf{x} \mid \theta_0) > 0\}$. Clearly, among all such tests a (the) most powerful test assigns $\varphi(\mathbf{x}) = 1$ on the rest of the observation space $\{\mathbf{x} : f(\mathbf{x} \mid \theta_0) = 0\}$.

(c) Let $0 < \alpha \leq 1$. Write $Y = f(\mathbf{X} \mid \theta_1)/f(\mathbf{X} \mid \theta_0)$. Note that $P_{\theta_0}(Y = \infty) = P_{\theta_0}(\{\mathbf{x} : f(\mathbf{x} \mid \theta_0) = 0\}) = 0$. Hence, under P_{θ_0}, $0 \leq Y < \infty$, a.s. We need to determine k and γ such that $P_{\theta_0}(Y > k) + \gamma P_{\theta_0}(Y = k) = \alpha$, or

$$P_{\theta_0}(Y \leq k) - \gamma P_{\theta_0}(Y = k) = 1 - \alpha. \tag{5.12}$$

If there exists k such that $P_{\theta_0}(Y \leq k) = 1 - \alpha$, then use this k and take $\gamma = 0$. If not, there exists k_0 such that $P_{\theta_0}(Y < k_0) < 1 - \alpha$ and $P_{\theta_0}(Y \leq k_0) > 1 - \alpha$. In this case, (5.12) is solved by taking $k = k_0$ and

$$\gamma = \frac{P_{\theta_0}(Y \leq k_0) - (1 - \alpha)}{P_{\theta_0}(Y = k_0)} = \frac{\alpha - P_{\theta_0}(Y > k_0)}{P_{\theta_0}(Y = k_0)}.$$

That is, $\alpha = P_{\theta_0}(Y > k_0) + \gamma P_{\theta_0}(Y = k_0)$. $\qquad\square$

Remark 5.1. The use of densities $f(\mathbf{x} \mid \theta_0)$, $f(\mathbf{x} \mid \theta_1)$ of the distributions P_{θ_0}, P_{θ_1} on $\mathscr{X}$ is not a restriction. For, one may always take the dominating measure $\mu = P_{\theta_0} + P_{\theta_1}$. In most of the examples that we deal with in this course, P_θ has density w.r.t. Lebesgue measure (on $\mathbb{R}^n$), or w.r.t. the counting measure (on S^n where S is countable).

5.3 Examples

Consider a one-parameter exponential family of distributions with density (w.r.t. a σ-finite measure μ on the observation space $\mathscr{X}$)

$$f(\mathbf{x} \mid \theta) = \overline{C}(\theta)h(\mathbf{x})\exp\{\pi(\theta)T(\mathbf{x})\}, \tag{5.13}$$

which may be written in natural parameter form as

$$\tilde{f}(\mathbf{x} \mid \pi) = \widetilde{C}(\pi)h(\mathbf{x})\exp\{\pi T(\mathbf{x})\}. \tag{5.14}$$

Assume the natural parameter space Π is a nonempty open interval, $\Pi = (a, b)$. We want to test

$$H_0 : \pi = \pi_0 \quad \text{against} \quad H_1 : \pi = \pi_1. \tag{5.15}$$

The most powerful test φ^* of a given size α is of the form given by the N–P Lemma:

$$\varphi^*(\mathbf{x}) = \begin{cases} 1 & \text{if } \tilde{f}(\mathbf{x} \mid \pi_1)/\tilde{f}(\mathbf{x} \mid \pi_0) > k \\ \gamma & \text{if } \tilde{f}(\mathbf{x} \mid \pi_1)/\tilde{f}(\mathbf{x} \mid \pi_0) = k \\ 0 & \text{if } \tilde{f}(\mathbf{x} \mid \pi_1)/\tilde{f}(\mathbf{x} \mid \pi_0) < k. \end{cases} \tag{5.16}$$

That is,

$$\varphi^*(\mathbf{x}) = \begin{cases} 1 & \text{if } \exp\{(\pi_1 - \pi_0)T(\mathbf{x})\} > k_1 \equiv k\widetilde{C}(\pi_0)/\widetilde{C}(\pi_1) \\ \gamma & \text{if } \exp\{(\pi_1 - \pi_0)T(\mathbf{x})\} = k_1 \\ 0 & \text{if } \exp\{(\pi_1 - \pi_0)T(\mathbf{x})\} < k_1. \end{cases}$$

Suppose $\pi_1 > \pi_0$. Then

$$\varphi^*(\mathbf{x}) = \begin{cases} 1 & \text{if } T(\mathbf{x}) > k_2 = (\ln k_1)/(\pi_1 - \pi_0) \\ \gamma & \text{if } T(\mathbf{x}) = k_2 \\ 0 & \text{if } T(\mathbf{x}) < k_2. \end{cases} \tag{5.17}$$

If, on the other hand, $\pi_1 < \pi_0$, then the most powerful test is

$$\varphi^*(\mathbf{x}) = \begin{cases} 1 & \text{if } T(\mathbf{x}) < k_2 \\ \gamma & \text{if } T(\mathbf{x}) = k_2 \\ 0 & \text{if } T(\mathbf{x}) > k_2. \end{cases} \tag{5.18}$$

Note that k_2 and γ are determined *only by the size of the test*. For example, in the case $\pi_1 > \pi_0$, one has

$$\alpha = P_{\pi_0}(T > k_2) + \gamma P_{\pi_0}(T = k_2). \tag{5.19}$$

Example 5.1. Let $\mathscr{X} = \mathbb{R}^n$, P_θ the joint distribution of n i.i.d. variables, each with distribution $N(\theta, 1)$, so that

$$f(\mathbf{x} \mid \theta) = (2\pi)^{-n/2} \exp\left\{ -\sum_{j=1}^{n}(x_j - \theta)^2/2 \right\}$$

$$\tilde{f}(\mathbf{x} \mid \pi) = (2\pi)^{-n/2} e^{-n\theta^2/2} \exp\left\{ -\sum_{j=1}^{n} x_j^2/2 \right\} \exp\left\{ \theta \sum_{j=1}^{n} x_j \right\}$$

$$= \widetilde{C}(\pi)h(\mathbf{x}) \exp\{\pi T(\mathbf{x})\}, \qquad (\mathbf{x} \in \mathscr{X})$$

where $\pi = \theta \in \Pi = \mathbb{R}$, and $T(\mathbf{x}) = \sum_{j=1}^{n} x_j$. The most powerful test for

$$H_0 : \theta = \theta_0 \quad (\text{or } \pi = \pi_0) \qquad \text{against} \qquad H_1 : \theta = \theta_1 (\pi = \pi_1),$$

with $\theta_1 > \theta_0$, is given by

$$\varphi^*(\mathbf{x}) = \begin{cases} 1 & \text{if } T(\mathbf{x}) > k_2 \\ \gamma & \text{if } T(\mathbf{x}) = k_2 \\ 0 & \text{if } T(\mathbf{x}) < k_2. \end{cases}$$

To determine k_2 and γ, using (5.19), note first that $P_{\pi_0}(T = k_2) = 0$, since the P_{π_0}-distribution of $T = \sum_{1}^{n} X_j$ is $N(n\theta_0, n)$ (Normal with mean $n\theta_0$ and variance n), which assigns probability 0 to every singleton $\{k_2\}$. Hence we can take the test to be nonrandomized,

$$\varphi^*(\mathbf{x}) = \begin{cases} 1 & \text{if } T(\mathbf{x}) > k_2, \\ 0 & \text{if } T(\mathbf{x}) \leq k_2, \end{cases} \tag{5.20}$$

where, noting that $(T - n\theta_0)/\sqrt{n}$ is $\mathbf{N}(0,1)$, one has

$$\alpha = P_{\theta_0}(T > k_2) = P_{\theta_0}\left(\frac{T - n\theta_0}{\sqrt{n}} > \frac{k_2 - n\theta_0}{\sqrt{n}}\right) = 1 - \Phi\left(\frac{(k_2 - n\theta_0)}{\sqrt{n}}\right) \quad (5.21)$$

i.e.,

$$\frac{k_2 - n\theta_0}{\sqrt{n}} = \Phi^{-1}(1 - \alpha) \quad (k_2 = n\theta_0 + \sqrt{n}\,\Phi^{-1}(1 - \alpha)), \quad\quad (5.22)$$

Φ being the (cumulative) distribution function of $\mathbf{N}(0,1)$. A standard way of expressing (5.21)–(5.22) is:

$$\varphi^*(\mathbf{x}) = \begin{cases} 1 \text{ if } \sqrt{n}(\overline{x} - \theta_0) > z_{1-\alpha} \equiv \Phi^{-1}(1 - \alpha) \\ 0 \text{ if } \sqrt{n}(\overline{x} - \theta_0) \leq z_{1-\alpha}. \end{cases} \quad\quad (5.23)$$

Note that (5.23) is of the form (5.20), and this test has size α (since $\sqrt{n}(\overline{x} - \theta_0)$ has the distribution $\mathbf{N}(0,1)$ if θ_0 is the true parameter value).

Example 5.2. Let $\mathscr{X} = \{0,1\}^n$, and the probability mass function (i.e., density w.r.t. counting measure μ on $\mathscr{X}$) is

$$p(\mathbf{x} \mid \theta) = \theta^{\sum_1^n x_j}(1 - \theta)^{n - \sum_1^n x_j} = (1 - \theta)^n \left(\frac{\theta}{1 - \theta}\right)^{\sum_1^n x_j}$$

$$= \widetilde{C}(\pi)e^{\pi \sum_1^n x_j} = \widetilde{C}(\pi)e^{\pi T(\mathbf{x})}, \quad \text{say, } (T = \textstyle\sum_{j=1}^n X_j),$$

with the natural parameter $\pi = \ln\left(\frac{\theta}{1-\theta}\right) \in \Pi = \mathbb{R}$ (corresponding to $\theta \in \Theta = (0,1)$). Suppose we wish to test

$$H_0 : \theta = \theta_0 \quad \text{against} \quad H_1 : \theta = \theta_1 \quad (\theta_1 > \theta_0). \quad\quad (5.24)$$

Since $\theta \to \pi(\theta)$ is strictly increasing, one has $\pi_0 \equiv \ln\left(\frac{\theta_0}{1-\theta_0}\right) < \pi_1 = \ln\left(\frac{\theta_1}{1-\theta_1}\right)$, since $\theta_1 > \theta_0$. The best test φ^* of a given size $\alpha \in (0,1)$ is then of the form

$$\varphi^*(\mathbf{x}) = \begin{cases} 1 & \text{if } \sum_1^n x_j > k_2 \\ \gamma & \text{if } \sum_1^n x_j = k_2 \\ 0 & \text{if } \sum_1^n x_j < k_2 \end{cases}$$

where k_2 (integer) and γ are determined by

$$\alpha = P_{\theta_0}\left(\sum_1^n X_j > k_2\right) + \gamma P_{\theta_0}\left(\sum_1^n X_j = k_2\right)$$

$$= \sum_{r=k_2+1}^n \binom{n}{r}\theta_0^r(1 - \theta_0)^{n-r} + \gamma\binom{n}{k_2}\theta_0^{k_2}(1 - \theta_0)^{n-k_2}. \quad\quad (5.25)$$

For example, let $n = 20$, $\theta_0 = 0.20$, $\alpha = 0.10$. Then the solution of (5.25) is given by (Look up Binomial tables with $n = 20$, $\theta = 0.20$)

$$k_2 = 6,$$

$$\gamma = \frac{\alpha - P_{\theta_0}\left(\sum_1^n X_j > 6\right)}{P_{\theta_0}\left(\sum X_j = 6\right)} = \frac{0.10 - 0.0867}{0.1091} = 0.1219.$$

Here, $P_{\theta_0}(T > 6) = 0.0867(P_{\theta_0}(T > 5) > 0.10)$, and $P_{\theta_0}(T = 6) = 0.1091$. Hence

$$P_{\theta_0}(T > 6) + \gamma P_{\theta_0}(T = 6) = 0.0867 + \left(\frac{0.10 - 0.0867}{0.1091}\right)0.1091$$
$$= 0.0867 + 0.10 - 0.0867 = 0.10 = \alpha.$$

Three important remarks may now be made.

Remark 5.2 (Uniformly Most Powerful Tests). It follows from (5.13)–(5.19) (as illustrated by Examples 5.1 and 5.2 above) that the test (5.18) for $H_0 : \pi = \pi_0$ satisfying the size restriction (5.19), is *most powerful of size α for every alternative* $H_1 : \pi = \pi_1$, *as long as* $\pi_1 > \pi_0$. Hence this test is *uniformly most powerful of size α for testing $H_0 : \pi = \pi_0$ against $H_1 : \pi > \pi_0$.* Now compare this test φ^* against the *test φ_α* given by $\varphi_\alpha(\mathbf{x}) = \alpha \ \forall \ \mathbf{x} \in \mathscr{X}$ (i.e., whatever be the observation $\mathbf{x}$, this test rejects H_0 with probability α, and accepts it with probability $1 - \alpha$). Since this test is of size α, it follows that $E_\pi \varphi^*(\mathbf{X}) \geq E_\pi \varphi_\alpha(\mathbf{X}) = \alpha \ \forall \pi > \pi_0$. This property of a test, namely, that its power is at least as large as its size (i.e., the probability of rejecting H_0 when it is false is at least as large as the probability of rejecting H_0 when H_0 is true), is called *unbiasedness* (of the test). We now argue that φ^* is *uniformly most powerful (UMP)* of its size, say α, for testing $H_0 : \pi \leq \pi_0$ against $H_1 : \pi > \pi_0$. Since it is UMP of size α among all tests of size α or less for testing $H_0 : \pi = \pi_0$ (against $H_1 : \pi > \pi_0$), and since every test of size α or less for $H_0 : \pi \leq \pi_0$ is also of size α or less for testing $H_0 : \pi = \pi_0$, we only need to show that φ^* is of size α for testing $H_0 : \pi \leq \pi_0$. Fix $\pi_- < \pi_0$. The test φ^* is most powerful of its size, say α_-, for testing $H_0 : \pi = \pi_-$ against $H_1 : \pi = \pi_0$ (since $\pi_0 > \pi_-$). In particular, it is at least as powerful as the test φ_{α_-} (of the same size α_-). But the power of φ^* (for testing $H_0 : \pi = \pi_-$, $H_1 : \pi = \pi_0$) is $E_{\pi_0}\varphi^*(\mathbf{X}) = \alpha$, and that of φ_{α_-} is $E_{\pi_0}\varphi_{\alpha_-}(\mathbf{X}) = \alpha_-$. Therefore, $\alpha_- \leq \alpha$. This shows that $E_\pi \varphi^*(\mathbf{X}) \leq \alpha \ \forall \pi < \pi_0$. Hence, for testing $H_0 : \pi \leq \pi_0$ against $H_1 : \pi > \pi_0$, the size of φ^* is α. Incidentally, we have shown that $E_\pi \varphi^*(\mathbf{X})$ *is monotone increasing with π.*

Finally, reversing the inequalities in the proof of the N–P Lemma, one shows that *for testing $H_0 : \pi = \pi_0$, against $H_1 : \pi < \pi_0$, φ^* is the least powerful of all tests of its size.*

One may also consider the problem of testing $H_0 : \sigma^2 \leq \sigma_0^2$ against $H_1 : \sigma^2 > \sigma_0^2$ based on i.i.d. observations $X_1, \ldots, X_n$ from $N(\mu_0, \sigma^2)$, with μ_0 known (Exercise 5.9(a)). One may think of this as a statistical test to determine if a new equipment for measuring length is at least as accurate as a standard equipment, by taking measurements with the new equipment. Here σ_0^2 is the known variance for the standard equipment.

Remark 5.3. Note that for testing $H_0 : \pi = \pi_0$ against $H_1 : \pi = \pi_1$, where $\pi_1 < \pi_0$, the most powerful test, given by the N–P Lemma in the case of a one-parameter exponential family (5.13)–(5.16) becomes instead of (5.17))

$$\varphi^*(\mathbf{x}) = \begin{cases} 1 & \text{if } T(\mathbf{x}) < k_2 \ (= (\ell_n k_1)/(\pi_1 - \pi_0)) \\ \gamma & \text{if } T(\mathbf{x}) = k_2 \\ 0 & \text{if } T(\mathbf{x}) > k_2. \end{cases} \tag{5.26}$$

Arguments entirely analogous to those made for the case $\pi_1 > \pi_0$ show that the test (5.26) is UMP of its size for testing $H_0 : \pi \geq \pi_0$ (for any given π_0) against

$H_1 : \pi < \pi_0$. One may also derive this fact by changing the parameter $\pi \to -\pi$ and the statistic $T \to -T$ (so that $\pi_1 < \pi_0$ becomes $-\pi_1 > -\pi_0$). (Exercise 5.1.)

Remark 5.4. Although one-parameter exponential families constitute the most important examples for the existence of uniformly most powerful tests for hypotheses of the kind $H_0 : \theta \le \theta_0$ against $H_1 : \theta > \theta_0$, the property that is actually used is that of the *monotone likelihood ratio* $f(\mathbf{x} \mid \theta_1)/f(\mathbf{x} \mid \theta_0)$ as a function of a (sufficient) statistic T : If $\theta_1 > \theta_0$, $f(\mathbf{x} \mid \theta_1)/f(\mathbf{x} \mid \theta_0)$ is a monotone increasing (or, monotone decreasing) function of T (Exercise 5.4).

Finally, in practice, statisticians generally do not rigidly fix the size α of a test. Instead, they compute the *P-value* of the test, namely, given the observation, the *smallest value of α for which H_0 would be rejected* (in favor of H_1). In Example 5.2 above, if $n = 20$, $\theta_0 = 0.20$, ($H_0 : \theta = \theta_0$, $H_1 : \theta > \theta_0$) and if the observed $T(\mathbf{x}) \equiv \sum_1^n x_j = 7$, then the P-value of the test is $P_{\theta_0}(T > 6) = P_{\theta_0}(T \ge 7) = 0.0867$. One would reject H_0 here if the size, or level of significance, α is larger than 0.0867 (e.g., $\alpha = 0.10$). In general, a test would reject H_0 for every $\alpha \ge P$-value. Hence the smaller the P-value, the stronger is the evidence against H_0.

5.4 The Generalized N–P Lemma and UMP Unbiased Tests

In this section we consider the problem of testing of a null hypothesis against a *two-sided alternative*
$$H_0 : \theta = \theta_0, \quad H_1 : \theta \ne \theta_0,$$
where θ is a real parameter. It is clear that for this, a UMP test of any given size $\alpha \in (0, 1)$ does not exist in general. As pointed out in the preceding section, even in the case of a one-parameter exponential family, a UMP test for $H_0 : \theta = \theta_0$ against $H_1 : \theta > \theta_0$ is *uniformly least powerful* (among all tests of the same size) against alternatives $\theta < \theta_0$ (and vice versa). The following diagram schematically illustrates the situation (Fig. 5.1): We will, therefore, restrict our attention to the class $\mathscr{G}$ of *unbiased tests* φ, i.e., *power of $\varphi \ge$ size of φ*, or

$$E_\theta \varphi(\mathbf{X}) \ge \alpha = \sup_{\theta \in \Theta_0} E_\theta(\mathbf{X}) \quad \forall\, \theta \in \Theta_1. \tag{5.27}$$

$[\inf_{\theta \in \Theta_1} E_\theta \varphi(\mathbf{X}) \ge \sup_{\theta \in \Theta_0} E_\theta \varphi(\mathbf{X}).]$ Note that there always exist unbiased tests of any given size. For example, *the test $\varphi_\alpha : \varphi_\alpha(\mathbf{x}) = \alpha \,\forall \mathbf{x} \in \mathscr{X}$*, is of size α and is unbiased (since $E_\theta \varphi_\alpha(\mathbf{X}) = \alpha \,\forall \theta$). The following theorem shows that, for

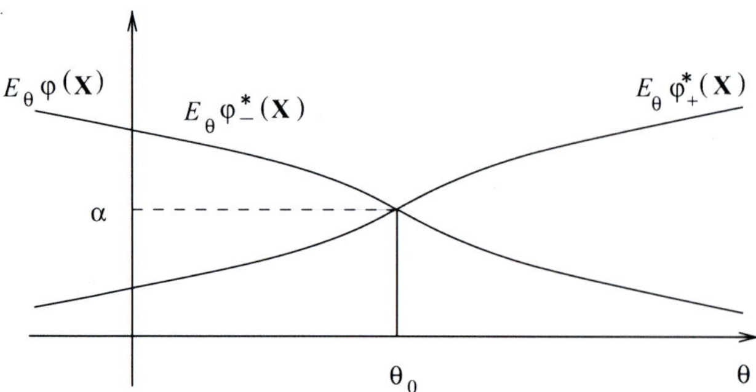

Fig. 5.1 φ_+^*, φ_-^* are UMP for $H_1 : \theta > \theta_0$, $H_1 : \theta < \theta_0$, respectively

one-parameter exponential families, a uniformly most powerful test exists in the class of all unbiased tests of any specified size. Such tests are said to be *uniformly most powerful unbiased* or *UMPU*.

Theorem 5.2. *Consider the one-parameter exponential family (5.14), in natural parameter form, with Π an open interval. A UMP unbiased test φ^* of size α, for testing $H_0 : \pi = \pi_0$ against $H_1 : \pi \neq \pi_0$ is given by*

$$\varphi^*(\mathbf{x}) = \begin{cases} 1 & \text{if } T(\mathbf{x}) < t_1 \text{ or } > t_2, \\ \gamma_i & \text{if } T(\mathbf{x}) = t_i \ (i = 1, 2) \\ 0 & \text{if } t_1 < T(\mathbf{x}) < t_2, \end{cases} \tag{5.28}$$

where $t_1 < t_2$ and γ_i $(i = 1, 2)$ satisfy the equations

$$\alpha = E_{\pi_0}\varphi^*(\mathbf{X}) = P_{\pi_0}(T < t_1) + P_{\pi_0}(T > t_2) + \gamma_1 P_{\pi_0}(T = t_1) + \gamma_2 P_{\pi_0}(T = t_2), \tag{5.29}$$

and all tests φ (including φ^*) in $\mathscr{G}$ satisfy

$$0 = \left(\frac{d}{d\pi} E_\pi \varphi(\mathbf{X}) \right)_{\pi = \pi_0} \qquad (\textit{Unbiasedness}), \tag{5.30}$$

or, equivalently, the relation (5.36) below, namely.

$$E_{\pi_0}\varphi(\mathbf{X})T(\mathbf{X}) = \alpha E_{\pi_0}T(\mathbf{X}).$$

The following generalized version of the N–P Lemma is needed for the proof.

Theorem 5.3 (Generalized N–P Lemma). *Let f_i, $1 \leq i \leq m+1$, be functions on $\mathscr{X}$, and $\mathscr{G}$ the class of tests φ, $0 \leq \varphi \leq 1$, satisfying*

$$\int_{\mathscr{X}} f_i(\mathbf{x})\varphi(\mathbf{x})d\mu(\mathbf{x}) = c_i \qquad (1 \leq i \leq m). \tag{5.31}$$

If φ^* is in $\mathscr{G}$ and is of the form

$$\varphi^*(\mathbf{x}) = \begin{cases} 1 & \text{if } f_{m+1}(\mathbf{x}) > \sum_{i=1}^{m} k_i f_i(\mathbf{x}), \\ \gamma(\mathbf{x}) & \text{if } f_{m+1}(\mathbf{x}) = \sum_{i=1}^{m} k_i f_i(\mathbf{x}), \\ 0 & \text{if } f_{m+1}(\mathbf{x}) < \sum_{i=1}^{m} k_i f_i(\mathbf{x}), \end{cases} \tag{5.32}$$

for some constants k_i $(1 \leq i \leq m)$ and some measurable $\gamma(\mathbf{x})$, $0 \leq \gamma(\mathbf{x}) \leq 1$, then

$$\int_{\mathscr{X}} \varphi^*(\mathbf{x})f_{m+1}(\mathbf{x})d\mu(\mathbf{x}) = \sup_{\varphi \in \mathscr{G}} \int_{\mathscr{X}} \phi(\mathbf{x})f_{m+1}(\mathbf{x})d\mu(\mathbf{x}). \tag{5.33}$$

Proof. Let $\varphi \in \mathscr{G}$. Then the function $(\varphi^*(\mathbf{x}) - \varphi(\mathbf{x}))(f_{m+1}(\mathbf{x}) - \sum_{i=1}^{m} k_i f(\mathbf{x}))$ is nonnegative on $\mathscr{X}$. Therefore, by (5.31),

$$0 \leq \int_{\mathscr{X}} (\varphi^*(\mathbf{x}) - \varphi(\mathbf{x}))(f_{m+1}(\mathbf{x}) - \sum_{i=1}^{m} k_i f_i(\mathbf{x}))d\mu(\mathbf{x})$$

$$= \int_{\mathscr{X}} (\varphi^*(\mathbf{x}) - \varphi(\mathbf{x}))(f_{m+1}(\mathbf{x}))d\mu(\mathbf{x})$$

$$= \int_{\mathscr{X}} \varphi^*(\mathbf{x})f_{m+1}(\mathbf{x})d\mu(\mathbf{x}) - \int_{\mathscr{X}} \varphi(\mathbf{x})f_{m+1}(\mathbf{x})d\mu(\mathbf{x}).$$

$\square$

Next note that, for the case of a one-parameter exponential family with Π an open interval, one can differentiate under the integral sign (Exercise 5.4(a)):

$$
\begin{aligned}
\frac{d}{d\pi} E_\pi \varphi(\mathbf{X}) &= \frac{d}{d\pi} \int_{\mathscr{X}} \varphi(\mathbf{x}) \mathbf{C}(\pi) h(\mathbf{x}) e^{\pi T(\mathbf{x})} d\mu(\mathbf{x}) \\
&= \mathbf{C}'(\pi) \int_{\mathscr{X}} \varphi(\mathbf{x}) h(\mathbf{x}) e^{\pi T(\mathbf{x})} d\mu(\mathbf{x}) \\
&\quad + \int_{\mathscr{X}} \varphi(\mathbf{x}) \mathbf{C}(\pi) h(\mathbf{x}) T(\mathbf{x}) e^{\pi T(\mathbf{x})} d\mu(\mathbf{x}) \\
&= \frac{\mathbf{C}'(\pi)}{\mathbf{C}(\pi)} E_\pi \varphi(\mathbf{X}) + E_\pi \varphi(\mathbf{X}) T(\mathbf{X}).
\end{aligned}
\tag{5.34}
$$

Now

$$
\begin{aligned}
0 &= \frac{d}{d\pi} \int_{\mathscr{X}} \mathbf{C}(\pi) h(\mathbf{x}) e^{\pi T(\mathbf{x})} d\mu(\mathbf{x}) \\
&= \mathbf{C}'(\pi) \int_{\mathscr{X}} h(\mathbf{x}) e^{\pi T(\mathbf{x})} d\mu(\mathbf{x}) + \int_{\mathscr{X}} \mathbf{C}(\pi) h(\mathbf{x}) T(\mathbf{x}) e^{\pi T(\mathbf{x})} d\mu(\mathbf{x}) \\
&= \frac{\mathbf{C}'(\pi)}{\mathbf{C}(\pi)} + E_\pi T(\mathbf{X}),
\end{aligned}
$$

or,

$$
E_\pi T(\mathbf{X}) = -\frac{\mathbf{C}'(\pi)}{\mathbf{C}(\pi)}.
\tag{5.35}
$$

Hence the condition (5.30) may be expressed as

$$
\begin{aligned}
E_{\pi_0} \varphi(\mathbf{X}) T(\mathbf{X}) &= E_{\pi_0} \varphi(\mathbf{X}) \cdot E_{\pi_0} T(\mathbf{X}) \\
&= \alpha E_{\pi_0} T(\mathbf{X}).
\end{aligned}
\tag{5.36}
$$

To prove Theorem 5.2, use the Generalized N–P Lemma with $f_1(\mathbf{x}) = f(\mathbf{x} \mid \pi_0)$, $f_2(\mathbf{x}) = T(\mathbf{x}) f(\mathbf{x} \mid \pi_0)$, $c_1 = \alpha$, $c_2 = \alpha E_{\pi_0} T(\mathbf{X})$, and $f_3(\mathbf{x}) = f(\mathbf{x} \mid \pi)$ for some $\pi \neq \pi_0$.

Case A. Fix π, π_0 with $\pi > \pi_0$. The inequality $f_3(\mathbf{x}) > k_1 f_1(\mathbf{x}) + k_2 f_2(\mathbf{x})$ may be written as $\mathbf{C}(\pi) \exp\{\pi T(\mathbf{x})\} h(\mathbf{x}) > k_1 \mathbf{C}(\pi_0) \exp\{\pi_0 T(\mathbf{x})\} h(\mathbf{x}) + k_2 \mathbf{C}(\pi_0) \exp\{\pi_0 T(\mathbf{x})\} h(\mathbf{x}) T(\mathbf{x})$, or, using different constants, $\exp\{\pi T(\mathbf{x})\} > k'_1 \exp\{\pi_0 T(\mathbf{x})\} + k'_2 \exp\{\pi_0 T(\mathbf{x})\} T(\mathbf{x})$, or

$$
\exp\{(\pi - \pi_0) T(\mathbf{x})\} > k'_1 + k'_2 T(\mathbf{x}), \quad [k'_i = k_i \mathbf{C}(\pi_0)/\mathbf{C}(\pi)].
\tag{5.37}
$$

For a given pair $t_1 < t_2$, one may find k'_1, $k'_2 > 0$ such that the function $g(t) := \exp\{(\pi - \pi_0)t\}$ satisfies $g(t) = k'_1 + k'_2 t$ for the values $t = t_1, t = t_2$ [i.e., $t \to k'_1 + k'_2 t$ is the line passing through the points $(t_1, g(t_1))$ and $(t_2, g(t_2))$.] See Fig. 5.2a.

Then "$t < t_1$ or $t > t_2$" is equivalent to "$g(t) > k'_1 + k'_2 t$", and "$t_1 < t < t_2$" means "$g(t) < k'_1 + k'_2 t$", "$t = t_i$ for $i = 1$ or 2" means "$g(t) = k'_1 + k'_2 t$". Hence the test φ^* given by (5.28) may be expressed in the form (5.32) which, by the Generalized N–P Lemma, is uniformly most powerful in the class of all tests φ satisfying (1) $E_{\pi_0} \varphi(\mathbf{X}) \equiv \int_{\mathscr{X}} \varphi(\mathbf{x}) f_1(\mathbf{x}) d\mu(\mathbf{x}) = \alpha$, (2) $\alpha E_{\pi_0} T(\mathbf{X}) \equiv \int_{\mathscr{X}} \varphi(\mathbf{x}) f_2(\mathbf{x}) d\mu(\mathbf{x})$. Since (5.30) is equivalent to (5.36), this class of tests is precisely the class of all tests of $H_0 : \pi = \pi_0$, $H_1 : \pi \neq \pi_0$ of size α, satisfying (5.30). Since this last class includes the class of all unbiased tests of size α, the proof of Theorem 5.2 is complete in the Case A. The *Case B: $\pi < \pi_0$*, is entirely analogous. In this case one takes $k'_2 < 0$, and Fig. 5.2b applies.

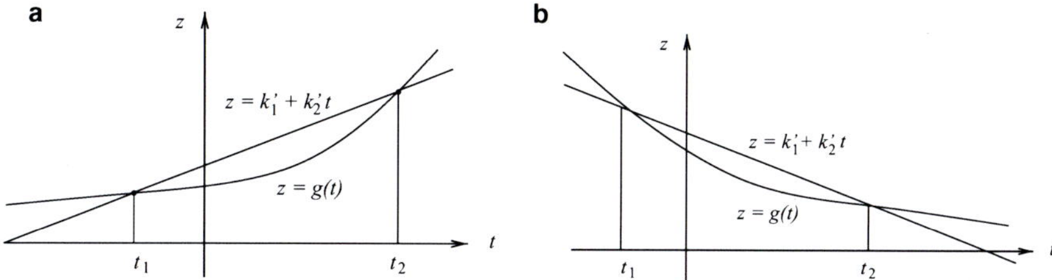

Fig. 5.2 Case A: $\pi > \pi_0$ $(k_2' > 0)$. Case B: $\pi < \pi_0$ $(k_2' < 0)$

It remains to show that for any $\alpha \in (0,1)$ there exists a test of the form (5.28) with $t_1 < t_2$, γ_i $(i = 1,2)$ satisfying (5.29) and (5.30). The proof in the present generality is long[1] and is omitted. Instead, Examples 5.3–5.5 and a number of Exercises illustrate the construction of UMPU tests for arbitrary sizes α, $0 < \alpha < 1$. $\square$

Example 5.3. Let $\mathscr{X} = \mathbb{R}^n$, $f(\mathbf{x} \mid \theta) = f(\mathbf{x} \mid \pi) = (2\pi)^{-n/2} \exp\{-\frac{1}{2}\sum_1^n (x_j - \pi)^2\} = \mathbf{C}(\pi)h(\mathbf{x}) \exp\{\pi \sum_{j=1}^n x_j\}$, where $h(\mathbf{x}) = \exp\{-\frac{1}{2}\sum_{j=1}^n x_j^2\}$. That is, $\mathbf{X} = (X_1, \ldots, X_n)$, where X_i are i.i.d. Normal $\mathbf{N}(\theta, 1) \equiv \mathbf{N}(\pi, 1)$. For a given $\pi_0 \in \mathbb{R}$, we will construct the UMPU test of size α for

$$H_0 : \theta = 0, \quad H_1 : \theta \neq 0. \tag{5.38}$$

By Theorem 5.2, the desired test is of the form

$$\varphi^*(\mathbf{X}) = \begin{cases} 1 & \text{if } T \equiv \sum_1^n X_j < t_1 \text{ or } > t_2 \\ 0 & \text{if } t_1 \leq T \leq t_2, \end{cases} \tag{5.39}$$

where

$$P_0(t_1 \leq T \leq t_2) = 1 - \alpha, \tag{5.40}$$

and

$$E_0\varphi^*(\mathbf{X})T = \alpha E_0 T = 0. \tag{5.41}$$

Now (5.40) is equivalent to

$$P_0\left(c_1 \leq \frac{T}{\sqrt{n}} \leq c_2\right) = 1 - \alpha \quad \left(c_i = \frac{t_i}{\sqrt{n}}, \ i = 1,2\right), \tag{5.42}$$

and (5.41) is equivalent to $E_0(1 - \varphi^*(\mathbf{X}))T = 0$, or $E_0(1 - \varphi^*(\mathbf{X}))\frac{T}{\sqrt{n}} = 0$, or, noting that $\frac{T}{\sqrt{n}}$ is $\mathbf{N}(0,1)$ under P_0,

$$\int_{c_1}^{c_2} z\varphi(z)dz = 0 \quad \left(z = \frac{t}{\sqrt{n}}\right). \tag{5.43}$$

Since $z\varphi(z)$ is an *odd function*, the last condition implies $c_1 = -c_2$ $(c_2 > 0)$. Then (5.42) fixes $c_2 = \Phi^{-1}\left(1 - \frac{\alpha}{2}\right)$, where $\Phi(x)$ is the cumulative distribution function of the standard Normal distribution. Hence the UMPU test of size α is given by

$$\text{Reject } H_0 \text{ iff } \left|\frac{\sum_{j=1}^n X_j}{\sqrt{n}}\right| > \Phi^{-1}\left(1 - \frac{\alpha}{2}\right) = z_{1-\frac{\alpha}{2}}, \text{ say.} \tag{5.44}$$

---

[1] See Ferguson (1967, pp. 215–221).

Example 5.4. Let $X_1, X_2, \ldots, X_n$ be i.i.d. gamma $\mathscr{G}(\theta, \beta)$, *with $\beta > 0$ known*, i.e., with density (w.r.t. Lebesgue measure on $(0, \infty)$) given by

$$f(x \mid \theta) = \frac{1}{\Gamma(\beta)\theta^\beta} \, e^{-x/\theta} x^{\beta-1}, \quad 0 < x < \infty.$$

The (joint) density of $\mathbf{X} = (X_1, \ldots, X_n)$ (w.r.t. Lebesgue measure on $(0, \infty)^n = \mathscr{X}$) is

$$\mathbf{f}(\mathbf{x} \mid \theta) = \left(\frac{1}{\Gamma(\beta)\theta^\beta}\right)^n e^{-\frac{1}{\theta}\sum_{j=1}^n x_j} \prod_{j=1}^n x_j^{\beta-1}$$

$$= C(\theta)h(\mathbf{x}) \exp\{\pi T\} \qquad \left(\pi = -\tfrac{1}{\theta}, \ T(\mathbf{x}) = \textstyle\sum_1^n x_j\right)$$

with the natural parameter $\pi = -\frac{1}{\theta}$, and the complete sufficient statistic $T = \sum_{j=1}^n X_j$ (for the one-parameter exponential family of probability measures $\{P_\pi : \pi = -\frac{1}{\theta} \in (-\infty, 0) = \Pi\}$). By Theorem 5.2, the UMPU test of size α for

$$H_0 : \theta = 1 \ \text{i.e.,} \ \pi = -1, \quad H_1 : \theta \neq 1 \tag{5.45}$$

is given by

$$\varphi^*(\mathbf{X}) = \begin{cases} 1 & \text{if } T < t_1 \text{ or } T > t_2, \\ 0 & \text{if } t_1 \leq T \leq t_2, \end{cases}$$

where, using the fact that, under P_0, T has the gamma distribution $\mathscr{G}(1, n\beta)$ with density

$$g(t) = \frac{1}{\Gamma(n\beta)} \, e^{-t} t^{n\beta-1}, \quad 0 < t < \infty, \tag{5.46}$$

one has

$$\int_{t_1}^{t_2} g(t)dt = 1 - \alpha, \quad \int_{t_1}^{t_2} tg(t)dt = (1-\alpha)E_0 T = (1-\alpha)(n\beta). \tag{5.47}$$

The second integral above is

$$(1-\alpha)n\beta = \int_{t_1}^{t_2} \frac{1}{\Gamma(n\beta)} \, t^{n\beta} d(-e^{-t})$$

$$= \frac{1}{\Gamma(n\beta)} \left\{ t_1^{n\beta} e^{-t_1} - t_2^{n\beta} e^{-t_2} \right\} + \frac{n\beta}{\Gamma(n\beta)} \int_{t_1}^{t_2} t^{n\beta-1} e^{-t} dt$$

$$= \frac{1}{\Gamma(n\beta)} \left\{ t_1^{n\beta} e^{-t_1} - t_2^{n\beta} e^{-t_2} \right\} + n\beta(1-\alpha),$$

that is, one has $t_1^{n\beta} e^{-t_1} - t_2^{n\beta} e^{-t_2} = 0$, which may be expressed as

$$\left(\frac{t_2}{t_1}\right)^{n\beta} = e^{t_2-t_1}. \tag{5.48}$$

Thus $t_1 < t_2$ are determined by the size restriction (or, the first integral in (5.47) and (5.48)), usually by trial and error, using interpolation and a table of gamma probabilities.

Example 5.5. Let X_j's $(1 \leq j \leq n)$ be i.i.d. $\mathbf{N}(0, \theta)$, $\theta \in \Theta = (0, \infty)$. Then the (joint) density of $\mathbf{X} = (X_1, \ldots, X_n)$ (w.r.t. Lebesgue measure on $\mathbb{R}^n$) is

$$\mathbf{f}(\mathbf{x} \mid \theta) = (2\pi\theta)^{-n/2} \exp\left\{ -\frac{1}{2\theta} \sum_{j=1}^{n} x_j^2 \right\},$$

which is the density of a one-parameter exponential family with natural parameter $\pi = -\frac{1}{\theta} \in \Pi = (-\infty, 0)$, and a complete sufficient statistic $T = \sum_{j=1}^{n} X_j^2/2$. The distribution of T is gamma $\mathcal{G}(\theta, n/2)$ (See the Appendix on Standard Distributions) with density (w.r.t. Lebesgue measure on $(0, \infty)$) given by

$$g(t) = \frac{1}{\Gamma\left(\frac{n}{2}\right)\theta^{n/2}} \, e^{-t/\theta} t^{n/2-1}, \quad 0 < t < \infty, \tag{5.49}$$

constituting a one-parameter exponential family, the same as in Example 5.4, with $\beta = 1/2$. Hence the UMPU test for $H_0 : \theta = 1$, $H_1 : \theta \neq 1$, is given by

$$\text{Reject } H_0 \text{ iff} \quad \frac{\sum_1^n X_j^2}{2} < t_1 \quad \text{or} \quad > t_2, \tag{5.50}$$

where t_1 and t_2 are determined by the first (size) condition in (5.47), and (5.48) with $\beta = \frac{1}{2}$.

Note that statistical tables are more readily available for the *chi-square distributions* $\text{Chi}(n) = \mathcal{G}(2, n/2)$ than for general gamma distributions. Hence the test is generally expressed in terms of the chi-square *random variable* $U = \sum_{j=1}^{n} X_j^2$ *with n degrees of freedom,* so that (5.50) becomes

$$\text{Reject } H_0 \text{ iff:} \quad U < u_1 \quad \text{or} \quad > u_2, \tag{5.51}$$

where $u_i = 2t_i$ $(i = 1, 2)$. That is,

$$P(u_1 \leq U \leq u_2) = 1 - \alpha, \quad \left(\frac{u_2}{u_1}\right)^{\frac{n}{2}} = e^{\frac{u_2}{2} - \frac{u_1}{2}}. \tag{5.52}$$

5.5 UMP Unbiased Tests in the Presence of Nuisance Parameters

Implicitly or explicitly, all optimal statistical procedures so far have been shown to be based on sufficient statistics. Since their considerations are particularly important in this section, we begin with a simple affirmation that it is enough to restrict attention to functions of sufficient statistics.

Proposition 5.1. *Let $T = t(\mathbf{X})$ be a sufficient statistic for $\{P_\theta : \theta \in \Theta\}$. Given a test $\varphi = \varphi(\mathbf{X})$ for $H_0 : \theta \in \Theta_0$, $H_1 : \theta \in \Theta_1 = \Theta \backslash \Theta_0$, there exists a test $\psi = \psi(T)$ which has the same performance as φ, in the sense $E_\theta \psi(T) = E_\theta \varphi(\mathbf{X}) \; \forall \theta \in \Theta$.*

Proof. Let $\psi(T) = E_\theta(\varphi(\mathbf{X}) \mid T)$. Then ψ does not depend on θ and is (hence) a statistic, $0 \leq \psi(T) \leq 1$ (almost surely $(P_\theta) \; \forall \theta$), and it satisfies $E_\theta \psi(T) = E_\theta[E_\theta(\varphi(\mathbf{X}) \mid T)] = E_\theta \varphi(\mathbf{X}) \; \forall \theta \in \Theta$. $\qquad \square$

Given a parameter set Θ which is a metric space, and nonempty subsets Θ_0, $\Theta_1 = \Theta \backslash \Theta_0$, let Θ_B be the *boundary* of Θ_0 (or, of Θ_1), i.e., $\Theta_B = \overline{\Theta}_0 \cap \overline{\Theta}_1$.

Proposition 5.2. *If the power function* $\theta \to E_\theta \varphi(\mathbf{X})$ *of a test is continuous on* Θ, *then every unbiased test* φ *of size* α *has the property*

$$E_\theta \varphi(\mathbf{X}) = \alpha \qquad \forall\, \theta \in \Theta_B. \tag{5.53}$$

Proof. Let $\theta \in \Theta_B$. Then $\exists$ sequences $\{\theta_n^{(i)} : n = 1, 2, \ldots\} \subset \Theta_i$ $(i = 0, 1)$ such that $\theta_n^{(i)} \to \theta$ as $n \to \infty$ $(i = 1, 2)$. If $i = 0$, then the size restriction of φ implies $E_{\theta_n^{(0)}} \varphi(\mathbf{X}) \leq \alpha\ \forall n$, so that, by continuity of the power function, $E_\theta \varphi(\mathbf{X}) \leq \alpha$. Similarly, unbiasedness of φ implies $E_{\theta_n^{(1)}} \varphi(\mathbf{X}) \geq \alpha\ \forall n$, and therefore $E_\theta \varphi(\mathbf{X}) \geq \alpha$. $\qquad\square$

Definition 5.4. A test φ satisfying (5.53) is called α-*similar*.

Proposition 5.3. *Suppose the power function* $\theta \to E_\theta \varphi(\mathbf{X})$ *is continuous on* Θ *for all tests* φ. *If* φ^* *is UMP in the class of all* α-*similar tests, and if* φ^* *is of size* α, *then* φ^* *is UMP in the class of all unbiased tests of size* α *of* $H_0 : \theta \in \Theta_0$, $H_1 : \theta \in \Theta_1$.

Proof. The test $\varphi_\alpha(\mathbf{x}) = \alpha\ \forall\ \mathbf{x}$ is α-similar. Therefore, $E_\theta \varphi^*(\mathbf{X}) \geq E_\theta \varphi_\alpha(\mathbf{X}) = \alpha\ \forall\, \theta \in \Theta_1$. Thus φ^* is unbiased (and of size α, by assumption). On the other hand, by Proposition 5.2, every unbiased test of size α is α-similar. Since φ^* is UMP in the (bigger) class of all α-similar tests, φ^* is UMP in the class of all unbiased tests of size α. $\qquad\square$

Consider now the problem of testing

$$H_0 : \theta_1 \leq \theta_1^0, \qquad H_1 : \theta_1 > \theta_1^0, \tag{5.54}$$

where θ_1 is the first coordinate of the parameter $\theta = (\theta_1, \theta_2, \ldots, \theta_k)$, parametrizing the family of probability measures $\{P_\theta : \theta \in \Theta\}$, $\Theta \subset \mathbb{R}^k$. Here $\Theta_0 = \{\theta \in \Theta : \theta_1 \leq \theta_1^0\}$, $\Theta_1 = \{\theta \in \Theta : \theta_1 > \theta_1^0\}$, and $\Theta_B = \{\theta \in \Theta : \theta_1 = \theta_1^0\}$. Suppose there exists a *sufficient statistic* T_B for the family $\{P_\theta : \theta \in \Theta_B\}$. Given any α-similar test $\varphi(\mathbf{X})$, the random variable

$$\psi(T_B) = E_\theta[\varphi(\mathbf{X}) \mid T_B] \qquad (\theta \in \Theta_B) \tag{5.55}$$

does not involve any unknown parameter and is, therefore, a test, which is α-similar.

Definition 5.5. An α-similar test $\varphi(\mathbf{X})$ is said to have the *Neyman structure* if $\psi(T_B) \equiv E_\theta[\varphi(\mathbf{X}) \mid T_B] = \alpha$ a.s. $P_\theta\ \forall\, \theta \in \Theta_B$.

Theorem 5.4. *Suppose* T_B *is a complete sufficient statistic for the family* $\{P_\theta : \theta \in \Theta_B\}$. *Then every* α-*similar test has the Neyman structure.*

Proof. Let $\varphi(\mathbf{X})$ be α-similar. Then, with $\psi(T_B) = E_\theta[\varphi(\mathbf{x}) \mid T_B]$ $(\theta \in \Theta_B)$, one has $E_\theta(\psi(T_B) - \alpha) = 0\ \forall\, \theta \in \Theta_B$. By the hypothesis of completeness, $P_\theta(\psi(T_B) = \alpha) = 1\ \forall\, \theta \in \Theta_B$. $\qquad\square$

Remark 5.5. Note that the proof only required that the *sufficient statistic* T (for $\{P_\theta : \theta \in \Theta_B\}$) be *boundedly complete*, i.e., if *for a bounded measurable function* g, $E_\theta g(T_B) = 0\ \forall\, \theta \in \Theta_B$, *then* $P_\theta(g(T_B) = 0) = 1\ \forall\, \theta \in \Theta_B$.

5.5.1 UMPU Tests in k-Parameter Exponential Families

For the problem of testing (5.54) in the case of a k-parameter exponential family with a complete sufficient statistic $T = (T_1, T_2, \ldots, T_k)$ with density (w.r.t. Lebesgue measure on an open subset $\mathcal{T}$ of $\mathbb{R}^k$)

$$g_T(\mathbf{t} \mid \theta) = C(\theta)h(\mathbf{t}) \exp\left\{\sum_{i=1}^{k} \theta_i t_i\right\}, \quad \mathbf{t} \in \mathcal{T} \subset \mathbb{R}^k, \ (h(\mathbf{t}) > 0 \ \forall\, \mathbf{t} \in \mathcal{T}), \quad (5.56)$$

with natural parameter $\theta = \pi \in \Theta = \Pi \subset \mathbb{R}^k$, the (marginal) density of $(T_2, \ldots, T_k)$ is

$$\tilde{g}(t_2, \ldots, t_k \mid \theta) = C(\theta) \exp\left\{\sum_{i=2}^{k} \theta_i t_i\right\} \int_{\mathbb{R}} h(t_1, t_2, \ldots, t_k) e^{\theta_1 t_1}\, dt_1$$

$$= C(\theta) h_1(t_2, \ldots, t_k, \theta_1) \exp\left\{\sum_{i=2}^{k} \theta_i t_i\right\}, \quad (5.57)$$

$$(t_2, \ldots, t_k) \in \widetilde{\mathcal{T}} = \{(t_2, \ldots, t_k) : h(t_1, t_2, \ldots, t_k) > 0 \text{ for some } t_1\}.$$

Dividing g_T by $\tilde{g}$ we obtain the conditional density of T_1, given $T_2 = t_2, \ldots, T_k = t_k$, as

$$g_1(t_1 \mid t_2, \ldots, t_k; \theta_1) = C_1(\theta_1; t_2, \ldots, t_k) h(t_1, t_2, \ldots, t_k) e^{\theta_1 t_1},$$
$$t_1 \in \mathcal{T}_1(t_2, \ldots, t_k) = \{t_1 : (t_1, t_2, \ldots, t_k) \in \mathcal{T}\}, \quad (5.58)$$

which is a one-parameter exponential family on $\mathcal{T}_1(t_2, \ldots, t_k)$. One may consider the problem of testing (5.54) for this one-parameter exponential family (conditionally given $T_2 = t_2, \ldots, T_k = t_k$). For this family a UMP test of size α is of the form

$$\varphi_1^*(T_1 \mid T_2 = t_2, \ldots, T_k = t_k) = \begin{cases} 1 & \text{if } T_1 > t_1(t_2, \ldots, t_k), \\ 0 & \text{if } T_1 \le t_1(t_2, \ldots, t_k), \end{cases} \quad (5.59)$$

where $t_1(t_2, \ldots, t_k)$ is determined by

$$\int_{t_1(t_2, \ldots, t_k)}^{\infty} g_1(t_1 \mid t_2, \ldots, t_k; \theta_1^0)\, dt_1 = \alpha. \quad (5.60)$$

Not that $T_B = (T_2, \ldots, T_k)$ is a complete sufficient statistic for the $(k-1)$-parameter exponential family $\{P_\theta : \theta \in \Theta_B\}$ with density [see (5.57)]

$$g_3(t_2, \ldots, t_k \mid \theta_2, \ldots, \theta_k) \equiv \tilde{g}(t_2, \ldots, t_k \mid \theta_1^0, \theta_2, \ldots, \theta_k)$$

$$= C(\theta_1^0, \theta_2, \ldots, \theta_k) h_1(t_2, \ldots, t_k, \theta_1^0) \exp\left\{\sum_{i=2}^{k} \theta_i t_i\right\}. \quad (5.61)$$

Hence, by Theorem 5.4, every α-similar test for (5.54) has the Neyman structure. That is, such a test $\varphi(T) = \varphi(T_1, T_2, \ldots, T_k)$ satisfies

$$\psi(T_B) \equiv E_\theta(\varphi(T) \mid T_2, \ldots, T_k) = \alpha \quad \text{a.s. } P_\theta \ \forall\, \theta \in \Theta_B. \quad (5.62)$$

But among all such tests $\varphi(T)$, $\varphi^*(T) \equiv \varphi_1^*(T_1 \mid T_2, \ldots, T_k)$ has the maximum conditional power:

$$E_\theta(\varphi^*(T) \mid T_2, \ldots, T_k) \geq E_\theta(\varphi(T) \mid T_2, \ldots, T_k) \quad (\text{a.s.}\,(P_\theta))\ \forall\,\theta \in \Theta_1,$$

for all unbiased $\varphi = \varphi(T)$ of size α. Taking expectation (E_θ) on both sides, one concludes that φ^* is UMP unbiased among all unbiased tests of size α.

Consider now a k-parameter exponential family on a countable observation space, and let (5.56) denote the probability mass function *(pmf)* of the complete sufficient statistic $T = (T_1, T_2, \ldots, T_k)$, with values in a (countable) set $\mathscr{T}$. Replacing integrations by summations in (5.57), one arrives at (5.58) as the pmf of the conditional distribution of T_1, given $T_2 = t_2, \ldots, T_k = t_k$, which is a one-parameter exponential family. Reasoning as above, the UMPU test of size α for (5.54) is given by

$$\varphi_1^*(T_1 \mid T_2 = t_2, \ldots, T_k = t_k) = \begin{cases} 1 & \text{if } T_1 > t_1(t_2, \ldots, t_k), \\ \gamma & \text{if } T_1 = t_1(t_2, \ldots, t_k), \\ 0 & \text{if } T_1 < t_1(t_2, \ldots, t_k), \end{cases} \qquad (5.63)$$

where $t_1(t_2, \ldots, t_k)$ and γ, $0 \leq \gamma < 1$, are determined by

$$\sum_{t_1 > t_1(t_2, \ldots, t_k)} g(t_1 \mid t_2, \ldots, t_k; \theta_1^0) + \gamma g(t_1(t_2, \ldots, t_k) \mid t_2, \ldots, t_k; \theta_1^0) = \alpha. \qquad (5.64)$$

Example 5.6. Let $X_1, \ldots, X_n$ be i.i.d. $\mathbf{N}(\mu, \sigma^2)$, $n \geq 2$, with joint density

$$(2\pi\sigma^2)^{-n/2} \exp\left\{ -\sum_{j=1}^{n} \frac{(x_j - \mu)^2}{2\sigma^2} \right\} = C(\boldsymbol{\theta}) e^{\theta_1 T_1(\mathbf{x}) + \theta_2 T_2(\mathbf{x})},$$

$$\theta_1 = \frac{\mu}{\sigma^2}, \qquad \theta_2 = -\frac{1}{2\sigma^2},$$

$$T_1(\mathbf{x}) = \sum_{j=1}^{n} x_j, \quad T_2(\mathbf{x}) = \sum_{j=1}^{n} x_j^2, \ C(\boldsymbol{\theta}) = (2\pi\sigma^2)^{-n/2} e^{n\mu^2/2\sigma^2}. \qquad (5.65)$$

Here the natural parameter is $\theta = (\theta_1, \theta_2) \in \mathbb{R} \times (-\infty, 0) = \Theta$. We wish to test

$$H_0 : \mu \leq 0, \quad H_1 : \mu > 0, \quad \text{or}$$
$$H_0 : \theta_1 \leq 0, \quad H_1 : \theta_1 > 0. \qquad (5.66)$$

By the preceding theory, the UMP unbiased test of size $\alpha \in (0, 1)$ is given by

$$\text{Reject } H_0 \text{ iff} \quad \sum_{1}^{n} X_j > \eta_1(T_2), \qquad (5.67)$$

where $\eta_1(t_2)$ is determined for each value $T_2 = t_2$ so that

$$P_\theta\left(\sum_{1}^{n} X_j > \eta_1(t_2) \mid T_2 = t_2 \right) = \alpha \text{ a.s. } (P_\theta) \qquad (5.68)$$

$$\forall\,\theta \in \Theta_B = \{(0, \theta_2) : \theta_2 < 0\} = \{(0, \sigma^2) : \sigma^2 > 0\}.$$

Since the conditional distribution in (5.68) (w.r.t. P_θ, $\theta \in \Theta_B$) depends only on $\theta_1^0 = 0$, and not on θ_2 (or σ^2), one can find t_1 (t_2), for every $t_2 > 0$, so that (5.68) holds. However, to use standard statistical tables express the test (5.67) as

$$\text{Reject } H_0 \text{ iff } \quad \frac{\sqrt{n}\,\overline{X}}{\sqrt{\left[\frac{1}{n-1}\left(\sum X_j^2 - n\overline{X}^2\right)\right]}} \equiv \frac{\sqrt{n}\,\overline{X}}{s} > c_1(T_2), \tag{5.69}$$

where $c_1(t_2)$ is determined, for $T_2 = t_2$, to satisfy

$$P_\theta\left(\frac{\sqrt{n}\,\overline{X}}{s} > c_1(t_2) \mid T_2 = t_2\right) = \alpha \quad \text{a.s. } P_\theta \ \forall\, \theta = (0, \theta_2). \tag{5.70}$$

Note that, for each t_2, $y \equiv \sum_1^n x_j \to \frac{\sqrt{n}\,\overline{x}}{s} = \sqrt{\frac{n-1}{n}}\,\frac{y}{(t_2 - y^2/n)^{1/2}} = h(y)$, say, is a strictly increasing function of y (for $y^2/n < t_2$), since the derivative of the function $h(y)$ w.r.t. y is positive. Hence, one can find a unique $\eta_1(t_2)$ such that $\{y > \eta_1(t_2)\}$ is equivalent to $\{h(y) > c_1(t_2)\}$.

A second important fact about the *t-statistic* $\frac{\sqrt{n}\,\overline{X}}{s}$ is that, under H_0, i.e., w.r.t. P_θ, $\theta \in \Theta_B$, its distribution does not depend on $\theta \in \Theta_B$. From Basu's Theorem below, it then follows that $\sqrt{n}\,\overline{X}/s$ is, under H_0, independent of T_2. Hence (5.70) becomes

$$P_{(0,\sigma^2)}\left(\frac{\sqrt{n}\,\overline{X}}{s} > c_1(t_2)\right) = \alpha,$$

so that $c_1(t_2)$ does not depend on t_2 and is obtained from the standard t-table (with $n - 1$ d.f.). Then, finally, the UMP test among all unbiased tests of size α is given by

$$\text{Reject } H_0 \text{ iff } \quad \frac{\sqrt{n}\,\overline{X}}{s} > t_{1-\alpha}(n - 1), \tag{5.71}$$

where $t_{1-\alpha}(n - 1)$ is the $(1 - \alpha)$th quantile of the t-distribution with $n - 1$ d.f.

Example 5.7. Let $X_1, X_2, \ldots, X_n$ be i.i.d. $\mathbf{N}(\mu, \sigma^2)$, $n \geq 2$, $(\mu, \sigma^2) \in \mathbb{R} \times (0, \infty)$. We wish to test, for some given $\sigma_0^2 > 0$,

$$H_0 : \sigma^2 \leq \sigma_0^2, \quad H_1 : \sigma^2 > \sigma_0^2. \tag{5.72}$$

As in Example 5.6, the UMP unbiased test of size α is given by

$$\text{Reject } H_0 \text{ iff } \quad \sum_1^n X_j^2 > \eta_2(T_1),$$

where $\eta_2(t_1)$ is determined by

$$P_{(\mu,\sigma_0^2)}\left(\sum_1^n X_j^2 > \eta_2(t_1) \mid T_1 = t_1\right) = \alpha \quad \forall\, \mu \in \mathbb{R} \quad (\text{i.e., } \forall\, \theta \in \Theta_B).$$

Since $y = \sum_1^n x_j^2 \to \sum_1^n (x_j - \overline{x})^2/\sigma_0^2 = [\sum_1^n x_j^2 - n\overline{x}^2]/\sigma_0^2 = (y - t_1^2/n)\sigma_0^2$ is strictly increasing (for $y > 0$, $T_1 = t_1$), this test reduces to

$$\text{Reject } H_0 \text{ iff } \quad \sum_1^n (X_j - \overline{X})^2/\sigma_0^2 > c_1(t_1), \tag{5.73}$$

where $c_1(t_1)$ is determined so that

$$P_{(\mu,\sigma_0^2)}\left(\sum_1^n (X_j - \overline{X})^2/\sigma_0^2 > c_1(t_1) \mid T_1 = t_1\right) = \alpha \quad \forall\, \mu \in \mathbb{R}. \tag{5.74}$$

But, as is well known, $T_1 = n\overline{X}$ and $\sum_1^n (X_j - \overline{X})^2/\sigma_0^2$ are independent random variables ($\forall\, (\mu, \sigma_0^2)$), the latter having a chi-square distribution with $n-1$ degrees of freedom (See Proposition 2.1). Hence $c_1(t_1)$ does not depend on t_1 and is given by $\chi_{1-\alpha}^2(n-1)$—the $(1-\alpha)$th quantile of the chi-square distribution with $n-1$ d.f.:

$$\text{Reject } H_0 \text{ iff } \quad \sum_1^n (X_j - \overline{X})^2/\sigma_0^2 > \chi_{1-\alpha}^2(n-1). \tag{5.75}$$

5.6 Basu's Theorem

In this section we use a useful result of Basu (1959) to compute certain UMPU tests.

Theorem 5.5 (Basu's Theorem). *Let $\mathbf{X}$ have distribution P_θ, $\theta \in \widetilde{\Theta}$. Suppose T is a (boundedly) complete sufficient statistic for $\{P_\theta : \theta \in \widetilde{\Theta}\}$. If Y is a statistic whose distribution does not depend on $\theta \in \widetilde{\Theta}$ (i.e., it is the same for all $\theta \in \widetilde{\Theta}$), then T and Y are independent under $P_\theta \ \forall\, \theta \in \widetilde{\Theta}$.*

Proof. We need to show that for every pair of bounded real-valued measurable statistics $g(T)$ and $h(Y)$, depending only on T and Y, respectively, one has

$$E_\theta(g(T)h(Y)) = E_\theta(g(T)) \cdot E_\theta(h(Y)), \quad \forall\, \theta \in \widetilde{\Theta}. \tag{5.76}$$

Since $c \equiv E_\theta h(Y)$ does not depend on θ, and neither does $V(T) = E_\theta(h(Y) \mid T)$, we have $E_\theta(V(T) - c) = 0 \ \forall\, \theta \in \widetilde{\Theta}$, so that, by (bounded) completeness of T, $V(T) = c$ a.s. $(P_\theta) \ \forall\, \theta \in \widetilde{\Theta}$. Hence

$$E_\theta(g(T)(h(Y) - c)) = E_\theta[g(T)E_\theta(h(Y) - c \mid T)]$$
$$= E_\theta[g(T)(V(T) - c)] = 0 \quad \forall\, \theta \in \widetilde{\Theta}.$$

Therefore, (5.76) holds. $\qquad\qquad\qquad\qquad\qquad\qquad\qquad\qquad\qquad\qquad\square$

We now apply the theory of this section to the so-called *two-sample problems*.

Example 5.8. Let $X_1, \ldots, X_m$, and $Y_1, Y_2, \ldots, Y_n$ be independent samples from $\mathbf{N}(\mu_1, \sigma_0^2)$ and $\mathbf{N}(\mu_2, \sigma_0^2)$, respectively, where $\sigma_0^2 > 0$ is given, and μ_1, μ_2 are unknown means. The (joint) density of these $m+n$ observations is

$$f(\mathbf{x}, \mathbf{y} \mid \mu_1, \mu_2) = (2\pi\sigma_0^2)^{-(m+n)/2} \exp\left\{-\frac{1}{2\sigma_0^2}\sum_1^m x_j^2 + \frac{m\mu_1\overline{x}}{\sigma_0^2} - \frac{m\mu_1^2}{2\sigma_0^2}\right\}$$

$$\cdot \exp\left\{-\frac{1}{2\sigma_0^2}\sum_1^n y_j^2 + \frac{n\mu_2\overline{y}}{\sigma_0^2} - \frac{n\mu_2^2}{2\sigma_0^2}\right\}$$

$$= C(\mu_1, \mu_2)h(\mathbf{x}, \mathbf{y})e^{(\mu_1/\sigma_0^2)\sum_1^m x_j + (\mu_2/\sigma_0^2)\sum_1^n y_j}$$

$$= C(\mu_1, \mu_2)h(\mathbf{x}, \mathbf{y})\exp\left\{\left(\frac{\mu_1}{\sigma_0^2} - \frac{\mu_2}{\sigma_0^2}\right)\sum_1^m x_j + \frac{\mu_2}{\sigma_0^2}\left(\sum_1^m x_j + \sum_1^n y_j\right)\right\}$$

$$= \widetilde{C}(\theta_1, \theta_2)h(\mathbf{x}, \mathbf{y})\exp\{\theta_1 T_1(\mathbf{x}) + \theta_2 T_2(\mathbf{x}, \mathbf{y})\}, \tag{5.77}$$

where $\theta_1 = (\mu_1 - \mu_2)/\sigma_0^2$, $\theta_2 = \mu_2/\sigma_0^2$, $T_1(\mathbf{x}) = \sum_1^m x_j$, $T_2(\mathbf{x}, \mathbf{y}) = \sum_1^m x_j + \sum_1^n y_j$. Thus $\{P_\theta : \theta \in \Theta\}$, with $\Theta = \mathbb{R}^2$, is a two-parameter exponential family, with natural parameter $\theta = (\theta_1, \theta_2)$. We wish to test

$$H_0 : \mu_1 \leq \mu_2, \quad H_1 : \mu_1 > \mu_2, \tag{5.78}$$

which may be cast as

$$H_0 : \theta_1 \leq 0, \quad H_1 : \theta_1 > 0.$$

By the preceding theory, a UMP unbiased test of size α is given by

$$\text{Reject } H_0 \text{ iff} \quad T_1 > \eta_1(T_2), \tag{5.79}$$

where $\eta_1(t_2)$ is determined, for every value t_2 of T_2, such that

$$P_\theta(T_1 > \eta_1(t_2) \mid T_2 = t_2) = \alpha \quad \forall \theta \in \Theta_B = \{(0, \theta_2) : \theta_2 \in \mathbb{R}\}.$$

Since $\overline{X} - \overline{Y} = \left(\frac{1}{m} + \frac{1}{n}\right)m\overline{X} - \frac{1}{n}(m\overline{X} + n\overline{Y}) = \left(\frac{1}{m} + \frac{1}{n}\right)T_1 - \frac{1}{n}T_2$, which is a strictly increasing function of T_1, for any given value of T_2, the test may be expressed as

$$\textit{Reject } H_0 \textit{ iff} \quad \frac{\overline{X} - \overline{Y}}{\sigma_0\sqrt{\frac{1}{n} + \frac{1}{m}}} > c(T_2), \tag{5.80}$$

where

$$P_\theta\left(\frac{\overline{X} - \overline{Y}}{\sigma_0\sqrt{\frac{1}{m} + \frac{1}{n}}} > c(T_2) \mid T_2\right) = \alpha \quad \forall \theta = (0, \theta_2) \in \Theta_B.$$

But the P_θ-distribution of $\frac{\overline{X} - \overline{Y}}{\sigma_0\sqrt{\frac{1}{m} + \frac{1}{n}}}$ is $\mathbf{N}(0, 1)$, which does not involve θ, $\forall \theta \in \Theta_B$. Hence, by Basu's Theorem (or, checking that this Normal random variable and the Normal random variable T_2 are uncorrelated $\forall \theta$), $\frac{\overline{X} - \overline{Y}}{\sigma_0\sqrt{\frac{1}{m} + \frac{1}{n}}}$ is independent of T_2. Therefore, $c(T_2) = z_{1-\alpha} = \Phi^{-1}(1 - \alpha)$, and the UMP unbiased test is given by

$$\text{Reject } H_0 \text{ iff} \quad \frac{\overline{X} - \overline{Y}}{\sigma_0\sqrt{\frac{1}{m} + \frac{1}{n}}} > z_{1-\alpha}. \tag{5.81}$$

Example 5.9. Let $X_1, X_2, \ldots, X_m$ and $Y_1, Y_2, \ldots, Y_n$ be independent random samples from $\mathbf{N}(\mu_1, \sigma^2)$ and $\mathbf{N}(\mu_2, \sigma^2)$, respectively, where μ_1, μ_2, σ^2 are all unknown. We wish to find the UMP unbiased test of size α for

$$H_0 : \mu_1 \leq \mu_2, \quad H_1 : \mu_1 > \mu_2. \tag{5.82}$$

The joint density of the observations maybe expressed as [see (5.77)]

$$f(\mathbf{x}, \boldsymbol{\mu} \mid \mu_1, \mu_2, \sigma^2) = C(\mu_1, \mu_2, \sigma^2) \exp\{\theta_1 T_1(\mathbf{x}) + \theta_2 T_2(\mathbf{x}) + \theta_3 T_3(\mathbf{x})\}, \qquad (5.83)$$

where $\theta_1 = \frac{\mu_1}{\sigma^2} - \frac{\mu_2}{\sigma^2}$, $\theta_2 = \frac{\mu_2}{\sigma^2}$, $\theta_3 = -\frac{1}{2\sigma^2}$, and $T_1(\mathbf{x}) = \sum_1^m x_j$, $T_2(\mathbf{x}) = \sum_1^m x_j + \sum_1^n y_j$, $T_3(\mathbf{x}) = \sum_1^m x_j^2 + \sum_1^n y_j^2$. $\theta = (\theta_1, \theta_2, \theta_3)$ is the natural parameter, $\theta \in \Theta = \mathbb{R} \times \mathbb{R} \times (-\infty, 0)$, and (T_1, T_2, T_3) is a complete sufficient statistic for $\{P_\theta : \theta \in \Theta\}$, while (T_2, T_3) is a complete sufficient statistic for $\{P_\theta : \theta \in \Theta_B\}$ with $\Theta_B = \{(\theta_1, \theta_2, \theta_3) \in \Theta : \theta_1 = 0\}$. Hence the UMP unbiased test of size α for (5.82) (i.e., for $H_0 : \theta_1 \leq 0$, $H_1 : \theta_1 > 0$) is given by

$$\text{Reject } H_0 \text{ iff } \quad T_1 > \eta_1(T_2, T_3), \qquad (5.84)$$

where $\eta_1(T_2, T_3)$ is determined by

$$P_\theta(T_1 > \eta_1(T_2, T_3) \mid T_2, T_3) = \alpha \quad \forall\, \theta \in \Theta_B. \qquad (5.85)$$

We will show that this is equivalent to the classical test which replaces the σ_0 in (5.80) (which is here unknown) by its *pooled estimate*

$$s_p = \sqrt{\frac{\sum(X_j - \overline{X})^2 + \sum(Y_j - \overline{Y})^2}{m + n - 2}} = \sqrt{\frac{\sum X_j^2 - \frac{(\sum X_j)^2}{m} + \sum Y_j^2 - \frac{(\sum Y_j)^2}{n}}{m + n - 2}}$$

$$= \sqrt{\frac{T_3 - \frac{T_1^2}{m} - \frac{(T_2 - T_1)^2}{n}}{m + n - 2}}. \qquad (5.86)$$

We have shown in Example 5.8 that $(\overline{X} - \overline{Y})/\sqrt{\frac{1}{m} + \frac{1}{n}}$ is an increasing function of T_1, for any given T_2. It is easy to check, using (5.86), that the derivative of s_p^2 w.r.t. T_1 (for fixed T_2, T_3) is a decreasing function of T_1. Therefore,

$$T_1 \longrightarrow \tau = \frac{\overline{X} - \overline{Y}}{s_p\sqrt{\frac{1}{m} + \frac{1}{n}}} \text{ is an increasing function of } T_1, \text{ for given } T_2 \text{ and } T_3.$$

Hence the test (5.84) is equivalent to

$$\text{Reject } H_0 \text{ iff } \quad \tau = \frac{\overline{X} - \overline{Y}}{s_p\sqrt{\frac{1}{m} + \frac{1}{n}}} > c(T_2, T_3)$$

where $c(T_2, T_3)$ is determined from

$$P_\theta(\tau > c(T_2, T_3) \mid T_2, T_3) = \alpha \quad \forall\, \theta \in \Theta_B. \qquad (5.87)$$

But for $\theta \in \Theta_B$ (i.e., for $\mu_1 - \mu_2 = 0$), the distribution of τ is that of a Student's t with $m + n - 2$ d.f.,[2] and is independent of $\theta \in \Theta_B$. Hence, by Basu's Theorem, τ is independent of (T_2, T_3) under P_θ, for all $\theta \in \Theta_B$. Thus $c(T_2, T_3)$ does not

---

[2] Note. Since $(\overline{X} - \overline{Y})/\sigma\sqrt{\frac{1}{m} + \frac{1}{n}}$ is $\mathbf{N}(0, 1)$ and independent of the chi-square random variables $U = \sum(X_j - \overline{X})^2/\sigma^2 + \sum(Y_i - \overline{Y})^2/\sigma^2$ (having d.f. $m - 1 + n - 2 = m + n - 2$), by definition, $[(\overline{X} - \overline{Y})/\sigma\sqrt{\frac{1}{m} - \frac{1}{n}}]/\sqrt{U/(m + n - 2)}$ has Student's distribution with $m + n - 2$ d.f.

depend on (T_2, T_3) and is equal to $t_{1-\alpha}(m + n - 2)$—the $1 - \alpha$th quantile of the t-distribution with $m + n - 2$ d.f. The UMP unbiased test of size α for (5.82) is given by

$$\text{Reject } H_0 \text{ iff } \quad \frac{\overline{X} - \overline{Y}}{s_p \sqrt{\frac{1}{m} + \frac{1}{n}}} > t_{1-\alpha}(m + n - 2). \tag{5.88}$$

Remark 5.6. There does not exist a UMP unbiased test of size $\alpha \in (0, 1)$ for $H_0 : \mu_1 \leq \mu_2$ against $H_1 : \mu_1 > \mu_2$ if the unknown variance σ_1^2 of X_i's is not equal to the unknown variance σ_2^2 of Y_j's, in Example 5.9. The problem of obtaining a "good" test in this case is known as the *Behrens-Fisher problem.*

Example 5.10. Let $X_1, \ldots, X_m$ and $Y_1, \ldots, Y_n$ $(m \geq 2,\ n \geq 2)$ be independent random samples from $\mathbf{N}(\mu_1, \sigma_1^2)$ and $\mathbf{N}(\mu_2, \sigma_2^2)$ with all the parameters $\mu_i \in \mathbb{R}$ $(i = 1, 2)$, $\sigma_i^2 > 0$ $(i = 1, 2)$ unknown. We wish to test, at a level of significance $\alpha \in (0, 1)$,

$$H_0 : \sigma_1^2 \leq \sigma_2^2 \quad \text{against} \quad H_1 : \sigma_1^2 > \sigma_2^2. \tag{5.89}$$

The (joint) density of the observation vector $(X_1, X_2, \ldots, X_m, Y_1, Y_2, \ldots, Y_n)$ may be expressed as [see (5.83)]

$$c(\mu_1, \mu_2, \sigma_1^2, \sigma_2^2) \exp\left\{ -\frac{1}{2\sigma_1^2} \sum_1^m x_i^2 - \frac{1}{2\sigma_2^2} \sum_1^n y_j^2 + \frac{\mu_1}{\sigma_1^2} \left(\sum x_i \right) + \frac{\mu_2}{\sigma_2^2} \left(\sum y_j \right) \right\}$$

$$= c_1(\theta_1, \theta_2, \theta_3, \theta_4) \exp\{\theta_1 T_1(\mathbf{x}) + \theta_2 T_2(\mathbf{x}, \mathbf{y}) + \theta_3 T_3(\mathbf{x}) + \theta_4 T_4(\mathbf{y})\}, \tag{5.90}$$

where

$$\theta_2 = -\frac{1}{2\sigma_2^2}, \quad T_2(\mathbf{x}, \mathbf{y}) = \sum x_i^2 + \sum y_j^2, \quad \theta_1 = \frac{1}{2\sigma_2^2} \left(1 - \frac{\sigma_2^2}{\sigma_1^2} \right), \quad T_1(\mathbf{x}) = \sum x_i^2,$$

$$\theta_3 = \frac{\mu_1}{\sigma_1^2}, \quad T_3(\mathbf{x}) = \sum x_i, \quad \theta_4 = \frac{\mu_2}{\sigma_2^2}, \quad T_4(\mathbf{y}) = \sum y_j. \tag{5.91}$$

Then (5.89) may be expressed as

$$H_0 : \theta_1 \leq 0, \quad H_1 : \theta_1 > 0. \tag{5.92}$$

Hence the UMP unbiased test is of the form

$$\text{Reject } H_0 : \text{ iff } \quad T_1 > \eta_1(T_2, T_3, T_4), \tag{5.93}$$

where $\eta(t_2, t_3, t_4)$ satisfies

$$P_\theta(T_1 > \eta_1(t_2, t_3, t_4) \mid T_2 = t_2, T_3 = t_3, T_4 = t_4) = \alpha \quad \forall \theta \in \Theta_B,$$

$$\Theta_B = \{(0, \theta_2, \theta_3, \theta_4) : \theta_2 < 0,\ \theta_3 \in \mathbb{R},\ \theta_4 \in \mathbb{R}\}. \tag{5.94}$$

This probability does not depend on $\theta \in \Theta_B$. Consider the statistic

$$F = \frac{\sum(X_i - \overline{X})^2/(m - 1)}{\sum(Y_j - \overline{Y})^2/(n - 1)} = \frac{\left(T_1 - \frac{T_3^2}{m} \right)/(m - 1)}{\left(T_2 - T_1 - \frac{T_4^2}{n} \right)/(n - 1)}, \tag{5.95}$$

which is easily seen to be a strictly increasing function of T_1, for every set of given values of T_2, T_3, T_4 that may occur a.s. (P_θ), for all $\theta \in \Theta_B$. Hence (5.93) is equivalent to

$$\text{Reject } H_0 \text{ iff } \quad F > \eta_2(T_2, T_3, T_4),$$

for an appropriate η_2. But

$$F = \frac{\sum(X_i - \overline{X})^2/\sigma_1^2}{\sum(Y_j - \overline{Y})^2/\sigma_1^2} \cdot \frac{n-1}{m-1} = \frac{U}{V} \cdot \frac{n-1}{m-1} \, , \tag{5.96}$$

where, $\forall\, \theta \in \Theta_B$, U and V are two independent chi-square random variables. Hence the distribution of F does not depend on $\theta \in \Theta_B$. It follows from Basu's Theorem that F is independent of T_2, T_3, T_4, under P_θ, $\theta \in \Theta_B$. Therefore, the UMP unbiased test is given by

$$\text{Reject } H_0 \text{ iff } \quad F > F_{1-\alpha}, \tag{5.97}$$

where $F_{1-\alpha}$ is the $1-\alpha$th quantile of the so-called *F-distribution with d.f.* $(m-1, n-1)$. The distribution of F is computed in Appendix A, Example A.VIII.

Example 5.11 (Match Pair Experiments). In order to test the effectiveness of an exercise regimen in reducing systolic blood pressures of people with moderately elevated blood pressure belonging to a certain age group, a random sample of n individuals is chosen from this population. For each individual the (systolic) blood pressure x before the start of the exercise regimen is recorded, as well as the blood pressure y after the completion of the exercise regimen. The model assumed is that (X_i, Y_i), $i = 1, \ldots, n$, are i.i.d. Normal $N((\mu_x, \mu_y)^t \sum)$ where $(\mu_x, \mu_y) \in \mathbb{R}^2$, and $\sum$ is a positive definite covariance matrix with $\sigma_x^2 = \text{var}(X_i)$, $\sigma_y^2 = \text{var}(Y_i)$, $\sigma_{x,y} = \text{covar}(X_i, Y_i) = p\sigma_x\sigma_y$. All the parameters are unknown. One wishes to test $H_0 : \mu_x = \mu_y$ against $H_1 : \mu_x > \mu_y$. Just as in the Behrens-Fisher problem, there is no uniformly most powerful unbiased (UMPU) test in this generality. However, if one assumes that $\sigma_x^2 = \sigma_y^2$, the UMPU test of size α exists by the theory of Sects. 5.5, 5.6, and it is given by (Exercise 5.14(a)): Reject H_0 iff $\overline{d} > t_{n-1}(1-\alpha)(s_d/\sqrt{n})$ where, writing $d_i = X_i - Y_i$, $\overline{d} = \sum_{1 \le i \le n} d_i/n$, $s_d^2 = \sum_{1 \le i \le n}(d_i - \overline{d})^2/(n-1)$, and $t_{n-1}(1-\alpha)$ is the $(1-\alpha)$-th quantile of the t-distribution with $n-1$ degrees of freedom.

Consider now an alternative design where a random sample of size n is chosen from the target population, and their blood pressures X_i are recorded $(i = 1, \ldots, n)$; this is the so-called *control group*. Another random sample of n individuals is drawn independently from this population and subjected to the exercise regimen. This is the *treatment group*. Let Y_i $(i = 1, \ldots, n)$ be the blood pressures of these individuals measured after the exercise regimen is completed. Suppose X_i's are $N(\mu_x, \sigma_x^2)$ and Y_i's are $N(\mu_y, \sigma_y^2)$, with μ_x, σ_x^2, μ_y, σ_y^2, the same as in the preceding paragraph. Assume that $\sigma_x^2 = \sigma_y^2$, and consider the UMPU test provided by Example 5.8 for $H_0 : \mu_x = \mu_y$ against $H_1 : \mu_x > \mu_y$. Give an argument to show that the match pair design is more efficient than this independent samples design if $p > 0$ and n is sufficiently large (Exercise 5.14(b)).

5.7 Duality Between Tests and Confidence Regions

Given a family of non-randomized tests $\{\varphi_{\theta_0} : \theta_0 \in \Theta\}$ for testing $H_0 : \theta = \theta_0$ against $H_1 : \theta \in \Theta_1^{\theta_0} \subset \Theta \backslash \{\theta_0\}$, there exists a 'confidence region' for the unknown parameter θ given by

$$S(\mathbf{x}) := \{\theta_0 \in \Theta : \varphi_{\theta_0}(\mathbf{x}) = 0\}, \quad \mathbf{x} \in \mathscr{X}. \tag{5.98}$$

That is, $S(\mathbf{x})$ is the set of values of $\theta = \theta_0$ which are accepted by the test when presented as the null hypothesis $H_0 : \theta = \theta_0$. If the family of tests is of size α, then

$$P_{\theta_0}\left(\theta_0 \in S(\mathbf{X})\right) = P_{\theta_0}\left(\varphi_{\theta_0}(\mathbf{X}) = 0\right) = 1 - \alpha \ \forall\, \theta_0 \in \Theta. \tag{5.99}$$

One expresses this as: '$S(\mathbf{X})$ is a confidence region for θ of confidence level $1 - \alpha$'. Conversely, suppose one is given a confidence region S of confidence level $1 - \alpha$, that is, $\mathbf{x} \to S(\mathbf{x})$ is a map on $\mathscr{X}$ into the class of subsets of Θ such that

$$P_{\theta_0}(\theta_0 \in S(\mathbf{X})) = 1 - \alpha \ \forall\, \theta_0 \in \Theta. \tag{5.100}$$

Then the family of tests φ_{θ_0} given by

$$\varphi_{\theta_0}(\mathbf{x}) = \begin{cases} 1 \text{ if } \theta_0 \notin S(\mathbf{x}), & \mathbf{x} \in \mathscr{X},\, \theta_0 \in \Theta, \\ 0 \text{ if } \theta_0 \in S(\mathbf{x}), \end{cases} \tag{5.101}$$

is of size α.

We now consider the more general situation of the estimation of, and confidence regions for, functions $f(\theta)$ of θ.

Definition 5.6. Let $\{P_\theta : \theta \in \Theta\}$ be a family of distributions on the observation space $\mathscr{X}$. Let $f(\theta)$ be a function on Θ into some space Γ. Suppose $\mathbf{x} \to S(\mathbf{x})$ is a map on $\mathscr{X}$ into the class of subsets of Γ such that, (i) for each $\theta \in \Theta$, the set $\{\mathbf{x} : f(\theta) \in S(\mathbf{x})\}$ is a measurable subset of $\mathscr{X}$, and

$$(\text{ii}) \quad P_\theta(\{\mathbf{x} : f(\theta) \in S(\mathbf{x})) = 1 - \alpha \quad \forall\, \theta \in \Theta.$$

Then $S(\cdot)$ (or, $S(\mathbf{X})$), where $\mathbf{X}$ has distribution P_θ, θ being the unknown 'true' parameter value) is called a *confidence region* for $f(\theta)$ having a *confidence level* $1 - \alpha$.

Remark 5.7. To avoid confusion, $\mathbf{X}$ may be taken to be the identity map $\mathbf{x} \to \mathbf{x}$ on $\mathscr{X}$.

Notation Let $\overline{H}(f(\theta_0))$ be the space of alternatives against $H_0 : f(\theta) = f(\theta_0)$.

Definition 5.7. A confidence region $S^*(\cdot)$ of confidence level $1 - \alpha$ is said to be *uniformly most accurate (UMA)* of level $1 - \alpha$, if, $\forall\, \theta_0 \in \Theta$ and $\forall\, \theta \in \overline{H}(f(\theta_0))$,

$$P_\theta(\{\mathbf{x} \in \mathscr{X} : f(\theta_0) \in S^*(\mathbf{x})\}) \le P_\theta(\{\mathbf{x} \in \mathscr{X}; f(\theta_0) \in S(\mathbf{x})\}), \tag{5.102}$$

holds for every confidence region $S(\cdot)$ for $f(\theta)$ of confidence level $1 - \alpha$. A confidence region $S(\cdot)$ for $f(\theta)$ is said to be *unbiased of level* $1 - \alpha$ if, $\forall\, \theta_0 \in \Theta$,

$$P_\theta(\{\mathbf{x} \in \mathscr{X} : f(\theta_0) \in S(\mathbf{x})\}) \le 1 - \alpha \quad \forall\, \theta \in \overline{H}(f(\theta_0)). \tag{5.103}$$

A confidence region $S^*(\cdot)$ is said to be *UMA unbiased of level* $1 - \alpha$, if among all unbiased confidence regions $S(\cdot)$ of level $1 - \alpha$, it is the most accurate, i.e., (5.102) holds.

Theorem 5.6 (Duality Between Confidence Regions and Tests).

(a) Given a family of non-randomized tests $\{\varphi_\theta : \theta \in \Theta\}$ of size α, where φ_{θ_0} is a test (of size α) for $H_0 : f(\theta) = f(\theta_0)$ $(\theta_0 \in \Theta)$,

$$S(\mathbf{x}) := \{f(\theta) : \theta \in \Theta, \varphi_\theta(\mathbf{x}) = 0\}, \quad \mathbf{x} \in \mathscr{X}, \tag{5.104}$$

is a confidence region for $f(\theta)$ of confidence level $1 - \alpha$.

(b) Conversely, given a confidence region $S(\cdot)$ for $f(\theta)$ of confidence level $1 - \alpha$, consider the family of tests $\{\varphi_\theta : \theta \in \Theta\}$ defined by

$$\varphi_\theta(\mathbf{x}) := \begin{cases} 1 & \text{if } f(\theta) \notin S(\mathbf{x}), \\ 0 & \text{if } f(\theta) \in S(\mathbf{x}), \end{cases} \quad (\mathbf{x} \in \mathscr{X}). \tag{5.105}$$

Then, for each $\theta_0 \in \Theta$, φ_{θ_0} is a test of size α for $H_0 : f(\theta) = f(\theta_0)$.

(c) If a family of tests $\{\varphi_\theta^* : \theta \in \Theta\}$ is given such that, for each $\theta_0 \in \Theta$, $\varphi_{\theta_0}^*$ is UMP of size α for testing $H_0 : f(\theta) = f(\theta_0)$, against $H_1 : \theta \in \overline{H}(f(\theta_0))$, then $S^*(\cdot)$ defined by (5.104), with φ_θ replaced by φ_θ^*, is UMA of size $1 - \alpha$.

(d) If $\varphi_{\theta_0}^*$ is UMA unbiased of size α for testing $H_0 : f(\theta) = f(\theta_0)$, against $H_1 : \theta \in \overline{H}(f(\theta_0))$ $(\forall\, \theta_0 \in \Theta)$, then the corresponding $S^*(\cdot)$ is UMA unbiased of level $1 - \alpha$.

Proof. (a) By (5.104), we have equality of the events

$$\{\mathbf{x} \in \mathscr{X} : f(\theta) \in S(\mathbf{x})\} = \{\mathbf{x} \in \mathscr{X} : \varphi_\theta(\mathbf{x}) = 0\}, \quad \forall\, \theta \in \Theta. \tag{5.106}$$

Hence

$$P_\theta(\{\mathbf{x} \in \mathscr{X} : f(\theta) \in S(\mathbf{x})\}) = P_\theta(\{\mathbf{x} \in \mathscr{X} : \varphi_\theta(\mathbf{x}) = 0\}) = 1 - \alpha \quad \forall\, \theta \in \Theta. \tag{5.107}$$

(b) if $S(\cdot)$ is a confidence region for $f(\theta)$ of confidence level $1 - \alpha$, then for the family of tests $\{\varphi_\theta : \theta \in \Theta\}$ defined by (5.105), again (5.106) holds. Hence $P_\theta(\{\mathbf{x} \in \mathscr{X} : \varphi_\theta(\mathbf{x}) = 1\}) = 1 - P_\theta(\{\mathbf{x} \in \mathscr{X} : \varphi_\theta(\mathbf{x}) = 0\}) = 1 - P_\theta(\{\mathbf{x} \in \mathscr{X} : f(\theta) \in S(\mathbf{x})\}) = 1 - (1 - \alpha) = \alpha$ $(\forall\, \theta \in \Theta)$.

(c) Suppose $\varphi_{\theta_0}^*$ is UMP of size α for testing $H_0 : f(\theta) = f(\theta_0)$, against $H_1 : \theta \in \overline{H}(f(\theta_0))$, for every $\theta_0 \in \Theta$. Then if $S^*(\cdot)$ is the corresponding confidence region for $f(\theta)$ and $S(\cdot)$ is any other confidence region of level $1 - \alpha$ for $f(\theta)$, with corresponding family of tests defined by (5.105), then, $\forall\, \theta \in \overline{H}(f(\theta_0))$,

$$\begin{aligned} P_\theta(&\{\mathbf{x} \in \mathscr{X} : f(\theta_0) \in S^*(\mathbf{x})\}) \\ &= P_\theta(\{\mathbf{x} \in \mathscr{X} : \varphi_{\theta_0}^*(\mathbf{x}) = 0\}) = 1 - P_\theta(\{\mathbf{x} : \varphi_{\theta_0}^*(\mathbf{x}) = 1\}) \\ &= 1 - (\text{Power of the test } \varphi_{\theta_0}^* \text{ at } \theta) \leq 1 - (\text{Power of the test } \varphi_{\theta_0} \text{ at } \theta) \\ &= 1 - P_\theta(\{\mathbf{x} \in \mathscr{X} : \varphi_{\theta_0}(\mathbf{x}) = 1\}) = P_\theta(\{\mathbf{x} \in \mathscr{X} : \varphi_{\theta_0}(\mathbf{x}) = 0\}) \\ &= P_\theta(\{\mathbf{x} \in \mathscr{X} : f(\theta_0) \in S(\mathbf{x})\}). \end{aligned} \tag{5.108}$$

(d) This follows from (5.108), since the unbiasedness of the family of tests $\{\varphi_\theta : \theta \in \Theta\}$ means, $P_\theta(\{\mathbf{x} \in \mathscr{X} : \varphi_{\theta_0}(\mathbf{x}) = 1\}) \equiv (\text{Power of } \varphi_{\theta_0} \text{ at } \theta) \geq \alpha \ \forall\, \theta \in \overline{H}(f(\theta_0))$. Therefore, for $S(\cdot)$ defined by (5.103),

$$P_\theta(\{\mathbf{x} \in \mathscr{X} : f(\theta_0) \in S(\mathbf{x})\}) = P_\theta(\{\mathbf{x} \in \mathscr{X} : \varphi_{\theta_0}(\mathbf{x}) = 0\})$$
$$= 1 - P_\theta(\{\mathbf{x} \in \mathscr{X} : \varphi_{\theta_0}(\mathbf{x}) = 1\}) \le 1 - \alpha, \quad \forall\, \theta \in \overline{H}(f(\theta_0)).$$

Same is true for $S^*(\cdot)$. $\square$

Example 5.12. Let $X_1, \ldots, X_n$ be i.i.d. exponential with p.d.f.

$$f(x \mid \theta) = \frac{e^{-x/\theta}}{\theta}, \quad 0 \le x < \infty, \ \theta \in (0, \infty). \tag{5.109}$$

The uniformly most powerful test of size α for testing $H_0 : \theta \le \theta_0$, $H_1 : \theta > \theta_0$ is (See Example 5.4, with $\beta = 1$; but consider one-sided alternatives; or, see Exercise 5.2)

$$\varphi_{\theta_0}^*(\mathbf{x}) = \begin{cases} 1 & \text{if } \sum_1^n x_i > \theta_0 c_{1-\alpha}(n), \\ 0 & \text{if } \sum_1^n x_i \le \theta_0 c_{1-\alpha}(n), \end{cases}$$

where $c_{1-\alpha}(n)$ is the $(1 - \alpha)$th quantile of the gamma distribution $\mathscr{G}(1, n)$. The *UMA confidence region $S^*(\cdot)$ for θ of level $1 - \alpha$* is given by

$$S^*(\mathbf{x}) = \left\{ \theta_0 > 0 : \sum_1^n x_i \le \theta_0 c_{1-\alpha}(n) \right\} = \left\{ \theta_0 : \theta_0 \ge \frac{\sum_1^n x_i}{c_{1-\alpha}(n)} \right\}$$
$$= \left[\frac{\sum_1^n x_i}{c_{1-\alpha}(n)}, \infty \right).$$

In this example, $f(\theta) = \theta$, $\overline{H}(\theta_0) = \Theta \backslash \{\theta_0\}$.

Example 5.13. Consider Exercise 5.7(c). Here $X_1, \ldots, X_n$ are i.i.d. $N(\mu, \sigma^2)$ ($\mu \in \mathbb{R}$, $\sigma^2 > 0$ are unknown). The UMPU test, of size $\alpha \in (0, 1)$, for $H_0 : \mu = \mu_0$, $H_1 : \mu \ne \mu_0$ is: $\varphi^*(\mathbf{x}) = 0$ if $\left| \frac{\sqrt{n}\,(\overline{x} - \mu_0)}{s} \right| \le t_{1-\frac{\alpha}{2}}(n - 1)$, and $\varphi^*(\mathbf{x}) = 1$ otherwise. Hence the UMA unbiased confidence region S^* of level $1 - \alpha$ is given by $S^*(\mathbf{x}) = \left\{ \mu \in \mathbb{R} : \left| \frac{\sqrt{n}\,(\overline{x} - \mu)}{s} \right| \le t_{1-\frac{\alpha}{2}}(n - 1) \right\} = \left\{ \mu \in \mathbb{R} : |\overline{x} - \mu| \le t_{1-\frac{\alpha}{2}}(n - 1) \frac{s}{\sqrt{n-1}} \right\} = \left[\overline{x} - t_{1-\frac{\alpha}{2}}(n - 1) \frac{s}{\sqrt{n}}, \overline{x} + t_{1-\frac{\alpha}{2}}(n - 1) \frac{s}{\sqrt{n}} \right]$.

If σ^2 is known, $\sigma^2 = \sigma_0^2$, say, then the UMPU test φ^* of size α for the above hypotheses is given by $\varphi^*(\mathbf{x}) = 0$ iff $\left| \frac{\sqrt{n}(\overline{x} - \mu)}{\sigma_0} \right| \le z_{1-\frac{\alpha}{2}}$ (See Example 5.3). Hence the UMA unbiased confidence region (interval) for μ is $S^*(\mathbf{x}) = \left[\overline{x} - z_{1-\frac{\alpha}{2}} \frac{\sigma_0}{\sqrt{n}}, \overline{x} + z_{1-\frac{\alpha}{2}} \frac{\sigma_0}{\sqrt{n}} \right]$.

Note that here $f(\theta) = \theta_1$, where $\theta = (\theta_1, \theta_2) = (\mu, \sigma^2)$. Here $f(\theta_0) = \mu_0$, $\overline{H}(f(\theta_0)) = \{(\mu, \sigma^2) : \mu \ne \mu_0\}$. Also, unbiasedness of a confidence region $S(\cdot)$ of level $(1 - \alpha)$ here means

$$P_{\mu, \sigma^2}(\{\mathbf{x} : \mu_0 \in S(\mathbf{x})\}) \le 1 - \alpha \quad \forall\, \mu_0 \ne \mu, \ \sigma^2 > 0. \tag{5.110}$$

Example 5.14. (See Example 5.7, and Exercise 5.8(b)). Let $X_1, \ldots, X_n$ be i.i.d. $N(\mu, \sigma^2)$ with $\mu \in \mathbb{R}$, $\sigma^2 > 0$ both unknown. Consider, for each $\sigma_0^2 > 0$, the test $H_0 : \sigma^2 = \sigma_0^2$ against $H_1 : \sigma^2 \ne \sigma_0^2$. The *UMPU test* of size α is given by $\varphi^*(\mathbf{x}) = 0$ if $c_1 \sigma_0^2 \le (n - 1)s^2 \le c_2 \sigma_0^2$, and $\varphi^*(\mathbf{x}) = 1$ otherwise, where $c_1 < c_2$ are determined by the Eq. (5.179). Hence $S^*(\mathbf{x}) = \{\sigma^2 : c_1 \sigma^2 \le (n - 1)s^2 \le c_2 \sigma^2\} = \{\sigma^2 : c_1 \le \frac{(n-1)s^2}{\sigma^2} \le c_2\} = \{\sigma^2 : \frac{1}{c_2} \le \frac{\sigma^2}{(n-1)s^2} \le \frac{1}{c_1}\} = \left[\frac{(n-1)s^2}{c_2}, \frac{(n-1)s^2}{c_1} \right]$, is the UMA unbiased confidence region.

Remark 5.8. To avoid computation of c_1, c_2 numerically, one sometimes uses an equal-tailed test, or confidence region, choosing $c_1 = \chi^2_{\frac{\alpha}{2}} < c_2 = \chi^2_{1-\frac{\alpha}{2}}$, where χ^2_p is the pth quantile of the chi-square distribution with $n - 1$ d.f.

5.8 Invariant Tests, the Two-Sample Problem and Rank Tests

Consider a testing problem with observation $\mathbf{X}$ having distribution P_θ, $\theta \in \Theta$, and $H_0 : \theta \in \Theta_0$, $H_1 : \theta \in \Theta_1 = \Theta \backslash \Theta_0$. Denote by $\mathscr{X}$ the observation space in which $\mathbf{X}$ takes values. Assume $P_{\theta_1} \neq P_{\theta_2}$ if $\theta_1 \neq \theta_2$ *(Identifiability)*.

Let g be a (bi-measurable) *transformation* on $\mathscr{X}$, i.e., g is a one-to-one map on $\mathscr{X}$ onto $\mathscr{X}$, and g and g^{-1} are both measurable.

Suppose $\mathbf{Y} := g\mathbf{X}$ ($:= g(\mathbf{X})$) has a distribution $P_{\theta'}$ (in the family $\{P_\theta : \theta \in \Theta\}$) when $\mathbf{X}$ has distribution P_θ. Writing $\theta' = \overline{g}_\theta$ in this case, one has a map $\overline{g}$ (associated with g) on Θ into Θ. This map is *one to one*, i.e., if $\theta_1 \neq \theta_2$ then $\overline{g}\theta_1 \neq \overline{g}\theta_2$ (Exercise 5.15). Assume that this map is also *onto* Θ:

$$\overline{g}\Theta = \Theta. \tag{5.111}$$

Assume also that g leaves Θ_0 (and, therefore, also Θ_1) invariant:

$$\overline{g}\Theta_0 = \Theta_0, \quad \overline{g}\Theta_1 = \Theta_1. \tag{5.112}$$

We then say that the statistical testing problem is *invariant under g*. The reason for the nomenclature is that the testing problem stated for $\mathbf{X}$ is exactly the same when stated for $\mathbf{Y}$, namely, with the same observation space $\mathscr{X}$, the same parameter space Θ or family of distributions $\{P_\theta : \theta \in \Theta\}$, the same null hypothesis $H_0 : \theta \in \Theta_0$ and the same alternative hypothesis $H_1 : \theta \in \Theta_1$. Therefore, if φ is a reasonable test, one should have

$$\varphi(\mathbf{x}) = \varphi(g^{-1}\mathbf{x}) \qquad \mathbf{x} \in \mathscr{X}. \tag{5.113}$$

For, given any $\mathbf{x} \in \mathscr{X}$, the decision based on $\mathbf{X} = \mathbf{x}$ (namely, $\varphi(\mathbf{x})$) should be the same as that based on $\mathbf{Y} = \mathbf{x}$ (where $\mathbf{Y} = g\mathbf{X}$), i.e., based on $\mathbf{x} = g^{-1}\mathbf{x}$. If one replaces $\mathbf{x}$ by $g\mathbf{x}$ in (5.113) then one gets

$$\varphi(g\mathbf{x}) = \varphi(\mathbf{x}) \quad \forall \; \mathbf{x} \in \mathscr{X}. \tag{5.114}$$

Note that (5.113) and (5.114) are equivalent.

Let now $\mathscr{G}$ be a *group of transformations* with each $g \in \mathscr{G}$ of the above kind (The group operations are (1) $g \to g^{-1}$, in the usual sense of the inverse of a function, and (2) $g_1 g_2 = g_1(g_2)$, the composition of functions). This also gives rise to the corresponding group $\overline{\mathscr{G}}$ on the parameter space Θ. Then (5.113), (5.114) should hold for all $g \in \mathscr{G}$ for a reasonable test φ. In other words, *for each $\mathbf{x} \in \mathscr{X}$, φ should be constant on the orbit* of $\mathbf{x}$,

$$O(\mathbf{x}) := \{g\mathbf{x} : g \in \mathscr{G}\}. \tag{5.115}$$

The invariant tests φ are therefore *functions on the space* of orbits. In this sense the map $\mathbf{x} \to O(\mathbf{x})$, say, $T(\mathbf{x}) = O(\mathbf{x})$, is a *maximal invariant*,

(i) *(Invariance).* $T(g\mathbf{x}) = T(\mathbf{x}) \quad \forall \; g \in \mathscr{G}, \forall \; \mathbf{x} \in \mathscr{X}$,

(ii) *(Maximality).* If $T(\mathbf{x}_1) = T(\mathbf{x}_2)$, then $\exists \; g \in \mathscr{G}$ such that $g\mathbf{x}_1 = \mathbf{x}_2$. (5.116)

Since it is not generally convenient to directly use the map $\mathbf{x} \to O(\mathbf{x})$ on $\mathscr{X}$ onto the space $\mathscr{O}$ of orbits, we will consider any suitable (measurable) map T with the properties (1), (2) in (5.116) as a *maximal invariant,* e.g., T may be a continuous map on $\mathscr{X}$ onto a subset of an Euclidean space with these properties. Note that the latter set may be considered a *relabeling* of the orbit space, and the map T a *renaming* of the map $\mathbf{x} \to O(\mathbf{x})$. After a suitable maximal invariant T is identified, one may restrict one's attention to tests which are functions of T (since all invariant tests are functions of T).

Example 5.15. Let $\mathbf{X} = (X_1, \ldots, X_k)'$ be Normal $N(\boldsymbol{\mu}, I_k)$, where $\boldsymbol{\mu} = (\mu_1, \ldots, \mu_k)' \in \Theta = \mathbb{R}^k$, and I_k is the $k \times k$ identity matrix. We wish to test $H_0 : \boldsymbol{\mu} = \mathbf{O}$. against $H_1 : \boldsymbol{\mu} \neq \mathbf{O}$. Let O be a $k \times k$ orthogonal matrix (i.e., $O'O = I_k$), and consider the transformed observation vector $\mathbf{Y} = O\mathbf{X}$. Then $\mathbf{Y}$ is $N(\boldsymbol{\nu}, I_k)$, $\boldsymbol{\nu} \in \mathbb{R}^k = \Theta$, and H_0 and H_1 are expressed as $\boldsymbol{\nu} = \mathbf{o}$ and $\boldsymbol{\nu} \neq \mathbf{o}$, respectively. Thus the statistical problem is invariant under all orthogonal transformations O. Since $T(\mathbf{x}) = \|\mathbf{x}\|^2$ is a maximal invariant under orthogonal $O : \mathbb{R}^k \to \mathbb{R}^k$, every invariant test here is a function of $T = \|\mathbf{X}\|^2$. Now, under H_0, T has the chi-square distribution with k degrees of freedom and, under H_1, it has the *non-central chi-square* distribution with the *non-centrality parameter* $\Delta = \sum_{i=1}^{k} \mu_i^2 = \|\boldsymbol{\mu}\|^2$, with respective densities $t_0(t)$, $f_1(t; \Delta)$ (with respect to Lebesgue measure on $[0, \infty)$). From Appendix C it follows that $f_1(t; \Delta)/f_0(t)$ is an increasing function of t. Therefore (See Remark 5.4), the UMP invariant test is given by

$$\text{Reject } H_0 \text{ iff } T \equiv \|\mathbf{X}\|^2 > \chi_{1-\alpha}^2(k).$$

It follows, by simple translation, that the UMP invariant test for $H_0 : \boldsymbol{\mu} = \boldsymbol{\mu}_0$ against $H_1 : \boldsymbol{\mu} \neq \boldsymbol{\mu}_0$ is given by

$$\text{Reject } H_0 \text{ iff } \|\mathbf{X} - \boldsymbol{\mu}_0\|^2 > \chi_{1-\alpha}^2(k).$$

More generally, Let $\mathbf{X}$ be $N(\boldsymbol{\mu}, \Sigma)$, $\boldsymbol{\mu} \in \mathbb{R}^k = \Theta$, and Σ a *known* $k \times k$ positive matrix. Let B be a $k \times k$ non-singular matrix such that $B'B = \Sigma^{-1}$ (See Lemma in Appendix B). Instead of $\mathbf{X}$, one may equivalently consider $\mathbf{Z} = B\mathbf{X}$ which is Normal $N(\boldsymbol{\nu}, I_k)$ with $\boldsymbol{\nu} = B\boldsymbol{\mu}$. The test for $H_0 : \boldsymbol{\mu} = \boldsymbol{\mu}_0$, $H_1 : \boldsymbol{\mu} \neq \boldsymbol{\mu}_0$ is equivalent to that for $H_0 : \boldsymbol{\nu} = \boldsymbol{\nu}_0 \equiv B\boldsymbol{\mu}_0$, $H_1 : \boldsymbol{\nu} \neq \boldsymbol{\nu}_0$. By the preceding, the UMP test invariant under the group of orthogonal transformations is given by

$$\text{Reject } H_0 \text{ iff } \|\mathbf{Z} - \boldsymbol{\nu}_0\|^2 > \chi_{1-\alpha}^2(k).$$

Since $\|\mathbf{Z} - \boldsymbol{\nu}_0\|^2 = \langle B(\mathbf{X} - \boldsymbol{\mu}_0), B(\mathbf{X} - \boldsymbol{\mu}_0)\rangle = \langle \Sigma^{-1}(\mathbf{X} - \boldsymbol{\mu}_0), (\mathbf{X} - \boldsymbol{\mu}_0)\rangle$, the above test may be expressed as

$$\text{Reject } H_0 \text{ iff } (\mathbf{X} - \boldsymbol{\mu}_0)' \Sigma^{-1} (\mathbf{X} - \boldsymbol{\mu}_0) > \chi_{1-\alpha}^2(k).$$

Next let $\mathbf{X}_i = (X_{i1} \ldots, X_{ik})'$, $1 \leq i \leq n$, be a random sample from $N(\boldsymbol{\mu}, \Sigma)$. Assume first that Σ is known. Then $\overline{\mathbf{X}} = (\overline{X}_{0.1}, \ldots, \overline{X}_{.k})'$ is a complete sufficient statistic for $\boldsymbol{\mu}(\overline{X}_{.j} = \sum_{i=1}^{n} X_{ij}/n)$. Hence the UMP invariant test for $H_0 : \boldsymbol{\mu} = \boldsymbol{\mu}_0$ against $H_1 : \boldsymbol{\mu} \neq \boldsymbol{\mu}_0$ is to reject H_0 if $n(\overline{\mathbf{X}} - \boldsymbol{\mu}_0)' \Sigma^{-1} (\overline{\mathbf{X}} - \boldsymbol{\mu}_0) > \chi_{1-\alpha}^2(k)$. Note that, by Basu's theorem, $\overline{\mathbf{X}}$ is independent of the sample covariance matrix $S = ((s_{jj'}))$, where $s_{jj'} = \sum_{i=1}^{n}(X_{ij} - \overline{X}_{.j})(X_{ij'} - \overline{X}_{.j'})/(n-1)$, $1 \leq j \leq j' \leq k$. If Σ is unknown then a UMP unbiased invariant test is of the form: Reject H_0 if $W = n(\overline{\mathbf{X}} - \boldsymbol{\mu}_0)' S^{-1} (\overline{\mathbf{X}} - \boldsymbol{\mu}_0) > c$. (See Exercise 5.7(c) for the case $k = 1$, and

Lehmann 1959, p. 299, for $k > 1$). Under H_0 the distribution of W does not involve Σ, so that c can be computed. The statistic W, or some multiple of it, is referred to as *Hotelling's* T^2. It can be shown that $\frac{n-k}{k(n-1)}W$ has the F distribution $F_{k,n-k}$ (Lehmann, 1959, pp. 296–300).

Finally, consider the two-sample problem: $H_0 : \boldsymbol{\mu}_1 = \boldsymbol{\mu}_2$, $H_1 : \boldsymbol{\mu}_1 \neq \boldsymbol{\mu}_2$, where $\boldsymbol{\mu}_1$, $\boldsymbol{\mu}_2$ are the means of the k-dimensional Normal distributions $N(\boldsymbol{\mu}_1, \Sigma)$, $N(\boldsymbol{\mu}_2, \Sigma)$. Arguing as above one can show (See Lehmann, loc. cit.) that the UMP unbiased invariant test based on independent samples $\mathbf{X}_i = (X_{i1}, \dots, X_{ik})'$, $1 \leq i \leq n_1$, from $N(\boldsymbol{\mu}_1, \Sigma)$, and $\mathbf{Y}_{i'} = (Y_{i'1}, \dots, Y_{i'k})'$, $1 \leq i' \leq n_2$, from $N(\boldsymbol{\mu}_2, \Sigma)$, is to reject H_0 if

$$T^2 \equiv \frac{n - k - 1}{k(n - 2)}(\overline{\mathbf{X}} - \overline{\mathbf{Y}})' \left[S \left(\frac{1}{n_1} + \frac{1}{n_2} \right) \right]^{-1} (\overline{\mathbf{X}} - \overline{\mathbf{Y}}) > F_{k,n-k-1}(1 - \alpha),$$

where $n = n_1 + n_2$, $S = \frac{(n_1 - 1)S_1 + (n_2 - 1)S_2}{n_1 + n_2 - 2}$, S_1 and S_2 being the sample covariance matrices of $\mathbf{X}$ and $\mathbf{Y}$. This is the two-sample Hotelling's T^2 test. This provides the motivation for the large sample nonparametric t tests used in Chap. 8.

5.8.1 The Two-Sample Problem

In many statistical investigations one inquires whether one commodity (or brand) is better than another, whether, e.g., one brand of tires has a greater life length than another, or whether one diet reduces cholesterol level more than another diet, etc. Among the most common parametric models of this kind is that of two normal distributions $\mathbf{N}(\mu, \sigma^2)$ and $\mathbf{N}(\eta, \sigma^2)$, and one wishes to test $H_0 : \mu = \eta$ (or, $\mu \leq \eta$) against $H_1 : \mu > \eta$. Here μ, η, σ^2 are unknown. The uniformly most powerful unbiased test of size α for this model, based on two independent random samples X_i $(1 \leq i \leq m)$ and Y_j $(1 \leq j \leq n)$ from $\mathbf{N}(\mu, \sigma^2)$ and $\mathbf{N}(\eta, \sigma^2)$, respectively, is the t-test: reject H_0 iff $T \equiv \frac{\overline{X} - \overline{Y}}{s_p \sqrt{\frac{1}{m} + \frac{1}{n}}} > t_{1-\alpha}(m + n - 2)$ (See Example 5.9). One could also have a model in which X_i's are i.i.d. exponential with parameter (mean) θ_1 and Y_j's are i.i.d. exponential with parameter θ_0, and one wishes to test $H_0 : \theta_1 \leq \theta_2$ against $H_1 : \theta_1 > \theta_2$ (See Exercise 5.11). In the absence of a reliable parametric model one may use a nonparametric model based on the following notion of stochastic order.

Definition 5.8. A random variable X is said to be *stochastically larger* than a random variable Y if

$$P(X > x) \geq P(Y > x) \quad x \in \mathbb{R}, \tag{5.117}$$

with strict inequality for some x. Note that, if F and G are the cumulative distribution functions of X and Y, respectively, then (5.117) is equivalent to

$$F(X) \leq G(x) \qquad \forall\, x \in \mathbb{R},$$
$$\text{with strict inequality for at least one } x. \tag{5.118}$$

Thus we may talk about a *stochastic order* (a partial order, given by the first line in (5.118) among distributions on $\mathbb{R}$. In both parametric examples given above, (the distribution of) X is stochastically larger than (that of) Y under H_1. Since the interest in the problems mentioned earlier in these examples seems to concern stochastic order, we may formulate the testing *nonparametrically* as follows. Based on independent random samples X_i $(1 \le i \le m)$ from a distribution with d.f. *(distribution function)* F and Y_j $(1 \le j \le n)$ from a distribution with d.f. G, consider the testing problem

$$H_\theta : \theta \in \Theta_0 = \{(F,G) : F(x) = G(x) \,\forall\, x,\, F \text{ and } G \text{ continuous}\}$$
$$= \{(F,F) : F \text{ continuous}\}$$
$$H_1 : \theta \notin \Theta_0 = \{(F,G) : F(x) \le G(x) \,\forall\, x,\, F \ne G,\, F \text{ and } G \text{ continuous}\}$$
$$\Theta = \Theta_0 \cup \Theta_1 = \{(F,G) : F, G \text{ are continuous}, F(x) \le G(x) \,\forall\, x\}. \qquad (5.119)$$

Remark 5.9. To simplify the discussion we have chosen H_0 as above, instead of $\{(F,G) : F(x) \ge G(x) \,\forall\, x,\, F \text{ and } G \text{ continuous}\}$. The tests we derive, are valid for this more reasonable version of H_0.

We will *first reduce the data by sufficiency.* The (pair of) ordered statistics $T = ((X_{(1)}, X_{(2)}, \ldots, X_{(m)}), (Y_{(1)}, Y_{(2)}, \ldots, Y_{(n)}))$ *are sufficient for* $\theta = (F,G) \in \Theta$. Here $X_{(1)} < X_{(2)} < \cdots < X_{(m)}$ are the ordering of the X_i's among themselves, and similarly for Y_j's. To prove sufficiency of T, note that (because X_i's are i.i.d.) the $m!$ different orderings $X_{i_1} < X_{i_2} < \cdots < X_{i_m}$ (with $(i_1, i_2, \ldots, i_m)$ an arbitrary permutation of $(1, 2, \ldots, m)$) are all equally likely, so that each has probability $\frac{1}{m!}$. Similarly, the $n!$ different orderings $Y_{j_1} < Y_{j_2} < \cdots < Y_{j_n}$ all have the same probability $\frac{1}{n!}$. Thus, for each value of $T = (u_1, u_2, \ldots, u_m, v_1, v_2, \ldots, v_n)$ with $u_1 < u_2, \cdots, < u_m$, $v_1 < v_2 < \cdots < v_n$, one has

$$P_\theta(X_1 = x_1, X_2 = x_2, \ldots, X_m = x_m, Y_1 = y_1, Y_2 = y_2, \ldots, Y_n = y_n) \mid$$
$$T = (u_1, u_2, \ldots, u_m, v_1, v_2, \ldots, v_n))$$
$$= \begin{cases} \frac{1}{m!n!} \text{ if } (x_1, x_2, \ldots, x_m) \text{ is a permutation of } (u_1, u_2, \ldots, u_m) \text{ and} \\ \qquad (y_1, y_2, \ldots, y_n) \text{ is a permutation of } (v_1, v_2, \ldots, v_n), \\ 0 \quad \text{otherwise.} \end{cases} \qquad (5.120)$$

Thus the conditional distribution of $(X_1, X_2, \ldots, X_m, Y_1, \ldots, Y_n)$, given T, does not depend on $\theta \in \Theta$. This establishes the desired sufficiency of T. Since F and G are continuous and the $m + n$ random variables X_i's and Y_j's are independent, we may take the *observation space* (of the sufficient statistic T) to be

$$\mathscr{X} = \left\{ (\mathbf{u}, \mathbf{v}) \in \mathbb{R}^{m+n} : u_1 < u_2, \cdots < u_m, v_1 < v_2 < \cdots < v_n,\, u_i \ne v_j \,\forall\, i, j \right\},$$
$$(\mathbf{u} = (u_1, \ldots, u_m), \mathbf{v} = (v_1, \ldots, v_n)). \qquad (5.121)$$

Let P_θ denote the distribution of T, when $\theta = (F,G)$ is the true parameter value, $\theta \in \Theta$.

That is, $P_\theta(B) = \text{Prob}_\theta(T \in B)$, where Prob_θ is the probability measure on the underlying probability space on which X_i's and Y_j's are defined.

Let ψ denote a *strictly increasing continuous* function on $\mathbb{R}$ onto $\mathbb{R}$. Define, $\forall\, \mathbf{x} = (\mathbf{u}, \mathbf{v}) \in \mathscr{X}$, $g = g_\psi$ by

$$g_\psi \equiv g_\psi(\mathbf{u}, \mathbf{v}) = (\psi(u_1), \ldots, \psi(u_m), \psi(v_1), \ldots, \psi(v_n)). \qquad (5.122)$$

Then g_ψ is a one-to-one map on $\mathscr{X}$ onto $\mathscr{X}$, and g_ψ and g_ψ^{-1} are both continuous. Indeed, if ψ^{-1} is the inverse of ψ on $\mathbb{R}$ onto $\mathbb{R}$, then $g_\psi^{-1} \equiv g_{\psi^{-1}}$:

$$g_\psi^{-1}\mathbf{x} = (\psi^{-1}(u_1), \psi^{-1}(u_2), \ldots, \psi^{-1}(u_m), \psi^{-1}(v_1), \ldots, \psi^{-1}(v_n)).$$

Let $\mathscr{G}$ denote the group of all such transformations on $\mathscr{X}$ (corresponding to the group Ψ of all transformations ψ on $\mathbb{R}$ onto $\mathbb{R}$ such that ψ is strictly increasing and continuous). Then one obtains the following.

Proposition 5.4. *(a) The testing problem (5.119) based on*
$T = T(X_1, \ldots, X_m, Y_1, \ldots, Y_n) = (X_{(1)}, \ldots, X_{(m)}, Y_{(1)}, \ldots, Y_{(n)})$ *is invariant under $\mathscr{G}$.*

(b) Every invariant test is a function of

$$\mathbf{R} = (R_1, R_2, \ldots, R_m) \tag{5.123}$$

where R_i is the rank of $X_{(i)}$ among the $m + n$ values $X_{(1)}, \ldots, X_{(m)}$, $Y_{(1)}, \ldots, Y_{(n)}$ (or, equivalently, the rank of $X_{(i)}$ among $\{X_1, \ldots, X_m, Y_1, \ldots, Y_n\}$). That is, $\mathbf{R}$ is a maximal invariant under $\mathscr{G}$.

Proof. (a) Suppose $\theta = (F, F) \in \Theta_0$. That is, $X_1, \ldots, X_m$ are i.i.d. with (common) d.f. F and $Y_1, \ldots, Y_n$ are i.i.d with (common) d.f. F. Then, for any given $\psi \in \Psi$, $g_\psi X_1, \ldots, g_\psi X_m$ are i.i.d. with d.f. $F \circ \psi^{-1}$. In particular, the distribution of T under this transformation (i.e., the distribution of $T \circ \psi(X_1, \ldots, X_m, Y_1, \ldots, Y_n) = T(\psi X_1, \ldots, \psi X_m, \psi Y_1, \ldots, \psi Y_n)$) is $P_{\bar{g}_\psi \theta}$, with $\bar{g}_\psi \theta = (F \circ \psi^{-1}, F \circ \psi^{-1}) \in \Theta_0$, when the distribution of $T(X_1, \ldots, X_m, Y_1, \ldots, Y_n)$ is P_θ with $\theta = (F, F) \in \Theta_0$. Also, if $F(x) \leq G(x) \; \forall x$, with strict inequality for some $x = x_0$, say, then $F \circ \psi^{-1}(x) \leq G \circ \psi^{-1}(x) \; \forall x$, with strict inequality for $x = \psi(x_0)$. Thus the distribution P_θ of $T(X_1, \ldots, X_m, Y_1, \ldots, Y_n)$ under $\theta = (F, G) \in \Theta$ is transformed to the distribution $P_{\bar{g}_\psi \theta}$ of $T(\psi X_1, \ldots, \psi X_m, \psi Y_1, \ldots, \psi Y_n)$ under $\bar{g}_{\psi \theta} = (F \circ \psi^{-1}, G \circ \psi^{-1}) \in \Theta_1$.

(b) It is clear that $\mathbf{R}$ is invariant under $\mathscr{G}$, since a strictly increasing transformation ψ does not change the relative orders of numbers on the line. To prove that $\mathbf{R}$ is a maximal invariant, let $(\mathbf{u}, \mathbf{v})$, $(\mathbf{u}', \mathbf{v}') \in \mathscr{X}$ be such that $\mathbf{R}(\mathbf{u}, \mathbf{v}) = (r_1, \ldots, r_m) = \mathbf{R}(\mathbf{u}', \mathbf{v}')$. This means that the relative orders of u_i's and v_j's are the same as those of the u_i''s and v_j''s. For example, the number of v_j's smaller than u_1 is $r_1 - 1$, the same is true of the number of v_j''s smaller than u_1'. In general, the number of v_j's lying between u_i and u_{i+1} is $r_{i+1} - r_i - 1$, and the same is true of u_i''s and v_j''s. Now order $m + n$ numbers u_i's and v_j's as $w_1 < w_2 < \cdots < w_{m+n}$, and similarly order the u_i''s and v_j''s as $w_1' < w_2' < \cdots < w_{m+n}'$. From the above argument it follows that if $w_k = u_i$ then $w_k' = u_i'$, and if $w_k = v_j$ then $w_k' = v_j'$. Define the piecewise linear strictly increasing map ψ by

$$\psi(w_i) = w_i', \qquad 1 \leq i \leq m + n,$$

with linear interpolation in (w_i, w_{i+1}), $1 \leq i \leq m + n - 1$. For $x < w_1$, let $\psi(x)$ be defined by extending the line segment joining (w_1, w_1') and (w_2, w_2'). Similarly, for $x > w_{m+n}$, define $\psi(x)$ by extending the line segment joining (w_{m+n-1}, w_{m+n-1}') and (w_{m+n}, w_{m+n}'). Then $\psi(u_i) = u_i'$ and $\psi(v_j) = v_j' \; \forall \, i, j$ and $g_\psi(\mathbf{u}, \mathbf{v}) = (\mathbf{u}', \mathbf{v}')$. $\qquad \square$

Among commonly used rank tests for the two-sample problem (5.119) are the following.

Example 5.16 (Wilcoxon, or Mann-Whitney Test).

$$\phi(r_1, \ldots, r_m) = \begin{cases} 1 \text{ if } \sum_{i=1}^{m} r_i > c, \\ \gamma \text{ if } \sum_{i=1}^{m} r_i = c, \\ 0 \text{ if } \sum_{i=1}^{m} r_i < c, \end{cases} \tag{5.124}$$

where $r_1 < \cdots < r_m$ are the values of $R_1 < R_2 < \cdots < R_m$. The positive integer c and $\gamma \in (0,1)$ are chosen so that the size is α.

Example 5.17 (The Fisher-Yates Test).

$$\phi(r_1, \ldots, r_m) = \begin{cases} 1 \text{ if } \sum_{i=1}^{m} \eta_{(r_i)}^{m+n} > c, \\ \gamma \text{ if } \sum_{i=1}^{m} \eta_{(r_i)}^{m+n} = c, \\ 0 \text{ if } \sum_{i=1}^{m} \eta_{(r_i)}^{m+n} < c, \end{cases} \tag{5.125}$$

where $\eta_{(r)}^{m+n}$ is the expected value of the r-th order statistic of a random sample of size $m + n$ from a standard Normal distribution $\mathbf{N}(0,1)$.

There is, of course, no uniformly most powerful test for (5.119). We may, however, look for *admissible tests* in the class of all rank tests. Both tests (5.124), (5.125) may be shown to be admissible in this sense. We will later discuss asymptotic (as $m \to \infty$, $n \to \infty$) superiority of these tests compared to the classical t-test based on $\overline{X} - \overline{Y}$, in the location model: $F(x) = F_0(x)$, $G(x) = F_0(x + \theta)$, $\theta < 0$. Here F_0 is a given continuous distribution function.

5.9 Linear Models

5.9.1 The Gauss-Markov Theorem

Consider the *linear model*

$$\mathbf{X} = A\boldsymbol{\theta} + \boldsymbol{\varepsilon}, \tag{5.126}$$

where $\mathbf{X} = (X_1, \ldots, X_n)'$ is the *observation vector*, $\boldsymbol{\theta} = (\theta_1, \ldots, \theta_k)' \in \mathbb{R}^k$ is the vector of *unknown parameters*, A is the $(n \times k)$ known *coefficient matrix*

$$A = \begin{pmatrix} a_{11} & a_{12} & \cdots & a_{1k} \\ a_{21} & a_{22} & \cdots & a_{2k} \\ \vdots & \vdots & & \vdots \\ a_{n1} & a_{n2} & \cdots & a_{nk} \end{pmatrix} = [\boldsymbol{\alpha}_1 \boldsymbol{\alpha}_2 \cdots \boldsymbol{\alpha}_k] = \begin{bmatrix} \boldsymbol{\beta}_1 \\ \boldsymbol{\beta}_2 \\ \vdots \\ \boldsymbol{\beta}_n \end{bmatrix}, \tag{5.127}$$

and $\boldsymbol{\varepsilon} = (\varepsilon_1, \varepsilon_2, \ldots, \varepsilon_n)'$ is the vector of *unobservable errors*,

$$E\varepsilon_i = 0, \quad E\varepsilon_i\varepsilon_j = \sigma^2 \delta_{ij} \quad (1 \leq i, j \leq n), \tag{5.128}$$

$0 < \sigma^2 < \infty$, σ^2 unknown.

Definition 5.9. A linear parametric function $\ell'\theta = \sum_{j=1}^{k} \ell_j \theta_j$ is said to be *estimable* if there exists a linear function of the observations $\mathbf{d}'\mathbf{X} = \sum_{i=1}^{n} d_i X_i$ such that

$$E_{(\theta,\sigma^2)}\mathbf{d}'\mathbf{X} = \ell'\theta \qquad \forall\, \theta \in \mathbb{R}^k,\ \sigma^2 > 0. \tag{5.129}$$

A linear unbiased estimator $\mathbf{d}_0'\mathbf{X}$ of $\ell'\theta$ is said to be the *best linear unbiased estimator* or *BLUE,* if

$$E_{(\theta,\sigma^2)}(\mathbf{d}_0'\mathbf{X} - \ell'\theta)^2 = \inf E_{(\theta,\sigma^2)}(\mathbf{d}'\mathbf{X} - \ell'\theta)^2 \quad \forall\, \theta \in \mathbb{R}^k,\ \sigma^2 > 0, \tag{5.130}$$

where the infimum is taken over the class $\mathbf{d}'\mathbf{X}$ of all linear unbiased estimators of $\ell'\theta$.

For the present subsection, nothing is assumed about the distribution of the errors other than (5.128). Hence the notation $E_{(\theta,\sigma^2)}$ here simply means expectation under an arbitrary distribution satisfying (5.128) and a given θ. In later subsections, ε's are also assumed to be Normal, so that the pair (θ,σ^2) specifies a distribution of $\mathbf{X}$.

Theorem 5.7 (Gauss-Markov Theorem).

(a) $\ell'\theta$ is estimable if and only if $\ell \in \mathscr{L}_r$—the vector space spanned by the rows $\boldsymbol{\beta}_i$ $(1 \leq i \leq n)$ of A.

(b) Suppose $\ell'\theta$ is estimable and $\mathbf{d}'\mathbf{X}$ is an unbiased estimator of $\ell'\theta$. Then the unique BLUE of $\ell'\theta$ is $\mathbf{d}_0'\mathbf{X}$, where $\mathbf{d}_0$ is the orthogonal projection of $\mathbf{d}$ on the vector space $\mathscr{L}_c$ spanned by the column vectors $\boldsymbol{\alpha}_j$ $(1 \leq j \leq k)$ of A.

Proof. (a) One has $E_{(\theta,\sigma^2)}\mathbf{d}'\mathbf{X} = \ell'\theta$ for some $\mathbf{d} \in \mathbb{R}^n$ if and only if

$$\ell'\theta \equiv \sum_{j=1}^{k} \ell_j \theta_j = E_{(\theta,\sigma^2)}\mathbf{d}'\mathbf{X} = \mathbf{d}'A\theta \equiv \sum_{j=1}^{k} \left(\sum_{i=1}^{n} d_i \boldsymbol{\beta}_i \right)_j \theta_j \quad \forall\, \theta \in \mathbb{R}^k,$$

i.e., iff $\ell = \sum_{i=1}^{n} d_i \boldsymbol{\beta}_i$ for some $\mathbf{d} \in \mathbb{R}^n$.

(b) Let $\mathbf{d}'\mathbf{X}$ be an unbiased estimator of $\ell'\theta$, and $\mathbf{d}_0$ the orthogonal projection of $\mathbf{d}$ onto $\mathscr{L}_c$. Then $E_{(\theta,\sigma^2)}(\mathbf{d} - \mathbf{d}_0)'\mathbf{X} = (\mathbf{d} - \mathbf{d}_0)'\, A\theta = (\mathbf{d} - \mathbf{d}_0)'\sum_{j=1}^{k} \theta_j \boldsymbol{\alpha}_j = 0$, since $\mathbf{d} - \mathbf{d}_0$ is orthogonal to $\sum_{j=1}^{k} \theta_j \boldsymbol{\alpha}_j \in \mathscr{L}_c$. It follows that $\mathbf{d}_0'\mathbf{X}$ is an unbiased estimator of $\ell'\theta$:

$$E_{(\theta,\sigma^2)}\mathbf{d}_0'\mathbf{X} = E_{(\theta,\sigma^2)}[\mathbf{d}'\mathbf{X} - (\mathbf{d} - \mathbf{d}_0)'\mathbf{X}] = E_{(\theta,\sigma^2)}\mathbf{d}'\mathbf{X} = \ell'\theta.$$

Also, $E_{(\theta,\sigma^2)}(\mathbf{d}_0'\mathbf{X} - \ell'\theta)(\mathbf{d}'\mathbf{X} - \mathbf{d}_0'\mathbf{X}) = E_{(\theta,\sigma^2)}(\mathbf{d}_0'\mathbf{X})(\mathbf{d}'\mathbf{X} - \mathbf{d}_0'\mathbf{X}) = \sigma^2 \mathbf{d}_0'(\mathbf{d} - \mathbf{d}_0) = 0$, so that

$$E_{(\theta,\sigma^2)}(\mathbf{d}'\mathbf{X} - \ell'\theta)^2 = E_{(\theta,\sigma^2)}(\mathbf{d}_0'\mathbf{X} - \ell'\theta + \mathbf{d}'\mathbf{X} - \mathbf{d}_0'\mathbf{X})^2$$

$$= E_{(\theta,\sigma^2)}(\mathbf{d}_0'\mathbf{X} - \ell'\theta)^2 + E_{(\theta,\sigma^2)}(\mathbf{d}'\mathbf{X} - \mathbf{d}_0'\mathbf{X})^2 \geq E_{(\theta,\sigma^2)}(\mathbf{d}_0'\mathbf{X} - \ell'\theta)^2,$$

with a strict inequality unless $E_{(\theta,\sigma^2)}(\mathbf{d}'\mathbf{X} - \mathbf{d}_0'\mathbf{X})^2 = 0$, i.e., unless $\sigma^2|\mathbf{d} - \mathbf{d}_0|^2 = 0$, i.e., unless $\mathbf{d}_0 = \mathbf{d}$, or $\mathbf{d} \in \mathscr{L}_c$.

To prove uniqueness, let $\boldsymbol{\gamma}'\mathbf{X}$ be another unbiased estimator of $\ell'\theta$, and let $\boldsymbol{\gamma}_0$ be the orthogonal projection of $\boldsymbol{\gamma}$ on $\mathscr{L}_c$. Then $E_{(\theta,\sigma^2)}(\boldsymbol{\gamma}_0'\mathbf{X} - \mathbf{d}_0'\mathbf{X}) = 0 \,\forall\, \theta$, i.e., $(\boldsymbol{\gamma}_0 - \mathbf{d}_0) \cdot \sum_{i=1}^{k} \theta_i \boldsymbol{\alpha}_i = 0 \,\forall\, \theta \in \mathbb{R}^k$. This implies $\boldsymbol{\gamma}_0 - \mathbf{d}_0$ is orthogonal to $\mathscr{L}_c$. But $\boldsymbol{\gamma}_0 - \mathbf{d}_0 \in \mathscr{L}_c$. Hence $\boldsymbol{\gamma}_0 - \mathbf{d}_0 = 0$. $\qquad\square$

Remark 5.10. Suppose $n \geq k$, and A is of full rank k. Then the vector space $\mathscr{L}_r$ is of rank k, so that every ℓ in $\mathbb{R}^k$ belongs to $\mathscr{L}_r$. Therefore, *all linear parametric functions are estimable,* by part (a) of the Gauss-Markov Theorem (Theorem 5.7). Also, in this case, $A'A$ is a non-singular $k \times k$ matrix, and

$$E_{(\boldsymbol{\theta},\sigma^2)}A'\mathbf{X} = A'A\boldsymbol{\theta}, \quad \text{or } E_{(\boldsymbol{\theta},\sigma^2)}(A'A)^{-1}A'\mathbf{X} = \boldsymbol{\theta} \quad \forall \; \boldsymbol{\theta} \in \mathbb{R}^k.$$

Hence an unbiased estimator of the vector $\boldsymbol{\theta}$ is given by

$$\widehat{\boldsymbol{\theta}} = (A'A)^{-1}A'\mathbf{X} = D_0'\mathbf{X}, \quad \text{cov}\,\widehat{\boldsymbol{\theta}} = \sigma^2 D_0'D_0 = \sigma^2(A'A)^{-1}, \quad (5.131)$$

$$D_0 := A(A'A)^{-1} = [\alpha_1\alpha_2\cdots\alpha_k]\begin{bmatrix} b_{11} & b_{12} & \cdots & b_{1k} \\ b_{21} & b_{22} & \cdots & b_{2k} \\ \vdots & \vdots & & \vdots \\ b_{k1} & b_{k2} & \cdots & b_{kk} \end{bmatrix}, \text{ say.}$$

The j-th column of D_0 is

$$\mathbf{d}_{0,j} = \sum_{j=1}^{k} b_{j'j}\alpha_{j'} \in \mathscr{L}_c,$$

and

$$\hat{\theta}_j = \mathbf{d}_{0,j}'\mathbf{X}$$

is, therefore, the best linear unbiased estimator of θ_j $(1 \leq j \leq k)$, as is $\sum_1^k \ell_j\hat{\theta}_j = \sum_1^k(\ell_j d_{0,j})'\mathbf{X}$ of $\sum_1^k \ell_j\theta_j$, whatever be $\ell \in \mathbb{R}^k$.

In view of part (b) of the Gauss-Markov Theorem, the best linear unbiased estimator of an estimable function $\ell'\boldsymbol{\theta}$ is also called its *least squares estimator.*

5.9.2 Testing in Linear Models

In addition to (5.126), (5.128), assume

$$\varepsilon_i \; \text{are i.i.d. } \mathbf{N}(0,\sigma^2) \qquad (1 \leq i \leq n)\,(0 < \sigma^2 < \infty).$$

In fact, for simplicity let us assume for the time being that we have the *canonical linear model* for the observations $Y_1,\ldots,Y_n$, i.e.,

$$\mathbf{Y} = \begin{pmatrix} \mu_1 \\ \mu_2 \\ \vdots \\ \mu_k \\ 0 \\ 0 \\ \vdots \\ 0 \end{pmatrix} + \begin{pmatrix} \varepsilon_1 \\ \varepsilon_2 \\ \vdots \\ \varepsilon_n \end{pmatrix} \qquad (5.132)$$

where $(\mu_1, \mu_2, \ldots, \mu_k)' \in \mathbb{R}^k$, $\Theta = \{(\mu_1, \ldots, \mu_k, \sigma^2) : \mu_i \in \mathbb{R}, 1 \leq i \leq k, 0 < \sigma^2 < \infty\} = \mathbb{R}^k \times (0, \infty)$ and ε_i's are i.i.d. $N(0, \sigma^2)$. We want to test

$$H_0 : \mu_1 = \mu_2 = \cdots = \mu_r = 0 \quad (\Theta_0 = \{(0, \ldots, 0)\} \times \mathbb{R}^{k-r} \times (0, \infty))$$

against

$$H_1 : H_0 \text{ is not true} \quad (\Theta_1 = \Theta \backslash \Theta_0), \tag{5.133}$$

where $1 \leq r \leq k$. For $r = k$, $\Theta_0 = \{\mathbf{O}\} \times (0, \infty)$.

The testing problem is invariant under each of the following three groups of transformations:

$$\mathscr{G}_1 : g_1(\mathbf{y}) = \mathbf{z}, \quad z_i = y_i + c_i \quad (r + 1 \leq i \leq k)$$
$$z_i = y_i \quad \text{for } i \notin \{r + 1, r + 2, \ldots, r + k\}.$$

c_i $(r + 1 \leq i \leq k)$ are arbitrary reals.

$$\mathscr{G}_2 : g_2(\mathbf{y}) = \mathbf{z}, \quad \begin{pmatrix} z_1 \\ z_2 \\ \vdots \\ z_r \end{pmatrix} = O_r \begin{pmatrix} y_1 \\ y_2 \\ \vdots \\ y_r \end{pmatrix}, \quad z_i = y_i \quad \text{for } r + 1 \leq i \leq n,$$

where O_r is an arbitrary $r \times r$ orthogonal matrix.

$$\mathscr{G}_3 : g_3(\mathbf{y}) = b\mathbf{y}, \quad b > 0.$$

Let $\mathscr{G}$ be the group of transformations generated by $\Gamma = \mathscr{G}_1 \cup \mathscr{G}_2 \cup \mathscr{G}_3$.

Proposition 5.5. *The statistic* $T(\mathbf{y}) \equiv \frac{\sum_1^r y_i^2}{\sum_{k+1}^n y_i^2}$ *is a maximal invariant function of the sufficient statistic*

$$S(\mathbf{y}) \equiv \left(y_1, \ldots, y_k, \sum_{i=k+1}^n y_i^2 \right). \tag{5.134}$$

Proof. The joint p.d.f. of $Y_1, \ldots, Y_n$ is

$$\left(\frac{1}{\sqrt{2\pi\sigma^2}} \right)^n e^{-\frac{1}{2\sigma^2} \sum_{i=1}^k (y_i - \mu_i)^2 - \frac{1}{2\sigma^2} \sum_{i=k+1}^n y_i^2}, \tag{5.135}$$

implying $S(\mathbf{y})$ is indeed a sufficient statistic for $(\mu_1, \ldots, \mu_k, \sigma^2)$. (Use the Factorization Theorem). We need to show that a function $T(\mathbf{y})$, of $S(\mathbf{y})$, which is also invariant under $\mathscr{G}$, is a function of $T(\mathbf{y})$.

Now a maximal invariant under $\mathscr{G}_1$ is

$$W_1(\mathbf{y}) = (y_1, \ldots, y_r, y_{k+1}, \ldots, y_n) \tag{5.136}$$

Observe that $W_1(g_1\mathbf{y}) = W_1(\mathbf{y})$, since W_1 does not involve the coordinates $y_{r+1}, \ldots, y_k$ which are the only ones affected by g_1; also, if $W_1(\mathbf{y}) = W_1(\mathbf{z})$, then $\mathbf{z} = (y_1, \ldots, y_r, y_{r+1} + (z_{r+1} - y_{r+1}), \ldots, y_k + (z_k - y_k), y_{k+1}, \ldots, y_n) = g_1(\mathbf{y})$ with $c_i = z_i - y_i$ for $r + 1 \leq i \leq k$. Thus $\mathbf{y}$ and $\mathbf{z}$ belong to the same orbit under $\mathscr{G}_1$ proving that W_1 is a maximal invariant under $\mathscr{G}_1$. Hence T_1 must be a function of W_1, say,

$$T_1(\mathbf{y}) = f_1(y_1, \ldots, y_r, y_{k+1}, \ldots, y_n). \tag{5.137}$$

A maximal invariant under $\mathscr{G}_2$ is

$$W_2(\mathbf{y}) = \left(\sum_{i=1}^{r} y_i^2, y_{k+1}, \ldots, y_n \right) \tag{5.138}$$

For, $W_2(g\mathbf{y}) = W_2(\mathbf{y})$, since $\left\| O_r \begin{pmatrix} y_1 \\ \vdots \\ y_r \end{pmatrix} \right\|^2 = \left\| \begin{pmatrix} y_1 \\ \vdots \\ y_r \end{pmatrix} \right\|^2$. Also, given two vectors $(z_1, \ldots, z_r)'$ and $(y_1, y_2, \ldots, y_r)'$ of equal length there exists an orthogonal transformation O_r such that $O_r(z_1, \ldots, z_r) = (y_1, \ldots, y_r)$. Hence $W_2(\mathbf{y}) = W_2(\mathbf{z})$ implies that $\mathbf{z} = g(\mathbf{y})$ for some $g \in \mathscr{G}_2$. Therefore, $T_1(\mathbf{y})$ depends on $y_1, \ldots, y_r$ only through $\sum_1^r y_i^2$, i.e., T_1 is of the form [see (5.137)]

$$T_1(\mathbf{y}) = f_2 \left(\sum_{i=1}^{r} y_i^2, \, y_{k+1}, \ldots, y_n \right). \tag{5.139}$$

Now since $T_1(\mathbf{y})$ is a function of $S(\mathbf{y})$, it must be of the form (using (5.134) and (5.139))

$$T_1(\mathbf{y}) = f_3 \left(\sum_{i=1}^{r} y_i^2, \sum_{k+1}^{n} y_i^2 \right). \tag{5.140}$$

In view of invariance under $\mathscr{G}_3$, one must have

$$T_1(b\mathbf{y}) = T_1(\mathbf{y}) \qquad \forall \, b > 0, \tag{5.141}$$

i.e.,

$$f_3 \left(b^2 \sum_1^r y_i^2, b^2 \sum_{k+1}^n y_i^2 \right) = f_3 \left(\sum_1^r y_i^2, \sum_{k+1}^n y_i^2 \right). \tag{5.142}$$

This implies T_1 is of the form

$$T_1(\mathbf{y}) = f_4 \left(\sum_1^r y_i^2 \Big/ \sum_{k+1}^n y_i^2 \right) \quad \text{if } \sum_{k+1}^n y_i^2 \neq 0, \tag{5.143}$$

where f_4 is one-to-one, i.e., strictly monotone on $(0, \infty)$. For *if this is not the case* then there exist $\mathbf{y}$, $\mathbf{z}$ such that $\sum_1^r y_i^2 / \sum_{k+1}^n y_i^2 = \sum_1^r z_i^2 / \sum_{k+1}^n z_i^2$, but $T_1(\mathbf{y}) \neq T_1(\mathbf{z})$. Write $b = (\sum_{k+1}^n z_i^2 / \sum_{k+1}^n y_i^2)^{1/2}$ (assuming $\sum_{k+1}^n z_i^2 > 0$, by (5.143)). Then, by invariance under $\mathscr{G}_3$,

$$T_1(b\mathbf{y}) = T_1(\mathbf{y})$$

so that

$$\begin{aligned} T_1(\mathbf{z}) &= f_3 \left(\sum_1^r z_i^2, \sum_{k+1}^n z_i^2 \right) = f_3 \left(\frac{\sum_1^r z_i^2}{\sum_{k+1}^n z_i^2} \cdot \sum_{k+1}^n z_i^2, b^2 \sum_{k+1}^n y_i^2 \right) \\ &= f_3 \left(\frac{\sum_1^r y_i^2}{\sum_{k+1}^n y_i^2} \cdot \sum_{k+1}^n z_i^2, b^2 \sum_{k+1}^n y_i^2 \right) = f_3 \left(b^2 \sum_1^r y_i^2, b^2 \sum_{k+1}^n y_i^2 \right) \\ &= T_1(b\mathbf{y}) = T_1(\mathbf{y}), \end{aligned}$$

which is a contradiction. Hence (5.143) holds. $\qquad\square$

Theorem 5.8. *For the canonical linear model, the U.M.P. invariant (under $\mathcal{G}$) test of H_0 against H_1 (of size α) is given by*

$$\varphi(\mathbf{y}) = \begin{cases} 1 \text{ if } \dfrac{\sum_1^r y_i^2}{r} \Big/ \dfrac{\sum_{k+1}^n y_i^2}{n-k} > F_\alpha(r, n-k) \\[2ex] 0 \text{ if } \dfrac{\sum_1^r y_i^2}{r} \Big/ \dfrac{\sum_{k+1}^n y_i^2}{n-k} \leq F_\alpha(r, n-k) \end{cases} \tag{5.144}$$

where $F_{1-\alpha}(r, n-k)$ is the $(1-\alpha)$th quantile of the F-distribution with d.f.s r for the numerator, and $n-k$ for the denominator.

Proof. We know that the UMP invariant test must be based on $\sum_1^r y_i^2 / \sum_{k+1}^n y_i^2$. We need to prove that among all tests which are functions of this ratio, the test (5.144) is UMP. This follows from the Neyman–Pearson Lemma and the fact that the ratio $h_{\gamma^2}(u)/h_0(u)$ of the p.d.f. of $\left(\sum_1^r y_i^2/r\right) / \left(\sum_{k+1}^n y_i^2/(n-k)\right)$ under H_1 and under H_0 is monotone increasing in u (see Remark 5.4 and Example C.6). Here $\gamma^2 = \sum_{i=1}^r \mu_i^2/\sigma^2$. $\qquad\square$

Remark 5.11. For (5.143) note that $\mathrm{Prob}_\mu(\sum_{k+1}^n Y_i^2 = 0) = 0$, so that one can restrict the observation space to exclude the set $\{\sum_{k+1}^n y_i^2 = 0\}$.

Reduction of the General Case to the Canonical Model The general linear hypothesis in the linear model (5.126) is of the form

$$H_0 : B\boldsymbol{\theta} = 0, \tag{5.145}$$

where $B = ((b_{ij}))$ is an $r \times k$ matrix of full rank r ($1 \leq r \leq k$). Without loss of generality one may assume that the first r columns of B are linearly independent. Then the matrix B_r, which is $r \times r$ and is composed of the first r columns of B, is nonsingular, and one may rewrite (5.145) as

$$B_r \begin{pmatrix} \theta_1 \\ \vdots \\ \theta_r \end{pmatrix} = - \begin{pmatrix} b_{1\,r+1} & \cdots & b_{1k} \\ \vdots & & \vdots \\ b_{r\,r+1} & \cdots & b_{rk} \end{pmatrix} \begin{pmatrix} \theta_{r+1} \\ \vdots \\ \theta_k \end{pmatrix},$$

or,

$$\begin{pmatrix} \theta_1 \\ \vdots \\ \theta_r \end{pmatrix} = -B_r^{-1} \begin{pmatrix} b_{1\,r+1} & \cdots & b_{1k} \\ \vdots & & \vdots \\ b_{r\,r+1} & \cdots & b_{rk} \end{pmatrix} \begin{pmatrix} \theta_{r+1} \\ \vdots \\ \theta_k \end{pmatrix} = \begin{pmatrix} d_{1\,r+1} & \cdots & d_{1k} \\ \vdots & & \vdots \\ d_{r\,r+1} & \cdots & d_{rk} \end{pmatrix} \begin{pmatrix} \theta_{r+1} \\ \vdots \\ \theta_k \end{pmatrix}, \tag{5.146}$$

say.

Observe that in the linear model (5.126), we assume that $\mathbf{X}$ is Normal with mean (vector) $A\boldsymbol{\theta}$ and dispersion matrix $\sigma^2 I_n$, and A is of full rank k. Here $\boldsymbol{\theta}$ is unknown (i.e., θ_i's can take arbitrary real values), and $\sigma^2 > 0$ is unknown; and I_n is the $n \times n$ identity matrix. Thus $E\mathbf{X} \equiv \sum_1^k \theta_i \boldsymbol{\alpha}_i$ ($\boldsymbol{\alpha}_i$ being the ith column of A) spans the subspace of $\mathcal{L}_c$ of dimension k of an n-dimensional Euclidean space E_n, say, which may be identified with $\mathbb{R}^n$ (w.r.t. the standard orthonormal basis). Under H_0, $E\mathbf{X}$ lies in a $(k-r)$-dimensional subspace, say $\mathcal{L}_{c,0}$, of $\mathcal{L}_c$ (expressing $\theta_1, \ldots, \theta_r$ in terms of $\theta_{r+1}, \ldots, \theta_k$, by (5.146)). We will now construct an $n \times n$ orthogonal matrix $\mathbf{O}$ such that $\mathbf{Y} = \mathbf{OX}$ is in the form (5.132) of a canonical model, with H_0 and H_1 given by (5.133). For this (1) choose the first r rows of $\mathbf{O}$ to lie in $\mathscr{L}_c$, but orthogonal to $\mathscr{L}_{c,0}$, (2) choose the next $k - r$ rows of $\mathbf{O}$ to span $\mathscr{L}_{c,0}$, and (3) choose the last $n - k$ rows of $\mathbf{O}$ orthogonal to $\mathscr{L}_c$. Because of the orthogonality of $\mathbf{O}$, the error $\mathbf{O}\,\epsilon$ remains $N(\mathbf{O}, \sigma^2 I_n)$, and $\boldsymbol{\mu} \equiv E\mathbf{Y} = \mathbf{O}\,A\boldsymbol{\theta}$ span $\mathscr{L}$. Also, under H_0, the first r elements of $\boldsymbol{\mu}$ are zero, leaving the next $k - r$ elements arbitrary (unknown).

Before turning to Examples, let us note that, because of orthogonality of $\mathbf{O}$ in the relation $\mathbf{Y} = \mathbf{O}\,\mathbf{X}$, one has $\|\mathbf{X} - E\mathbf{X}\|^2 = \|\mathbf{Y} - E\mathbf{Y}\|^2$. Hence

$$\sum_{i=k+1}^{n} Y_i^2 = \min_{\boldsymbol{\mu}} \left\{ \sum_{i=1}^{k} (Y_i - \mu_i)^2 + \sum_{i=k+1}^{n} Y_i^2 \right\} = \min_{E\mathbf{Y}} \|\mathbf{Y} - E\mathbf{Y}\|^2$$

$$= \min_{\mathbf{m} \equiv E\mathbf{X} \in \mathscr{L}_c} \|\mathbf{X} - E\mathbf{X}\|^2 = \|\mathbf{X} - \hat{\mathbf{m}}\|^2, \tag{5.147}$$

where $\hat{\mathbf{m}} = A\hat{\boldsymbol{\theta}}$ is the projection of $\mathbf{X}$ on $\mathscr{L}_c$. Similarly, writing $\boldsymbol{\mu}^0 = (0, \ldots, 0, \mu_{r+1}, \ldots, \mu_k)$, one has

$$\sum_{i=1}^{r} Y_i^2 + \sum_{i=k+1}^{n} Y_i^2 = \min_{\boldsymbol{\mu}^0} \|\mathbf{Y} - E\mathbf{Y}\|^2 = \min_{\mathbf{m} \in \mathscr{L}_{c,0}} \|\mathbf{X} - \mathbf{m}\|^2 = \|\mathbf{X} - \hat{\hat{\boldsymbol{m}}}\|^2,$$

where $\hat{\hat{\boldsymbol{m}}}$ is the projection of $\mathbf{X}$ on $\mathscr{L}_{c,0}$, $\hat{\hat{\boldsymbol{m}}} = A\hat{\hat{\boldsymbol{\theta}}}$, say. Hence, by Pythagoras,

$$\sum_{i=1}^{r} Y_i^2 = \|\mathbf{X} - \hat{\hat{\mathbf{m}}}\|^2 - \|\mathbf{X} - \hat{\mathbf{m}}\|^2 = \|\hat{\mathbf{m}} - \hat{\hat{\mathbf{m}}}\|^2. \tag{5.148}$$

Therefore, the UMP invariant test of size α for $H_0 : B\boldsymbol{\theta} = 0$ against $H_1 : B\boldsymbol{\theta} \neq 0$ is given by (See Theorem 5.8)

$$\text{Reject } H_0 \text{ iff } \quad \frac{\|\hat{\mathbf{m}} - \hat{\hat{\mathbf{m}}}\|^2 / r}{\|\mathbf{X} - \hat{\mathbf{m}}\|^2 / (n-k)} \equiv \frac{\|A(\hat{\boldsymbol{\theta}} - \hat{\hat{\boldsymbol{\theta}}})\|^2 / r}{\|\mathbf{X} - A\hat{\boldsymbol{\theta}}\|^2 / (n-k)} > F_{1-\alpha}(r, n-k),$$

$$\tag{5.149}$$

where $\hat{\boldsymbol{\theta}}, \hat{\hat{\boldsymbol{\theta}}}$ are, respectively, the least squares estimators of $\boldsymbol{\theta}$ under the linear model (5.126) and under H_0.

Example 5.18 (One-Way Layout, or the k-Sample Problem). For $k \geq 2$, let X_{ji} be independent $N(\theta_i, \sigma^2)$ random variables ($1 \leq j \leq n_i$, $1 \leq i \leq k$). We wish to test $H_0 : \theta_1 = \theta_2 = \cdots = \theta_k$ against H_1 which says H_0 is not true. Here $n = \sum_1^k n_i$, $r = k - 1$, and $\mathbf{X} = A\boldsymbol{\theta} + \boldsymbol{\epsilon}$, where $a_{ji} = 1$ for $n_1 + \cdots + n_{i-1} < j \leq n_1 + \cdots + n_i$ ($2 \leq i \leq k$), $a_{j1} = 1$ for $1 \leq j \leq n_1$: $a_{ji} = 0$ otherwise. The minimum of $\|\mathbf{X} - A\boldsymbol{\theta}\|^2 \equiv \sum_{i=1}^{k} \sum_{j=1}^{n_i} (X_{ji} - \theta_i)^2$ over all $\boldsymbol{\theta} \in \mathbb{R}^k$ is attained by taking $\theta_i = \hat{\theta}_i = \overline{X}_{\cdot i} \equiv \frac{1}{n_i} \sum_{j=1}^{n_i} X_{ji}$ ($1 \leq i \leq k$). Under H_0, letting θ_0 denote the common value of θ_i's, the minimum value of $\sum_{i=1}^{k} \sum_{j=1}^{n_i} (X_{ji} - \theta_0)^2$ is attained by $\theta_0 = \hat{\theta}_0 = \frac{1}{n} \sum_{j,i} X_{ji} = \overline{X}_{\cdot\cdot}$, say. Hence the test (5.149) becomes, in view of (5.147), (5.148),

$$\text{Reject } H_0 \text{ iff } \quad \frac{\sum_{i=1}^{k} n_i (\overline{X}_{\cdot i} - \overline{X}_{\cdot\cdot})^2 / (k-1)}{\sum_{i=1}^{k} \sum_{j=1}^{n_i} (X_{ji} - \overline{X}_{\cdot i})^2 / (n-k)} > F_{1-\alpha}(k-1, n-k). \tag{5.150}$$

Example 5.19 (Two-Way Layout). Consider an agricultural field experiment in which certain *varieties* of a crop, say wheat, is grown on plots of equal size treated with different *fertilizers*. The objective is to study the effect of variety and fertilizer on the yield of the crop. Let X_{ijq} denote the yield of the q-th plot treated with fertilizer j on which the ith variety is grown ($q = 1, \ldots, S; j = 1, 2, \ldots, J; i = 1, 2, \ldots, I$), $I > 1$, $J > 1$, $S > 1$. Assume the linear model

$$X_{ijq} = \theta_{ij} + \varepsilon_{ijq}, \tag{5.151}$$

where ε_{ijq} are independent $N(0, \sigma^2)$ random variables. To test various hypotheses, it is useful to express θ_{ij} as

$$\theta_{ij} = \mu + \delta_i + \gamma_j + \eta_{ij} \qquad (1 \le i \le I, 1 \le j \le J), \tag{5.152}$$

where with $\overline{\theta}_{..} = \frac{1}{IJ} \sum_{i=1}^{I} \sum_{j=1}^{J} \theta_{ij}$, $\overline{\theta}_{i.} = \frac{1}{J} \sum_{j=1}^{J} \theta_{ij}$, $\overline{\theta}_{.j} = \frac{1}{I} \sum_{i=1}^{I} \theta_{ij}$,

$$\mu = \overline{\theta}_{..}, \quad \delta_i = \overline{\theta}_{i.} - \overline{\theta}_{..}, \quad \gamma_j = \overline{\theta}_{.j} - \overline{\theta}_{..}, \quad \eta_{ij} = \theta_{ij} - \overline{\theta}_{i.} - \overline{\theta}_{.j} + \overline{\theta}_{..}. \tag{5.153}$$

The quantities δ_i, γ_j are called the *main effects* (of variety i and fertilizer j, respectively), and η_{ij} are the *interactions*. The presence of non-zero interactions indicate that some varieties yield more in the presence of certain fertilizers than with others even after the main effects are accounted for. In analogy with continuous variables x, y (in place of i, j), the presence of interactions indicates a *nonlinear dependence* of the mean yield θ on the two variables—variety and fertilizer. Note that, in the parametrization (5.152), the following restrictions hold:

$$\mu \in \mathbb{R}, \quad \sum_{i=1}^{I} \delta_i = 0, \quad \sum_{j=1}^{J} \gamma_j = 0, \quad \sum_{j=1}^{J} \eta_{ij} = 0 \quad \forall\, i,$$

$$\sum_{i=1}^{I} \eta_{ij} = 0 \quad \forall\, j. \tag{5.154}$$

We consider several tests of interest.

(a) *Test of Equality of all θ_{ij}*: $H_0 : \theta_{ij} = \mu \ \forall\, i, j$. In this case, under H_0, the minimum value of $\sum_{i,j,q}(X_{ijq} - \mu)^2$ is attained by setting $\mu = \overline{X}_{...} = \frac{1}{n} \sum_{i,j,q} X_{ijq}$, where $n = SIJ$. Also write $\overline{X}_{i..} = \frac{1}{SJ} \sum_{j,q} X_{ijq}$, $\overline{X}_{.j.} = \frac{1}{SI} \sum_{i,q} X_{ijq}$.
   Under the general model (5.151) (or, (5.152)), the minimum value of $\sum_{i,j,q}(X_{ijq} - \theta_{ij})^2$ is attained by taking $\theta_{ij} = \hat{\theta}_{ij} = \overline{X}_{ij.} \equiv \frac{1}{S} \sum_q X_{ijq}$. Hence, with $k = IJ$ and $r = IJ - 1$, the test is to

$$\text{Reject } H_0 \text{ iff } \quad \frac{\sum_{i=1}^{I} \sum_{j=1}^{J} (\overline{X}_{ij.} - \overline{X}_{...})^2/(IJ - 1)}{\sum_{i,j,q}(X_{ijq} - \overline{X}_{ij.})^2/(n - IJ)} > F_{1-\alpha}(IJ - 1, n - IJ).$$

$$\tag{5.155}$$

   Note that this is the same test as the k-sample test of the preceding example, with $k = IJ$.

(b) *Test of Absence of Variety Main Effect:* $H_0 : \delta_i = 0 \ \forall\, i$. For this, and for the cases (c), (d) below, it is convenient to express $\sum_{i,j,q}(X_{ijq} - \theta_{ij})^2$ as

$$\|\mathbf{X} - \mathbf{m}\|^2 = \sum_{i,j,q}(X_{ijq} - \mu - \delta_i - \gamma_j - \eta_{ij})^2$$

$$= \sum_{i,j,q}(X_{ijq} - \overline{X}_{ij.})^2 + \sum_{i,j,q}(\overline{X}_{ij.} - \overline{X}_{i..} - \overline{X}_{.j.} + \overline{X}_{...} - \eta_{ij})^2$$

$$+ \sum_{i,j,q}(\overline{X}_{i..} - \overline{X}_{...} - \delta_i)^2 + \sum_{i,j,q}(\overline{X}_{.j.} - \overline{X}_{...} - \gamma_j)^2 + \sum_{i,j,q}(\overline{X}_{...} - \mu)^2.$$

$$(5.156)$$

That (5.156) is an identity follows from the representation

$$X_{ijq} - \mu - \delta_i - \gamma_j - \eta_{ij} = (X_{ijq} - \overline{X}_{ij.}) + (\overline{X}_{ij.} - \overline{X}_{i..} - \overline{X}_{.j.} + \overline{X}_{...} - \eta_{ij})$$
$$+ (\overline{X}_{i..} - \overline{X}_{...} - \delta_i) + (\overline{X}_{.j.} - \overline{X}_{...} - \gamma_j) + (\overline{X}_{...} - \mu), \quad (5.157)$$

noting that the sum (over i, j, q) of products of any two terms among the five terms on the right vanishes. Now the minimum of the sum of squares in (5.156), subject to $\delta_i = 0 \; \forall \; i$, is attained by setting $\eta_{ij} = \overline{X}_{ij.} - \overline{X}_{i..} - \overline{X}_{.j.} + \overline{X}_{...}$, $\gamma_j = \overline{X}_{.j.} - \overline{X}_{...}$, $\mu = \overline{X}_{...}$. Thus the minimum sum of squares under H_0 is attained by setting

$$\theta_{ij} = \hat{\hat{\theta}}_{ij} = \overline{X}_{...} + \overline{X}_{.j.} - \overline{X}_{...} + \overline{X}_{ij.} - \overline{X}_{i..} - \overline{X}_{.j.} + \overline{X}_{...}$$
$$= \overline{X}_{ij.} - \overline{X}_{i..} + \overline{X}_{...} \qquad (1 \le i \le I, \; 1 \le j \le J).$$

Hence

$$\|\hat{\boldsymbol{m}} - \hat{\hat{\boldsymbol{m}}}\|^2 = \sum_{i,j,q}(\hat{\theta}_{ij} - \hat{\hat{\theta}}_{ij})^2 = SJ\sum_{i=1}^{I}(\overline{X}_{i..} - \overline{X}_{...})^2.$$

Thus the test is:

Reject H_0 iff $\dfrac{\sum_{i=1}^{I} SJ(\overline{X}_{i..} - \overline{X}_{...})^2/(I-1)}{\sum_{i,j,q}(X_{ijq} - \overline{X}_{ij.})^2/(n-IJ)} > F_{1-\alpha}(I-1, n-IJ).$ (5.158)

(c) The test for $H_0 : \gamma_j = 0 \; \forall \; j$ is entirely analogous to the case (b):

Reject H_0 iff $\dfrac{\sum_{j=1}^{J} SI(\overline{X}_{.j.} - \overline{X}_{...})^2/(J-1)}{\sum_{i,j,q}(X_{ijq} - \overline{X}_{ij.})^2/(n-IJ)} > F_{1-\alpha}(J-1, n-IJ).$ (5.159)

(d) Finally, we consider the test for the *absence of interactions:* $H_0 : \eta_{ij} = 0 \; \forall \; i, j$. The minimum value of the sum of squares (5.156) is attained under H_0 by setting $\delta_i = \overline{X}_{i..} - \overline{X}_{...}$, $\gamma_j = \overline{X}_{.j.} - \overline{X}_{...}$, $\mu = \overline{X}_{...}$. That is, $\hat{\hat{\theta}}_{ij} = \overline{X}_{...} + \overline{X}_{i..} - \overline{X}_{...} + \overline{X}_{.j.} - \overline{X}_{...} = \overline{X}_{i..} + \overline{X}_{.j.} - \overline{X}_{...}$. Hence

$$\|\hat{\boldsymbol{m}} - \hat{\hat{\boldsymbol{m}}}\|^2 = S\sum_{i,j}(\overline{X}_{ij.} - \overline{X}_{i..} - \overline{X}_{.j.} + \overline{X}_{...})^2,$$

so that the test is:

Reject H_0 iff $\dfrac{S\sum_{i,j}(\overline{X}_{ij.} - \overline{X}_{i..} - \overline{X}_{.j.} + \overline{X}_{...})^2/(I-1)(J-1)}{\sum_{i,j,q}(X_{ijq} - \overline{X}_{ij.})^2/(n-IJ)}.$

Note that there are $(I-1)(J-1)$ linearly independent functions of $\boldsymbol{\theta} = (\theta_{ij} : 1 \le i \le I, \; 1 \le j \le J)$ among $\{\eta_{ij}, 1 \le i \le I, \; 1 \le j \le J\}$.

Remark 5.12. If in the two-way layout above $s = 1$, then the denominators of the tests (a)–(d) all vanish. This is due to the fact that in this case the least squares estimators $\hat{\theta}_{ij} = X_{ij}$ take up all the observations and it is not possible to estimate σ^2. If one assumes $\eta_{ij} = 0 \; \forall \; i, j$, then one can still find UMP invariant tests for (a)–(c) (Exercise 5.17).

Example 5.20 (UMPU Invariant Tests for Regression). Consider the multiple regression model

$$X_i = \alpha + \sum_{j=1}^{p} \beta_j z_{ji} + \varepsilon_i \quad (i = 1, \ldots, n), \; p + 1 < n, \tag{5.160}$$

where $\beta_1, \ldots, \beta_p$ are unknown regression coefficients, $\beta_i \in \mathbb{R} \; \forall i$, and $Z = ((z_{ji}))_{1 \leq j \leq p, 1 \leq i \leq n}$ a known design matrix of full rank p. As usual, assume ε_i's are i.i.d. $N(0, \sigma^2)$, $\sigma^2 > 0$ is unknown. We wish to explore the effects of the p *regressors* or *independent variables*, z_j, $1 \leq j \leq p$, on the *predictor* or *dependent variable* X, assuming that the relationship is linear.

(a) **Test $H_0 : \beta_1 = \cdots = \beta_p = 0$,** i.e., the regressors have no effect on X. To simplify the computation it is best to rewrite the model (5.160) as the equivalent model

$$X_i = \delta + \sum_{j=1}^{p} \beta_j (z_{ji} - \overline{z}_{j.}) + \varepsilon_i, \tag{5.161}$$

$$\delta := \alpha + \sum_{j=1}^{p} \beta_j \overline{z}_{j.}, \overline{z}_{j.} = \frac{1}{n} \sum_{i=1}^{n} z_{ji} \quad (1 \leq j \leq p).$$

Here $k = p + 1$, and to minimize

$$\|\mathbf{X} - E\mathbf{X}\|^2 = \sum_{i=1}^{n}(X_i - \delta - \sum_{j=1}^{p} \beta_j (z_{ji} - \overline{z}_{j.}))^2,$$

differentiate with respect to these parameters to obtain

$$\sum_{i=1}^{n}(X_i - \delta - \sum_{j=1}^{p} \beta_j (z_{ji} - \overline{z}_{j.})) = 0,$$

$$\sum_{i=1}^{n}(z_{j'i} - \overline{z}_{j'.}) \left(\sum_{i=1}^{n} X_i - \delta - \sum_{j=1}^{p} \beta_j (z_{ji} - \overline{z}_{j.}) \right) = 0, \; 1 \leq j' \leq p. \tag{5.162}$$

The first equation yields the solution

$$\delta = \hat{\delta} = \overline{X} = \frac{1}{n} \sum_{i=1}^{n} X_i, \tag{5.163}$$

while the remaining p equations yield

$$S_{xj} \equiv \sum_{i=1}^{n}(z_{j'i} - \overline{z}_{j'.})(X_i - \overline{X}) = \sum_{j=1}^{p} S_{j'j}\beta_{j'},$$

$$S_{j'j} := \sum_{i=1}^{n}(z_{j'i} - \overline{z}_{j'.})(z_{ji} - \overline{z}_{j.}), \quad 1 \le j' \le p. \tag{5.164}$$

Hence, in matrix notation, the solution $\hat{\boldsymbol{\beta}}$ of the test p equations may be expressed as

$$\hat{\boldsymbol{\beta}} = \left(\hat{\beta}_1, \ldots, \hat{\beta}_p\right)' = S^{-1}\mathbf{S}_x,$$

$$S = ((S_{j'j})), \quad \mathbf{S}_x = (S_{x1}, \ldots, S_{xp})'. \tag{5.165}$$

Together (5.161) and (5.165) provide the solution to α as

$$\hat{\alpha} = \hat{\delta} - \sum_{j=1}^{p} \hat{\beta}_j \overline{z}_{j.} = \overline{X} - \sum_{j=1}^{p} \hat{\beta}_j \overline{Z}_{j.}. \tag{5.166}$$

Hence, writing $S_{xx} = \sum_{i=1}^{n}(X_i - \overline{X})^2$,

$$\begin{aligned}
\min_{EX} \|\mathbf{X} - E\mathbf{X}\|^2 &= \sum_{i=1}^{n}(X_i - \overline{X})^2 + \sum_{j',j=1}^{p} \hat{\beta}_{j'}\hat{\beta}_j S_{j'j} - 2\sum_{j=1}^{p} \hat{\beta}_j S_{xj} \\
&= \sum_{i=1}^{n}(X_i - \overline{X})^2 + \hat{\beta}' S\hat{\beta} - 2\hat{\beta}'\mathbf{S}_x \\
&= \sum_{i=1}^{n}(X_i - \overline{X})^2 + \mathbf{S}_x' S^{-1}\mathbf{S}_x - 2\mathbf{S}_x' S^{-1}\mathbf{S}_x \\
&= S_{xx} - \mathbf{S}_x S^{-1}\mathbf{S}_x \\
&= \sum_{i=1}^{n}(X_i - \overline{X})^2 - \sum_{j,j'=1}^{p} S_{jj'}\hat{\beta}_j\hat{\beta}_{j'}.
\end{aligned} \tag{5.167}$$

Next, under H_0, $EX_x = \delta = \alpha$, and $\|\mathbf{X} - E\mathbf{X}\|^2 = \sum_{i=1}^{n}(X_i - \delta)^2$ is minimized by $\delta = \hat{\delta} = \overline{X}$, so that

$$\min_{E\mathbf{X} \text{ under } H_0} \|\mathbf{X} - E\mathbf{X}\|^2 = \sum_{i=1}^{n}(X_i - \overline{X})^2 = S_{xx}. \tag{5.168}$$

Hence the UMPU invariant test for (a) is to

reject H_0 iff $\quad \dfrac{\mathbf{S}_x' S^{-1}\mathbf{S}_x / p}{[S_{xx} - \mathbf{S}_x' S^{-1}\mathbf{S}_x]/(n - p - 1)} \to F_{1-\alpha}(p, n - p - 1). \tag{5.169}$

It may be noted that $\mathbf{S}_x' S^{-1}\mathbf{S}_x = \sum_{j,j'=1}^{p} S^{jj'}\hat{\beta}_j\hat{\beta}_{j'}$, where $((S^{jj'})) = S^{-1}$.

(b) Test $\boldsymbol{H_0 : \beta_{q+1} = \cdots = \beta_p = 0}$, where $1 \le q \le p - 1$. This is to test if some of the independent variables Z_j have no effect on the predictor (or dependent) variable X. Proceeding as in (5.160)–(5.167), but with p replaced by q, one has the least squares estimate $\hat{\boldsymbol{\beta}}_1^q = (\hat{\beta}_1, \ldots, \hat{\beta}_q)'$ given by

$$\hat{\boldsymbol{\beta}}_1^q = (S_1^q)^{-1}\mathbf{S}_{x,1}^q, \quad S_1^q := ((S_{j'j}))_{1\le j,j'\le q}, \quad \mathbf{S}_{x,1}^q := (S_{x,1}, \ldots, S_{x,q})'. \tag{5.170}$$

Also, $\hat{\hat{\delta}} = \overline{X}$, and

$$\hat{\hat{\alpha}} = \overline{X} - \sum_{j=1}^{q} \hat{\hat{\beta}}_j \overline{z}_j. \tag{5.171}$$

Then

$$\min_{E\mathbf{X} \text{ under } H_0} \|\mathbf{X} - E\mathbf{X}\|^2 = S_{xx} - (\mathbf{S}_{x,1}^q)'(S_1^q)^{-1}\mathbf{S}_{x,1}^q. \tag{5.172}$$

The F test is now to

$$\text{reject } H_0 \text{ iff } \frac{[\mathbf{S}_x'S^{-1}\mathbf{S}_x - (\mathbf{S}_{x,1}^q)'(S_1^q)^{-1}\mathbf{S}_{x,1}^q]/(p-q)}{[S_{xx} - \mathbf{S}_x'S^{-1}\mathbf{S}_x]/(n-p-1)} > F_{1-\alpha}(p-q, n-p-1).$$

(c) For the test $\boldsymbol{H_0 : \alpha = 0}$, the sum of squares under H_0 to be minimized is

$$\|\mathbf{X} - E\mathbf{X}\|^2 = \sum_{i=1}^{n}(X_i - \sum_{j=1}^{p}\beta_j z_{ji})^2, \tag{5.173}$$

and the minimizer is $\hat{\hat{\beta}} = (\hat{\hat{\beta}}_1, \ldots, \hat{\hat{\beta}}_p)'$ obtained by setting zero to the derivatives of (5.173) with respect to β_j $(1 \le j \le p)$. This gives

$$\hat{\hat{\beta}} = S_0^{-1}\mathbf{S}_{0x}, \quad S_0 := ((S_{0jj'}))_{1 \le j,j' \le p}, \quad \mathbf{S}_{0x} := (X_{0x_1}, \ldots, S_{0x_p})', \tag{5.174}$$

where

$$S_{0jj'} = \sum_{i=1}^{n} Z_{ji}Z_{j'i}, \quad S_{0xj} = \sum_{i=1}^{n} X_i z_{ji} \ (1 \le j \le p). \tag{5.175}$$

Therefore, writing $S_{0xx} = \sum_{i=1}^{n} X_i^2$, $S_0^{-1} = ((S_0^{jj'}))$, one has

$$\min_{E\mathbf{X} \text{ under } H_0} \|\mathbf{X} - E\mathbf{X}\|^2 = S_{Oxx} - \mathbf{S}_{Ox}'S_0^{-1}\mathbf{S}_{Ox} = \sum_{i=1}^{n} X_i^2 - \sum_{j,j'=1}^{p} S_O^{jj'} S_{Oxj}S_{Oxj'}. \tag{5.176}$$

This leads to the UMPU invariant test for H_0 which would

$$\text{reject } H_0 \text{ iff } \frac{n\overline{x}^2 + \mathbf{S}_x'S^{-1}\mathbf{S}_x - S_{Ox}'S_O^{-1}\mathbf{S}_{Ox}}{[S_{xx} - \mathbf{S}_x'S^{-1}\mathbf{S}_x]/(n-p-1)} > F_{1-\alpha}(1, n-p-1). \tag{5.177}$$

Note that one may similarly test (a)$_e$ $H_0 : \beta_j = c_j$ for arbitrarily given constants c_j $(1 \le j \le p)$, (b)$_d$ $H_0 : \beta_j = d_j$ for given d_j $(1 \le j \le q)$, and (c)$_{a_0}$ $H_0 : \alpha = a_0$ for given a_0.

Remark 5.13 (ANOVA—Analysis of Variance). In Examples 5.18, 5.19, a convenient and intuitively meaningful tabulation of calculations of the various tests is by the so-called ANOVA-table, ANOVA being the abbreviation for analysis of variance. The idea is to consider the *total variability* in the data, namely, the *total sum of squares* (Total SS) of all the observed X-values around the *grand total*. This Total SS into sums of squares of orthogonal components. For example, for Example 5.18, the ANOVA Table is

Source	DF	SS	MS	F
Treatments	$k-1$	SST	$MST = SST/(k-1)$	MST/MSE
Error	$N-k$	SSE	$MSE = SSE(N-k)$	
Total	$N-1$	Total SS		

Here DF is *degrees of freedom*, i.e., the number of linearly independent contrasts or comparisons which can be attributed to the particular source of variation; SS is the sum of squares and MS stands for mean squares. The *treatment sum of squares* SSI equals $\sum_{i=1}^{k} n_i(\overline{X}_{i\cdot} - \overline{X}_{\cdot\cdot})^2$, the Total sum of squares is $\sum_{i=1}^{k}\sum_{j=1}^{n_i}(X_{ij} - \overline{X}_{\cdot\cdot})^2$, and the error sum of squares is $SSE = $ Total $SS - SST = \sum_{i=1}^{k}\sum_{j=1}^{n_i}(X_{ij} - \overline{X}_{i\cdot})^2$.

For the two-way layout of Example 5.19, the ANOVA table is

Source	DF	SS	MS	F
Variety	$I-1$	SSV	$SSV/(I-1)$	MSV/MSE
Fertilizer	$J-I$	SSF	$SSF/(J-1)$	MSF/MSE
Interaction	$(I-1)\times(J-1)$	SSI	$SSI(I-1)(J-1)$	MSI/MSE
Error	$n-IJ$	SSE	$SSE/(n-IJ)$	
Total	$IFS-1=n-1$	Total		

Here $SSV = JS\sum_{i=1}^{I}(\overline{X}_{i\cdot\cdot} - \overline{X}_{\cdots})^2$, $SSF = IS\sum_{j=1}^{J}(\overline{X}_{\cdot j\cdot} - \overline{X}_{\cdots})^2$, $SSI = S\sum_{i=1}^{I}\sum_{j=1}^{J}(\overline{X}_{ij\cdot} - \overline{X}_{i\cdot\cdot} - \overline{X}_{\cdot j\cdot} + \overline{X}_{\cdots})^2$, Total $SS = \sum_{i,j,q}(X_{ijq} - \overline{X}_{\cdots})^2$, and $SSE = $ Total $SS - SSV - SSF - SSI$.

5.10 Notes and References

The presentation in this chapter follows (Ferguson, 1967, Chap. 5), which in turn is strongly influenced by Lehmann (1959). The theory in Sects. 5.1–5.7 is mostly due to Neyman and Pearson (1933, 1936, 1938). Theorem 5.5 is due to Basu (1955). Pitman (1939) introduced the notion of invariance in problems of estimation of location and scale, and this was generalized in broader estimation problems by Kiefer (1957). As mentioned by Lehmann (1959, p. 261), the general theory of invariant tests is due to Hunt and Stein (1946). Analysis of variance introduced and presented in Fisher (1925, 1935) has had a profound impact on parametric statistics especially in the study of so-called linear models and multivariate analysis. Rao (1952), Scheffé (1959) and Anderson (1958) are early standard references on these topics. Fisher's efforts at construction of designs that allow proper and/or optimal statistical analysis of agricultural experiments in the 1930s and 1940s led to the creation of a theory by him and other statisticians with wide applications to the construction of error correcting codes in information theory, such as the famous *Bose–Chaudhuri–Hocquenghem code* (Bose and Ray-Chaudhuri (1960), Hocquenghem (1959)). Many of its applications to combinatorial mathematics included the settling in 1959 in the negative by R.C. Bose, S.S. Shrikhande and, independently, by E.T. Parker (Bose et al. (1960)), of the famous conjecture of Euler on the construction of certain *Latin squares*.

Exercises for Chap. 5

Ex. 5.1. Using the theory of UMP tests for $H_0 : \pi \leq \pi_0$ against $H_1 : \pi > \pi_0$ in one-parameter exponential families (See Sect. 5.3, and Remarks 5.2, 5.3), show that the test (5.26) is UMP of its size for testing $H_0 : \pi \geq \pi_0$ against $H_1 : \pi < \pi_0$.

Ex. 5.2. Given $\mathbf{X} = (X_1, \ldots, X_n)$ with X_j's i.i.d. gamma $\mathscr{G}(\theta, \beta)$, $\beta > 0$ known (as in Example 5.4), find the UMP test of size α ($0 < a < 1$) for $H_0 : \theta \leq \theta_0$ against $H_1 : \theta > \theta_0$, where $\theta_0 > 0$ is a given (threshold) number.

Ex. 5.3. Use Remark 5.4 to find UMP tests of size α for testing $H_0 : \theta \leq \theta_0$ against $H_1 : \theta > \theta_0$ in the following examples, based on i.i.d. observations $X_1, \ldots, X_n$.

(a) X_j has the double exponential distribution with p.d.f. $f(x \mid \theta) = (2\alpha)^{-1} \exp\{-|x - \theta|/\alpha\}$ ($x \in \mathbb{R}$), $\theta \in \mathbb{R} = \Theta$ ($\alpha > 0$ known).
(b) X_j has the shifted exponential distribution with p.d.f. $f(x \mid \theta) = \exp\{-x(x - \theta)\}$ ($x > \theta$), and $f(x \mid \theta) = 0$ ($x \leq \theta$). Here $\Theta = \mathbb{R}$.
(c) X_j has the uniform distribution $\mathscr{U}(0, \theta)$ on $(0, \theta)$, $\theta \in (0, \infty) = \Theta$.
(d) X_j has the uniform distribution $\mathscr{U}(\theta, \theta + 1)$ on $(\theta, \theta + 1)$, $\theta \in \Theta = \mathbb{R}$.

Ex. 5.4. (a) Justify the interchange of the order of differentiation (w.r.t. π) and integration (w.r.t. μ) in (5.34).
(b) Show that the UMP unbiased test in Example 5.1 for

$$H_0 : \theta = \theta_0, \quad H_1 : \theta \neq \theta_0 \tag{5.178}$$

is given by

$$\text{Reject } H_0 \text{ iff } \quad \left| \frac{\sum_1^n X_j - n\theta_0}{\sqrt{n}} \right| > z_{1-\frac{\alpha}{2}} .$$

More generally, if the model is $\mathbf{N}(\theta, \sigma_0^2)$, *with* $\sigma_0^2 > 0$ *known*, then the UMP unbiased test for (5.178) is given by

$$\text{Reject } H_0 \text{ iff } \quad \left| \frac{\sqrt{n}(\overline{X} - \theta_0)}{\sigma_0} \right| > z_{1-\frac{\alpha}{2}} .$$

[Hint: Take $T = \frac{\sum_1^n X_j - n\theta_0}{\sigma_0}$, and apply (5.42), (5.43).]

Ex. 5.5. In the Normal example $\mathbf{N}(\mu_0, \theta)$, show that the UMP unbiased test of size α for $H_0 : \theta = \sigma_0^2$, $H_1 : \theta \neq \sigma_0^2$, is given by

$$\text{Reject } H_0 \text{ iff } \quad \sum_{j=1}^n \frac{(X_j - \mu_0)^2}{2\sigma_0^2} < t_1 \quad \text{or} \quad > t_2,$$

where t_1 and t_2 are determined by the first condition in (5.47) and (5.48) (both with $\beta = \frac{1}{2}$). [Hint: Consider observations $(X_j - \mu_0)/\sigma_0$ ($1 \leq j \leq n$), and apply Example 5.8.]

Ex. 5.6. Let X_j ($1 \leq j \leq n$) be i.i.d. with common density (w.r.t. Lebesgue measure on $(0, 1)$)

$$f(x \mid \theta) = \theta x^{\theta - 1} \quad 0 < x < 1, \quad \theta \in \Theta = (0, \infty).$$

Find the UMP unbiased test of size α, $0 < \alpha < 1$, for testing $H_0 : \theta = 1$ against $H_1 : \theta \neq 1$.

Ex. 5.7. In Example 5.6, show that

(a) the UMP unbiased test of size α for $H_0 : \mu \leq \mu_0$, $H_1 : \mu > \mu_0$, is given by:
Reject H_0 iff $\frac{\sqrt{n}(\overline{X} - \mu_0)}{s} > t_{1-\alpha}(n-1)$,

(b) the UMP unbiased test of size α for $H_0 : \mu \geq \mu_0$, $H_1 : \mu < \mu_0$, is given by:
Reject H_0 iff $\frac{\sqrt{n}(\overline{X} - \mu_0)}{s} < -t_{1-\alpha}(n-1)$,

(c) the UMP unbiased test of size α for $H_0 : \mu = \mu_0$, $H_1 : \mu \neq \mu_0$, is given by:
Reject H_0 if $\left| \frac{\sqrt{n}(\overline{X} - \mu_0)}{s} \right| > t_{1-\frac{\alpha}{2}}(n-1)$.

Ex. 5.8. (a) Suppose $X_1, \ldots, X_n$ are i.i.d. observations from $N(\mu_0, \sigma^2)$, with μ_0 known. Find the UMP test of size α for testing $H_0 : \sigma^2 \leq \sigma_0^2$ against $H_1 : \sigma^2 > \sigma_0^2$, where $\sigma_0^2 > 0$ is given.

In Example 5.7, show that

(b) the UMP unbiased test of size α for $H_0 : \sigma^2 \geq \sigma_0^2$, $H_1 : \sigma^2 < \sigma_0^2$ is given by: Reject H_0 iff $\sum_1^n (X_j - \overline{X})^2 / \sigma_0^2 < \chi_\alpha^2$, the αth quantile of the chi-square distribution with $n-1$ d.f.

(c) the UMP unbiased test of size α for $H_0 : \sigma^2 = \sigma_0^2$, $H_1 : \sigma^2 \neq \sigma_0^2$ is given by: Reject H_0 iff $\sum_1^n (X_j - \overline{X})^2 < c_1 \sigma_0^2$ or $> c_2 \sigma_0^2$ where $0 < c_1 < c_2$ are determined by the equations (see Sect. 5.4, Examples 5.4, 5.5)

$$\int_{C_1}^{C_2} g(t)dt = 1 - \alpha, \qquad \left(\frac{c_2}{c_1} \right)^{(n-1)/2} = e^{(c_2 - c_1)/2}, \qquad (5.179)$$

where $g(t)$ is the density of the chi-square distribution with $n-1$ d.f.

Ex. 5.9. Let $X_1, \ldots, X_m$ and $Y_1, Y_2, \ldots, Y_n$ be independent random samples from exponential distributions with means θ_1, θ_2 respectively (i.e., from $\mathscr{G}(\theta_1, 1)$ and $\mathscr{G}(\theta_2, 1)$). Find the UMP unbiased test of size α for $H_0 : \theta_1 \leq \theta_2$ against $H_1 : \theta_1 > \theta_2$.

Ex. 5.10. Let U_1 and U_2 be independent gamma random variables $\mathscr{G}(\theta, m)$ and $\mathscr{G}(\theta, n)$. Prove that $Z_1 \equiv U_1/(U_1 + U_2)$ and $Z_2 \equiv U_1 + U_2$ are independent random variables with Z_1 having the beta distribution $\text{Beta}(m, n)$ and Z_2 have the gamma distribution $\mathscr{G}(\theta, m+n)$.

Ex. 5.11. (a) Let $X_1, \ldots, X_m$ and $Y_1, Y_2, \ldots, Y_n$ be independent random samples from $\mathbf{N}(\mu_1, \sigma_1^2)$ and $\mathbf{N}(\mu_2, \sigma_2^2)$, respectively, where $\sigma_i^2 > 0$ $(i = 1, 2)$ are known, and $(\mu_1, \mu_2) \in \mathbb{R}^2$. Find a UMP unbiased test of size $\alpha \in (0, 1)$ for (5.78).

(b) In the context of Example 5.8, find the UMP unbiased test of size α for (i) $H_0 : \mu_1 \geq \mu_2$, $H_1 : \mu_1 < \mu_2$, and for (ii) $H_0 : \mu_1 = \mu_2$, $H_1 : \mu_1 \neq \mu_2$.

(c) Extend (b) to the case (a).

Ex. 5.12. In Example 5.10, find the UMP unbiased test of size $\alpha \in (0, 1)$ for

$$H_0 : \frac{\sigma_2^2}{\sigma_1^2} \leq \gamma_0, \quad H_1 : \frac{\sigma_2^2}{\sigma_1^2} > \gamma_0,$$

for a given $\gamma_0 > 0$. [Hint: change Y_j to Y_j/γ_0, $1 \leq j \leq n$, in Example 5.10.]

Ex. 5.13. Let $\mathbf{X} = (X_1, \ldots, X_{20})$, where X_j, $1 \leq j \leq 20$, are i.i.d. Bernoulli $\mathscr{B}(\theta)$, $\theta \in \Theta = (0,1)$. Construct the UMPU test of size α for $H_0 : \theta = 0.5$, against $H_1 : \theta \neq 0.5$. [Hint: By symmetry, the test should be of the form: $\varphi(\mathbf{x}) = 1$ if $\sum_1^{20} x_j < 10 - r$ or $> 10 + r$, $\varphi(\mathbf{x}) = \gamma$ if $\sum_1^{20} x_j = 10 - r$ or $10 + r$, and $\varphi(\mathbf{x}) = 0$ if $10 - r < \sum_1^{20} x_j < 10 + r$. Here r is a positive integer, and $0 \leq \gamma < 1$.]

Ex. 5.14 (Match Pair Test). Assume that (X_i, Y_i), $1 \leq i \leq n$, are i.i.d. Normal $N((\mu_x, \mu_y)^t, \Sigma)$, where Σ is positive definite and all the parameters are unknown.

(a) Prove that under the assumption $\sigma_x^2 = \sigma_y$, the match pair test for $H_0 : \mu_x = \mu_y$, $H_1 : \mu_x > \mu_y$ described in Example 5.11 is UMPU of size α.
(b) Prove the assertion that the match pair design is more efficient than the independent samples design, at least for sufficiently large n, provided $\rho > 0$.

Ex. 5.15. (a) Show that the map $\overline{g} : \Theta \to \Theta$ defined in Sect. 5.8 is one-to-one.
(b) (i) Show that the two-sample problem in Example 5.8 is invariant under the group $\mathscr{G}$ of all *translations* $g_a : \mathbb{R}^m \times \mathbb{R}^n \longrightarrow \mathbb{R}^m \times \mathbb{R}^n$, given by $g_a((x_1, \ldots, x_m, y_1, \ldots, y_n)) = (x_1 + a, \ldots, x_m + a, y_1 + a, \ldots, y_n + a)$, $a \in \mathbb{R}$.
(ii) Also, show that the UMPU test is also an invariant test.
(c) Show that the two-sample problem in Example 5.9 is invariant under the group $\mathscr{G}$ of transformations $g_{a,c} : \mathbb{R}^m \times \mathbb{R}^n \to \mathbb{R}^m \times \mathbb{R}^n$ given by $g_{a,c}((x_1, \ldots, x_m, y_1, \ldots, y_n)) = \left(\frac{x_1 + a}{c}, \ldots, \frac{x_m + a}{c}, \frac{y_1 + a}{c}, \ldots, \frac{y_n + a}{c}\right)$, $a \in \mathbb{R}$, $c > 0$. Show also that the UMPU test is invariant.

Ex. 5.16. Let $\mathbf{X}$ be k-dimensional Normal $N(\boldsymbol{\mu}, \Sigma)$ with Σ a known positive definite matrix.

(a) Find the UMA invariant confidence region for $\boldsymbol{\mu}$ using the UMP invariant test in Example 5.15.
(b) Find a UMA invariant confidence region for $\boldsymbol{\mu}$ (under the group as in (a)) based on n i.i.d. observations $\mathbf{X}_1, \ldots, \mathbf{X}_n$ with common distribution $N(\boldsymbol{\mu}, \Sigma)$).

Ex. 5.17 (Two-way Layout with One Observation Per Cell). In Example 5.19, let $S = 1$, and assume $\eta_{ij} = 0 \ \forall \ i, j$ (in addition to the other assumptions). Find the UMP invariant tests (a)–(c).

Ex. 5.18. In Example 5.19, let the number of replications, say S_{ij}, vary for different pairs (i, j), with $S_{ij} \geq 2 \ \forall \ (i, j)$. Carry out the tests (a)–(d) in this case.

References

Anderson, T. W. (1958). *An introduction to multivariate analysis* (1st ed.). New York: Wiley.

Basu, D. (1955). On statistics independent of a complete sufficient statistic. *Sankhya, 15,* 377–380.

Basu, D. (1959). The family of ancillary statistics. *Sankhya, 21,* 247–256.

Bose, R. C., & Ray-Chaudhuri, D. K. (1960). On a class of error-correcting binary codes. *Information and Control, 3,* 68–79.

Bose, R. C., Shrikhande, S. S., & Parker, E. T. (1960). Further results on the construction of mutually orthogonal Latin squares and the falsity of Euler's conjecture. *Canadian Journal of Mathematics, 12*, 189–203.

Ferguson, T. (1967). *Mathematical statistics: A decision theoretic approach.* Boston: Academic.

Fisher, R. A. (1925). *Statistical methods for research workers* (1st ed.). Edinburgh: Oliver and Boyd.

Fisher, R. A. (1935). *The design of experiments* (1st ed.). Edinburgh: Oliver and Boyd.

Hocquenghem, A. (1959). Codes correcteurs d'erreurs. *Chiffres (in French) (Paris), 2*, 147–156.

Hunt, G., & Stein, C. (1946). Most stringent tests of statistical hypotheses. Unpublished manuscript.

Kiefer, J. (1957). Invariance, minimax sequential estimation, and continuous time processes. *Annals of Mathematical Statistics, 28*, 573–601.

Lehmann, E. L. (1959). *Testing statistical hypotheses* (1st ed.). New York: Wiley.

Neyman, J., & Pearson, E. S. (1933). On the problem of the most efficient tests of statistical hypotheses. *Philosophical Transactions of the Royal Society of London. Series A, 231*, 289–337.

Neyman, J., & Pearson, E. S. (1936). Contributions to the theory of testing statistical hypotheses. Part I. Unbiased critical regions of type A and type A_1. *Statistical Research Memoirs, 1*, 1–37.

Neyman, J., & Pearson, E. S. (1938). Contributions to the theory of testing statistical hypotheses. Part II. Certain theorems on unbiased critical regions of type A. Part III. Unbiased tests of simple statistical hypotheses specifying the values of more than one unknown parameter. *Statistical Research Memoirs, 2*, 25–57.

Pitman, E. J. G. (1939). The estimation of location and scale parameters of a continuous population of any given form. *Biometrika, 30*, 391–421.

Rao, C. R. (1952). *Advanced statistical methods in biometric research.* New York: Wiley.

Scheffé, H. (1959). *Analysis of variance.* New York: Wiley.

Part II
Mathematical Statistics: Large Sample Theory

Chapter 6
Consistency and Asymptotic Distributions of Statistics

Abstract Notions of convergence for large sample theory are introduced: almost sure convergence, convergence in probability and convergence in distribution. Consistency of several classes of estimators and their asymptotic distributions are derived, including those of sample moments, quantiles and linear regression coefficients.

6.1 Introduction

Unlike the so-called exact sampling theory, where one needs to look for optimal estimators separately for each parametric function in every individual parametric family, in large sample theory there is a remarkable unity in the methodology for optimal estimators. Also, unlike the former, the optimality criterion is essentially uniquely and unambiguously specified and this (asymptotic) optimality is achieved under fairly general hypotheses. Indeed, one may say that the maximum likelihood estimator (MLE) is optimal in an asymptotic sense under broad assumptions. In addition, if nothing is known about the form of the distribution except perhaps that it has certain finite moments, or that it has a continuous and positive density over the effective range, one may still construct reasonably good estimators for important classes of population indices such as moments or quantiles. For example, one may use sample moments and sample quantiles as the respective estimators. The present chapter introduces the basic notions of convergence in large sample theory and develops some of its main tools. Asymptotics of sample moments and quantiles, and of semiparametric linear regression, are derived here, which allow one to construct nonparametric or semiparametric confidence regions or tests for the corresponding population parameters.

6.2 Almost Sure Convergence, Convergence in Probability and Consistency of Estimators

A minimal requirement of any reasonable estimator $U_n := U_n(X_1, \ldots, X_n)$ of a population parameter γ is that of *consistency*.

© Springer-Verlag New York 2016
R. Bhattacharya et al., *A Course in Mathematical Statistics and Large Sample Theory*, Springer Texts in Statistics, DOI 10.1007/978-1-4939-4032-5_6

Definition 6.1. A statistic U_n $(n \geq 1)$ is said to be a *consistent estimator* of a parameter γ, if U_n converges to γ in probability: $U_n \xrightarrow{P} \gamma$, i.e.,

$$P(|U_n - \gamma| > \varepsilon) \longrightarrow 0 \quad \text{as } n \to \infty, \text{ for every } \varepsilon > 0. \tag{6.1}$$

More generally, a sequence of random variables Y_n is said to *converge to a random variable Y in probability*, $Y_n \xrightarrow{P} Y$, if

$$P(|Y_n - Y| > \varepsilon) \longrightarrow 0 \quad \text{as } n \to \infty, \text{ for every } \varepsilon > 0. \tag{6.2}$$

In (6.1), Y is the constant random variable $Y = \gamma$.

A common method for proving consistency is the following.

Proposition 6.1. *(a) If, for some $r > 0, E|U_n - \gamma|^r \to 0$, then U_n is a consistent estimator of γ. (b) If U_n is an unbiased estimator of γ and $\mathrm{var}(U_n) \to 0$, then U_n is a consistent estimator of γ.*

Proof. (a) By Chebychev's Inequality (see note below), for every $\varepsilon > 0$,

$$P(|U_n - \gamma| \geq \varepsilon) \leq \frac{E|U_n - \gamma|^r}{\varepsilon^r} \longrightarrow 0 \quad \text{as } n \to \infty. \tag{6.3}$$

(b) is a special case of (a) with $r = 2$, and $\mathrm{var}(U_n) = E|U_n - \gamma|^2$. $\qquad\square$

Note: Let X be a random variable (e.g., $X = U_n - \gamma$) such that $E|X|^r < \infty$ for some $r > 0$, then writing $E(Y : A)$ for the *expectation of Y on the set A*, i.e., $E(Y : A) = \int_A Y dP$,

$$\begin{aligned}
E|X|^r &= E\left[|X|^r : |X| < \varepsilon\right] + E\left[|X|^r : |X| \geq \varepsilon\right] \\
&\geq E\left[|X|^r : |X| \geq \varepsilon\right] \geq \varepsilon^r P(|X| \geq \varepsilon),
\end{aligned} \tag{6.4}$$

which gives *Chebyshev's Inequality*

$$P(|X| \geq \varepsilon) \leq \frac{E|X|^r}{\varepsilon^r}. \tag{6.5}$$

Proposition 6.2. *Suppose U_n and V_n are two sequences of random variables such that $U_n \xrightarrow{P} a$, $V_n \xrightarrow{P} b$. If $g(u, v)$ is a function (of two variables) which is continuous at (a, b), then $g(U_n, V_n) \xrightarrow{P} g(a, b)$.*

Proof. Fix $\varepsilon > 0$. There exists $\delta = \delta(\varepsilon)$ such that if $|u - a| \leq \delta$ and $|v - b| \leq \delta$ then $|g(u, v) - g(a, b)| \leq \varepsilon$. Now

$$\begin{aligned}
&P(|g(U_n, V_n) - g(a, b)| > \varepsilon) = \\
&= P(\{|U_n - a| > \delta \text{ or } |V_n - b| > \delta\} \cap \{|g(U_n, V_n) - g(a, b)| > \varepsilon\}) \\
&\quad + P(\{|U_n - a| \leq \delta \text{ and } |V_n - b| \leq \delta\} \cap \{|g(U_n, V_n) - g(a, b)| > \varepsilon\}) \\
&\leq P(|U_n - a| > \delta) + P(|V_n - b| > \delta) \to 0.
\end{aligned} \tag{6.6}$$

Note that the set $\{|U_n - a| \leq \delta, |V_n - b| \leq \delta\} \cap \{g(U_n, V_n) - g(a, b)| > \varepsilon\}$ is empty, and has therefore zero probability—a fact used for the last inequality. $\qquad\square$

Remark 6.1. Proposition 6.2 extends to any fixed number, say k, of sequences $U_n^{(i)} \xrightarrow{P} a_i, 1 \leq i \leq k$, and a function $g(u_1, u_2, \ldots, u_k)$ of k variables which is continuous at $(a_1, \ldots, a_k)$, yielding: $g(U_n^{(1)}, U_n^{(2)}, \ldots, U_n^{(k)}) \xrightarrow{P} g(a_1, a_2, \ldots, a_k)$. The proof is entirely analogous (Exercise 6.1).

Corollary 6.1. *If* $U_n \xrightarrow{P} a$, $V_n \xrightarrow{P} b$ *then (i)* $U_n + V_n \xrightarrow{P} a + b$, *(ii)* $U_n V_n \xrightarrow{P} ab$, *and (iii) assuming* $b \neq 0$, $U_n/V_n \xrightarrow{P} a/b$.

Proof. Use Proposition 6.2 with (i) $g(u, v) = u + v$, (ii) $g(u, v) = uv$ and (iii) $g(u, v) = u/v$. $\qquad\square$

Proposition 6.2 and Remark 6.1 extend to vector-valued random variables (i.e., random vectors) U_n, V_n and vector-valued functions $g(u, v)$. One needs to use the definition for convergence in probability in Eq. (6.2) with $|\cdot|$ denoting Euclidean norm: $|x| = (\sum_{i=1}^{k} x_i^2)^{1/2}$ (Exercise 6.1).

A stronger form of consistency than that considered above involves the notion of *almost sure convergence*. A sequence Y_n $(n \geq 1)$ of random variables (vectors) *converges almost surely* to a random variable (vector) Y, denoted $Y_n \to Y$ a.s., or $Y_n \xrightarrow{\text{a.s.}} Y$, if

$$\lim_{n \to \infty} Y_n = Y \qquad (\text{or,} \quad \lim_{n \to \infty} |Y_n - Y| = 0) \tag{6.7}$$

holds with probability one. That is, $Y_n \to Y$ a.s. if (6.7) holds for all sample points ω in the underlying probability space $(\Omega, \mathcal{F}, P)$ on which Y_n $(n \geq 1)$, Y are defined, except for a set N of ω's with $P(N) = 0$. It is a standard result in probability that almost sure convergence implies convergence in probability.[1] The main tool for proving almost sure convergence is the following.

Proposition 6.3 (Strong Law of Large Numbers). [2] *Let* X_n $(n \geq 1)$ *be a sequence of independent and identically distributed (i.i.d.) random variables having a finite mean* $\mu = EX_n$. *Then* $(X_1 + \cdots + X_n)/n$ *converges almost surely to* μ.

We will often refer to this result by the abbreviation *SLLN*.

Definition 6.2. A statistic U_n $(n \geq 1)$ is said to be a *strongly consistent estimator* of a parameter γ if $U_n \to \gamma$ a.s.

Proposition 6.4. *If* $U_n \to a$ *a.s. and* $V_n \to b$ *a.s., then* $g(U_n, V_n) \longrightarrow g(a, b)$ *a.s. for every function* g *of two variables which is continuous at the point* (a, b).

We leave the proof of Proposition 6.4, as well as that of the a.s. convergence version of Corollary 6.1 to Exercise 6.2.

6.3 Consistency of Sample Moments and Regression Coefficients

Example 6.1 (Consistency of the Sample Mean). Let $X_1, \ldots, X_n$ be independent observations from an unknown distribution of which we assume a finite variance σ^2.

---

[1] See, e.g., Bhattacharya and Waymire (2007, pp. 7, 179, 180) or Billingsley (1986, p. 274).

[2] See, e.g., Bhattacharya and Waymire (2007, pp. 7, 50–53) or Billingsley (1986, p. 80).

Let $U_n(X) = \overline{X} = \frac{X_1 + \cdots + X_n}{n}$ be used as an estimator of the unknown population mean μ. Since $E\overline{X} = \mu$, and $\mathrm{var}(\overline{X}) = \sigma^2/n \to 0$ as $n \to \infty$, it follows from Proposition 6.2, that $\overline{X}$ is a consistent estimator of μ.

One may actually prove strong consistency of $\overline{X}$ under the assumption of finiteness of μ, using the SLLN.

Example 6.2 (Consistency of Sample Moments). Suppose a random sample X_1, $\ldots$, X_n is taken from a distribution with a finite k-th moment, for some $k \geq 1$ (i.e., $E|X_j|^k < \infty$). Then it can be shown by the strong law of large numbers (SLLN) that the sample moments $\hat{m}_r$ are strongly consistent estimators of population moments m_r for $r = 1, \ldots, k$,

$$\hat{m}_r = \frac{1}{n} \sum_{j=1}^{n} X_j^r \xrightarrow{\text{a.s.}} E(X_1^r) = m_r \qquad \text{(the r-th 'population moment')}$$

$$r = 1, 2, \ldots, k. \tag{6.8}$$

Note that $\hat{m}_r$ is an unbiased estimator of m_r ($r = 1, 2, \ldots, k$). Hence if $EX_1^{2k} \equiv m_{2k} < \infty$, then it follows from Proposition 6.1 that $\hat{m}_r$ is a consistent estimator of m_r ($r = 1, 2, \ldots, k$). The SLLN implies that it is enough to assume $E|X_1|^k < \infty$. Next consider the centered population moments $\mu_r = E(X_1 - m_1)^r$, where $m_1 = \mu$ is the mean of the distribution (population). A natural estimator of μ_r is the (corresponding) centered r-th sample moment $\hat{\mu}_r = \frac{1}{n} \sum_{j=1}^{n} (X_j - \overline{X})^r$. Note that by the binomial expansion,

$$\hat{\mu}_r = \frac{1}{n} \sum_{j=1}^{n} \left\{ X_j^r - \binom{r}{1} X_j^{r-1} \overline{X} + - \cdots + (-1)^t \binom{r}{t} X_j^{r-t} \overline{X}^t \pm \cdots + (-1)^r \overline{X}^r \right\}$$

$$= \frac{1}{n} \sum_{j=1}^{n} \sum_{t=0}^{r} (-1)^t \binom{r}{t} X_j^{r-t} \overline{X}^t = \sum_{t=0}^{r} (-1)^t \binom{r}{t} \overline{X}^t \hat{m}_{r-t} \cdot$$

$$= \sum_{t=0}^{r} (-1)^t \binom{r}{t} \hat{m}_{r-t} \hat{m}_1^t. \tag{6.9}$$

By repeated application of Proposition 6.2 or Corollary 6.1 it follows that the last sum converges in probability to

$$\sum_{t=0}^{r} (-1)^t \binom{r}{t} m_{r-t} m_1^t = E(X_1 - m_1)^r, \tag{6.10}$$

provided $\hat{m}_{r'} \xrightarrow{p} m_{r'}$ as $n \to \infty$ ($r' = 1, \ldots, r$). The latter is assured for all $r' = 1, \ldots, r$ if $EX_1^{2r} < \infty$ (by Proposition 6.1). Once again this last requirement may be relaxed to $E|X_1^r| < \infty$, by the SLLN. Thus *sample moments*, raw as well as centered, are strongly *consistent estimators* of the *corresponding population moments* if the corresponding population moments are finite.

Remark 6.2. Although the 'raw' sample moments $\hat{m}_r$ are unbiased estimators of the corresponding population moments m_r (if m_r is finite), this is not true for centered sample moments μ_r, $r \geq 2$. For example, writing $\mu = m_1$, $\sigma^2 = \mu_2$, and assuming $\sigma^2 > 0$, one has

$$
\begin{aligned}
E\hat{\mu}_2 &= E\left\{\frac{1}{n}\sum_{j=1}^{n}(X_j - \overline{X})^2\right\} = E\left\{\frac{1}{n}\sum_{j=1}^{n}[X_j - \mu - (\overline{X} - \mu)]^2\right\} \\
&= E\left\{\frac{1}{n}\sum_{j=1}^{n}(X_j - \mu)^2 + (\overline{X} - \mu)^2 - 2(\overline{X} - \mu)^2\right\} \\
&= E\left\{\frac{1}{n}\sum_{j=1}^{n}(X_j - \mu)^2 - (\overline{X} - \mu)^2\right\} = \mu_2 - \frac{\mu_2}{n} = \sigma^2 - \frac{\sigma^2}{n} \\
&= \left(1 - \frac{1}{n}\right)\sigma^2 \neq \mu_2.
\end{aligned}
\tag{6.11}
$$

Example 6.3 (Linear Regression). Consider the semiparametric regression model

$$
Y_j = \alpha + \beta X_j + \varepsilon_j \qquad (1 \leq j \leq n),
\tag{6.12}
$$

where the *response variable* Y and the nonstochastic *explanatory variable* X are observable, while the random errors ε_j are not. Assume that ε_j are i.i.d. with mean 0 and finite variance σ^2. Consider the least squares estimators $\hat{\alpha}$, $\hat{\beta}$ of α, β, i.e., values of α, β which minimize $\sum_1^n(Y_j - \alpha - \beta X_j)^2$. For ease of computation, let $\delta = \alpha + \beta\overline{X}$, and express (6.12) in terms of the new parameters δ and β as

$$
Y_j = \delta + \beta(X_j - \overline{X}) + \varepsilon_j.
\tag{6.13}
$$

Then, by calculus, $\hat{\delta} = \overline{Y}$ and

$$
\hat{\beta} = \frac{\sum_{j=1}^{n}(X_j - \overline{X})Y_j}{\sum_{j=1}^{n}(X_j - \overline{X})^2} = \frac{\sum_{j=1}^{n}(X_j - \overline{X})(\alpha + \beta X_j + \varepsilon_j)}{\sum_{j=1}^{n}(X_j - \overline{X})^2} = \beta + \frac{\sum_{j=1}^{n}\varepsilon_j(X_j - \overline{X})}{\sum_{j=1}^{n}(X_j - \overline{X})^2}.
\tag{6.14}
$$

In particular,

$$
\begin{aligned}
E\hat{\delta} &= \delta, \qquad \operatorname{var}\hat{\delta} = \sigma^2/n, \\
E\hat{\beta} &= \beta, \qquad \operatorname{var}\hat{\beta} = \frac{\sigma^2\sum_{j=1}^{n}(X_j - \overline{X})^2}{[\sum_{j=1}^{n}(X_j - \overline{X}^2)]^2} = \frac{\sigma^2}{\sum_{j=1}^{n}(X_j - \overline{X})^2}, \\
\operatorname{cov}(\hat{\delta}, \hat{\beta}) &= \operatorname{cov}(\overline{\varepsilon}, \hat{\beta}) = 0.
\end{aligned}
\tag{6.15}
$$

Hence, by Proposition 6.1, $\hat{\beta}$ is a consistent estimator of β if $\sum_{j=1}^{n}(X_j - \overline{X})^2 \to \infty$ as $n \to \infty$. Next,

$$
\hat{\alpha} = \overline{Y} - \hat{\beta}\overline{X} = \alpha + \beta\overline{X} + \frac{1}{n}\sum_{j=1}^{n}\varepsilon_j - \hat{\beta}\overline{X} = \alpha - (\hat{\beta} - \beta)\overline{X} + \frac{1}{n}\sum_{j=1}^{n}\varepsilon_j
$$

$$
E\hat{\alpha} = \alpha, \ \operatorname{var}\hat{\alpha} = \overline{X}^2(\operatorname{var}\hat{\beta}) + \frac{\sigma^2}{n} - \frac{2\overline{X}}{n}\operatorname{cov}\left\{\frac{\sum_{j=1}^{n}\varepsilon_j(X_j - \overline{X})}{\sum_{j=1}^{n}(X_j - \overline{X})^2}, \sum_{j=1}^{n}\varepsilon_j\right\}
$$

$$
= \overline{X}^2(\operatorname{var}\hat{\beta}) + \frac{\sigma^2}{n} - \frac{2}{n}(0)
$$

$$
= \sigma^2\left\{\frac{1}{n} + \frac{\overline{X}^2}{\sum_{j=1}^{n}(X_j - \overline{X})^2}\right\}, \ \operatorname{cov}(\hat{\alpha}, \hat{\beta}) = \operatorname{cov}(\hat{\delta} - \hat{\beta}\overline{X}, \hat{\beta}) = -\overline{X}\operatorname{var}\hat{\beta}.
\tag{6.16}
$$

Hence $\hat\alpha$, $\hat\beta$ are consistent estimators of α, β, respectively, if $\sum_1^n (X_j - \overline{X})^2 \to \infty$ and $\overline{X}^2 / \sum_1^n (X_j - \overline{X})^2 \to 0$ as $n \to \infty$ (also see Exercise 6.5).

Remark 6.3 (Consistency of $\hat\alpha$, $\hat\beta$ Under Dependence). Note that the calculations (6.14)–(6.16) only required that

$$E\varepsilon_j = 0, \qquad E\varepsilon_j^2 = \sigma^2, \qquad E\varepsilon_j \varepsilon_{j'} = 0 \quad \text{for } j \neq j'. \tag{6.17}$$

Thus $\hat\alpha$, $\hat\beta$ are consistent estimators of α, β if ε_j are mean-zero uncorrelated (but not necessarily independent) random variables having a common variance σ^2, and $\sum_1^n (X_j - \overline{X})^2 \to \infty$, $\sum_1^n (X_j - \overline{X})^2 / \overline{X}^2 \to \infty$.

Remark 6.4 (Consistency of $\hat\alpha$, $\hat\beta$ under Heteroscedasticity). Assume again that ε_j are mean-zero and uncorrelated, but heteroscedastic, i.e.,

$$E\varepsilon_j^2 = \sigma_j^2 \qquad (j = 1, 2, \dots). \tag{6.18}$$

Then $\hat\delta = \overline{Y}$, $\hat\beta$ are still unbiased estimators, as is $\hat\alpha$. (This only requires $E\varepsilon_j = 0$ for all j). But

$$\mathrm{var}\hat\delta = \sum_{j=1}^n \sigma_j^2 / n,$$

$$\mathrm{var}\hat\beta = \frac{\Sigma_{j=1}^n \sigma_j^2 (X_j - \overline{X})^2}{\left[\Sigma_{j=1}^n (X_j - \overline{X})^2\right]^2},$$

$$\mathrm{var}\,\hat\alpha = \overline{X}^2 (\mathrm{var}\,\hat\beta) + \frac{\Sigma_{j=1}^n \sigma_j^2}{n^2} - \frac{2\Sigma \sigma_j^2 (X_j - \overline{X})}{n \Sigma_{j=1}^n (X_j - \overline{X})^2} \overline{X}$$

$$\mathrm{cov}(\hat\delta, \hat\beta) = \mathrm{cov}\left(\overline{\varepsilon}, \frac{\sum_1^n (X_j - \overline{X})\varepsilon_j}{\sum_1^n (X_j - \overline{X})^2}\right) = \frac{\sum_1^n \sigma_j^2 (X_j - \overline{X})^2}{\sum_1^n (X_j - \overline{X})^2}. \tag{6.19}$$

Check that $\mathrm{var}\,\hat\beta$ and $\mathrm{var}\,\hat\alpha$ both go to zero as $n \to \infty$, provided the following conditions hold:

$$\sigma_j^2 \leq c < \infty \quad \text{for all } j, \qquad \sum_{j=1}^n (X_j - \overline{X})^2 \longrightarrow \infty,$$

$$\sum_1^n (X_j - \overline{X})^2 / \overline{X}^2 \longrightarrow \infty \quad \text{as } n \to \infty. \tag{6.20}$$

For the last term in $\mathrm{var}\,\hat\alpha$, use $|\frac{1}{n}\sum_{j=1}^n \sigma_j^2 (X_j - \overline{X})| \leq (\frac{1}{n}\sum_{j=1}^n \sigma_j^4)^{1/2}(\frac{1}{n}\sum_{j=1}^n (X_j - \overline{X})^2)^{1/2} \leq c^2 (\frac{1}{n}\sum_{j=1}^n (X_j - \overline{X})^2)^{1/2}$. One may relax the assumption of boundedness of σ_j^2 in (6.18) even further (Exercise 6.5).

The estimators $\hat\delta$, $\hat\alpha$, $\hat\beta$ here are called *ordinary least squares estimators* or *OLS* as opposed to *weighted least squares estimators* considered later in Example 6.8.

Consider now the problem of *predicting Y from a future observation X*, based on past observations (Y_j, X_j), $1 \leq j \leq n$, satisfying (6.12), with i.i.d. errors ε_j. The natural predictor $\widehat{Y} = \hat\alpha + \hat\beta X$ has the prediction error $\widehat{Y} - Y$ with expected squared error $\operatorname{var}\hat{\alpha} + X^2\operatorname{var}\hat{\beta} + 2X\operatorname{cov}(\hat{\alpha},\hat{\beta}) + \sigma^2$ which may be estimated by replacing σ^2 by $\hat{\sigma}^2 = \sum_{j=1}^{n}(Y_j - \hat{\alpha} - \hat{\beta}X_j)^2/(n-2)$ [See (6.87), (6.88)].

Example 6.4 (Autoregressive Model). A commonly used time series model is the linear autoregressive model of order $k \geq 1$. For the case $k = 1$, it takes the form

$$Y_j = \alpha + \beta Y_{j-1} + \varepsilon_j \qquad (j = 1, 2, \ldots, n), \tag{6.21}$$

where $Y_0, Y_1, \ldots, Y_n$ are observed, α and β are unknown parameters to be estimated and ε_j, $1 \leq j \leq n$, are uncorrelated mean zero random variables with common variance $\sigma^2 > 0$ (unknown). Once again the least squares estimators are (see (6.14), (6.16), with $X_j = Y_{j-1}$)

$$\hat{\alpha} = \overline{Y}_{1,n} - \hat{\beta}\overline{Y}_{0,n-1}, \qquad \hat{\beta} = \beta + \frac{\sum_{r=1}^{n}\varepsilon_r(Y_{r-1} - \overline{Y}_{0,n-1})}{\sum_{r=1}^{n}(Y_{r-1} - \overline{Y}_{0,n-1})^2}, \tag{6.22}$$

where $\overline{Y}_{1,n} = \frac{1}{n}\sum_{r=1}^{n}Y_r$, $\overline{Y}_{0,n-1} = \frac{1}{n}\sum_{r=0}^{n-1}Y_r = \overline{Y}_{1,n} + (Y_0 - Y_n)/n$. Assume the stability condition

$$|\beta| < 1. \tag{6.23}$$

One also assumes that Y_0 is uncorrelated with of $\{\varepsilon_1, \ldots, \varepsilon_n\}$, and $E\varepsilon_j^4 < \infty$, $EY_0^2 < \infty$. It may be shown that, $\hat{\alpha}, \hat{\beta}$ are consistent estimators of α, β, respectively:

$$\hat{\beta} \xrightarrow{P} \beta, \ \hat{\alpha} \xrightarrow{P} \alpha \quad \text{as } n \to \infty. \tag{6.24}$$

Proof of (6.24). First, by iteration of (6.21),

$$Y_1 = \alpha + \beta Y_0 + \varepsilon_1, Y_2 = \alpha + \beta Y_1 + \varepsilon_2 = \alpha + \alpha\beta + \beta^2 Y_0 + \beta\varepsilon_1 + \varepsilon_2, \ldots,$$
$$Y_r = \alpha + \alpha\beta + \cdots + \alpha\beta^{r-1} + \beta^r Y_0 + \beta^{r-1}\varepsilon_1 + \beta^{r-2}\varepsilon_2 + \cdots + \beta\varepsilon_{r-1} + \varepsilon_r$$
$$= \alpha(1 - \beta^r)/(1 - \beta) + \beta^r Y_0 + \sum_{j=1}^{r}\beta^{r-j}\varepsilon_j. \tag{6.25}$$

Hence

$$Y_r - \frac{\alpha}{1 - \beta} = \frac{-\alpha\beta^r}{1 - \beta} + \beta^r Y_0 + \sum_{j=1}^{r}\beta^{r-j}\varepsilon_j,$$

$$\overline{Y}_{0,n-1} - \frac{\alpha}{1 - \beta} = \frac{-\alpha(1 - \beta)}{n(1 - \beta)} + \frac{1 - \beta^n}{n(1 - \beta)}Y_0 + \frac{1}{n}\sum_{r=1}^{n}\sum_{j=1}^{r}\beta^{r-j}\varepsilon_j,$$

$$E\left(\overline{Y}_{0,n-1} - \frac{\alpha}{1 - \beta}\right)^2 = E\frac{\left[\frac{-\alpha(1-\beta^n)}{(1-\beta)} + \frac{1-\beta^n}{1-\beta}Y_0 + \sum_{j=1}^{n-1}\frac{1-\beta^{n-j}}{1-\beta}\varepsilon_j\right]^2}{n^2}$$

$$\leq \frac{3\left[\frac{\alpha^2(1-\beta^n)^2}{(1-\beta)^2} + \frac{(1-\beta^n)^2}{(1x-\beta)^2}EY_0^2 + E(\sum_{j=1}^{n-1}\frac{1-\beta^{n-j}}{1-\beta}\varepsilon_j)^2\right]}{n^2}$$

$$\text{(for, } (a + b + c)^2 \leq 3a^2 + 3b^2 + 3c^2)$$

$$= O\left(\frac{1}{n^2}\right) + \frac{3}{n^2}\sum_{j=1}^{n-1}\frac{(1 - \beta^{n-j})^2}{(1 - \beta)^2}\sigma^2$$

$$= O\left(\frac{1}{n^2}\right) + O\left(\frac{1}{n}\right) \longrightarrow 0. \tag{6.26}$$

Next, writing $Y_{r-1} - \overline{Y}_{0,n-1} = Y_{r-1} - \frac{\alpha}{1-\beta} - (\overline{Y}_{0,n-1} - \frac{\alpha}{1-\beta})$ one gets

$$\frac{1}{n}\sum_{r=1}^{n}(Y_{r-1} - \overline{Y}_{0,n-1})^2$$

$$= \frac{1}{n}\sum_{r=1}^{n}\left(Y_{r-1} - \frac{\alpha}{1-\beta}\right)^2 - \frac{2}{n}\sum_{r=1}^{n}\left(\overline{Y}_{0,n-1} - \frac{\alpha}{1-\beta}\right)\left(\overline{Y}_{0,n-1} - \frac{\alpha}{1-\beta}\right)$$

$$+ \left(\overline{Y}_{0,n-1} - \frac{\alpha}{1-\beta}\right)^2$$

$$= \frac{1}{n}\sum_{r=1}^{n}\left(Y_{r-1} - \frac{\alpha}{1-\beta}\right)^2 - \left(\overline{Y}_{0,n-1} - \frac{\alpha}{1-\beta}\right)^2,$$

so that, by (6.26),

$$\frac{1}{n}\sum_{r=1}^{n}(Y_{r-1} - \overline{Y}_{0,n-1})^2 - \frac{1}{n}\sum_{r=1}^{n}(Y_{r-1} - \frac{\alpha}{1-\beta})^2 \xrightarrow{P} 0. \tag{6.27}$$

Now, using the first relation in (6.26),

$$\left(Y_{r-1} - \frac{\alpha}{1-\beta}\right)^2 - \left(\sum_{j=1}^{r-1}\beta^{r-j-1}\varepsilon_j\right)^2$$

$$= \left(\frac{-\alpha\beta^{r-1}}{1-\beta} + \beta^{r-1}Y_0\right)^2 + 2\left(\frac{-\alpha\beta^{r-1}}{1-\beta} + \beta^{r-1}Y_0\right)\cdot\left(\sum_{j=1}^{r-1}\beta^{r-j-1}\varepsilon_j\right),$$

so that, by the Cauchy-Schwartz Inequality,

$$\left|\sum_{r=1}^{n}\left(Y_{r-1} - \frac{\alpha}{1-\beta}\right)^2 - \sum_{r=1}^{n}\left(\sum_{j=1}^{r-1}\beta^{r-j-1}\varepsilon_j\right)^2\right|$$

$$\le \sum_{r=1}^{n}\left(\frac{-\alpha\beta^{r-1}}{1-\beta} + \beta^{r-1}Y_0\right)^2 + 2\left[\left(\sum_{r=1}^{n}\left\{\frac{-\alpha\beta^{r-1}}{1-\beta} + \beta^{r-1}Y_0\right\}^2\right)\right]^{1/2}$$

$$\cdot\left[\sum_{r=1}^{n}\left(\sum_{j=1}^{r-1}\beta^{r-j-1}\varepsilon_j\right)^2\right]^{1/2},$$

$$E\left|\left[\frac{\sum_{r=1}^{n}(Y_{r-1} - \frac{\alpha}{1-\beta})^2}{n} - \frac{1}{n}\sum_{r=1}^{n}\left(\sum_{j=1}^{r-1}\beta^{r-j-1}\varepsilon_j\right)^2\right]\right|$$

$$\le \frac{2\alpha^2}{n(1-\beta)^2}\frac{1-\beta^{2n}}{1-\beta^2} + \frac{2(1-\beta^{2n})}{n(1-\beta^2)}EY_0^2$$

$$+ \frac{2}{n}\left[E\left(\sum_{r=1}^{n}\left\{\frac{-\alpha\beta^{r-1}}{1-\beta} + \beta^{r-1}Y_0\right\}^2\right)\cdot E\left(\sum_{r=1}^{n}\sum_{j=1}^{r-1}\beta^{r-j-1}\varepsilon_j\right)^2\right]^{1/2}$$

$$= \mathrm{O}\left(\frac{1}{n}\right) + \frac{2}{n} \cdot \mathrm{O}(1) \left(E\left[\sum_{j=1}^{n-1} \frac{1-\beta^{n-j}}{1-\beta}\varepsilon_j\right]^2 \right)^{1/2}$$

$$= \mathrm{O}\left(\frac{1}{n}\right) + \frac{2}{n} \cdot \mathrm{O}(1) \left[\sum_{j=1}^{n-1} \frac{(1-\beta^{n-j})^2}{(1-\beta)^2}\sigma^2\right]^{1/2} = \mathrm{O}(n^{-1/2}) \longrightarrow 0. \quad (6.28)$$

Therefore, by Chebyshev's inequality (with $p=1$),

$$\frac{\sum_{r=1}^{n}(Y_{r-1}-\overline{Y}_{0,n-1})^2}{n} - \frac{1}{n}\sum_{r=1}^{n}\left(\sum_{j=1}^{r-1}\beta^{r-j-1}\varepsilon_j\right)^2 \xrightarrow{P} 0. \quad (6.29)$$

Also,

$$E\,\frac{1}{n}\sum_{r=1}^{n}\left(\sum_{j=1}^{r-1}\beta^{r-j-1}\varepsilon_j\right)^2$$

$$= \frac{\sigma^2}{n}\sum_{r=1}^{n}\sum_{j=1}^{r-1}\beta^{2(r-j-1)}$$

$$= \frac{\sigma^2}{n}\sum_{r=1}^{n}\frac{1-\beta^{2(r-1)}}{1-\beta^2} \longrightarrow \frac{\sigma^2}{1-\beta^2}. \quad (6.30)$$

A little extra work shows that (Exercise 6.8)

$$\frac{1}{n}\sum_{r=1}^{n}\left(\sum_{j=1}^{r-1}\beta^{r-j-1}\varepsilon_j\right)^2 \xrightarrow{P} \frac{\sigma^2}{1-\beta^2}. \quad (6.31)$$

Also, from the above estimates (6.28), one gets

$$E\left[\frac{1}{n}\sum_{r=1}^{n}\varepsilon_r(Y_{r-1}-\overline{Y}_{0,n-1})\right]^2 \longrightarrow 0 \quad \text{as } n \to \infty. \quad (6.32)$$

Combining (6.29), (6.31) and (6.32), we get

$$|\hat{\beta}-\beta| = \left|\frac{\frac{1}{n}\sum_{r=1}^{n}\varepsilon_r(Y_{r-1}-\overline{Y}_{0,n-1})}{\frac{1}{n}\sum_{r=1}^{n}(Y_{r-1}-\overline{Y}_{0,n-1})^2}\right| \xrightarrow{P} \frac{0}{\frac{\sigma^2}{1-\beta}} = 0.$$

That is, $\hat{\beta} \xrightarrow{P} \beta$. Clearly then

$$\hat{\alpha} = \overline{Y}_{1,n} - \hat{\beta}\overline{Y}_{0,n-1} \xrightarrow{P} \frac{\alpha}{1-\beta} - \frac{\beta\alpha}{1-\beta} = \alpha.$$

$$\square$$

Remark 6.5. To prove (6.24) one may relax the assumption of common variance of ε_j by "$\sigma_j^2 := E\varepsilon_j^2$ $(j=1,2,\dots)$ is a bounded sequence".

Remark 6.6. Suppose $Y_n \xrightarrow{P} Y$. Although this does not in general imply that $Y_n \xrightarrow{\text{a.s.}} Y$, it is a useful fact that there exists a subsequence Y_{n_k} $(k \geq 1)$, $n_1 < n_2 < \cdots$, such that $Y_{n_k} \xrightarrow{\text{a.s.}} Y$.[3]

6.4 Consistency of Sample Quantiles

As a general definition, a *p-th quantile* of a random variable X, or of its distribution Q, is a number c such that $P(X \leq c) \geq p$, $P(X < c) \leq p$, i.e., $F(c) \geq p$, $F(c-) \leq p$, where $F(c-) = \lim_{x \downarrow c\ (x<c)} F(x)$ is the *left-hand limit* of $F(x)$ as $x \downarrow c$. Note that if F is continuous at $x = c$, then $F(c-) = F(c)$ but otherwise $F(c-) < F(c)$, i.e., there is a jump in F at $x = c$, $P(X = c) > 0$. One may also write the requirement as $F(c) \geq p$, $1 - F(c-) \equiv P(X \geq c) \geq 1 - p$.

Let $X_1, X_2, \ldots, X_n$ be a random sample from Q. Ordering them from the smallest to the largest, one may arrange them as $X_{(1)} \leq X_{(2)}, \leq \cdots \leq X_{(n)}$. Note that if we add another observation X_{n+1}, then the orderings (even of the first n order statistics) generally change. For example, consider the samples $\{2, 7, 5\}$ $(n = 3)$, and $\{2, 7, 5, 3\}$ $(n = 4)$. For the first sample $X_{(1)} = 2$, $X_{(2)} = 5$, $X_{(3)} = 7$. When a fourth observation is added one has $X_{(1)} = 2$, $X_{(2)} = 3$, $X_{(3)} = 5$, $X_{(4)} = 7$. For this reason, one should write the order statistics as $X_{(1):n}$, $X_{(2):n}$, $X_{(n):n}$, indicating that this ordering is based on the first n observations. When there is little chance of confusion, we will continue to write $X_{(j)}$ in place of $X_{(j):n}$, keeping in mind that for any given n, $X_{(j)}$ depends on n, as the sample size is increased.

Definition 6.3. For a random sample $\{X_1, \ldots, X_n\}$ the *sample p-th quantile* $\hat{\xi}_p$ $(0 < p < 1)$ is defined either as $X_{([np])}$ or $X_{([np]+1)}$, where $[np]$ is the *integer part* of np. One may think of (or define) the sample p-th quantile as the p-th quantile of *the empirical distribution* Q_n, which assigns mass $\frac{1}{n}$ to each of the observed n points $X_1, X_2, \ldots, X_n$. The empirical *distribution function* (d.f.) of Q_n is:

$$F_n(x) = \frac{1}{n} \, \# \{j : 1 \leq j \leq n, \, X_j \leq x\} \qquad (x \in \mathbb{R}). \tag{6.33}$$

If F is continuous then, with probability one, $X_{(1)} < X_{(2)} < \cdots < X_{(n)}$, i.e., there are no ties.

It follows from the SLLN that, for a given x, $\widehat{F}_n(x) \to F(x)$ almost surely, i.e., $\widehat{F}_n(x)$ is a strongly consistent estimator of $F(x)$. Indeed, by the so-called Gilvenko–Cantelli Theorem[4] $\sup\{|\widehat{F}_n(x) - F(x)| : x \in \mathbb{R}\} \to 0$ almost surely as $n \to \infty$.

Proposition 6.5 (Consistency of Sample Quantiles). *Let X_j, $1 \leq j \leq n$, be i.i.d. real-valued random variables with the common distribution function F. Suppose, for a given $p \in (0, 1)$, the p-th quantile ξ_p is uniquely defined, i.e., there is a unique solution of $F(x) = p$. Then $\hat{\xi}_p$ is a consistent estimator of ξ_p.*

Proof. The hypothesis on F implies

$$F(x) < p \text{ for all } x < \xi_p \text{ and } F(x) > p \text{ for all } x > \xi_p. \tag{6.34}$$

---

[3] See Bhattacharya and Waymire (2007, pp. 179, 180) or Billingsley (1986, p. 274).
[4] See Billingsley (1986, pp. 275, 276).

Fix any $\varepsilon > 0$. Then, writing $Y_n = \Sigma_{j=1}^n 1_{\{X_j \leq \xi_p - \varepsilon\}}$,

$$P(\hat{\xi}_p \leq \xi_p - \varepsilon) = P(X_{([np])} \leq \xi_p - \varepsilon) = P(Y_n \geq [np])$$

$$= P\left(\frac{Y_n}{n} - F(\xi_p - \varepsilon) \geq \frac{[np]}{n} - F(\xi_p - \varepsilon)\right)$$

$$\leq P\left(\left|\frac{Y_n}{n} - F(\xi_p - \varepsilon)\right| \geq \frac{[np]}{n} - F(\xi_p - \varepsilon)\right). \qquad (6.35)$$

In view of (6.34) and the fact that $\frac{[np]}{n} \to p$ as $n \to \infty$ (indeed, $|\frac{[np]}{n} - p| \leq \frac{1}{n}$), it follows that for all sufficiently large n (e.g., $n \geq 2/\delta(\varepsilon)$)

$$P(\hat{\xi}_p \leq \xi_p - \varepsilon) \leq P\left(\left|\frac{Y_n}{n} - F(\xi_p - \varepsilon)\right| \geq \frac{\delta(\varepsilon)}{2}\right), \qquad (6.36)$$

where

$$\delta(\varepsilon) = p - F(\xi_p - \varepsilon) > 0.$$

By Chebyshev's Inequality (6.5) applied to $Y_n/n - F(\xi_p - \varepsilon) = Y_n/n - E(Y_n/n)$, one gets from (6.36) the inequality

$$P\left(\hat{\xi}_p \leq \xi_p - \varepsilon\right) \leq \frac{\text{var}(Y_n/n)}{(\delta(\varepsilon)/2)^2} = \frac{F(\xi_p - \varepsilon)(1 - F(\xi_p - \varepsilon))}{n(\delta(\varepsilon/2)^2)} \longrightarrow 0 \text{ as } n \to \infty.$$

$$(6.37)$$

Similarly, writing $Z_n = \sum_{j=1}^n 1_{\{X_j \leq \xi_p + \varepsilon\}}$,

$$P\left(\hat{\xi}_p > \xi_p + \varepsilon\right) = P(Z_n < [np])$$

$$= P\left(\frac{Z_n}{n} - F(\xi_p + \varepsilon) < \frac{[np]}{n} - F(\xi_p + \varepsilon)\right) \leq P\left(\frac{Z_n}{n} - F(\xi_p + \varepsilon) < -\delta'(\varepsilon)\right)$$

$$\leq P\left(\left|\frac{Z_n}{n} - F(\xi_p + \varepsilon)\right| > \delta'(\varepsilon)\right) \qquad (6.38)$$

where, by (6.34),

$$\delta'(\varepsilon) := F(\xi_p + \varepsilon) - p > 0.$$

Note that $\frac{[np]}{n} - F(\xi_p + \varepsilon) < p - F(\xi_p + \varepsilon) = -\delta'(\varepsilon)$. It now follows from (6.38) that, as $n \to \infty$,

$$P\left(\hat{\xi}_p > \xi_p + \varepsilon\right) \leq \frac{\text{var}(\frac{Z_n}{n})}{(\delta'(\varepsilon))^2} = \frac{F(\xi_p + \varepsilon)(1 - F(\xi_p + \varepsilon))}{n(\delta'(\varepsilon))^2} \longrightarrow 0. \qquad (6.39)$$

The inequalities (6.37) and (6.39) imply $P(|\hat{\xi}_p - \xi_p| > \varepsilon) \to 0$, as $n \to \infty$.

$$\square$$

Note that $\xi_{1/2}$, if it is uniquely defined, is called the *median* of F.

Remark 6.7. Suppose $U_n \xrightarrow{P} U$, and U_n^2 $(n \geq 1)$ are *uniformly integrable* i.e., $\sup_n E[U_n^2 : |U_n| \geq \lambda] \to 0$ as $\lambda \to \infty$, then $E(U_n - U)^2 \to 0$.[5] This allows one to sometimes derive $E(U_n - \theta)^2 \to 0$ from $U_n \xrightarrow{P} \theta$.

---

[5] See Bhattacharya and Waymire (2007, p. 12).

6.5 Convergence in Distribution or in Law (or Weak Convergence): The Central Limit Theorem

Definition 6.4. A sequence of random variables Z_n (real-valued) is said to *converge in distribution* to a probability measure P on $\mathbb{R}$ if

$$Ef(Z_n) \longrightarrow \int_{-\infty}^{\infty} f\,dP \quad \text{as } n \to \infty \tag{6.40}$$

for all bounded real-valued continuous functions f on $\mathbb{R}$. If the *distribution* of Z_n is P_n (i.e., $\text{Prob}(Z_n \in A) = P_n(A)$ for all Borel sets, A), then one also describes this convergence as "P_n *converges weakly* to P". Sometimes instead of "convergence in distribution" one uses the terminology "*convergence in law*," and uses the notation

$$Z_n \xrightarrow{\mathscr{L}} P. \tag{6.41}$$

Recall that the distribution function F of a random variable Z is defined by

$$F(x) = \text{Prob}(Z \le x) \ < -\infty < x < \infty. \tag{6.42}$$

One may show that the convergence in distribution (6.41) is equivalent to convergence of the distribution functions $F_n(x)$ of Z_n to $F(x) = P((-\infty, x])$ at all points of continuity of F.

Convergence in distribution of Z_n to a probability law P is also equivalent to the convergence of characteristic functions of Z_n to the characteristic function of P:

$$Ee^{i\xi Z_n} \longrightarrow \int_{-\infty}^{\infty} e^{i\xi z}\,dP(z), \qquad \xi \in \mathbb{R}. \tag{6.43}$$

The above definitions and results extend word for word when Z_n are vector-valued.[6]

The most important convergence theorem in law is the following theorem, abbreviated as *CLT*.[7]

Proposition 6.6 (Central Limit Theorem). *If Z_n is a sequence of i.i.d. random variables (or vectors) with common mean (or mean vector) zero and a finite common variance σ^2 (or dispersion matrix Σ), then*

$$n^{-\frac{1}{2}}(Z_1 + Z_2 + \cdots + Z_n) \xrightarrow{\mathscr{L}} N(0, \sigma^2) \quad (\text{or } N(0, \Sigma)). \tag{6.44}$$

Here $N(0, \sigma^2)$ is the *Normal distribution* having mean zero and variance σ^2. For convenience we allow the possibility $\sigma^2 = 0$ and in this case interpret $N(0,0)$ as the probability measure *degenerate* at 0 (which assigns probability one to the singleton $\{0\}$). Similarly, Σ is in general a nonnegative definite matrix.

We will sometimes use the alternative notation Φ_{0,σ^2} (or $\Phi_{0,\Sigma}$) for $N(0,\sigma^2)$ (or $N(0,\Sigma)$), and denote the corresponding distribution function by $\Phi_{0,\sigma^2}(x)$, and the density function by $\phi_{0,\sigma^2}(x)$. According to the criterion stated immediately after (6.42), the central limit theorem (CLT) says: if $\sigma^2 > 0$.

---

[6] See, e.g., Bhattacharya and Waymire (2007, pp. 62, 63, 86).

[7] For a proof, see Bhattacharya and Waymire (2007, pp. 92, 93) or Billingsley (1986, pp. 398, 399).

$$Prob\left(\frac{Z_1 + Z_2 + \cdots + Z_n}{n^{\frac{1}{2}}} \leq x\right) \longrightarrow \Phi_{0,\sigma^2}(x) = \frac{1}{\sqrt{2\pi\sigma^2}} \int_{-\infty}^{x} e^{-y^2/2\sigma^2}\,dy \quad (6.45)$$

for all x. An old result of Polya implies that actually the convergence in (6.45) *is uniform* over all x:

$$\sup_{x}\left|P\left(\frac{Z_1 + \cdots + Z_n}{\sqrt{n}} \leq x\right) - \Phi_{0,\sigma^2}(x)\right| \longrightarrow 0 \quad \text{as } n \to \infty. \quad (6.46)$$

For a precise statement and proof of Polya's Theorem, see the Appendix to this chapter (Appendix D).

An immediate application of the *CLT* is that the rth *sample moment* $(X_1^r + \cdots + X_n^r)/n$ *is asymptotically normal if* $EX_1^{2r} < \infty$. (Here r is a positive integer). The italicized statement means:

$$\sqrt{n}\left(\frac{1}{n}\sum_{i=1}^{n} X_i^r - EX_1^r\right) \xrightarrow{\mathcal{L}} N(0,\sigma^2), \quad (6.47)$$

where $\sigma^2 = \text{var}(X_1^r) = EX_1^{2r} - (EX_1^r)^2$.

Remark 6.8. The law of large numbers shows that for a sequence of i.i.d. random variables Z_n, $\frac{1}{n}\sum_{i=1}^{n} Z_i \simeq \mu\ (= EZ_1)$, i.e., the difference between the two sides in $\simeq$ goes to zero as $n \to \infty$. Under the additional assumption $EZ_1^2 < \infty$, Chebyshev's Inequality strengthens this approximation by showing that the difference is of the order of $n^{\frac{1}{2}}$ in probability:

$$\frac{1}{n}\sum_{i=1}^{n} Z_i - \mu = O_p\left(\frac{1}{\sqrt{n}}\right). \quad (6.48)$$

This means that, given $\varepsilon < 0$ there exists $A > 0$ such that

$$\text{Prob}\left(\left|\frac{1}{n}\sum_{i=1}^{n} Z_i - \mu\right| > \frac{A}{\sqrt{n}}\right) < \varepsilon \quad (6.49)$$

for all sufficiently large n. The CLT is of course a more precise statement than (6.48) [or (6.49)]. In particular, it provides a computation of the left side of (6.49) as

$$2\Phi_{0,1}\left(-\frac{A}{\sigma}\right) + \delta_n = \frac{2}{\sqrt{2\pi}}\int_{-\infty}^{-A/\sigma} e^{-y^2/2}dy + \delta_n, \quad (6.50)$$

where $\delta_n \to 0$ as $n \to \infty$ (uniformly w.r.t. A). (See Exercise 6.12.)

Before we may consider more important applications of the CLT we need some useful elementary facts.

First, we shall use the notation

$$Z_n \xrightarrow{\mathcal{L}} Z \quad (6.51)$$

sometimes, in place of (6.41), and say that Z_n *converges in distribution* (or *law*) to Z, with the understanding that Z is a random variable (or random vector) whose distribution is P. Although the definition (6.41) does not imply (and in most situations of interest it is not true) that there exists a limiting random variable in the sense of convergence in probability or a.s. convergence, the notation (6.51) and the corresponding language has some advantage—as the following results show.

Proposition 6.7. *Suppose* $Z_n \xrightarrow{\mathcal{L}} Z$ *and* g *is a continuous function. Then* $g(Z_n) \xrightarrow{\mathcal{L}} g(Z)$.

Proof. Use Definition 6.4 (6.40) (Exercise 6.14). $\square$

Proposition 6.8 (Slutsky's Lemma). *Suppose* U_n, V_n, W_n $(n \geq 1)$ *are three sequences of real-valued random variables such that* $U_n \xrightarrow{P} a$, $V_n \xrightarrow{P} b$, $W_n \xrightarrow{\mathcal{L}} W$. *Let* $h(u, v, w)$ *be a function which is continuous on* $[a - \delta_1, a + \delta_1] \times [b - \delta_2, b + \delta_2] \times (-\infty, \infty)$ *for some* $\delta_1 > 0$, $\delta_2 > 0$. *Then* $h(U_n, V_n, W_n) \xrightarrow{\mathcal{L}} h(a, b, W)$.

Proof. We will first prove that $h(U_n, V_n, W_n) - h(a, b, W_n) \xrightarrow{P} 0$. The desired result would follow from this, using the continuity of $w \to h(a, b, w)$ (see Proposition 6.4). Fix $\varepsilon > 0$ and $\theta > 0$, however small. In view of the convergence in distribution of W_n there exists $A = A(\theta)$ such that $P(|W_n| > A) < \theta/3$ for all n (Exercise 6.15). In view of (uniform) continuity of h on the compact set $[a - \delta_1, a + \delta_1] \times [b - \delta_2, b + \delta_2] \times [-A, A]$, there exists $\delta = \delta(\varepsilon) > 0$ such that $|h(u, v, w) - h(a, b, w)| \leq \varepsilon$ for all (u, v, w) satisfying $|u - a| \leq \delta$, $|v - b| \leq \delta$ and $|w| \leq A$. Now since $U_n \xrightarrow{P} a$, $V_n \xrightarrow{P} b$, there exists a positive integer $n(\theta, \varepsilon)$ such that

$$P(|U_n - a| > \delta) < \frac{\theta}{3}, \quad P(|V_n - b| > \delta) < \frac{\theta}{3} \quad \forall\, n \geq n(\theta, \varepsilon). \tag{6.52}$$

Hence

$$\begin{aligned}
&P(|h(U_n, V_n, W_n) - h(a, b, W_n)| > \varepsilon) \\
&\leq P(|U_n - a| > \delta) + P(|V_n - b| > \delta) + P(|W_n| > A) \\
&\quad + P(\{|U_n - a| \leq \delta, |V_n - b| \leq \delta, |W_n| \\
&\leq A, |h(U_n, V_n, W_n) - h(a, b, W_n)| > \varepsilon\}) \\
&\leq 3\frac{\theta}{3} = \theta,
\end{aligned} \tag{6.53}$$

since the set within curly brackets in (6.53) is empty and has, therefore, probability zero. $\square$

Remark 6.9. Proposition 6.8 easily extends to the case of k sequences $U_{ni} \xrightarrow{P}$, a_i $(i = 1, \ldots, k)$, $W_n \xrightarrow{\mathcal{L}} W$ $\mathbb{R}^p$-valued, $h(u_1, \ldots, u_k, w)$ continuous on $O \times \mathbb{R}^p$, where O is an open neighborhood of $(a_1, \ldots, a_k)$, $h(U_{n1}, \ldots, U_{nk}, W_n) \xrightarrow{\mathcal{L}} h(a_1, \ldots, a_k, W)$. Indeed, the function h may also be vector-valued.

The following simple result is widely used in large sample theory.

Theorem 6.1. *Suppose* W_n *is a sequence of random variables and* $g(n)$ *a sequence of constants,* $g(n) \uparrow \infty$, *such that* $g(n)(W_n - \mu) \xrightarrow{\mathcal{L}} V$. *Then for every function* H *which is continuously differentiable in a neighborhood of* μ, *one has*

$$g(n)[H(W_n) - H(\mu)] \xrightarrow{\mathcal{L}} H'(\mu)V. \tag{6.54}$$

(b)

Proof. By the mean value theorem there exists μ^* between μ and W_n such that the left side of (6.54) is

$$g(n)(W_n - \mu)H'(\mu^*) = g(n)(W_n - \mu)\{H'(\mu) + o_p(1)\} \qquad (6.55)$$

where $o_p(1) \to 0$ in probability. Note that this is a consequence of the fact that $W_n - \mu \xrightarrow{P} 0$ [since $g(n) \uparrow \infty$ and $g(n)(W_n - \mu) \xrightarrow{\mathscr{L}} V$ (Exercise 6.13(c))]. The convergence (6.54) now follows from Proposition 6.8. $\qquad \square$

The most commonly used consequence of Theorem 6.1 is the following, called the delta method.

Corollary 6.2 (The Delta Method). *Suppose Z_j, $j \geq 1$, are i.i.d. random variables with common mean and variance μ and $\sigma^2 < \infty$. If H is continuously differentiable in a neighborhood of μ, then*

$$\sqrt{n}[H(\overline{Z}) - H(\mu)] \xrightarrow{\mathscr{L}} H'(\mu)V \stackrel{\mathscr{L}}{=} N(0, (H'(\mu))^2\sigma^2), \qquad (6.56)$$

where $\overline{Z} = \overline{Z}_n = \sum_{j=1}^{n} Z_j/n$, and V is $N(0, \sigma^2)$.

For future reference and notational convenience, the following definition is useful.

Definition 6.5. A *statistic* $T = T_n$ (i.e., a function of observations $X_1, \ldots, X_n$) is said to be *asymptotically Normal* with mean θ and variance $\frac{\sigma^2}{n}$, or $AN(\theta, \frac{\sigma^2}{n})$, if $\sqrt{n}(T_n - \theta) \xrightarrow{\mathscr{L}} N(0, \sigma^2)$. A vector valued $\mathbf{T}_n$ is said to be $AN(\boldsymbol{\theta}, \frac{1}{n}\Sigma)$ if $\sqrt{n}(T_n - \theta) \xrightarrow{\mathscr{L}} N(0, \Sigma)$.

Remark 6.10. Corollary 6.2 (and, indeed, Theorem 6.1) extends to the case where Z_j, $j \geq 1$, are i.i.d. k-dimensional random vectors with mean vector μ and covariance matrix $\Sigma = ((\sigma_{ij}))$, while H is real-valued and continuously differentiable (as a function of k variables) in a neighborhood of μ. In this case

$$\sqrt{n}[H(\overline{Z}) - H(\mu)] \xrightarrow{\mathscr{L}} \text{Grad } H(\mu) \cdot V \stackrel{\mathscr{L}}{=} N(0, \sum_{i,j=1}^{k} \ell_i\ell_j\sigma_{ij}), \qquad (6.57)$$

with $\ell_i := (D_iH)(\mu) = (\partial H(z)/\partial z_i)_{z=\mu}$, and V is $N(0, \sigma)$.

Example 6.5 (t-Statistic). Let Y_n be a sequence of i.i.d. one-dimensional random variables and consider the *t*-statistic

$$t_n = \frac{\sqrt{n}(\overline{Y} - \mu)}{[(\sum_{j=1}^{n} Y_j^2 - n\overline{Y}^2)/(n-1)]^{\frac{1}{2}}} = \frac{\sqrt{n-1}(\overline{Y} - \mu)}{(\frac{1}{n}\sum_1^n Y_j^2 - \overline{Y}^2)^{\frac{1}{2}}}$$

$$= \sqrt{\frac{n-1}{n}}\,\bar{t}_n, \qquad \bar{t}_n = \frac{\sqrt{n}(\overline{Y} - \mu)}{(\frac{1}{n}\sum_1^n Y_j^2 - \overline{Y}^2)^{\frac{1}{2}}}. \qquad (6.58)$$

Here $EY_n = \mu$, $\text{var } Y_n = \sigma^2 > 0$ (finite). Note that

$$\sqrt{n}(\overline{Y} - \mu) \xrightarrow{\mathscr{L}} N(0, \sigma^2),$$

$$\frac{1}{n}\sum_1^n Y_j^2 - \overline{Y}^2 \xrightarrow{\text{a.s.}} EY_1^2 - \mu^2 = \sigma^2 > 0,$$

so that Proposition 6.8 applies to show that $t_n \xrightarrow{\mathscr{L}} N(0, 1)$.

Example 6.6 (Sample Correlation Coefficient). Let (X_j, Y_j), $j \geq 1$, be i.i.d. two-dimensional random vectors with $EX_j = \mu_x$, $EY_j = \mu_y$, $\text{var}\,X_j = \sigma_x^2 > 0$, $\text{var}\,Y_j = \sigma_y^2 > 0$, $\text{cov}(X_j, Y_j) = \sigma_{xy}$. The population coefficient of correlation is $\rho = \rho_{xy} = \sigma_{xy}/\sigma_x\sigma_y$. The sample coefficient correlation based on n observations is

$$r = r_{xy} = \frac{\frac{1}{n}\sum_{j=1}^n (X_j - \overline{X})(Y_j - \overline{Y})}{\left[\frac{1}{n}\sum_{j=1}^n (X_j - \overline{X})^2\right]^{1/2}\left[\frac{1}{n}\sum_{j=1}^n (Y_j - \overline{Y})^2\right]^{1/2}}. \tag{6.59}$$

Since the sample and population correlation coefficients are invariant under translation and scale change, we will replace X_j and Y_j in (6.59) by $U_j = (X_j - \mu_x)/\sigma_x$ and $V_j = (Y_j - \mu_y)/\sigma_y$, respectively. Then

$$
\begin{aligned}
r = r_{uv} &= \frac{\frac{1}{n}\sum_{j=1}^n (U_j - \overline{U})(V_j - \overline{V})}{\left[\frac{1}{n}\sum_{j=1}^n (U_j - \overline{U})^2\right]^{1/2}\left[\frac{1}{n}\sum_{j=1}^n (V_j - \overline{V})^2\right]^{1/2}} \\[2mm]
&= \frac{\frac{1}{n}\sum_{j=1}^n U_j V_j - \overline{U}\,\overline{V}}{\left[\frac{1}{n}\sum_{j=1}^n U_j^2 - \overline{U}^2\right]^{1/2}\left[\frac{1}{n}\sum_{j=1}^n V_j^2 - \overline{V}^2\right]^{1/2}}, \\[2mm]
\rho = \rho_{uv} &= E\,U_j V_j.
\end{aligned}
\tag{6.60}
$$

We will show that, under the assumption $E\,U_j^4 < \infty$ and $E\,V_j^4 < \infty$, one has

$$\sqrt{n}(r - \rho) \xrightarrow{\mathscr{L}} N(0, \sigma^2), \tag{6.61}$$

for some $\sigma^2 > 0$, to be computed later. We will apply the delta method in the form (6.57) with $k = 5$ and

$$Z_j = (U_j, V_j, U_j^2, V_j^2, U_j V_j) \quad (j \geq 1), \quad \mu = EZ_j = (0, 0, 1, 1, \rho),$$

$$\overline{Z} = \left(\overline{U}, \overline{V}, \frac{1}{n}\sum_1^n U_j^2, \frac{1}{n}\sum_1^n V_j^2, \frac{1}{n}\sum_1^n U_j V_j\right), \tag{6.62}$$

and

$$H(z_1, z_2, z_3, z_4, z_5) = \frac{z_5 - z_1 z_2}{(z_3 - z_1^2)^{\frac{1}{2}}(z_4 - z_2^2)^{\frac{1}{2}}}, \tag{6.63}$$

defined and continuously differentiable on the open set $(\subset \mathbb{R}^5)$

$$\left\{(z_1, z_2, z_3, z_4, z_5) : z_3 > z_1^2, z_4 > z_2^2\right\}.$$

Note that $H(\overline{Z}) = r$ and $H(\mu) = \rho$, so that $\sqrt{n}(r - \rho) = \sqrt{n}[H(\overline{Z}) - H(\mu)]$. To apply (6.57) we need to compute the five partial derivatives of H evaluated at μ. It is simple to check

$$\ell_1 = \left(\frac{\partial H}{\partial z_1}\right)_{z=\mu} = 0, \quad \ell_2 = \left(\frac{\partial H}{\partial z_2}\right)_{z=\mu} = 0, \quad \ell_3 \left(\frac{\partial H}{\partial z_3}\right)_{z=\mu} = -\frac{1}{2}\rho,$$

$$\ell_4 = \left(\frac{\partial H}{\partial z_4}\right)_{z=\mu} = -\frac{1}{2}\rho, \quad \ell_5 = \left(\frac{\partial H}{\partial z_5}\right)_{z=\mu} = 1. \tag{6.64}$$

Then, since $\ell_1 = 0 = \ell_2$,

$$\sigma^2 = \sum_{i,j=1}^{5} \ell_i \ell_j \sigma_{ij} = \sum_{i,j=3}^{5} \ell_i \ell_j \sigma_{ij}. \tag{6.65}$$

Now

$$
\begin{aligned}
\sigma_{33} &= \mathrm{var}(U_j^2) = EU_j^4 - 1, & \sigma_{44} &= \mathrm{var}(V_j^2) = EV_j^4 - 1, \\
\sigma_{55} &= \mathrm{var}(U_j V_j) = EU_j^2 V_j^2 - \rho^2, & \sigma_{34} &= \sigma_{43} = \mathrm{cov}(U_j^2, V_j^2) = EU_j^2 V_j^2 - 1, \\
\sigma_{35} &= \sigma_{53} = \mathrm{cov}(U_j^2, U_j V_j) & \sigma_{45} &= \sigma_{54} = EV_j^3 U_j - \rho. \\
&= EU_j^3 V_j - \rho,
\end{aligned}
\tag{6.66}
$$

From (6.64) to (6.66) we get

$$
\begin{aligned}
\sigma^2 &= \left(-\frac{1}{2}\rho\right)^2 (EU_j^4 - 1) + \left(-\frac{1}{2}\rho\right)^2 (EV_j^4 - 1) + 1^2(EU_j^2 V_j^2 - \rho^2) \\
&\quad + 2\left(-\frac{1}{2}\rho\right)^2 (EU_j^2 V_j^2 - 1) + 2\left(-\frac{1}{2}\rho\right)(1)(EU_j^3 V_j - \rho) \\
&\quad + 2\left(-\frac{1}{2}\rho\right)(1)(EV_j^3 U_j - \rho) \\
&= \rho^2 \left[\frac{EU_j^4}{4} + \frac{EV_j^4}{4} + \frac{EU_j^2 V_j^2}{2}\right] - \rho\left[EU_j^3 V_j + EV_j^3 U_j\right] + EU_j^2 V_j^2. \tag{6.67}
\end{aligned}
$$

In particular, if (X_j, Y_j) are normal, then (Exercise 6.16)

$$\sigma^2 = \rho^2 \left[\frac{3}{4} + \frac{3}{4} + \frac{1}{2} + \rho^2\right] - \rho[3\rho + 3\rho] + 1 + 2\rho^2 = 1 - 2\rho^2 + \rho^4 = (1-\rho^2)^2, \tag{6.68}$$

using the facts $EU^2 = EV^2 = 1$, $EU^4 = EV^4 = 3$, the conditional distribution of U, given V, is $N(\rho V, 1 - \rho^2)$ and, similarly, the conditional distribution of V, given U, is $N(\rho U, 1 - \rho^2)$.

Remark 6.11. If one attempts to derive (6.61) directly without using the delta method, then one arrives at

$$
\begin{aligned}
\sqrt{n}(r - \rho) = {}& \frac{\sqrt{n}\left[\frac{1}{n}\sum_{j=1}^{n}(U_j V_j - \rho) - \overline{U}\,\overline{V}\right]}{\left[\frac{1}{n}\sum_{j=1}^{n}(U_j - \overline{U}^2)\right]^{1/2}\left[\frac{1}{n}\sum_{j=1}^{n}(V_j - \overline{V})^2\right]^{1/2}} \\
& - \sqrt{n}\rho \left(1 - \frac{1}{\left[\frac{1}{n}\sum_{j=1}^{n}(U_j - \overline{U})^2\right]^{1/2}\left[\frac{1}{n}\sum_{j=1}^{n}(V_j - \overline{V})^2\right]^{1/2}}\right).
\end{aligned}
\tag{6.69}
$$

Using the fact that $\sqrt{n}\,\overline{U}\,\overline{V} \xrightarrow{P} 0$, and the denominator of the first term on the right in (6.69) converges to 1 in probability, one arrives at the fact that this term converges in distribution to $N(0, EU_1^2 V_1^2 - \rho^2)$. One may similarly prove that the second term on the right converges in distribution to some Normal distribution $N(0, \delta)$, say. However, from these facts alone one can not conclude convergence of the sum to a Normal law and/or derive the asymptotic variance of $\sqrt{n}(r - \rho)$. For problems such as these, Efron's, percentile *bootstrap method* for estimating the distribution of the statistic is very effective (see Chap. 9). Diaconis and Efron (1983) may be consulted for an example.

6.6 Asymptotics of Linear Regression

Consider the semiparametric regression model described in Example 6.3, Sect. 6.3. Write

$$b_n = \left[\sum_{j=1}^{n}(X_j - \overline{X})^2\right]^{\frac{1}{2}}. \tag{6.70}$$

We will first prove that

$$b_n(\hat{\beta} - \beta) \equiv \left(\sum_{j'=1}^{n}(X_{j'} - \overline{X})^2\right)^{\frac{1}{2}}(\hat{\beta} - \beta) \xrightarrow{\mathscr{L}} N(0, \sigma^2) \quad \text{as } n \to \infty, \tag{6.71}$$

if

$$\delta_n := \max_{1 \leq j \leq n} \frac{(X_j - \overline{X})^2}{\sum_{j' \leq 1}^{n}(X_{j'} - \overline{X})^2} \longrightarrow 0 \quad \text{as } n \to \infty. \tag{6.72}$$

Now the left side of (6.71) may be expressed as [see (6.14)]

$$\sum_{j=1}^{n}\varepsilon_{j,n}, \qquad \varepsilon_{j,n} =: \frac{(X_j - \overline{X})\varepsilon_j}{(\sum_{j'=1}^{n}(X_{j'} - \overline{X})^2)^{\frac{1}{2}}}, \tag{6.73}$$

so that

$$E\varepsilon_{j,n} = 0 \qquad E\varepsilon_{j,n}^2 = \frac{(X_j - \overline{X})^2}{\sum_{j'=1}^{n}(X_{j'} - \overline{X})^2}\sigma^2 \leq \delta_n\sigma^2,$$

$$\sum_{j=1}^{n}E\varepsilon_{j,n}^2 = \sigma^2. \tag{6.74}$$

By the Lindeberg central limit theorem (see Appendix D), it is enough to prove that for every $\eta > 0$,

$$\gamma_n := E\sum_{j=1}^{n}\varepsilon_{j,n}^2 \mathbf{1}_{[|\varepsilon_{j,n}|>\eta]} \longrightarrow 0 \quad \text{as } n \to \infty. \tag{6.75}$$

But the expectation on the left satisfies

$$\gamma_n \leq \sum_{j=1}^{n}\frac{(X_j - \overline{X})^2}{\sum_{j'=1}^{n}(X_{j'} - \overline{X})^2}\, E\varepsilon_j^2 \mathbf{1}_{[\varepsilon_j^2>\eta^2/\delta_n]}$$

$$= E\varepsilon_1^2 \mathbf{1}_{[\varepsilon_1^2>\eta^2/\delta_n]} \longrightarrow 0. \tag{6.76}$$

Note that $\mathbf{1}_{[\varepsilon_1^2>\eta^2/\delta_n]} \xrightarrow{\text{a.s.}} 0$ since $\eta^2/\delta_n \to \infty$, and $\varepsilon_1^2\mathbf{1}_{[\varepsilon_1^2>\eta^2/\delta_n]} \leq \varepsilon_1^2$, so that one may apply Lebesgue's dominated convergence theorem to obtain the last relation in (6.76). Using (6.16) one may write

$$a_n(\hat{\alpha} - \alpha) = \sum_{j=1}^{n} \tilde{\varepsilon}_{j,n} \tag{6.77}$$

where

$$a_n = \left\{ \frac{1}{n} + \frac{\overline{X}^2}{\sum_{j=1}^{n}(X_j - \overline{X})^2} \right\}^{-\frac{1}{2}} = n^{\frac{1}{2}} \left[\frac{\sum_{j=1}^{n}(X_j - \overline{X})^2}{\sum_{j=1}^{n} X_j^2} \right]^{\frac{1}{2}} = \frac{b_n}{m_2^{\frac{1}{2}}}, \tag{6.78}$$

so that

$$\tilde{\varepsilon}_{j,n} = a_n \left\{ \frac{-(X_j - \overline{X})\overline{X}}{\sum_{j=1}^{n}(X_j - \overline{X})^2} + \frac{1}{n} \right\} \varepsilon_j = \frac{1}{m_2^{\frac{1}{2}}} \left\{ \frac{-(X_j - \overline{X})\overline{X}}{b_n} + \frac{b_n}{n} \right\} \varepsilon_j. \tag{6.79}$$

Here

$$m_2 = \frac{1}{n} \sum_{j=1}^{n} X_j^2. \tag{6.80}$$

Then $\tilde{\varepsilon}_{j,n}$, $1 \leq j \leq n$, are independent,

$$E\tilde{\varepsilon}_{j,n} = 0, \quad \tilde{\varepsilon}_{j,n}^2 \leq 2a_n^2 \left\{ \frac{(X_j - \overline{X})^2 \overline{X}^2}{\left[\sum_{j=1}^{n}(X_j - \overline{X})^2 \right]^2} + \frac{1}{n^2} \right\} \varepsilon_j^2 = \theta_{j,n} \varepsilon_j^2, \text{ say,}$$

$$\sum_{j=1}^{n} E\tilde{\varepsilon}_{j,n}^2 = \sigma^2. \tag{6.81}$$

Now, noting that $n\overline{X}^2 \leq \sum_{j=1}^{n} X_j^2$, one has

$$\theta_{j,n} = \frac{2n\overline{X}^2(X_j - \overline{X})^2}{\left(\sum_{j=1}^{n}(X_j - \overline{X})^2 \right) \left(\sum_{j=1}^{n} X_j^2 \right)} + \frac{2 \sum_{j=1}^{n}(X_j - \overline{X})^2}{n \sum_{j=1}^{n} X_j^2}$$

$$\leq 2\delta_n + \frac{2}{n} \longrightarrow 0, \quad (1 \leq j \leq n);$$

$$\sum_{1}^{n} \theta_{jn} = \frac{2n\overline{X}^2}{\sum_{j=1}^{n} X_j^2} + \frac{2 \sum_{j=1}^{n}(X_j - \overline{X})^2}{\sum_{j=1}^{n} X_j^2}$$

$$\leq 2 + 2 = 4. \tag{6.82}$$

Hence

$$E \sum_{j=1}^{n} \tilde{\varepsilon}_{j,n}^2 \mathbf{1}_{\{\tilde{\varepsilon}_{j,n}^2 > \eta^2\}} \leq E \sum_{j=1}^{n} \theta_{j,n} \varepsilon_j^2 \mathbf{1}_{\{\varepsilon_j^2 > \eta^2/(2\delta_n + \frac{2}{n})\}}$$

$$= \sum_{j=1}^{n} \theta_{j,n} E\varepsilon_1^2 \mathbf{1}_{\{\varepsilon_1^2 > \eta^2/(2\delta_n + \frac{2}{n})\}} \longrightarrow 0 \quad \text{as } n \to \infty, \tag{6.83}$$

for every $\eta > 0$. Hence, by the Lindeberg CLT,

$$a_n(\hat{\alpha} - \alpha) \xrightarrow{\mathscr{L}} N(0, \sigma^2), \quad \text{as } n \to \infty. \tag{6.84}$$

From (6.72), (6.84) one can construct confidence intervals for α and β, separately, for any desired asymptotic level (or, confidence coefficient), provided a consistent estimator of σ^2 can be found. Such an estimator is

$$\hat{\sigma}^2 = \frac{1}{n-2} \sum_{j=1}^{n} \left(Y_j - \hat{\alpha} - \hat{\beta} X_j \right)^2 . \tag{6.85}$$

To see this, let $Z_n = \sum_{j=1}^{n}(X_j - \overline{X})\varepsilon_j$ and express the right side of (6.85) as

$$\frac{1}{n-2} \sum_{j=1}^{n} \left[-\frac{Z_n}{b_n^2}(X_j - \overline{X}) + \varepsilon_j - \overline{\varepsilon} \right]^2 = \frac{1}{n-2} \left[\frac{Z_n^2}{b_n^2} + \sum_{j=1}^{n}(\varepsilon_j - \overline{\varepsilon})^2 - 2\frac{Z_n^2}{b_n^2} \right]$$

$$= \frac{1}{n-2} \sum_{j=1}^{n} \varepsilon_j^2 - \frac{n}{n-2}\overline{\varepsilon}^2 - \frac{Z_n^2}{b_n^2(n-2)} . \tag{6.86}$$

By the SLLN, the first two terms on the extreme right side converge to σ^2 and 0 a.s.; also, $EZ_n^2/(b_n^2(n-2)) = \sigma^2/(n-2) \to 0$, so that the last term in (6.86) converges in probability to 0. Hence

$$\hat{\sigma}^2 \longrightarrow \sigma^2 \quad \text{in probability, as } n \to \infty. \tag{6.87}$$

In addition, $\hat{\sigma}^2$ is an *unbiased* estimator of σ^2. For

$$E\hat{\sigma}^2 = \frac{1}{n-2} E \left[\sum_{j=1}^{n}(\varepsilon_j - \overline{\varepsilon})^2 - \frac{Z_n^2}{b_n^2} \right] = \frac{1}{n-2} \left[(n-1)\sigma^2 - \sigma^2 \right]$$

$$= \sigma^2. \tag{6.88}$$

Thus

$$\hat{\alpha} \pm \frac{\hat{\sigma}}{a_n} z_{1-\frac{\theta}{2}}, \quad \hat{\beta} \pm \frac{\hat{\sigma}}{b_n} z_{1-\frac{\theta}{2}} \tag{6.89}$$

are confidence intervals for α and β, each of asymptotic confidence coefficient $1-\theta$. Here z_α is the α-th quantile of the standard Normal distribution $N(0,1)$,—that is, the $N(0,1)$—probability of this set of values larger than $z_{\theta/2}$ is $\theta/2$. Although, a *Bonferroni-type confidence region* for the pair (α, β) may be given by the rectangle

$$\left\{ (\alpha, \beta) : \hat{\alpha} - \frac{\hat{\sigma}}{a_n} z_{1-\frac{\theta}{4}} \leq \alpha \leq \hat{\alpha} + \frac{\hat{\sigma}}{a_n} z_{1-\frac{\theta}{4}}, \ \hat{\beta} - \frac{\hat{\sigma}}{b_n} z_{1-\frac{\theta}{4}} \leq \beta \leq \hat{\beta} + \frac{\hat{\sigma}}{b_n} z_{1-\frac{\theta}{4}} \right\},$$
$$\tag{6.90}$$

having an asymptotic confidence coefficient of at least $(1 - \theta)$, for a better and more precise confidence region for (α, β), we need to look at the joint distribution of $(\hat{\alpha}, \hat{\beta})$. The $k \times k$ *identity matrix* is denoted by I_k.

Theorem 6.2 (Asymptotic Joint Distribution of Regression Coefficients). *Consider the linear regression model (6.12) with i.i.d. errors ε_j satisfying $E\varepsilon_j = 0$, $0 < E\varepsilon_j^2 \equiv \sigma^2 < \infty$. If the quantity δ_n in (6.72) goes to zero, then $(\hat{\alpha}, \hat{\beta})$ is asymptotically Normal, i.e.,*

$$(U_n, V_n)' \equiv \Gamma_n(a_n(\hat{\alpha} - \alpha), b_n(\hat{\beta} - \beta))' \xrightarrow{\mathscr{L}} N \left(\begin{pmatrix} 0 \\ 0 \end{pmatrix}, \sigma^2 I_2 \right), \tag{6.91}$$

where $\Gamma_n = ((\gamma_{ii'}))$ is the symmetric positive definite matrix satisfying

$$\Gamma_n^2 = \begin{bmatrix} 1 & \dfrac{-\overline{X}}{m_2^{1/2}} \\[2mm] \dfrac{-\overline{X}}{m_2^{1/2}} & 1 \end{bmatrix}^{-1} \equiv \dfrac{1}{1 - \dfrac{\overline{X}^2}{m_2}} \begin{bmatrix} 1 & \dfrac{\overline{X}}{m_2^{1/2}} \\[2mm] \dfrac{\overline{X}}{m_2^{1/2}} & 1 \end{bmatrix}. \tag{6.92}$$

Proof. Note that the covariance matrix of $(a_n(\hat{\alpha} - \alpha), b_n(\hat{\beta} - \beta))'$ is

$$\Sigma_n = \sigma^2 \begin{bmatrix} 1 & \dfrac{-\overline{X}}{m_2^{1/2}} \\[2mm] \dfrac{-\overline{X}}{m_2^{1/2}} & 1 \end{bmatrix}, \tag{6.93}$$

so that the covariance matrix of the left side of (6.91) is $\sigma^2 I_2$, where I_2 is the 2×2 identity matrix. Let a, b be arbitrary reals. We will show that $aU_n + bV_n \xrightarrow{\mathscr{L}} N(0, \sigma^2)$. For this write

$$aU_n + bV_n = \sum_{j=1}^{n} \zeta_{j,n}, \quad \zeta_{j,n} := (a\gamma_{11} + b\gamma_{21})\tilde{\varepsilon}_{j,n} + (a\gamma_{12} + b\gamma_{22})\varepsilon_{j,n} \tag{6.94}$$

Then $\zeta_{j,n}$ are independent, $E\zeta_{j,n} = 0$, and [see (6.74), (6.90)]

$$E\zeta_{j,n}^2 \le 2\sigma^2 \left[(a\gamma_{11} + b\gamma_{21})^2 \left(2\delta_n + \frac{2}{n} \right) + (a\gamma_{12} + b\gamma_{22})^2 \, \delta_n \right]$$

$$\le 2\sigma^2 \left(2\delta_n + \frac{2}{n} \right) \left| \Gamma_n \begin{pmatrix} a \\ b \end{pmatrix} \right|^2 \longrightarrow 0 \ \text{ as } n \to \infty. \tag{6.95}$$

To prove the convergence to 0 in (6.95), check that

$$\gamma_{11} = \gamma_{22} = \frac{1}{2\sqrt{1 - \dfrac{\overline{X}^2}{m_2}}} \left[\sqrt{1 + \frac{\overline{X}}{m_2^{1/2}}} + \sqrt{1 - \frac{\overline{X}}{m_2^{1/2}}} \, \right],$$

$$\gamma_{12} = \gamma_{21} = \frac{\overline{X}}{2\sqrt{1 - \dfrac{\overline{X}^2}{m_2}}} \left[\sqrt{1 + \frac{\overline{X}}{m_2^{1/2}}} - \sqrt{1 - \frac{\overline{X}}{m_2^{1/2}}} \, \right]. \tag{6.96}$$

Now it is easy to verify (6.95). Also, $\sum_{j=1}^{n} E\zeta_{j,n}^2 = \sigma^2(a^2 + b^2)$, using the fact that the covariance matrix of $(U_n, V_n)'$ is $\Gamma_n \Sigma_n \Gamma_n = \sigma^2 I_2$. The proof of $\sum_{j=1}^{n} E\zeta_{j,n}^2 \mathbf{1}_{\{\zeta_{j,n}^2 > \eta^2\}} \to 0$, as $n \to \infty$, for every $\eta > 0$, now follows exactly as in (6.76), or (6.83). Thus for all (a, b), $aU_n + bV_n \xrightarrow{\mathscr{L}} N(0, (a^2 + b^2)\sigma^2)$. By the so-called Cramér–Wold device,[8] it now follows that $(U_n, V_n)' \xrightarrow{\mathscr{L}} N\left(\binom{0}{0}, \sigma^2 I_2\right)$.
$\qquad\qquad\qquad\qquad\qquad\qquad\qquad\qquad\qquad\qquad\qquad\qquad\qquad\qquad\quad \square$

It follows from Theorem 6.2 that, if (6.74) holds, *a confidence region for (α, β), with asymptotic confidence coefficient $1 - \theta$, is given by the ellipse* (Exercise 6.21).

$$C_n := \Big\{ (\alpha, \beta) : p(a_n(\hat{\alpha} - \alpha))^2$$

$$+ q(b_n(\hat{\beta} - \beta))^2 + \gamma a_n b_n(\hat{\alpha} - \alpha)(\hat{\beta} - \beta) \le \hat{\sigma}^2 \chi_{1-\theta}^2(2) \Big\}, \tag{6.97}$$

---

[8] See Bhattacharya and Waymire (2007, p. 105) or Billingsley (1986, p. 397).

where

$$p = \frac{1}{1 - \frac{\overline{X}^2}{m_2}} = \frac{m_2}{\frac{b_n^2}{n}} = q, \quad r = \frac{2\frac{\overline{X}}{m_2^{1/2}}}{1 - \frac{\overline{X}^2}{m_2}}, \tag{6.98}$$

and $\chi_\alpha^2(2)$ is the α-th quantile of the chi square distribution with 2 degrees of freedom.

Remark 6.12. In Sect. 6.9, Theorem 6.2 is extended to the case of multiple linear regression, allowing more than one explanatory real-valued variable X. It is also shown there in particular that the condition $\delta_n \to 0$ in (6.72) is also *necessary* from the asymptotic Normality of $(\hat{\alpha}, \hat{\beta})$.

Remark 6.13 (Parametric Linear Regression with Normal Errors). Assume that ε_j's in (6.12) are i.i.d. $N(0, \sigma^2)$. Then the least squares estimators $\hat{\alpha}$, $\hat{\beta}$ above are also the maximum likelihood estimators of α, β. It follows, from the fact that $\hat{\alpha} - \alpha$, $\hat{\beta} - \beta$ are linear combinations of ε_j's, that they are Normal, (individually and jointly, with zero means and variances given by (6.15), (6.16), and covariance $-\sigma^2\overline{X}/b_n^2$. Note that this is *true for all* $n \geq 3$, assuming X_j's $(1 \leq j \leq n)$ are not all the same:

$$\left(\hat{\alpha} - \alpha, \hat{\beta} - \beta\right)' \overset{\mathscr{L}}{=} N\left(\begin{pmatrix} 0 \\ 0 \end{pmatrix}, \sigma^2 \begin{bmatrix} \frac{1}{n} + \frac{\overline{X}^2}{b_n^2} & -\frac{\overline{X}}{b_n^2} \\ -\frac{\overline{X}}{b_n^2} & b_n^{-2} \end{bmatrix}\right) \tag{6.99}$$

From the theory of Linear Models in Part I, Sect. 5.9, it is known that $\hat{\sigma}^2$ is independent of $(\hat{\alpha}, \hat{\beta})$, and that the left side of (6.91) has the Normal distribution on the right side for all $n \geq 3$. In particular, the quadratic form appearing on the left of the inequality within curly brackets in (6.97) is σ^2-times chi-square random variable with degrees of freedom 2, and

$$\widetilde{D}_n := \left\{(\alpha, \beta) : n\left[\hat{\alpha} - \alpha + (\hat{\beta} - \beta)\overline{X}\right]^2 + b_n^2(\hat{\beta} - \beta)^2 \leq 2\hat{\sigma}^2 F_{1-\theta}(2, n-2)\right\} \tag{6.100}$$

is a confidence region for (α, β) of *exact* confidence coefficient $1 - \theta$, for all $n \geq 3$ (assuming X_j, $1 \leq j \leq n$, are not all the same) (Exercise 6.22). Here $F_\alpha(r, s)$ is the α-th quantile of the F-distribution with numerator d.f. r and denominator d.f. s.

Example 6.7 (A Heteroscedastic Linear Regression Model with Known Error Variances). Consider the linear regression

$$Y_j := \alpha + \beta X_j + \varepsilon_j \quad (1 \leq j \leq n),$$

ε_j's are independent,

$$E\varepsilon_j = 0, \quad 0 < \sigma_j^2 = E\varepsilon_j^2 < \infty. \tag{6.101}$$

Assume that σ_j^2 are known $(1 \leq j \leq n)$. If one also assumes that ε_j's are Normal $N(0, \sigma_j^2)$, then the M.L.E.'s $\hat{\alpha}$, $\hat{\beta}$ are the solutions of

$$\sum_{j=1}^n \frac{1}{\sigma_j^2}(Y_j - \alpha - \beta X_j) = 0, \quad \sum_{j=1}^n \frac{1}{\sigma_j^2} X_j(Y_j - \alpha - \beta X_j) = 0. \tag{6.102}$$

Rewrite these as

$$\sum_{j=1}^{n} \frac{1}{\sigma_j^2}\left(Y_j - \{\alpha + \beta\overline{X}_\omega\} - \beta\{X_j - \overline{X}_\omega\}\right) = 0,$$

$$\sum_{j=1}^{n} \frac{1}{\sigma_j^2}X_j\left(Y_j - \{\alpha - \beta\overline{X}_\omega\} - \beta\{X_j - \overline{X}_\omega\}\right) = 0, \tag{6.103}$$

where $\overline{X}_\omega$ is the weighted mean of X_j $(1 \le j \le n)$,

$$\overline{X}_\omega = \frac{\sum_{j=1}^{n}\frac{1}{\sigma_j^2}X_j}{\sum_{j=1}^{n}\frac{1}{\sigma_j^2}}. \tag{6.104}$$

Writing $\alpha + \beta\overline{X}_\omega = \delta$, one obtains

$$\hat{\delta} = \overline{Y}_\omega \equiv \frac{\sum_{j=1}^{n}\frac{Y_j}{\sigma_j^2}}{\sum_{j=1}^{n}\frac{1}{\sigma_j^2}}, \quad \hat{\beta} = \frac{\sum_{1}^{n}(Y_j - \overline{Y}_\omega)X_j/\sigma_j^2}{\sum_{1}^{n}(X_j - \overline{X}_\omega)X_j/\sigma_j^2}, \tag{6.105}$$

$$\hat{\alpha} = \hat{\delta} - \hat{\beta}\overline{X}_\omega = \overline{Y}_\omega - \hat{\beta}\overline{X}_\omega. \tag{6.106}$$

Note that

$$\hat{\beta} = \frac{\sum_{1}^{n}Y_j(X_j - \overline{X}_\omega)/\sigma_j^2}{\sum_{1}^{n}X_j(X_j - \overline{X}_\omega)/\sigma_j^2} = \beta + \sum_{j=1}^{n}\frac{(X_j - \overline{X}_\omega)/\sigma_j^2}{\sum_{1}^{n}X_j(\overline{X}_j - \overline{X}_\omega)/\sigma_j^2}\,\varepsilon_j,$$

$$\hat{\delta} = \alpha + \beta\overline{X}_\omega + \sum_{j=1}^{n}\frac{1/\sigma_j^2}{\sum_{j=1}^{n}1/\sigma_j^2}\,\varepsilon_j$$

$$= \delta + \sum_{j=1}^{n}\frac{1/\sigma_j^2}{\sum_{j=1}^{n}1/\sigma_j^2}\,\varepsilon_j. \tag{6.107}$$

Thus, for Normal ε_j's, $(\hat{\delta},\hat{\beta})'$ is Normal $N\left(\binom{\delta}{\beta}, \widetilde{\Sigma}\right)$, where $\widetilde{\Sigma} = ((\tilde{\sigma}_{ii'}))$ is given by

$$\sigma_{\hat{\delta}}^2 = \tilde{\sigma}_{11} = \left(\sum_{j=1}^{n}\frac{1}{\sigma_j^2}\right)^{-1}, \quad \sigma_{\hat{\beta}}^2 = \tilde{\sigma}_{22} = \left(\sum_{1}^{n}(X_j - \overline{X}_\omega)^2/\sigma_j^2\right)^{-1},$$

$$\tilde{\sigma}_{12} = \tilde{\sigma}_{21} = 0. \tag{6.108}$$

Also, $(\hat{\alpha},\hat{\beta})'$ is Normal $N\left(\binom{\alpha}{\beta}, \Sigma\right)$ with $\Sigma = ((\sigma_{ii'}))$, given by

$$\sigma_{11} = \tilde{\sigma}_{11} + \overline{X}_\omega^2\tilde{\sigma}_{22}, \quad \sigma_{22} = \tilde{\sigma}_{22}, \quad \sigma_{12} = \sigma_{21} = -\overline{X}_\omega\tilde{\sigma}_{22}. \tag{6.109}$$

Consider next the general case (6.101), ε_j's not necessarily Normal, but σ_j^2 $(1 \le j \le n)$ are known. Writing the *weighted least squares estimators* as

$$\hat{\delta} = \delta + \sum_{1}^{n}\omega_j\varepsilon_j, \quad \hat{\beta} = \beta + \sum_{1}^{n}\gamma_j\varepsilon_j, \quad \omega_j = \frac{\frac{1}{\sigma_j^2}}{\sum_{1}^{n}\frac{1}{\sigma_j^2}}. \tag{6.110}$$

one has

$$\sigma_{\hat{\delta}}^{-1}(\hat{\delta} - \delta) = \sum_1^n \varepsilon_{j,n}, \quad \sigma_{\hat{\beta}}^{-1}(\hat{\beta} - \beta) = \sum_1^n \xi_{j,n},$$

$$\varepsilon_{j,n} = \left\{ \frac{\frac{1}{\sigma_j^2}}{\left(\sum_1^n \frac{1}{\sigma_j^2}\right)^{\frac{1}{2}}} \right\} \varepsilon_j, \quad \xi_{j,n} = \frac{\frac{(X_j - \overline{X}_\omega)}{\sigma_j^2}}{\left[\sum_1^n (X_j - \overline{X}_\omega)^2 / \sigma_j^2\right]^{\frac{1}{2}}} \varepsilon_j. \qquad (6.111)$$

Note first that $\varepsilon_{j,n}$ $(1 \leq j \leq n)$ are independent, and

$$E\varepsilon_{j,n} = 0, \quad E\varepsilon_{j,n}^2 = (1/\sigma_j^2)/\sum_1^n 1/\sigma_j^2 = \omega_j,$$

$$\sum_{j=1}^n E\varepsilon_{j,n}^2 = 1. \qquad (6.112)$$

If

(i) ε_j/σ_j $(j = 1, 2, \dots)$ are uniformly integrable and

(ii) $\theta_n := \max\{\omega_j : 1 \leq j \leq n\} \longrightarrow 0,$ \qquad (6.113)

then it is simple to check that the Lindeberg conditions hold for the (triangular) sequence $\{\varepsilon_{j,n} : 1 \leq j \leq n, \, n \geq 1\}$ (Exercise 6.23). Hence, if (6.20) holds,

$$\sigma_{\hat{\delta}}^{-1}(\hat{\delta} - \delta) \xrightarrow{\mathscr{L}} N(0,1). \qquad (6.114)$$

Similarly, if

$$\tilde{\delta}_n := \frac{\max\left\{(X_j - \overline{X}_\omega)^2 / \sigma_j^2 : 1 \leq j \leq n\right\}}{\sum_1^n (X_j - \overline{X}_\omega)^2 / \sigma_j^2} \longrightarrow 0, \qquad (6.115)$$

then one can show that (Exercise 6.23)

$$\sigma_{\hat{\beta}}^{-1}(\hat{\beta} - \beta) \xrightarrow{\mathscr{L}} N(0,1). \qquad (6.116)$$

More generally, if (6.20), (6.115) hold, then (Exercise 6.23)

$$\begin{pmatrix} \sigma_{\hat{\delta}}^{-1}(\hat{\delta} - \delta) \\ \sigma_{\hat{\beta}}^{-1}(\hat{\beta} - \beta) \end{pmatrix} \xrightarrow{\mathscr{L}} N\left(\begin{pmatrix} 0 \\ 0 \end{pmatrix}, I_2\right). \qquad (6.117)$$

Just as argued in Remark 6.12 in the case of homoscedastic linear regression (6.115), for the present heteroscedastic case, a confidence region for (α, β) with asymptotic confidence coefficient $1 - \theta$ is given by the ellipse (Exercise 6.23)

$$\tilde{D}_n := \left\{(\alpha, \beta) : (\sigma_{\hat{\delta}}^2)^{-1}\{(\hat{\alpha} - \alpha) + (\hat{\beta} - \beta)\overline{X}_\omega\}^2 + \left(\sigma_{\hat{\beta}}^2\right)^{-1}(\hat{\beta} - \beta)^2 \leq \chi_{1-\theta}^2(2)\right\}. \qquad (6.118)$$

For an application, consider the commonly arising problem[9] of estimating the postulated relationship $D = \alpha + \beta T^{-1}$ between the diffusion coefficient D of a substance in a given medium and the temperature T of the medium (Ito and Ganguly 2006). Under isothermal conditions at each of temperatures $T_1, \dots, T_n,$

---

[9] Communicated by Professor Jiba Ganguly, University of Arizona.

the diffusion coefficient D is estimated in n independent experiments, perhaps by observing concentration profiles. These experiments also yield estimates of the standard deviations σ_j of the estimates y_j of D_j ($1 \leq j \leq n$). Writing $x = T^{-1}$, one then has the model (6.101).

Remark 6.14. Suppose that σ_j^2 in (6.101), is known up to a positive scalar, i.e., $\sigma_j^2 = \eta_j^2 \sigma^2$, where η_j^2 is known ($1 \leq j \leq n$), but $\sigma^2 > 0$ is not. For example, one may postulate that σ_j^2 is proportional to X_j (for a positive explanatory variable), in which case $\sigma_j^2 = X_j \sigma^2 (\eta_j^2 = X_j)$. If the ε_j's are also Normal, then the Eqs. (6.102)–(6.107), (6.110), all hold if σ_j^2 *is replaced by* η_j^2. For the variances and covariances in (6.108), (6.109), a multiplier σ^2 is to be used to each of $\tilde{\sigma}_{ii'}$, $\sigma_{ii'}$. If ε_j's are not necessarily Normal, the arguments leading to (6.12), (6.15), (6.117) remain the same. However, for statistical inference using the CLT one needs to have a consistent estimator of σ^2. This is given by

$$\hat{\sigma}^2 = \frac{1}{n-2} \sum_{j=1}^{n} (Y_j - \hat{\alpha} - \hat{\beta} X_j)^2 / \eta_j^2. \tag{6.119}$$

First note that $\hat{\sigma}^2$ is *unbiased* (Exercise 6.24). Next

$$\hat{\sigma}^2 = \frac{1}{n-2} \sum_{j=1}^{n} \frac{1}{\eta_j^2} \left[\hat{\delta} - \delta + (\hat{\beta} - \beta)(X_j - \overline{X}_\omega) - \varepsilon_j \right]^2$$

$$= \frac{1}{n-2} \left(\sum_{j=1}^{n} \frac{1}{\eta_j^2} \right) (\hat{\delta} - \delta)^2 + \frac{1}{n-2} (\hat{\beta} - \beta)^2 \sum_{1}^{n} (X_j - \overline{X}_\omega)^2 / \eta_j^2$$

$$+ \frac{1}{n-2} \sum_{j=1}^{n} \varepsilon_j^2 / \eta_j^2 - \frac{2}{n-2} (\hat{\delta} - \delta) \sum_{j=1}^{n} \varepsilon_j / \eta_j^2$$

$$- \frac{2}{n-2} (\hat{\beta} - \beta) \sum_{j=1}^{n} \varepsilon_j (X_j - \overline{X}_\omega) / \eta_j^2 - \frac{1}{n-2} (\hat{\delta} - \delta)(\hat{\beta} - \beta) \sum_{j=1}^{n} (X_j - \overline{X}_\omega) / \eta_j^2$$

$$= \frac{\sigma^2}{n-2} \sigma_{\hat{\delta}}^{-2} (\hat{\delta} - \delta)^2 + \frac{\sigma^2}{n-2} \sigma_{\hat{\beta}}^{-2} (\hat{\beta} - \beta)^2 + \frac{1}{n-2} \sum_{j=1}^{n} \varepsilon_j^2 / \eta_j^2$$

$$- \frac{2\sigma^2}{n-2} \sigma_{\hat{\delta}}^{-2} (\hat{\delta} - \delta)^2 - 2\sigma^2 \sigma_{\hat{\beta}}^{-2} \frac{(\hat{\beta} - \beta)^2}{n-2} - \frac{2\sigma^2}{n-2} \sigma_{\hat{\delta}}^{-1} \sigma_{\hat{\beta}}^{1} (\hat{\delta} - \delta)(\hat{\beta} - \beta). \tag{6.120}$$

Since $\sigma_{\hat{\delta}}^{-1}(\hat{\delta} - \delta) \xrightarrow{\mathscr{L}} N(0,1)$, $\sigma_{\hat{\beta}}^{-1}(\hat{\beta} - \beta) \xrightarrow{\mathscr{L}} N(0,1)$, these quantities are bounded in probability. Hence all the terms on the extreme right of (6.120) go to zero in probability, except for the term $\frac{1}{n-2} \sum_{j=1}^{n} \varepsilon_j^2 / \eta_j^2 = J_n$, say. Now $E \varepsilon_j^2 / \eta_j^2 = \sigma^2$. So if either ε_j / η_j are i.i.d. (as in the Normal case), or if their variances are bounded, i.e.,

$$\sup_j E \left(\varepsilon_j / \eta_j \right)^4 < \infty, \tag{6.121}$$

then J_n converges in probability to σ^2. Hence $\hat{\sigma}^2$ is an unbiased consistent estimator of σ^2 if, in addition to (6.113) and (6.115), (6.121) holds. In particular, it then follows that a confidence region for (α, β) with asymptotic confidence coefficient $1 - \theta$ is given by the modified version of (6.118) obtained by replacing σ^2 by $\hat{\sigma}^2$ as given in (6.119), provided (6.113), (6.115), (6.121) hold.

The estimates in Example 6.7 are special cases of what are called *weighted least squares estimates*.

Remark 6.15. If one simply assumes (6.101) and that the ε_j's are uncorrelated (instead of being independent), then the OLS estimator (6.14) is unbiased with variance $\sum[(X_j - \overline{X})^2\sigma_j^2 / \sum(X_j - \overline{X})^2]$. If the latter goes to zero as $n \to \infty$, then the OLS is a consistent estimator of β.

Remark 6.16 (Confidence Region for Regression Lines). Given a confidence region D for (α, β) with (asymptotic) confidence coefficient $1 - \theta$, the set L of lines $y = a + bx$ with $(a, b) \in D$ is a confidence region for the true regression line $y = \alpha + \beta x$ with (asymptotic) confidence coefficient $1 - \theta$. However, for a given x, the confidence bound for the regression is $\hat{\alpha} + \hat{\beta}x \pm z_{1-\frac{\theta}{2}} SE(\hat{\alpha} + \hat{\beta}x)$, where the *standard error* $SE(\hat{\alpha} + \hat{\beta}x)$ is obtained from (6.99) with σ^2 replaced by $\hat{\sigma}^2$ [see (6.85)]. By a *confidence band* (with confidence level $1 - \theta$) one usually means the regions between the upper and lower curves constructed as above (for all x).

The asymptotic probability that this confidence band contains the entire regression line is in general *smaller* than $1 - \theta$. On the other hand, an application of *Scheffé's method of multiple comparison*[10] provides a confidence band containing the regression line with a probability at least $1 - \theta$. Scheffé's method is based on the simple fact that for a random vector U and a constant $c > 0$, the events $\{\|\mathbf{U}\|^2 \leq c^2\}$ and $\{\langle\mathbf{U}, \boldsymbol{\gamma}\rangle^2 \leq \|\boldsymbol{\gamma}\|^2 c^2 \ \forall \ \text{vectors} \ \boldsymbol{\gamma}\}$ are equivalent and, therefore, have the same probability.

Proposition 6.9 (Simultaneous Confidence Region for Regression). *Under the hypothesis of Theorem 6.2, the asymptotic probability that*

$$|\hat{\alpha} + \hat{\beta}x - (\alpha + \beta x)| \leq \frac{\hat{\sigma}}{b_n}\left[m_2 - 2x\overline{X} + x^2\right]^{\frac{1}{2}} \sqrt{\chi_{1-\theta}^2(2)} \quad \text{for all } x \qquad (6.122)$$

is at least $1 - \theta$.

Proof. By (6.93) or (6.16), the covariance matrix of $(\hat{\alpha}, \hat{\beta})$ is

$$V_n = \frac{\sigma^2}{b_n^2}\begin{bmatrix} m_2 & -\overline{X} \\ -\overline{X} & 1 \end{bmatrix},$$

and, by Theorem 6.2, $Q \equiv (\hat{\alpha} - \alpha, \hat{\beta} - \beta)V_n^{-1}(\hat{\alpha} - \alpha, \hat{\beta} - \beta)'$ converges in distribution to the chi-square distribution with two degrees of freedom, as $n \to \infty$. Now $Q = \|\mathbf{U}\|^2$ where $\mathbf{U} = V_n^{-1/2}(\hat{\alpha} - \alpha, \hat{\beta} - \beta)'$, $V_n^{1/2}$ being the symmetric positive definite matrix such that $V_n^{1/2}V_n^{1/2} = V_n$ and $V_n^{-1/2} = (V_n^{1/2})^{-1}$. One has $P(\|\mathbf{U}\|^2 \leq \chi_{1-\theta}^2(2)) \longrightarrow 1 - \theta$ as $n \to \infty$. Now $\{\|\mathbf{U}\|^2 \leq \chi_{1-\theta}^2(2)\} = \{\langle\mathbf{U}, V_n^{1/2}\boldsymbol{\gamma}\rangle^2 \leq \|V_n^{1/2}\boldsymbol{\gamma}\|^2\chi_{1-\theta}^2(2) \ \forall \ \boldsymbol{\gamma} \in \mathbb{R}^2\} = \{\langle(\hat{\alpha} - \alpha, \hat{\beta} - \beta)', \boldsymbol{\gamma}\rangle^2 \leq \langle\boldsymbol{\gamma}, V_n\boldsymbol{\gamma}\rangle\chi_{1-\theta}^2(2) \ \forall \ \boldsymbol{\gamma} \in \mathbb{R}^2\}$. In particular taking $\boldsymbol{\gamma} = (1, x)'$, one gets $\{\|\mathbf{U}\|^2 \leq \chi_{1-\theta}^2(2)\} \subset \{|(\hat{\alpha} + \hat{\beta}x - (\alpha + \beta x)|^2 \leq \frac{\sigma^2}{b_n^2}[m_2 - 2n\overline{X} + x^2]\chi_{1-\theta}^2(2) \forall \ x\}$. Hence the last event has asymptotic probability at least $1 - \theta$. Replacing σ^2 by $\hat{\sigma}^2$ one obtains the desired result.

---

[10] H. Scheffé (1959).

Often the explanatory variable X is random. In the so-called *correlation model* one assumes that, in (6.12), X_j's are i.i.d. with finite positive variance ($j \geq 1$) and that they are independent of the error sequence $\{\varepsilon_j : j \geq 1\}$, the latter being i.i.d. with mean zero and variance $\sigma^2 > 0$ as assumed above. One may then apply the above arguments verbatim, *conditionally given* X_j's ($1 \leq j \leq n$), noting that in this case (6.72) holds with probability one. Hence (6.91) holds, conditionally given X_j's ($j \geq 1$). Since the *nonrandom* limit does not depend on the X_j's it follows that (6.91) holds unconditionally as well. We state this as a corollary to Theorem 6.2. (Exercise 6.25).

Corollary 6.3 (The Correlation Model). *Suppose that in (6.12), X_j's are also i.i.d., independent of the i.i.d. ε_j's, and that $0 < \mathrm{var}\, X_j < \infty$. Then (6.91) holds.*

As a final remark on convergence in distribution, we state the following useful fact.[11]

Remark 6.17. Suppose P_n ($n \geq 1$), P are probability measures on $(\mathbb{R}, \mathscr{B}(\mathbb{R}))$ such that P_n converges weakly to P. If f is a bounded measurable function whose points of discontinuity comprise a set D with $P(D) = 0$, then $\int f dP_n \to \int f dP$.

6.7 Asymptotic Distribution of Sample Quantiles, Order Statistics

Throughout this section it will be assumed that $X_1, X_2, \ldots$ is an i.i.d. sequence of random variables whose common *distribution function* F *is continuous on* $(-\infty, \infty)$. For each fixed n, let $X_{(1)} < X_{(2)} < \cdots < X_{(n)}$ be an ordering of $X_1, X_2, \ldots, X_n$. Note that $\mathrm{Prob}(X_i = X_j) = \int_{-\infty}^{\infty} \mathrm{Prob}(X_i = x/X_j = x)dF(x) = \int_{-\infty}^{\infty} \mathrm{Prob}(X_i = x)dF(x) = \int_{-\infty}^{\infty} 0 dF(x) = 0$, for each pair (i, j) with $i \neq j$. Hence one may assume *strict* ordering among $X_1, \ldots, X_n$. Observe also that one should write $X_{(1):n} < X_{(2):n} < \cdots < X_{(n):n}$, in order to emphasize that $X_{(i)}$, for any i ($1 \leq i \leq n$), depends on n. However, we will write $X_{(1)}, \ldots, X_{(n)}$ for the n *order statistics* to simplify notations. With this notation the distribution function of the rth *order statistic* is easily seen to be

$$\mathrm{Prob}(X_{(r)} \leq x) = \mathrm{Prob}(\text{At least } r \text{ of the } n \text{ random variables } X_1, \ldots, X_n \text{ are } \leq x)$$

$$= \sum_{j=r}^{n} \mathrm{Prob}(\text{Exactly } j \text{ of the random variables are } \leq x)$$

$$= \sum_{j=r}^{n} \binom{n}{j} (F(x))^j (1 - F(x))^{n-j}, \qquad [1 \leq r \leq n]. \qquad (6.123)$$

Fix $0 < p < 1$. Define the *pth quantile* ξ_p of F as the solution of

$$F(x) = p \qquad (6.124)$$

if this solution is unique; else let a *pth quantile* be any solution of (6.124). By the sample *pth quantile* $\hat{\zeta}_p$ we shall mean the order statistic $X_{([np])}$. (Once again, a more appropriate notation would be $\hat{\zeta}_{p:n}$).

---

[11] See Bhattacharya and Waymire (2007, Theorem 5.2, p. 62).

Theorem 6.3. *(a)* *Fix p, $0 < p < 1$. Assume that the solution ζ_p of (6.124) is unique, F is continuously differentiable in a neighborhood of ζ_p and $F'(\zeta_p) = f(\zeta_p) > 0$. Then*

$$\sqrt{n}(\hat{\zeta}_p - \zeta_p) \xrightarrow{\mathscr{L}} N\left(0, \frac{p(1-p)}{f^2(\zeta_p)}\right) \qquad \text{as } n \to \infty. \tag{6.125}$$

(b) *Let $0 < p_1 < p_2 < \cdots < p_k$, $k > 1$, and assume that the hypothesis of (a) holds for $p = p_i$ for all $i = 1, 2, \ldots, k$. Then*

$$\sqrt{n}\left(\hat{\xi}_{p_1} - \xi_{p_1}, \hat{\xi}_{p_2} - \xi_{p_2}, \ldots, \hat{\xi}_{p_k} - \xi_{p_k}\right) \xrightarrow{\mathscr{L}} N(\mathbf{0}, \Sigma), \tag{6.126}$$

where $\Sigma = ((\sigma_{ij}))$, $\sigma_{ij} = p_i(1 - p_j)/f(\xi_{p_i})f(\xi_{p_j})$ for $1 \le i \le j < k$.

Proof. (a) Fix $z \in (-\infty, \infty)$. Then, writing $\mathbf{1}_A$ for the *indicator* of A,

$$\text{Prob}\left(\sqrt{n}(\hat{\zeta}_p - \zeta_p) \le z\right)$$

$$= \text{Prob}\left(\hat{\zeta}_p \le \zeta_p + \frac{z}{\sqrt{n}}\right)$$

$$= \text{Prob}\left(\# \text{ of observations among } X_1, \ldots, X_n \text{ which are } \le \zeta_p + \frac{z}{\sqrt{n}} \text{ is } \ge \lceil np \rceil\right)$$

$$= \text{Prob}\left(\sum_{j=1}^{n} \mathbf{1}_{\{X_j \le \zeta_p + z/\sqrt{n}\}} \ge \lceil np \rceil\right)$$

$$= \text{Prob}\left[\frac{1}{\sqrt{n}}\left(\sum_{j=1}^{n} \mathbf{1}_{\{X_j \le \zeta_p + \frac{z}{\sqrt{n}}\}} - nF\left(\zeta_p + \frac{z}{\sqrt{n}}\right)\right)\right.$$

$$\left. \ge \frac{1}{\sqrt{n}}\left(\lceil np \rceil - nF\left(\zeta_p + \frac{z}{\sqrt{n}}\right)\right)\right] = P(Z_n \ge C_n), \tag{6.127}$$

where $Z_n = \frac{1}{\sqrt{n}}(\sum_{j=1}^{n} \mathbf{1}_{\{X_j \le \xi_p + z/\sqrt{n}\}} - nF(\xi_p + z/\sqrt{n}))$, and $c_n = \frac{1}{\sqrt{n}}(\lceil np \rceil - nF(\xi_p + z/\sqrt{n}))$. Let

$$W_n = n^{\frac{1}{2}}(F_n(\xi_p) - F(\xi_p))$$

$$= \frac{1}{\sqrt{n}} \sum_{j=1}^{n} \left(\mathbf{1}_{\{X_j \le \xi_p\}} - F(\xi_p)\right). \tag{6.128}$$

Then $EZ_n = 0$, $EW_n = 0$, and

$$\text{var}(Z_n - W_n) = \text{var}\left[\frac{1}{\sqrt{n}} \sum_{j=1}^{n} \left(\mathbf{1}_{\{\xi_p < X_j \le \xi_p + z/\sqrt{n}\}}\right)\right]$$

$$= \left(F(\xi_p + \frac{z}{\sqrt{n}}) - F(\xi_p)\right)\left(1 - F(\xi_p + \frac{z}{\sqrt{n}}) + F(\xi_p)\right)$$

$$\to 0 \text{ as } n \to \infty. \tag{6.129}$$

Therefore $Z_n - W_n$ converges in probability to zero. By the classical CLT, W_n converges in distribution to $N(0, p(1 - p))$. By (6.129), Z_n converges in distribution to the same limit. Also, $c_n \to -zf(\zeta_p)$. Hence (6.127) yields

$$\lim_{n \to \infty} \text{Prob}(\sqrt{n}(\hat{\zeta}_p - \zeta_p) \le z) = \text{Prob}(W \ge -zf(\zeta_p))$$

$$= \text{Prob}(W \le zf(\zeta_p)) = \text{Prob}(W/f(\zeta_p) \le z),$$

where W has the Normal distribution $N(0, p(1 - p))$. This proves (6.125).

(b) From the argument above applied to $p = p_i$ $(1 \leq i \leq k)$, it follows that

$$\mathbf{Z}_n - \tilde{\mathbf{W}}_n = (Z_n^{(1)} - W_n^{(1)}, \ldots, Z_n^{(k)} - W_n^{(k)}) \text{ converges in probability to } \mathbf{0},$$
(6.130)

where $Z_n^{(i)}$ and $W_n^{(i)}$ are the same as Z_n and W_n, but with $p = p_i$, and $\mathbf{Z}_n = (Z_n^{(1)}, \ldots, Z_n^{(k)})$, $\mathbf{W}_n = (W_n^{(1)}, \ldots, W_n^{(k)})$.

Now

$$\mathbf{W}_n = \frac{1}{\sqrt{n}} \sum_{j=1}^{n} \left(\mathbf{1}_{\{X_j \leq \zeta_{p_1}\}} - F(\xi_{p_1}), \ldots, \mathbf{1}_{\{X_j \leq \zeta_{p_k}\}} - F(\zeta_{p_k}) \right)$$

$$\xrightarrow{\mathscr{L}} N(\mathbf{0}, ((p_i(1 - p_j)))_{1 \leq i \leq j \leq k}), \quad \text{as } n \to \infty,$$
(6.131)

by the classical multivariate CLT. Writing $c_n^{(i)}$ for c_n and $z^{(i)}$ for z in (a), one has $c_j^{(i)} \to -z^{(i)} f(\zeta_{p_i})$. Therefore, as in the proof of (a),

$$\text{Prob}(\sqrt{n}(\hat{\zeta}_{p_1} - \zeta_p) \leq z^{(1)}, \ldots, \sqrt{n}(\hat{\zeta}_{p_k} - \zeta_{p_k}) \leq z^{(k)})$$
$$\longrightarrow \text{Prob}(W^{(1)} \leq z^{(1)} f(\zeta_{p_1}), \ldots, W^{(k)} \leq z^{(k)} f(\zeta_{p_k}),$$

where $\mathbf{W} = (W^{(1)}, \ldots, W^{(k)})$ is $N(\mathbf{0}, ((p_i(1 - p_j))))$. Let $0 < p_1 < p_2 < \cdots < p_k$, $k > 1$, *and assume that the hypothesis of (a) holds for $p = p_i$ for all $i = 1, 2, \ldots, k$. Then*

$$\sqrt{n} \left(\hat{\zeta}_{p_1} - \zeta_{p_1}, \hat{\zeta}_{p_2} - \zeta_{p_2}, \ldots, \hat{\zeta}_{p_k} - \zeta_{p_k} \right) \xrightarrow{\mathscr{L}} N(\mathbf{0}, \Sigma),$$
(6.132)

where $\Sigma = ((\sigma_{ij}))$, $\sigma_{ij} = p_i(1 - p_j)/f(\zeta_{p_i})f(\zeta_{p_j})$ for $1 \leq i \leq j < k$.

$$\text{Prob}(\sqrt{n}(\hat{\zeta}_{p_1} - \zeta_p) \leq z^{(1)}, \ldots, \sqrt{n}(\hat{\zeta}_{p_k} - \zeta_{p_k}) \leq z^{(k)})$$
$$= \text{Prob}(Z_n^{(1)} \geq c_n^{(1)}, \ldots, Z_n^{(k)} \geq c_n^{(k)})$$
$$\longrightarrow \text{Prob}(W^{(1)} \leq z^{(1)} f(\zeta_{p_1}), \ldots, W^{(k)} \leq z^{(k)} f(\zeta_{p_k})),$$

where $\mathbf{W} = (W^{(1)}, \ldots, W^{(k)})$ is $N(\mathbf{0}, ((p_i(1 - p_j)))_{1 \leq i \leq j \leq k})$.

$\square$

Precise asymptotic analysis of sample quantiles was spurred by the so-called *Bahadur representation*

$$\hat{\zeta}_p = \zeta_p + \frac{p - F_n(\zeta_p)}{f(\zeta_p)} + R_n$$
(6.133)

where $R_n \to 0$ almost surely as $n \to \infty$. This was derived by Bahadur (1966) under the additional assumption of twice differentiability of F at ζ_p, but with an estimation of the remainder term $R_n = O(n^{-3/4}(\log n)^{1/2}(\log \log n)^{1/4})$ a.s. as $n \to \infty$, and he suggested the problem of finding the precise rate of convergence of R_n to zero. Kiefer (1967) derived the precise rate given by

$$\overline{\lim}_{n \to \infty} \frac{n^{3/4} R_n}{(\log \log n)^{3/4}} = \frac{z^{3/4}[p(1 - p)]^{1/4}}{3^{3/4}}$$
(6.134)

with probability one. Our proof follows Ghosh (1971), who proved the representation (6.133) with the remainder satisfying $n^{\frac{1}{2}} R_n \to 0$ in probability with the help of a simple but useful lemma.

6.8 Asymptotics of Semiparametric Multiple Regression

In this section, we extend the results of Sect. 6.6 to *multiple regression* with p explanatory variables, $p \geq 1$. This is one of the most widely applied models in statistics. Consider the semiparametric regression model

$$\mathbf{y} = X\boldsymbol{\theta} + \boldsymbol{\epsilon} \quad \left[\text{i.e., } y_i = \sum_{j=1}^{p} x_{ij}\theta_j + \varepsilon_i \quad (1 \leq i \leq n)\right], \tag{6.135}$$

where $\mathbf{y}$ is the *observation vector*, $X = ((x_{ij}))_{1 \leq i \leq n,\, 1 \leq j \leq p}$ $(p \leq n)$ is a known *design matrix* of full rank p and ε $(1 \leq i \leq n)$ are i.i.d. (unobserved) *errors* satisfying

$$E\varepsilon_i = 0, \qquad E\varepsilon_i^2 = \sigma^2 \qquad (0 < \sigma^2 < \infty), \tag{6.136}$$

(σ^2 unknown) and $\boldsymbol{\theta} = (\theta_1, \theta_2, \ldots, \theta_p)'$ is the unknown *parameter vector* to be estimated. The *least squares estimator* of $\boldsymbol{\theta}$ is $\hat{\boldsymbol{\theta}}$ which minimizes

$$f(\boldsymbol{\theta}) := \sum_{i=1}^{n} \left(y_i - \sum_{j=1}^{p} x_{ij}\theta_j\right)^2. \tag{6.137}$$

Differentiating w.r.t. θ_k one gets the *normal equations* $\partial f(\boldsymbol{\theta})/\partial\theta_k = 0$, or

$$\sum_{i=1}^{n} x_{ik} \underbrace{\sum_{j=1}^{p} x_{ij}\theta_j}_{(X\boldsymbol{\theta})_i} = \underbrace{\sum_{i=1}^{n} x_{ik}y_i}_{(X'y)_k} \qquad (1 \leq k \leq p),$$

or, treating the left side as the k-th element of the column vector $X'X\boldsymbol{\theta}$ one has

$$X'X\boldsymbol{\theta} = X'\mathbf{y}, \tag{6.138}$$

so that the minimum of $f(\boldsymbol{\theta})$ is attained as

$$\hat{\boldsymbol{\theta}} = (X'X)^{-1}X'\mathbf{y}. \tag{6.139}$$

Substituting from (6.135), one has

$$\hat{\boldsymbol{\theta}} = \boldsymbol{\theta} + (X'X)^{-1}X'\boldsymbol{\epsilon}, \tag{6.140}$$

and

$$E\hat{\boldsymbol{\theta}} = \boldsymbol{\theta}, \qquad \text{cov}\hat{\boldsymbol{\theta}} = (X'X)^{-1}X'\sigma^2 I_n X(X'X)^{-1} = \sigma^2(X'X)^{-1}, \tag{6.141}$$

where I_n is the $n \times n$ identity matrix. It is well known, and not difficult to prove, that $\hat{\boldsymbol{\theta}}$ has the smallest expected square error $(E|\hat{\boldsymbol{\theta}} - \boldsymbol{\theta}|^2 = \sigma^2 \text{ trace } (X'X)^{-1})$ in the class of all linear unbiased estimators of $\boldsymbol{\theta}$ (i.e., among all estimators of the form $\boldsymbol{\theta}^* = b + A\mathbf{y}$). In fact this latter property holds even under the milder assumption on $\boldsymbol{\epsilon}$: ε_i satisfy (6.136) and are uncorrelated. (See the Gauss-Markov Theorem in Part I, Sect. 5.9.) Write $(X'X)^{1/2}$ as the positive definite symmetric matrix whose square is $X'X$.

If, in addition to the assumptions made above, ε_i's are Normal $N(0, \sigma^2)$, then it follows from (6.140) that $\hat{\boldsymbol{\theta}}$ is Normal $N(\boldsymbol{\theta}, \sigma^2(X'X)^{-1}))$. For this classical linear model an optimal confidence region for $\boldsymbol{\theta}$ is based on the F-statistic $\{(\hat{\boldsymbol{\theta}} - \boldsymbol{\theta})'X'X(\hat{\boldsymbol{\theta}} - \boldsymbol{\theta})/p\}/\hat{\sigma}^2$ where $\hat{\sigma}^2 = \|Y - X\hat{\boldsymbol{\theta}}\|^2/(n - p)$ (see Chap. 5, Example 5.20).

Theorem 6.4. *In the model (6.135), assume (6.136) where the ε_i's are i.i.d. and that the nonstochastic matrix X is of full rank p. Assume also that ε_i's are non-Normal. Then $(X'X)^{1/2}(\hat{\boldsymbol{\theta}} - \boldsymbol{\theta})$ converges in distribution to $N(0, \sigma^2 I_p)$ if and only if the maximum among the diagonal elements of the matrix $X(X'X)^{-1}X'$ goes to zero as $n \to \infty$.*

We first prove a lemma.

Lemma 6.1. *The $n \times n$ matrix $H = X(X'X)^{-1}X'$ is symmetric and idempotent, and it has the eigenvalue 1 of multiplicity p and the eigenvalue 0 of multiplicity $n - p$. Also, $0 \le H_{ii} \le 1 \ \forall \ i$.*

Proof. Symmetry and idempotence are easy to check. If λ is an eigenvalue with a corresponding eigenvector $\mathbf{x}$, then $H\mathbf{x} = \lambda\mathbf{x}$ and $H\mathbf{x} = HH\mathbf{x} = H\lambda\mathbf{x} = \lambda H\mathbf{x} = \lambda^2\mathbf{x}$, implying $\lambda = \lambda^2$, i.e. $\lambda(1 - \lambda) = 0$, or, $\lambda = 0$ or 1. Clearly, rank of H is no more than p (since $H\mathbf{x} = 0 \ \forall \ x$ orthogonal to the rows of X', i.e., columns of X). On the other hand, rank of H is no less than the rank of $HX = X$, which is p. Hence the rank of H is p. Note that H has p linearly independent row vectors and, therefore, exactly an $(n - p)$-dimensional subspace orthogonal to them. That is, the null space of H is of dimension $n - p$. Hence the multiplicity of the eigenvalue zero is $n - p$. It follows that the eigenvalue 1 is of multiplicity p. Since $H_{ii} \le$ max eigenvalue of H, and $H_{ii} \ge$ min eigenvalue of H, one gets $0 \le H_{ii} \le 1$. $\qquad\square$

Proof of Theorem. Let $\mathbf{a} \in \mathbb{R}^p$, $\mathbf{a} \ne 0$. Then

$$\mathbf{a}'\hat{\boldsymbol{\theta}} - \mathbf{a}'\boldsymbol{\theta} = \mathbf{a}'(X'X)^{-1}X'\boldsymbol{\epsilon} = \boldsymbol{\epsilon}'X(X'X)^{-1}\mathbf{a}$$

$$\text{var}(\mathbf{a}'\hat{\boldsymbol{\theta}}) = \mathbf{a}'(X'X)^{-1}X'\sigma^2 I_n X(X'X)^{-1}\mathbf{a}$$

$$= \sigma^2\mathbf{a}'(X'X)^{-1}\mathbf{a} = \gamma^2, \quad \text{say.} \tag{6.142}$$

$$\frac{\mathbf{a}'\hat{\boldsymbol{\theta}} - \mathbf{a}'\boldsymbol{\theta}}{\gamma} = \boldsymbol{\epsilon}'X(X'X)^{-\frac{1}{2}}\frac{\mathbf{b}}{\sigma} = \sum_1^n \frac{s_i}{\sigma}\varepsilon_i$$

where

$$\mathbf{b} = \frac{(X'X)^{-\frac{1}{2}}\mathbf{a}}{\|(X'X)^{-\frac{1}{2}}\mathbf{a}\|} \quad \text{is a unit vector in } \mathbb{R}^p,$$

and s_i is the i-th element of

$$X(X'X)^{-\frac{1}{2}}\mathbf{b} = G\mathbf{b}, \quad \text{say,} \quad s_i = \sum_{k=1}^p g_{ik}b_k. \tag{6.143}$$

Note that

$$s_i^2 = \left(\sum_{k=1}^{p} g_{ik} b_k\right)^2 \leq \sum_k g_{ik}^2 \cdot \sum_k b_k^2 = \sum_k g_{ik}^2 = (GG')_{ii}$$

$$= \left[X(X'X)^{-\frac{1}{2}}(X'X)^{-\frac{1}{2}}X'\right]_{ii} = H_{ii}. \tag{6.144}$$

Now

$$\frac{\mathbf{a}'\hat{\boldsymbol{\theta}} - \mathbf{a}'\hat{\boldsymbol{\theta}}}{\gamma} = \sum_{i=1}^{n} \frac{s_i}{\sigma}\varepsilon_i, \qquad E\left(\sum_{i=1}^{n} \frac{s_i}{\sigma}\varepsilon_i\right)^2 = \frac{\operatorname{var}(\mathbf{a}'\hat{\boldsymbol{\theta}})}{\gamma^2} = 1, \tag{6.145}$$

and, for every $\delta > 0$,

$$\sum_{i=1}^{n} E\left(\frac{s_i^2}{\sigma^2}\varepsilon_i^2 \mathbf{1}_{\left\{\left|\frac{s_i}{\sigma}\varepsilon_i\right|>\delta\right\}}\right) = \frac{1}{\sigma^2}\sum_{i=1}^{n} s_i^2 E\varepsilon_i^2 \mathbf{1}_{\{|s_i\varepsilon_i|>\delta\sigma\}}$$

$$\leq \frac{1}{\sigma^2}\sum_{i=1}^{n} s_i^2 E\varepsilon_i^2 \mathbf{1}_{\{|\varepsilon_i|>\frac{\delta\sigma}{s}\}}, \tag{6.146}$$

where

$$s^2 = \max_{1\leq i\leq n} s_i^2 \leq \max_i H_{ii} \longrightarrow 0 \qquad \text{by hypothesis.} \tag{6.147}$$

Thus

$$\sum_{i=1}^{n} E\left(\frac{s_i^2}{\sigma^2}\varepsilon_i^2 \mathbf{1}_{\left\{\left|\frac{s_i}{\sigma}\varepsilon_i\right|>\delta\right\}}\right)$$

$$\leq \frac{1}{\sigma^2}\sum_{i=1}^{n} s_i^2 E\varepsilon_1^2 \mathbf{1}_{\{|\varepsilon_1|>\frac{\delta\sigma}{s}\}}$$

$$= \frac{\sum_1^n s_i^2}{\sigma^2} E\varepsilon_1^2 \mathbf{1}_{\{|\varepsilon_1|>\frac{\delta\sigma}{s}\}} \tag{6.148}$$

Since [see (6.143)],

$$\sum_1^n s_i^2 = \langle G\mathbf{b}, G\mathbf{b}\rangle = \mathbf{b}'G'G\mathbf{b} = \mathbf{b}'\mathbf{b} = 1,$$

$$\left(G'G = (X'X)^{-\frac{1}{2}}X'X(X'X)^{-\frac{1}{2}} = I_p\right)$$

one gets

$$\sum_{i=1}^{n} E\left(\frac{s_i^2}{\sigma^2}\varepsilon_i^2 \mathbf{1}_{\left\{\frac{|s_i\varepsilon_i|}{\sigma}>\delta\right\}}\right) \leq \frac{1}{\sigma^2} E\varepsilon_1^2 \mathbf{1}_{\{|\varepsilon_1|>\frac{\delta\sigma}{s}\}} \longrightarrow 0$$

$$\text{as } n\to\infty \quad \left(\text{since } \frac{\delta\sigma}{s}\to\infty\right). \tag{6.149}$$

This proves the "sufficiency" part of the theorem, applying the Lindeberg-Feller CLT.

To prove the "necessary" part, choose $\mathbf{b}$ (and, therefore, $\mathbf{a}$) such that (6.144) is an equality, i.e., $b_k = g_{i_0 k}$ for the *particular* i_0 for which $\max_{i'} H_{i'i'} = H_{i_0 i_0}$. Then, if $H_{i_0 i_0} :\nrightarrow 0$, $E\left(\frac{s_{i_0}^2}{\sigma^2}\varepsilon_{i_0}^2\right) = s_{i_0}^2 \nrightarrow 0$ (as $n\to\infty$). This violates Feller's necessary condition for the Lindeberg-Feller CLT.

You may directly consider $\sum_{i=1}^{n} \frac{s_i}{\sigma} \varepsilon_i = \frac{s_{i_0}}{\sigma} \varepsilon_{i_0} + \sum_{i \neq i_0} \frac{s_i}{\sigma} \varepsilon_i$. There exists a subsequence n' of integers $n = 1, 2, \ldots$, such that as $n' \to \infty$, $\frac{s_{i_0}}{\sigma} \to \alpha > 0$, so that $\frac{s_{i_0}}{\sigma} \varepsilon_{i_0} \xrightarrow{\mathscr{L}} \alpha \varepsilon_1$ (which is *non-Normal*). Whatever may be the limit of $\sum_{i \neq i_0} \frac{s_i}{\sigma} \varepsilon_i$, the sum of two independent random variables, one of which is non-Normal, cannot be Normal (Exercise 6.30). $\qquad \square$

Remark 6.18. If ε_i's are i.i.d. $N(0, \sigma^2)$, then of course $\hat{\theta} = \theta + (X'X)^{-1}X'\epsilon$ is normal, no matter what X is (assuming full rank).

Corollary 6.4. *Let*

$$s_{ij} = \frac{1}{n} \sum_{k=1}^{n} x_{ki} x_{kj} \qquad (1 \leq i, j \leq p),$$

$$S = ((s_{ij})),$$

$$\Lambda = \text{largest eigenvalue of } S^{-1} = \frac{1}{\lambda}. \tag{6.150}$$

where $\lambda = $ *smallest eigenvalue of S. Thus* $\sqrt{n}\, S^{\frac{1}{2}}(\hat{\theta} - \theta) \xrightarrow{\mathscr{L}} N(0, \sigma^2 I_p)$ *if*

$$n^{-1} \Lambda \max_i \left(\sum_{j=1}^{p} x_{ij}^2 \right) \longrightarrow 0 \quad \text{as } n \to \infty. \tag{6.151}$$

Proof. Let $e_i\ (\in \mathbb{R}^n)$ have 1 in the i-th coordinate and 0 elsewhere. Then, writing $\langle\ ,\ \rangle_m$ for the Euclidean inner product in $\mathbb{R}^m$,

$$
\begin{aligned}
H_{ii} &= \langle He_i, e_i \rangle_n \equiv \langle X(X'X)^{-1}X'e_i, e_i \rangle_n \\
&= \langle (X'X)^{-1}X'e_i, X'e_i \rangle_p \qquad [\text{since } \langle \mathbf{B}x, y \rangle_n = \langle x, B'y \rangle] \\
&\leq \frac{\Lambda}{n} \|X'e_i\|^2 = \frac{\Lambda}{n} \|(x_{ij})_{1 \leq j \leq p}\|^2 = \frac{\Lambda}{n} \left(\sum_{j=1}^{p} x_{ij}^2 \right).
\end{aligned}
\tag{6.152}
$$

$\qquad \square$

Corollary 6.5. *Let*

$$\mathbf{y}_i = \alpha + \sum_{j=1}^{r} \beta_j Z_{ij} + \varepsilon_i \qquad (1 \leq i \leq n), \tag{6.153}$$

where $\mathbf{y} = (y_1, \ldots, y_n)'$ *is observed,* $\mathbf{Z}_i \equiv (Z_{i1}, \ldots, Z_{ir})'$, $i \geq 1$, *are i.i.d. r-dimensional random vectors with finite mean vector* $\boldsymbol{\mu} = (\mu_1, \ldots \mu_r)'$ *and nonsingular covariance matrix* $\Sigma = ((\sigma_{jk}))_{1 \leq j,k \leq r}$. *Assume* ε_i, $i \geq 1$, *are i.i.d. satisfying (6.136), and that the two families* $\{\mathbf{Z}_i : i \geq 1\}$ *and* $\{\varepsilon_i : i \geq 1\}$ *are independent.*

(a) Then the least squares estimators of $\alpha, \beta_1, \ldots, \beta_r$ *are*

$$
\begin{pmatrix} \hat{\beta}_1 \\ \vdots \\ \hat{\beta}_r \end{pmatrix} = \hat{\Sigma}^{-1} \begin{pmatrix} \hat{\sigma}_{y1} \\ \vdots \\ \hat{\sigma}_{yr} \end{pmatrix} \left[\begin{matrix} \hat{\Sigma} := ((\hat{\sigma}_{jk})),\ \hat{\sigma}_{jk} := \frac{1}{n} \sum_{i=1}^{n} (Z_{ij} - \overline{Z}_{\cdot j})(Z_{ik} - \overline{Z}_{\cdot k}), \\ \hat{\sigma}_{yj} := \frac{1}{n} \sum_{i=1}^{n} (y_i - \overline{y})(Z_{ij} - \overline{Z}_{\cdot j}) \end{matrix} \right],
$$

$$\hat{\alpha} = \overline{y} - \sum_{j=1}^{r} \hat{\beta}_j \overline{Z}_{\cdot j} \qquad (\overline{y} = \tfrac{1}{n}\sum_{i=1}^{n} y_i, \ \ \overline{Z}_{\cdot j} = \tfrac{1}{n}\sum_{i=1}^{n} Z_{ij}). \tag{6.154}$$

(b) Also,

$$\begin{pmatrix} \hat{\alpha} \\ \hat{\beta}_1 \\ \vdots \\ \hat{\beta}_r \end{pmatrix} \ \text{is} \ \ AN \left(\begin{pmatrix} \alpha \\ \beta_1 \\ \vdots \\ \beta_r \end{pmatrix}, \ \frac{\sigma^2}{n} \overbrace{\begin{bmatrix} a_0 & a_1 & \cdots & a_r \\ a_1 & & & \\ \vdots & & \Sigma^{-1} & \\ a_r & & & \end{bmatrix}}^{A} \right), \tag{6.155}$$

$$[a_0 := 1 + \textstyle\sum_{j,k} \sigma^{jk}\mu_j\mu_k, \ a_k := - \sum_{j'=1}^{r} \sigma^{kj'}\mu_{j'} \ (k \geq 1)].$$

Proof. (a) One may directly prove (6.154), by rewriting (6.153) as

$$y_i = \alpha_1 + \sum_{j=1}^{r} \beta_j (Z_{ij} - \overline{Z}_{\cdot j}) + \varepsilon_i \qquad (\alpha_1 := \alpha + \textstyle\sum_{j=1}^{r} \beta_j \overline{Z}_{\cdot j}), \tag{6.156}$$

and solve for $\alpha_1, \beta_1 \ldots \beta_r$ by differentiating w.r.t. $\alpha_1, \beta_1, \ldots, \beta_r$ the quantity $\sum_{i=1}^{n}(y_i - \alpha_1 - \sum_{j=1}^{r} \beta_j (Z_{ij} - \overline{Z}_{\cdot j}))^2$, and setting the derivatives equal to zero.
(b) For the proof of (b), write (6.153) in the form (6.135) with $p = r + 1$, and

$$X = \begin{bmatrix} 1 & Z_{11} & Z_{12} & \ldots & Z_{1r} \\ 1 & Z_{21} & Z_{22} & \ldots & Z_{2r} \\ \vdots & \vdots & \vdots & \ldots & \vdots \\ 1 & Z_{n1} & Z_{n2} & \ldots & Z_{nr} \end{bmatrix}. \tag{6.157}$$

Then

$$X'X = n \begin{bmatrix} 1 & \overline{Z}_{\cdot 1} & \overline{Z}_{\cdot 2} & \ldots & \overline{Z}_{\cdot r} \\ \overline{Z}_{\cdot 1} & \tilde{s}_{11} & \tilde{s}_{12} & \ldots & \tilde{s}_{2r} \\ \vdots & \vdots & \vdots & & \vdots \\ \overline{Z}_{\cdot r} & \tilde{s}_{r1} & \tilde{s}_{r2} & \ldots & \tilde{s}_{rr} \end{bmatrix} = n \begin{bmatrix} 1 & \overline{Z}_{\cdot 1} & \ldots & \overline{Z}_{\cdot r} \\ \overline{Z}_{\cdot 1} & & & \\ \vdots & & \tilde{S} & \\ \overline{Z}_{\cdot r} & & & \end{bmatrix},$$

$$\frac{X'X}{n} \ \xrightarrow{\text{a.s.}} \ \begin{bmatrix} 1 & \mu_1 & \ldots & \mu_r \\ \mu_1 & & & \\ \vdots & & \Gamma & \\ \mu_r & & & \end{bmatrix} = B, \ \text{say}, \ [\Gamma := ((\gamma_{jk}))], \tag{6.158}$$

where $\tilde{s}_{jk} = \tfrac{1}{n}\sum_{i=1}^{n} Z_{ij}Z_{ik}$, $\gamma_{jk} = E(Z_{ij}Z_{ik}) = E\tilde{s}_{jk}$. It is simple to check that $AB = I_{r+1}$. In particular, B is nonsingular so that the probability that $X'X$ is nonsingular ($\Longrightarrow X$ is of full rank) converges to 1 as $n \to \infty$. Thus, one may apply Corollary 6.4 above, conditionally given $\{\mathbf{Z}_i : i \geq 1\}$, to show that $(\hat{\alpha}, \hat{\beta}_1, \ldots, \hat{\beta}_r)'$ is

$$AN \left(\begin{pmatrix} \alpha \\ \beta_1 \\ \vdots \\ \beta_r \end{pmatrix}, \ \sigma^2 (X'X)^{-1} \right),$$

outside a set of sample points whose probability is zero. $\qquad\square$

Remark 6.19. The model in Corollary 6.5 is called the *correlation model*, to indicate that y_i and $\mathbf{Z}_i \equiv (Z_{i1}, \ldots, Z_{i,r})$ are correlated. The asymptotic normality can be proved directly in this case, using the expressions (6.154) for $\hat{\alpha}$ and $\hat{\boldsymbol{\beta}} \equiv (\hat{\beta}_1, \ldots, \hat{\beta}_r)'$, without invoking the general result contained in Theorem 6.4.

6.9 Asymptotic Relative Efficiency (ARE) of Estimators

We have seen in Sects. 6.5–6.8 that large classes of statistics T_n are asymptotically normal, i.e., $\sqrt{n}(T_n - g(\theta)) \xrightarrow{\mathscr{L}} N(0, \sigma^2(\theta))$, where T_n maybe taken to be an estimator of a parametric function $g(\theta)$.

Here we consider an index of asymptotic comparison of estimators of $g(\theta)$, θ being an unknown parameter which identifies the underlying probability distribution, say, P_θ, from which independent observations $X_1, \ldots, X_n$ are drawn.

Definition 6.6. If two estimators $T_n^{(1)}, T_n^{(2)}$ of $g(\theta)$ are both asymptotically normal with

$$
\begin{aligned}
\sqrt{n}(T_n^{(1)} - g(\theta)) &\xrightarrow{\mathscr{L}} N(0, \sigma_1^2(\theta)), \\
\sqrt{n}(T_n^{(2)} - g(\theta)) &\xrightarrow{\mathscr{L}} N(0, \sigma_2^2(\theta)),
\end{aligned}
\qquad \text{when } \theta \text{ is the true parameter value} \qquad (6.159)
$$

then the *asymptotic relative efficiency* (ARE) of $T_n^{(2)}$ with respect to $T_n^{(1)}$ is defined by

$$
e_{T^{(2)}, T^{(1)}} = \frac{\sigma_1^2(\theta)}{\sigma_2^2(\theta)}, \qquad (T^{(i)} = \{T_n^{(i)} : n = 1, 2, \ldots \}). \qquad (6.160)
$$

Thus if (for some value θ of the parameter) the above efficiency is 2/3 then σ_1^2 is two-thirds of σ_2^2, and adopting $T^{(1)}$ with a sample size $(2/3)n$ leads to the same accuracy (asymptotically, as $n \to \infty$) in the estimation of $g(\theta)$ as would be achieved by using $T^{(2)}$ with n observations.

Example 6.8 (Mean Versus Median in Normal and Cauchy Models). If $X_1, X_2, \ldots,$ X_n are i.i.d. $N(\theta, \sigma^2)$, $\theta \in (-\infty, \infty)$, $\sigma^2 \in (0, \infty)$. Consider $\overline{X} = \frac{X_1 + \cdots + X_n}{n}$ and $\hat{\zeta}_{\frac{1}{2}}$ = the sample median ($= \frac{1}{2}$-quantile) as estimators of θ. Now $\sqrt{n}(\overline{X} - \theta)$ is $N(0, \sigma^2)$ (and, therefore, $\sqrt{n}(\overline{X} - \theta) \xrightarrow{\mathscr{L}} N(0, \sigma^2)$, trivially) and, by Theorem 6.3, $\sqrt{n}(\hat{\zeta}_{\frac{1}{2}} - \theta) \xrightarrow{\mathscr{L}} N(0, \frac{2\pi\sigma^2}{4}) = N(0, \frac{\pi}{2}\sigma^2)$. Hence

$$
e_{\hat{\zeta}_{\frac{1}{2}}, \overline{X}} = \frac{\sigma^2}{\frac{\pi}{2}\sigma^2} = \frac{2}{\pi} \simeq 0.637. \qquad (6.161)
$$

Thus $\overline{X}$ is decidedly the better of the two estimators (no matter what θ, σ^2 may be).

When the same two estimators are used to estimate the median θ of a Cauchy distribution with density

$$
f(x; \theta) = \left(\frac{1}{\pi a}\right) \frac{1}{1 + \left(\frac{x-\theta}{a}\right)^2}, \qquad -\infty < x < \infty, \ (-\infty < \theta < \infty, a > 0), \qquad (6.162)
$$

then $\overline{X}$ is not even consistent, since the distribution of $\overline{X}$ is the same as that of a single observation (Exercise). But Theorem 6.3 applies to yield

$$\sqrt{n}(\hat{\zeta}_{\frac{1}{2}} - \theta) \xrightarrow{\mathscr{L}} N\left(0, \frac{\pi^2 a^2}{4}\right). \tag{6.163}$$

Although, strictly speaking, the ARE of $\hat{\zeta}_{\frac{1}{2}}$ w.r.t. $\overline{X}$ is not defined for this case, one may informally take it to be ∞. No matter how large a number A may be, $\overline{X}$ based on nA observations is worse than $\hat{\zeta}_{\frac{1}{2}}$ based on n observations.

To compare the two statistics as estimators of the median θ of a distribution which has a density symmetric about θ, let $f(x)$ be a p.d.f. (with respect to Lebesgue measure) which is an even function: $f(x) = f(-x)$. The common p.d.f. of the observations is $f(x;\theta) = f(x - \theta)$, $\theta \in (-\infty, \infty)$. (There may be other unknown parameters in $f(x;\theta)$). Here are some examples:

Example 6.9. The Logistic $L(0, a)$ has the density $f(x) = \frac{1}{a}e^{-x/a}/(1 + e^{-x/a})^2$, $a > 0$. For this case

$$\sqrt{n}(\hat{\zeta}_{\frac{1}{2}} - \theta) \xrightarrow{\mathscr{L}} N(0, 4a^2),$$

$$\sqrt{n}(\overline{X} - \theta) \xrightarrow{\mathscr{L}} N(0, \sigma^2),$$

where

$$\sigma^2 = \int_{-\infty}^{\infty} x^2 f(x)dx = a^2 \int_{-\infty}^{\infty} x^2 e^{-x}/(1 + e^{-x})^2 dx$$

$$= 2a^2 \int_0^{\infty} x^2 e^{-x}/(1 + e^{-x})^2 dx = (4a^2)\frac{\pi^2}{12}.$$

(Exercise 6.34). Hence

$$e_{\hat{\zeta}_{\frac{1}{2}},\overline{X}} = \frac{\pi^2}{12} \simeq 0.82.$$

Example 6.10 (Contaminated Normal or Tukey Model).

$$f(x) = (1 - \varepsilon)\varphi(x) + \varepsilon\varphi_\tau(x), \qquad (0 < \varepsilon < 1, \tau > 0)$$

where φ is the Normal p.d.f. with mean zero and variance one and φ_τ is the Normal p.d.f. with mean zero and variance τ. For $f(x - \theta)$ the median $\zeta_{\frac{1}{2}}$ is θ and the mean is also θ.

$$f(\zeta_{\frac{1}{2}};\theta) = f(\theta - \theta) = f(0) = \frac{1}{\sqrt{2\pi}}(1 - \varepsilon) + \varepsilon\left(\frac{1}{\sqrt{2\pi}\sqrt{\tau}}\right) = \frac{1}{\sqrt{2\pi}}\left(1 - \varepsilon + \frac{\varepsilon}{\sqrt{\tau}}\right).$$

The variance is $\sigma^2 = \int_{-\infty}^{\infty} x^2 f(x)dx = (1 - \varepsilon) + \varepsilon\tau$. Hence

$$e_{\hat{\zeta}_{\frac{1}{2}},\overline{X}} = \frac{(1 - \varepsilon) + \varepsilon\tau}{\frac{1}{4}/f^2(0)} = \frac{4\{(1 - \varepsilon) + \varepsilon\tau\}(1 - \varepsilon + \varepsilon/\sqrt{\tau})^2}{2\pi}.$$

6.10 Constructing (Nonparametric) Confidence Intervals

In this section we briefly consider constructing confidence regions for nonparametric functional estimates. Sections 6.6 and 6.8 dealt with what one might call semiparametric estimation of regression.

Corollary 6.2 may be used to construct an asymptotic *confidence interval* for the unknown quantity $H(\mu)$: such as $[H(\overline{Z}) - z_{1-\alpha/2}\frac{\hat{\sigma}}{\sqrt{n}}, H(\overline{Z}) + z_{1-\alpha/2}\frac{\hat{\sigma}}{\sqrt{n}}]$, where $z_{1-\alpha/2}$ is the $(1 - \frac{\alpha}{2})$th quantile of $N(0,1)$, and

$$\hat{\sigma}^2 = \sum_{i,j=1}^{k} \hat{\ell}_i \hat{\ell}_j \hat{b}_{ij}$$

with $\hat{\ell}_i$ obtained by replacing μ by $\overline{Z}$ in the expression for ℓ_i in (6.57), and letting $\hat{b}_{ij}$ be the sample covariance between $f_i(Y_1)$ and $f_j(Y_1)$:

$$\hat{b}_{ij} = \frac{1}{n} \sum_{r=1}^{n} f_i(Y_r)f_j(Y_r) - \left(\frac{1}{n}\sum_{r=1}^{n} f_i(Y_r)\right)\left(\frac{1}{n}\sum_{r=1}^{n} f_j(Y_r)\right).$$

By the SLLN, $\hat{\ell}_i$ and $\hat{b}_{ij}$ are consistent estimators of ℓ_i and b_{ij}, respectively ($1 \leq i$, $j \leq k$). Hence $\hat{\sigma}^2 \xrightarrow{P} \sigma^2$. It follows that (use Slutsky's Lemma and Corollary 6.2)

$$P\left(H(\overline{Z}) - z_{1-\alpha/2}\frac{\hat{\sigma}}{\sqrt{n}} \leq H(\mu) \leq H(\overline{Z}) + z_{1-\alpha/2}\frac{\tilde{\sigma}}{\sqrt{n}}\right)$$
$$= P\left(\left|\frac{\sqrt{n}(H(\overline{Z}) - H(\mu))}{\hat{\sigma}}\right| \leq z_{1-\alpha/2}\right) \longrightarrow 1 - \alpha.$$

$$(6.164)$$

One may similarly obtain an asymptotic confidence interval for ζ_p, under the hypotheses of Theorem 6.3, by estimating the density at ζ_p. A *nonparametric density estimation* is carried out in a later chapter. It is possible, however, to provide a different nonparametric confidence interval for ζ_p without resorting to density estimation. We describe this below.

For every s, $1 \leq s \leq n$, one has

$$P(\zeta_p < X_{(s)}) = P(\#\text{of observations } X_j, 1 \leq j \leq n, \text{ which are less than}$$
$$\text{or equal to } \zeta_p \text{ is less than } s)$$
$$= \sum_{m=0}^{s-1} \binom{n}{m} F^m(\zeta_p)(1 - F(\zeta_p))^{n-m} = \sum_{m=0}^{s-1} \binom{n}{m} p^m(1-p)^{n-m}.$$

$$(6.165)$$

For $1 \le r < s \le n$, one then has

$$P(X_{(r)} \le \zeta_p < X_{(s)}) = P(\zeta_p < X_{(s)}) - P(\zeta_p < X_{(r)})$$

$$= \sum_{m=r}^{s-1} \binom{n}{m} p^m (1-p)^{n-m}. \tag{6.166}$$

One now needs to look up *binomial tables* to find r and s such that (6.166) is approximately $1 - \alpha$. For the approximately equal tailed confidence interval, one finds s such that (6.165) is closest to $1 - \alpha/2$, and r such that $\sum_{m=0}^{r-1} \binom{n}{m} p^m (1-p)^{n-m}$ is closest to $\alpha/2$. If n is large and np, $n(1-p)$ are both moderately large then one may use *normal approximation* to the binomial:

$$s = \left[np + z_{1-\alpha/2} \sqrt{np(1-p)} \right], \qquad r = \left[np - z_{1-\alpha/2} \sqrt{np(1-p)} \right],$$

$$([y] := \text{integer part of } y).$$

$$\tag{6.167}$$

Addendum The Berry-Esséen bound for the classical CLT says: If X_j, $1 \le j \le n$, are i.i.d. with mean μ, variance σ^2 and a finite third absolute moment $\rho_3 = E|X_1|^3$, then

$$\sup_{x \in \mathbb{R}} \left| P(\sqrt{n}(\overline{X} - \mu) \le x) - \Phi_{0,\sigma^2}(x) \right| \le (0.5600) \frac{\rho_3}{\sigma^3 \sqrt{n}}. \tag{6.168}$$

The constant 0.7975 is due to Shevtsova (2010). The same inequality holds also for non-identically distributed, but independent, X_j ($1 \le j \le n$) with $\mu = \sum_{j=1}^{n} EX_j/n$, $\sigma^2 = 1/n \sum_{j=1}^{n} \text{var}(X_j)$, and $\rho_3 = \frac{1}{n} \sum_{j=1}^{n} E|X_j - E(X_j)|^3$.

For the multidimensional CLT there are Berry-Esséen type bounds available, although good numerical constants in the bound [such as given in (6.168)] are difficult to obtain. (Reference: Bhattacharya and Rang (2010).)

6.11 Errors in Variables Models

In standard regression models the independent or explanatory variables are assumed to be observed exactly, i.e., without errors. However, in many situations the independent variables can be contaminated, mismeasured or they cannot be directly observed. Under these circumstances the model is referred to as an *errors in variables model*. Consider for simplicity the linear regression model

$$Y_j = \alpha + \beta X_j^* + \varepsilon_j \qquad (j = 1, \ldots, n), \tag{6.169}$$

where ε_j has mean zero and a finite variance $\sigma_\varepsilon^2 > 0$. The "true" value X_j^* of the independent variable cannot be observed and, instead, one observes X_j subject to a random error,

$$X_j = X_j^* + \eta_j \qquad (j = 1, \ldots, n), \tag{6.170}$$

where η_j are mean zero i.i.d. random variables with finite variance $\sigma_\eta^2 > 0$. Consider first X_j^* to be random with a finite variance $\sigma_{x^*}^2 > 0$, and assume that X_j^*, η_j

and ε_j are independent. Also, assume the n observations (X_j, Y_j), $1 \leq j \leq n$, are independent. The ordinary least squares (OLS) estimate of β based on these observations is (see (6.14))

$$
\hat{\beta} = \frac{[\sum_{1 \leq j \leq n}(X_j - \overline{X})(Y_j - \overline{Y})]}{[\sum_{1 \leq j \leq n}(X_j - \overline{X})^2]}
$$

$$
= \frac{[\sum_{1 \leq j \leq n}(X_j^* - \overline{X}^* + \eta_j - \overline{\eta})(\beta(X_j^* - \overline{X}^*) + \varepsilon_j - \overline{\varepsilon})]}{[\sum_{1 \leq j \leq n}(X_j^* - \overline{X}^* + \eta_j - \overline{\eta})]^2}. \tag{6.171}
$$

When divided by n, the denominator converges in probability to $\sigma_{x^*}^2 + \sigma_\eta^2$, and the numerator to $\beta\sigma_{x^*}^2$ (Exercise 6.37), so that

$$
\hat{\beta} \xrightarrow{p} \frac{\beta}{(1 + \sigma_\eta^2/\sigma_{x^*}^2)} = \kappa\beta, \tag{6.172}
$$

where

$$
\kappa := \left(1 + \sigma_\eta^2/\sigma_{x^*}^2\right)^{-1} \tag{6.173}
$$

is the so-called *reliability ratio* determined by the *noise-to-signal ratio* $\sigma_\eta^2/\sigma_{x^*}^2$. Therefore, $\hat{\beta}$ is an *inconsistent underestimate* of β. This errors in variables model with a stochastic X^* is called a *structural model* in economics; here one stipulates an error free relation between the "true" random variables X^* and Y^* of the form

$$
\theta_1 Y^* + \theta_2 X^* = \alpha, \qquad (\theta_1 \neq 0, \theta_2 \neq 0), \tag{6.174}
$$

but with both X^* and Y^* subject to measuremental errors.

In the so-called *functional model*, X^* is non-stochastic, while the other assumptions above remain intact. If $n^{-1}\sum_{1 \leq j \leq n}(X_j^* - \overline{X}^*)^2$ converges to a positive quantity, say $\sigma_{x^*}^2$, as $n \to \infty$, then (6.172) still holds (see Exercise 6.37).

We now concentrate on the structural model. Suppose one knows, or has a consistent estimate of, κ (i.e. of the noise-to-signal ratio $\sigma_\eta^2/\sigma_{x^*}^2$). We assume that κ is known. Then

$$
\tilde{\beta} = \left(\frac{1}{\kappa}\right)\hat{\beta} = \left(1 + \sigma_\eta^2/\sigma_{x^*}^2\right)\hat{\beta} \tag{6.175}
$$

is a consistent estimate of β. Assume, for simplicity, that X^*, η and ε are Normal. Then (X, η) is bivariate Normal with $EX = \mu$, say, $\text{var}(X) = \sigma_{x^*}^2 + \sigma_\eta^2$, $E\eta = 0$, $\text{var}(\eta) = \sigma_\eta^2$, $\text{cov}(X, \eta) = \sigma_\eta^2$. Therefore, the conditional distribution of η, given X, is Normal with mean $E\eta + \left[\sigma_\eta^2/(\sigma_{x^*}^2 + \sigma_\eta^2)\right]^{1/2}\rho(X - \mu)$ and variance $\sigma_\eta^2(1 - \rho^2)$, where $\rho = \text{cov}(X, \eta)/(\text{var}(X)\text{var}(\eta))^{1/2} = (1 - \kappa)^{1/2}$ (see Appendix A.3, Exercise A.3). Now express $\hat{\beta}$ as [see (6.171)]

$$
\hat{\beta} = \sum(X_j - \overline{X})\frac{\left[\beta(X_j - \overline{X}) - \beta(\eta_i - \overline{\eta}) + \varepsilon_j - \overline{\varepsilon}\right]}{\sum(X_j - \overline{X})^2}
$$

$$
= \beta\left(1 - \frac{\sum(X_j - \overline{X})\eta_j}{\sum(X_j - \overline{X})^2}\right) + \frac{\sum(X_j - \overline{X})\varepsilon_j}{\sum(X_j - \overline{X})^2}. \tag{6.176}
$$

Hence, conditionally given $\mathbf{X} = (X_j; 1 \leq j \leq n)$, $\hat{\beta}$ is Normal with mean and variance given by (Exercise 6.38)

$$
E(\hat{\beta} \mid \mathbf{X}) = \kappa\beta, \quad \text{var}(\tilde{\beta} \mid \mathbf{X}) = \frac{\beta^2\sigma_{x^*}^2 + \sigma_\varepsilon^2}{[\sum_{1 \leq j \leq n}(X_j - \overline{X})^2]}. \tag{6.177}
$$

Since $[\sum_{1\leq j\leq n}(X_j - \overline{X})^2]/n \xrightarrow{P} \sigma_{x*}^2 + \sigma_\eta^2$, it follows that the conditional distribution of $\sqrt{n}(\hat{\beta} - \kappa\beta)$ converges in distribution to $N(0, \delta^2)$, where (Exercise 6.38)

$$\delta^2 = \frac{[\beta^2 \sigma_{x*}^2 + \sigma_\varepsilon^2]}{[\sigma_{x*}^2 + \sigma_\eta^2]} = \beta^2 \kappa + \frac{\sigma_\varepsilon^2}{\sigma_{x*}^2 + \sigma_\eta^2} = \beta^2 \kappa + \frac{\sigma_\varepsilon^2 \kappa}{\sigma_{x*}^2}. \tag{6.178}$$

This latter distribution does not involve $\mathbf{X}$; hence we have shown that

$$\sqrt{n}(\hat{\beta} - \kappa\beta) \xrightarrow{\mathscr{L}} N(0, \delta^2), \qquad \text{as } n \to \infty. \tag{6.179}$$

It follows that

$$\sqrt{n}(\tilde{\beta} - \beta) \xrightarrow{\mathscr{L}} N\left(0, \frac{\delta^2}{\kappa^2}\right). \tag{6.180}$$

One may similarly derive the asymptotic distribution of $\sqrt{n}(\tilde{\alpha} - \alpha, \tilde{\beta} - \beta)$.

Since the above argument is conditional on $\mathbf{X}$, the corresponding results also hold for the functional model, if

$$\sum_{1\leq j\leq n} \frac{X_j^*}{n} \longrightarrow \mu, \qquad \sum_{1\leq j\leq n} \frac{(X_j^* - \overline{X}^*)^2}{n} \longrightarrow \sigma_{x*}^2 \quad (\text{for some } \mu, \sigma_{x*}^2; \; 0 < \sigma_{x*}^2 < \infty). \tag{6.181}$$

Fuller (1987, pp. 18–20), describes an agricultural experiment in which a consistent estimate of σ_η^2 and, therefore, of κ, is obtained. In the absence of such additional information the parameter $\boldsymbol{\theta} = (\alpha, \beta, \mu = x^*, \sigma_{x*}^2, \sigma_\eta^2, \sigma_\varepsilon^2)$ is *not identifiable* in the bivariate Normal model for (X, Y) presented above [see (6.169), (6.170)]. In particular, β is *not identifiable*. That is, there are $\boldsymbol{\theta}_1$ and $\boldsymbol{\theta}_2$ with different values of β yielding the same bivariate Normal distribution for (X, Y). We give a proof of this fact due to Riersøl (1950). That $\boldsymbol{\theta}$ is not identifiable in the Normal model, of course, is easy to see. For the bivariate Normal is entirely determined by five parameters—two means, two variances and a covariance. But $\boldsymbol{\theta}$ has six functionally independent parameters, i.e., with a parameter space in $\mathbb{R}^6$ with a non-empty interior. On the other hand five parametric functions are identifiable and, therefore, can be estimated consistently. Unfortunately, the most important parameter β is not. To see this assume that the errors ε and η are independent and Normal with zero means and variances $\sigma_\varepsilon^2 > 0$, $\sigma_\eta^2 > 0$. Also assume that X^* is independent of ε and η. It will be shown that β is not identifiable precisely when X^* is Normal, i.e., if (X, Y) is bivariate Normal. Let there exist $\boldsymbol{\theta}_j = (\alpha_j, \beta_j, \mu_j, \sigma_{x*j}^2, \sigma_{\eta j}^2, \sigma_{\varepsilon j}^2)$, $j = 1, 2$, such that $\beta_1 \neq \beta_2$, but the distributions of (X, Y) the same under $\boldsymbol{\theta}_1$ and $\boldsymbol{\theta}_2$. The characteristic function of (X, Y) under $\boldsymbol{\theta}_j$ is

$$\begin{aligned}
\varphi(t_1, t_2) &= E_{\boldsymbol{\theta}_j} \exp\{it_1 X + it_2 Y\} \\
&= E_{\boldsymbol{\theta}_j} \exp\{it_1 X^* + it_1 \eta_j + it_2 \alpha_j + it_2 \beta_j X^* + it_2 \varepsilon_j\} \\
&= e^{it_2 \alpha_j} \exp\left\{-\frac{1}{2}\left(t_1^2 \sigma_{\eta j}^2 + t_2^2 \sigma_{\varepsilon j}^2\right)\right\} \psi_{\boldsymbol{\theta}_j}(t_1 + t_2 \beta_j) \quad (j = 1, 2),
\end{aligned} \tag{6.182}$$

where $\psi_{\boldsymbol{\theta}_j}$ is the characteristic function of X^* under $\boldsymbol{\theta}_j$. Fix $z \in \mathbb{R}$. There exist t_1, t_2 such that $t_1 + t_2 \beta_1 = z$ and $t_1 + t_2 \beta_2 = 0$, namely, $t_1 = -\beta_2 z/(\beta_1 - \beta_2)$, $t_2 = z/(\beta_1 - \beta_2)$. Then by (6.182), and equating φ_1 and φ_2, one has

$$\psi_{\boldsymbol{\theta}_1}(z) = \exp\left\{iz\frac{(\alpha_2 - \alpha_1)}{(\beta_1 - \beta_2)} - \frac{1}{2}\left(\frac{\beta_2^2}{(\beta_1 - \beta_2)^2}\right)\left(\sigma_{\eta 2}^2 - \sigma_{\eta 1}^2\right)z^2\right.$$
$$\left. - \frac{1}{2}(\beta_1 - \beta_2)^{-2}\left(\sigma_{\varepsilon 2}^2 - \sigma_{\varepsilon 1}^2\right)z^2\right\}$$
$$= \exp\left\{icz - \frac{1}{2}dz^2\right\}$$

for some constants c and d. In other words, the distribution of X^* under $\boldsymbol{\theta}_1$ (and, therefore, also $\boldsymbol{\theta}_2$) is Normal (or a constant, which may be taken as Normal). We have arrived at the fact that if the errors η and ε in the variables are Gaussian, then *β is unidentifiable if and only if X^* is Gaussian, that is, if and only if (X, Y) has the bivariate Normal.*

Before moving away from the bivariate Normal model for (X, Y) in the absence of additional information, note that the OLS $\hat{\beta}_{xy}$ for the regression of X on Y provides a lower estimate of $1/\beta$ and, therefore, $\hat{\beta}_{xy}$ is an upper estimate of β, just as $\hat{\beta} = \hat{\beta}_{yx}$ is a lower estimate of β.

It was pointed out by Berkson (1950) that in many applications observations X_j of the regressor are *controlled*. For example, in bioassay for different levels of *fixed* dosages X_j one finds the response Y_j. In this case there may still be some errors in the actual dosages x_j^* that are administered, but X_j is nonrandom, $X_j = x_j^* + \eta_j$. Here x_j^* and η_j are negatively correlated, but X_j is uncorrelated with them. The OLS $\hat{\beta}$ is now an unbiased and consistent estimator of β and the regression model is formally the same as the one without measuremental errors [see (6.14)–(6.16)]

$$Y_j = \alpha + \beta X_j = \gamma_j, \qquad \gamma_j := -\beta\eta_j + \varepsilon_j,$$
$$\hat{\beta} = \beta + \frac{\sum_{j=1}^{n}\gamma_j(X_j - \overline{X})}{\sum_{j=1}^{n}(X_j - \overline{X})^2}, \qquad \hat{\alpha} = \overline{Y} - \hat{\beta}\overline{X}. \tag{6.183}$$

Theorem 6.2 holds for the joint distribution of $\hat{\alpha}, \hat{\beta}$, where $\sigma_\gamma^2 = \beta^2\sigma_\eta^2 = \sigma_\varepsilon^2$ replaces σ^2.

As a final observation on the model (6.169), (6.170), express the equation (6.169) as

$$Y_j = \alpha + \beta X_j - \beta\eta_j + \varepsilon_j. \tag{6.184}$$

Since η is not observed, this can be thought of as a special case of linear regression with one regressor missing. See Bhattacharya and Bhattacharyya (1994).

6.12 Notes and References

A general reference to this chapter is Ferguson (1996). Serfling (1980, Chaps. 1, 2) and Bickel and Doksum (2001, Chap. 5 and Appendices A.14, A.15), contain many basic results and fine exercises. Dasgupta (2008, Chaps. 1, 5 and 7) may be consulted for many additional facts and references as well as a wealth of exercises.

Exercises for Chap. 6

Exercises for Sect. 6.2

Ex. 6.1. (a) Extend Proposition 6.1 to k sequences $U_n^{(i)} \xrightarrow{P} a_i$ $(1 \le i \le k)$ and a function g of k variables continuous at $(a_1, a_2, \ldots, a_k)$, as stated in Remark 6.1.
(b) Extend (a) to vector-valued sequences $U_n^{(i)}$, $1 \le i \le k$, and vector-valued functions $g(u, v)$.

Ex. 6.2. (a) Show that Corollary 6.1 holds if $\xrightarrow{P}$ is replaced by $\xrightarrow{\text{a.s.}}$.
(b) Prove Proposition 6.4.

Ex. 6.3. Let X_n $(n \ge 1)$ be a sequence of i.i.d. random variables, and assume that the infimum and supremum of values of X_1 are m and M, respectively. That is, $P(m \le X_1 \le M) = 1$, $P(X_1 < a) > 0 \; \forall \, a > m$, $P(X_1 > b) > 0 \; \forall \, b < M$ (Here X_n real-valued, but m and/or M may be infinite). Prove that $\max\{X_1, \ldots, X_n\} \xrightarrow{\text{a.s.}} M$ and $\min\{X_1, \ldots, X_n\} \xrightarrow{\text{a.s.}} m$.

Ex. 6.4. Let $Y_n \xrightarrow{P} Y$. Prove that there exists a subsequence Y_{n_k} $(k = 1, 2, \ldots)$ $(n_1 < n_2 < \cdots)$ such that $Y_{n_k} \xrightarrow{\text{a.s.}} Y$ as $k \to \infty$.

Exercises for Sect. 6.3

Ex. 6.5. (a) Let $\delta_n = \max\{(X_j - \overline{X})^2 : 1 \le j \le n\} / \sum_1^n (X_j - \overline{X})^2$. Prove that, under the assumptions of Example 6.3, $\hat{\alpha}$ and $\hat{\beta}$ are consistent estimators of α, β if $\delta_n \to 0$.
(b) Prove the statement in Remark 6.4 for consistency of $\hat{\alpha}$, $\hat{\beta}$ under the hypothesis (6.17), and assuming $\delta_n \to 0$ as $n \to \infty$.
(c) Extend (b) to the heteroscedastic case (6.18) with bounded σ_j^2 $(j \ge 1)$.
(d) Write $\gamma_n = \max\{\sigma_j^2 : 1 \le j \le n\}$. Extend (c) to the case of possibly unbounded sequences σ_j^2 $(j \ge 1)$, but satisfying $\gamma_n/n \longrightarrow 0$, $m_2 \gamma_n / \sum_{j=1}^n (X_j - \overline{X})^2 \to 0$, where $m_2 = \sum_1^n X_j^2 / n$.

Ex. 6.6. Give simpler proofs of consistency of $\hat{\alpha}$, $\hat{\beta}$ assuming ε_j are i.i.d. $N(0, \sigma^2)$ in

(a) Example 6.3,
(b) Example 6.4.

Ex. 6.7. (a) Let X_n $(n \ge 1)$ be uncorrelated random variables, $\sigma_n^2 = \mathrm{var}\, X_n$. Suppose $\frac{1}{n} \sum_{j=1}^n E X_j \to \mu$ as $n \to \infty$. Prove that $\overline{X} = \frac{1}{n} \sum_{j=1}^n$ is a consistent estimator of μ if $\frac{1}{n^2} \sum_{j=1}^n \sigma_j^2 \to 0$.
(b) In (a), assume $E X_j = \mu \; \forall \, j$, and while μ is unknown, $\sigma_j^2 > 0$ are known. If $X_1, \ldots, X_n$ are observed, show that the linear unbiased estimator $U_n = \sum_1^n \omega_j X_j$ of μ with the minimum expected squared error is obtained by taking $\omega_j = \frac{1}{\sigma_j^2} / \sum_1^n \frac{1}{\sigma_i^2}$ $(1 \le j \le n)$.
(c) Show that the optimal estimator U_n $(n \ge 1)$ of μ in (b) is consistent if $\sum_1^n \frac{1}{\sigma_i^2} \to \infty$ as $n \to \infty$.

Ex. 6.8. Prove (6.30) by showing that the second moment of the left side converges to $[\sigma^2/(1 - \beta^2)]^2$, and then using (6.31).

Exercises for Sect. 6.4

Ex. 6.9. Using the general definition of a quantile, prove the following assertions for $0 < p < 1$.

(a) If F is continuous and strictly increasing on (c, d) where $F(c) < p$ and $F(d) > p$, then F has a unique p-th quantile.
(b) If $F(x) = p$ for all $x \in (c, d)$, $F(x) < p \ \forall \ x < c$ and $F(x) > p \ \forall \ x > d$, then the set of p-th quantiles is $[c, d]$.

Ex. 6.10. Assume F is continuous.

(a) If np *is not an integer*, show that $X_{([np]+1)}$ is the unique p-th quantile of the empirical distribution.
(b) If np is an integer, show that the set of p-th quantiles is the interval $[X_{([np])}, X_{([np]+1)}]$.

Ex. 6.11. Let X be a discrete random variable with values $a_1 < a_2 < \cdots < a_k < \cdots$ (finite or denumerable sequence), $P(X = a_k) = \pi_k > 0 \ \forall \ k$. Its d.f. is

$$F(x) = \begin{cases} 0 & \text{if } x < a_1 \\ p_k & \text{if } a_k \le x < a_{k+1} \end{cases} \quad (k = 1, 2, \cdots) \tag{6.185}$$

where $p_k = \pi_1 + \pi_2 + \cdots + \pi_k \ (k \ge 1)$. Let $p_k < p < p_{k+1}$. Show that $X_{([np])}$ ($\equiv X_{([np]):n}$) converges in probability to the population p-th quantile a_{k+1}, as $n \to \infty$. [Hint: Let $N_k \ (\equiv N_{k:n}) = \#\{j : 1 \le j \le n, X_j \le a_k\}$. Then N_k is binomial $B(n, p_k)$, and $P(X_{([np])} = a_{k+1}) = P(N_k < [np] \le N_{k+1}) = P(N_k < [np]) - P(N_{k+1} < [np]) \longrightarrow 1 - 0 = 1.$]

Exercises for Sect. 6.5

Ex. 6.12 (Designing a Sampling Plan). For taking an opinion poll one wishes to know the size n of the random sample needed to ensure that the error of estimating the population proportion p by the sample proportion $\hat{p}$ be no more than 0.03 with a 95 % probability.

(a) Use Chebyshev's Inequality (6.5) with $r = 2$ to get $n = 5556$.
(b) Use Chebyshev's Inequality with $r = 4$ to get $n = 2153$.
(c) Use the CLT to get $n = 1067$. [Hint: Use the fact that $p(1-p)$ has the maximum value $\frac{1}{4}$ at $p = \frac{1}{2}$.]

Ex. 6.13. Let $P_n \ (n \ge 1)$, P be probability measures on $(\mathbb{R}, \mathscr{B}(\mathbb{R}))$ such that P_n converges weakly to P.

(a) Give an example to show that $P_n(B)$ need not converge to $P(B)$ for all Borel sets B.
(b) Give an example to show that the distribution function F_n of P_n may not converge to the distribution function F of P at every point x.
(c) Suppose $g_n Y_n \xrightarrow{\mathscr{L}} V$ for some random variable V, where $g_n \to \infty$ as $n \to \infty$. Show that $Y_n \xrightarrow{P} 0$.

Ex. 6.14. Extend (6.57) to vector valued H.

Ex. 6.15. Suppose P_n $(n \geq 1)$, P are probability measures on $(\mathbb{R}, \mathscr{B}(\mathbb{R}))$ such that P_n converges weakly to P. Show that $\{P_n : n \geq 1\}$ is *tight: for every $\varepsilon > 0$ there exists $A_\varepsilon > 0$ such that $P_n(\{x : |x| > A_\varepsilon\}) < \varepsilon$ for all n.*

[Hint: (i) Find points of continuity $-B_\varepsilon$, C_ε of the distribution function F of P such that $F(-B_\varepsilon) < \varepsilon/3$, $F(C_\varepsilon) > 1 - \varepsilon/3$.

(ii) Find N_ε such that $F_n(-B_\varepsilon) < \varepsilon/3$ and $F_n(C_\varepsilon) > 1 - \varepsilon/3$ for all $n \geq N_\varepsilon$, where F_n is the distribution function of P_n. Then $P_n([-B_\varepsilon, C_\varepsilon]) > 1 - \frac{2\varepsilon}{3}$ for all $n \geq N_\varepsilon$.

(iii) For $n = 1, \ldots, N_\varepsilon$, find $D_\varepsilon > 0$ such that $P_n([-D_\varepsilon, D_\varepsilon]) > 1 - \varepsilon$ $(1 \leq n \leq N_\varepsilon)$.

(iv) Let $A_\varepsilon = \max B_\varepsilon, C_\varepsilon, D_\varepsilon$ to get $P_n(\{x : |x| > A_\varepsilon\}) < \varepsilon$ for all n.]

Ex. 6.16. Suppose $\binom{U}{V}$ has the bivariate Normal distribution $N\left(\binom{0}{0}, \begin{bmatrix} 1 & \rho \\ \rho & 1 \end{bmatrix}\right)$.

(a) Prove that the conditional distributions of V, given U, is $N(\rho U, 1 - \rho^2)$.

(b) Show that $EU^2V^2 = 1 + 2\rho^2$, $EU^3V = EVU^3 = 3\rho$.

Ex. 6.17. Assume X_j, $j \geq 1$, are i.i.d. real-valued, with $EX_j = \mu$, $\text{var}(X_j) = \sigma^2 > 0$, $EX_j^4 < \infty$. Prove that

(a) $\sqrt{n}(s^2 - \sigma^2) \xrightarrow{\mathscr{L}} N(0, E(X_1 - \mu)^4 - \sigma^4)$, and

(b) $\sqrt{n}(\frac{1}{s} - \frac{1}{\sigma}) \xrightarrow{\mathscr{L}} N(0, [E(X_1 - \mu)^4 - \sigma^4] \cdot [1/4\sigma^6])$.

[Hint: (a) Consider $U_j = X_j - \mu$, $j \geq 1$, $s^2 = (\frac{n}{n-1})\frac{1}{n}\sum_{j=1}^{1}(U_j - \overline{U})^2 = (\frac{n}{n-1})\left[\frac{1}{n}(\sum_{j=1}^{n})U_j^2 - \overline{U}^2\right]$, so that $\sqrt{n}(s^2 - \sigma^2) - \sqrt{n}(\frac{1}{n}\sum_{j=1}^{n}(U_j^2 - \sigma^2)) \xrightarrow{P} 0$.

(b) $\sqrt{n}(\frac{1}{s} - \frac{1}{\sigma}) - \sqrt{n}[H(\overline{z}) - H(\sigma^2)] \xrightarrow{P} 0$, where $z_j = U_j^2$, $EZ_j = \sigma^2$, $H(z) = z^{-1/2}$, $H(\overline{Z}) = (\frac{1}{n}\sum_{j=1}^{n}U_j^2)^{-1/2}$, $H(\sigma^2) = 1/\sigma$. Apply Corollary 6.2.]

Ex. 6.18. (a) Let X_n have the discrete uniform distribution on $\{0, \frac{1}{n}, \frac{2}{n}, \ldots, 1\}$ (i.e., $P(X_n = \frac{k}{n}) = \frac{1}{n+1}$ $(k = 0, 1, \ldots, n)$. Show that X_n converges in distribution to the uniform distribution on $[0, 1]$ (with constant density 1).

(b) Use (a) to prove that $(1/n+1)\sum_{k=0}^{n} f(k/n) \to \int_0^1 f(x)dx$ for every continuous function f on $[0, 1]$.

(c) Extend (b) to the case of all bounded measurable f on $[0, 1]$ with a finite set of discontinuities.

Ex. 6.19. $Y_n = \min\{X_i : 1 \leq i \leq n\}$ where $X_1, X_2, \ldots$ are i.i.d. beta $\mathscr{B}_e(\alpha, 1)$ random variables.

(a) What is the distribution of Y_n?

(b) Find a value of α such that Y_n converges in distribution to a nondegenerate law.

Ex. 6.20 (Fieller's Method for the Estimation of a Ratio). Let θ_1, θ_2 be unknown parameters, $\theta_2 > 0$. The problem is to obtain a confidence interval for $\rho = \frac{c+d\theta_1}{\theta_2}$ $(d \neq 0)$, based on asymptotically jointly Normal estimators $\hat{\theta}_1$, $\hat{\theta}_2$

of θ_1, θ_2, with available consistent estimates $\widehat{W}_{11}$, $\widehat{W}_{22}$, $\widehat{W}_{12}$ of $W_{11} = \mathrm{var}(\hat{\theta}_1)$, $W_{22} = \mathrm{var}(\hat{\theta}_2)$, $W_{12} = \mathrm{cov}(\hat{\theta}_1, \hat{\theta}_2)$.

(a) Use the delta method to obtain a confidence interval for ρ of confidence level $1 - \alpha$.

(b) *(Fieller method.)* Let $W_2 = \mathrm{var}(c + d\hat{\theta}_1 - \rho\hat{\theta}_2) = d^2 W_{11} + \rho^2 W_{22} - 2d\rho W_{12}$. Let $\widehat{W}^2 = d^2 \widehat{W}_{11} + \rho^2 \widehat{W}_{22} - 2d\rho\widehat{W}_{12}$. Show that $(c + d\hat{\theta}_1 - \rho\hat{\theta}_2)/\widehat{W}$ is AN $N(0,1)$, so that $\{(c + d\hat{\theta}_1 - \rho\hat{\theta}_2)^2 \leq \widehat{W}^2 z^2_{1-\frac{\alpha}{2}}\}$ has asymptotic probability $1 - \alpha$ (z_β being the β-th quantile of $N(0,1)$). Use the last relation to obtain a quadratic equation in ρ with roots $\hat{\rho}_\ell < \hat{\rho}_u$, and argue that $[\hat{\rho}_\ell, \hat{\rho}_u]$ is a confidence interval for ρ with asymptotic level $1 - \alpha$. [See Fieller 1940.]

Exercises for Sect. 6.6

Ex. 6.21. (a) Prove that $\delta_n \to 0$ implies $m_2/b_n^2 \to 0$ (see (6.70)). [Hint: Let $|X_j - X_{j'}| = c > 0$ for some j, j'. Then $\delta_n \geq (c/2)^2/b_n^2 \ \forall \ n \geq \max\{j, j'\}$, since either $|X_j - \overline{X}| \geq c/2$ or $|X_{j'} - \overline{X}| \geq c/2$. Also, $X_i^2 \leq 2X_1^2 + 4\delta_n b_n^2$ $(1 \leq i \leq n)$, so that $m_2/b_n^2 \leq 2X_1^2/b_n^2 + 4\delta_n \to 0$.]

(b) Verify (6.95).

(c) Prove that the random ellipse given by (6.97) is a confidence region for (α, β) with asymptotic confidence coefficient $1 - \theta$, if (6.72) holds for the linear regression model (6.12).

Ex. 6.22 (Normal Correlation Model). Show that, under Normal errors ε_j in (6.12),

(a) $\hat{\alpha}$, $\hat{\beta}$ are M.L.E.'s, and

(b) the confidence region (6.100) for (α, β) has exact confidence coefficient $1 - \theta$, for all $n \geq 3$, assuming that X_j's, $1 \leq j \leq n$, are not all the same.

Ex. 6.23. For the heteroscedastic linear regression model (6.101) of Example 6.7,

(a) prove (6.114) under the assumption (6.113),

(b) prove (6.116) under the assumption (6.115),

(c) derive (6.117), assuming (6.113) and (6.115), and show that $\widetilde{D}_n$ in (6.118) is a confidence region for (α, β) with asymptotic confidence coefficient $1 - \theta$.

Ex. 6.24. For the heteroscedastic linear regression model considered under Remark 6.14 (i.e., in (6.101) with $E\varepsilon_j^2 = \eta_j^2 \sigma^2$ $(1 \leq j \leq n)$, where $\eta_j^2 > 0$ are known, but $\sigma^2 > 0$ is unknown), prove that $\hat{\sigma}^2$ defined in (6.119) is an unbiased estimator of σ^2.

Ex. 6.25. (a) Write out a detailed proof of Corollary 6.3.

(b) Show that the *hypothesis of the correlation model holds if* (X_j, Y_j), $j \geq 1$, *are i.i.d. observations from a bivariate normal distribution*, where $\alpha + \beta_j X_j$ is the conditional mean of Y_j, given X_j. [*Hint:* Use the fact that the conditional distribution of Y_j, given X_j, is Normal with mean $\alpha + \beta_j X_j$ (for appropriate α, β) and $\varepsilon_j := Y_j - \alpha - \beta X_j$ uncorrelated with X_j. This says that the conditional distribution of $\varepsilon_j := Y_j - \alpha - \beta X_j$, given X_j is $N(0, \sigma^2)$, so that ε_j is independent of X_j.]

Ex. 6.26 (Linear Regression Passing Through a Point).

(a) Consider the linear regression (6.12) with $\alpha = 0$ (i.e., the regression line passes through the origin $(0,0)$). Show that the least squares estimator of β is $\hat{\beta} = \sum_1^n X_j Y_j / \sum_1^n X_j^2$ and that $\hat{\beta}$ is asymptotically Normal $N(\beta, \sigma^2 / \sum_1^n X_j^2)$ if $\tilde{\delta}_n := \max\{X_j^2 / \sum_1^n X_j^2 : 1 \le j \le n\} \to 0$ as $n \to \infty$.

(b) Consider the linear regression passing through a given point (x_0, y_0), i.e., $Y_j - y_0 = \beta(X_j - x_0)$ (so that $\alpha = y_0 - \beta x_0$). Show that the least squares estimator of β is $\hat{\beta} = \sum_1^n (X_j - x_0)(Y_j - y_0) / \sum_1^n (X_j - x_0)^2$ and that $\hat{\beta}$ is asymptotically Normal $N(\beta, \sigma^2 / \sum_1^n (X_j - x_0)^2)$, provided $\bar{\delta}_n := \max\{(X_j - x_0)^2 / \sum_1^n (X_i - x_0)^2 : 1 \le j \le n\} \to 0$.

Ex. 6.27. Suppose in (6.12) X represents blood pressure and Y platelet calcium in 38 people with normal blood pressure. Let $\overline{X} = 84.5$, $\overline{Y} = 105.8$, $\sum (X_j - \overline{X})^2 = 2397.5$, $\sum (X_j - \overline{X}) Y_j = 2792.5$.

(a) Find the usual 90% confidence band for the regression line.
(b) Find the simultaneous confidence bound with asymptotic confidence level at least 90% using Scheffé's method (See Proposition 6.9).

Ex. 6.28 (Transformation to Linear Regression). Consider the relations (i) $y = \beta_0 \exp\{\beta_1 x\}$ ($\beta_0 > 0$, $\beta_1 \in \mathbb{R}$), $x \in \mathbb{R}$; (ii) $y = \beta_0 x^{\beta_1}$ ($\beta_0 >: 0$, $\beta_1 \in \mathbb{R}$), $x > 0$. In both (i) and (ii) $y > 0$ and a *multiplicative error* ζ may be more appropriate than a linear one.

(a) Find appropriate transformations of observations Y_j, X_j, $1 \le j \le n$, to make them linear regression models.
(b) Assuming that ζ_j are i.i.d. positive multiplicative errors, obtain appropriate estimates of β_0, β_1 and find their asymptotic distributions in each case (i), (ii).

Exercises for Sect. 6.7

Ex. 6.29. This exercise shows that Theorem 6.3(a) breaks down if the density f is not continuous (and positive) at ζ_p. Consider a density whose right-hand limit $f(\zeta_{\frac{1}{2}+})$ and left-hand limit $f(\zeta_{\frac{1}{2}-})$ are positive but unequal (at the median $\zeta_{\frac{1}{2}}$). Show that $\sqrt{n}(\hat{\zeta}_{\frac{1}{2}} - \zeta_{\frac{1}{2}})$ does not converge to a Normal distribution. [Hint: If $z > 0$, $\mathbf{1}_{\{X_i \le \zeta_{\frac{1}{2} + \frac{z}{\sqrt{n}}}\}}$ ($1 \le i \le n$) are i.i.d. Bernoulli, taking the value 1 with probability $\frac{1}{2} + \int_{\zeta_{1/2}}^{\zeta_{1/2} + \frac{z}{\sqrt{n}}} f(x)dx = \frac{1}{2} + \frac{z}{\sqrt{n}} f\left(\zeta_{\frac{1}{2}+}\right) + o(n^{-\frac{1}{2}})$. Hence, by following the steps (6.127)–(6.128), $P\left(\sqrt{n}\left(\hat{\zeta}_{1/2} - \zeta_{1/2}\right) \le z\right) \longrightarrow P\left(\dfrac{1}{2f\left(\zeta_{\frac{1}{2}+}\right)} Z \le z\right)$, where Z is standard Normal. But for $z < 0$, the same argument shows that $P(\sqrt{n}(\hat{\zeta}_{1/2} - \zeta_{1/2}) \le z) \longrightarrow P(\frac{1}{2f(\zeta_{\frac{1}{2}-})} Z \le z)$.]

Ex. 6.30. Fix $p \in (0,1)$. Assume that ζ_p is uniquely defined and F is three times continuously differentiable in a neighborhood of ζ_p, with $f(\zeta_p) \equiv F'(\zeta_p) = 0$, and $f''(\zeta_p) \ne 0$. Show that $P(n^{\frac{1}{6}}(\hat{\zeta}_p - \zeta_p) \le z) \to \Phi(cz^3)$ where Φ is the standard Normal distribution function and $c = \frac{1}{6} f''(\zeta_p) / \sqrt{p(1-p)}$.

Ex. 6.31. Find the asymptotic distribution of the sample *inter-quartile range* $\hat{\zeta}_{0.75} - \hat{\zeta}_{0.25}$ assuming that the hypothesis of Theorem 6.3(a) holds for $p = 0.75$ and $p = 0.25$.

Exercises for Sect. 6.8

Ex. 6.32. Prove that the sum of two independent mean-zero random variables, one of which is Normal while the other is non-Normal, can not be Normal. [Hint: Use characteristic functions.]

Ex. 6.33. Derive Theorem 6.2 as a corollary of Theorem 6.4, and show that the condition $\delta_n \to 0$ in (6.72) is also necessary for (6.91) to hold, if ε_i's are not Normal.

Exercises for Sect. 6.9

Ex. 6.34. Verify the computations in Example 6.9.

Ex. 6.35. In Example 6.10, compare the relative performances of $\hat{\zeta}_{1/2}$ and $\overline{X}$ as the parameters ε and τ vary over their respective ranges $(0 < \varepsilon < 1, \tau > 0)$. [Hint: Write $\sigma = \sqrt{\tau}$.] Let $h(\varepsilon, \sigma) := e_{\hat{\zeta}_{1/2}, \overline{X}}$, as given in Example 6.10. Fix $\varepsilon \in (0,1)$. Show that $\partial h/\partial \sigma$ is negative for $0 < \sigma < 1$, positive for $\sigma > 1$, and vanishes at $\sigma = 1$. Also, $h(\varepsilon, \sigma) \to \infty$ as $\sigma \downarrow 0$ and as $\sigma \uparrow \infty$. Thus,

(i) for every given $\varepsilon \in (0,1)$, $h(\varepsilon, \sigma)$ has the unique minimum value $h(\varepsilon, 1) = \frac{2}{\pi} <$ 1, and

(ii) there exist $\sigma_i(\varepsilon)$ $(i = 1, 2)$, $0 < \sigma_1(\varepsilon) < 1 < \sigma_2(\varepsilon)$ such that $h(\varepsilon, \sigma_i(\varepsilon)) = 1$ $(i = 1, 2)$, $h(\varepsilon, \sigma) > 1$ if $\sigma \in (0, \sigma_1(\varepsilon)) \cup (\sigma_2(\varepsilon), \infty)$, $h(\varepsilon, \sigma) > 1$ if $\sigma_1(\varepsilon) < \sigma < \sigma_2(\varepsilon)$. In particular, $h(\varepsilon, \sigma) \geq \frac{2}{\pi} \; \forall \; \varepsilon, \sigma$, but the lower bound of $e_{\overline{X}, \hat{\zeta}_{1/2}} = 1/h(\varepsilon, \sigma)$ is zero.]

Conclusion: $\hat{\xi}_{1/2}$ is *robust* compared to $\overline{X}$ in the Normal mixture model.

Exercise for Sect. 6.10

Ex. 6.36. Take a random sample of size $n = 50$ from the standard Normal distribution. Treating this as a random sample from an unknown population, find an approximate 95 % confidence interval for $\xi_{.25}$.

Exercise for Sect. 6.11

Ex. 6.37. (a) In the so-called *structural model* as considered in (6.169), (6.170), (6.174) with a stochastic X^*, prove (6.172). [Hint: Show that the quantities $n^{-1} \sum (X_j^* - \overline{X}^*)(\varepsilon_j - \overline{\varepsilon})$, $n^{-1} \sum (X_j^* - \overline{X}^*)(\eta_j - \overline{\eta})$ and $n^{-1} \sum (\eta_j - \overline{\eta})(\varepsilon_j - \overline{\varepsilon})$ have all zero means and variances converging to zero.]

(b) Prove (6.172) for the *functional model*, assuming $n^{-1} \sum_{1 \leq j \leq n} (X_j^* - \overline{X}^*)^2 \to \sigma_{x^*}^2 > 0$, as $n \to \infty$.

(c) In the functional model, show that $\hat{\beta}$ is a consistent estimator of β if and only if $n^{-1} \sum_{1 \leq j \leq n} (X_j^* - \overline{X}^*)^2 \to \infty$ as $n \to \infty$.

Ex. 6.38. Derive (6.177), (6.178) under the given assumptions of the structural bivariate Normal model.

Ex. 6.39. Carry out a simulation study of the structural bivariate Normal model with $\kappa = 0.9$, and compute $\hat{\beta} = \hat{\beta}_{yx}$ and $\hat{\beta}_{xy}^{-1}$.

References

Bahadur, R. R. (1966). A note on quantiles in large samples. *Annals of Mathematical Statistics, 37*(3), 577–580.

Berkson, J. (1950). Are there two regressions? *Journal of the American Statistical Association, 45*(250), 164–180.

Bhattacharya, R., & Bhattacharyya, D. K. (1994). Proxy versus instrumental variable methods in regression with one regressor missing. *Journal of Multivariate Analysis, 47*, 123–138.

Bhattacharya, R., & Ranga Rao, R. (2010). *Normal approximation and asymptatic expansions.* SIAM classics in applied mathematics (Vol. 64). Philadelphia: SIAM

Bhattacharya, R., & Waymire, E. (2007). *A basic course in probability theory.* New York: Springer.

Bickel, P. J., & Doksum, K. (2001). *Mathematical statistics* (2nd ed.). Upper Saddle River, NJ: Prentice Hall.

Billingsley, P. (1986). *Probability and measure.* Wiley.

Dasgupta, A. (2008). *Asymptotic theory of statistics and probability.* New York: Springer.

Diaconis, P., & Efron, B. (1983). Computer intensive methods in statistics. *Division of Biostatistics, 248*(5), 116–126.

Ferguson, T. S. (1996). *A course in large sample theory.* London: Taylor & Francis.

Fieller, E. C. (1940). The biological standardisation of insulin. *Journal of the Royal Statistical Society (Supplement), 1*, 1–54.

Fuller, W. A. (1987). *Measurement error models.* Wiley.

Ghosh, J. K. (1971). A new proof of the Bahadur representation of quantiles and an application. *Annals of Mathematical Statistics, 42*(6), 1957–1961.

Ito, M., & Ganguly, J. (2006). Diffusion kinetics of Cr in olivine and 53Mn-53Cr thermochronology of early solar system objects. *Geochimica et Cosmochimica Acta, 70*, 799–806.

Kiefer, J. (1967). On Bahadur's representation of sample quantiles. *Annals of Mathematical Statistics, 38*(5), 1323–1342.

Riersøl, O. (1950). Identifiability of a linear relation between variables which are subject to error. *Econometrica, 18*, 375–389.

Scheffé, H. (1959). *The analysis of variance.* New York: Wiley.

Serfling, R. (1980). *Approximation theorems of mathematical statistics.* New York: Wiley.

Shevtsova, I. G. (2010). An improvement of convergence rate estimates in the Lyapunov theorem. *Doklady Mathematics, 82*(3), 862–864.

Chapter 7
Large Sample Theory of Estimation in Parametric Models

Abstract The main focus of this chapter is the asymptotic Normality and optimality of the maximum likelihood estimator (MLE), under regularity conditions. The Cramér–Rao lower bound for the variance of unbiased estimators of parametric functions is shown to be achieved asymptotically by the MLE. Also derived are the asymptotic Normality of M-estimators and the asymptotic behavior of the Bayes posterior.

7.1 Introduction

We begin with the derivation of an important inequality known as the *Cramér-Rao bound* which gives a lower bound to the expected squared error (or, variance) of unbiased estimators of parametric functions under certain regularity conditions. This bound is attained by some estimators of special parametric functions in exponential families (see Chap. 4, Part I for definition of exponential families). In general the bound is rarely attained (exactly). We will see, however, that this lower bound is attained in an asymptotic sense by maximum likelihood estimators (MLEs) in large samples, provided certain regularity conditions hold. This shows that the MLEs are asymptotically optimal under these conditions.

Notation In order to avoid having to make the arguments separately for the discrete and absolutely continuous cases (or, for a mix of them), we will in this section write $g(\mathbf{x}; \theta)$ for the density of the observed random vector $\mathbf{X}$ w.r.t. a sigma finite measure μ on the set $\mathbf{x}$ of all values $\mathbf{x}$ in the range $\mathscr{X}$ of $\mathbf{X}$. In the absolutely continuous case, $g(\mathbf{x}; \theta)$ is the *probability density function (p.d.f.)* of $\mathbf{X}$ (and μ is Lebesgue measure: $\mu(dx) = dx$). In the discrete case $g(\mathbf{x}; \theta)$ is the *probability mass function (p.m.f.)* of $\mathbf{X}$ on a countable set $\mathscr{X}$ (and μ is the counting measure, $\int_{\mathscr{X}} h(\mathbf{x}) g(\mathbf{x}; \theta) \mu(d\mathbf{x}) = \sum_{\mathbf{x} \in \mathscr{X}} h(\mathbf{x}) g(\mathbf{x}; \theta)$). Sometimes, when $\mathbf{X} = (X_1, \ldots, X_n)$ with $X_1, \ldots, X_n$ i.i.d. (or when X_i's are n observations in a time series), we write $f_n(\mathbf{x}; \theta)$, instead of $g(\mathbf{x}; \theta)$, to indicate the *sample size*. In the i.i.d. case, $f(x; \theta)$ always indicates the density of a single observation X_i. The range of X_i is $\mathscr{X}$ in this case, while that of $\mathbf{X}$ is $\mathscr{X} = \mathscr{X}^n$.

© Springer-Verlag New York 2016
R. Bhattacharya et al., *A Course in Mathematical Statistics and Large Sample Theory*, Springer Texts in Statistics, DOI 10.1007/978-1-4939-4032-5_7

The random variables, or vectors, are all defined on a measurable space $(\Omega, \mathscr{F})$ and, for each value θ of the parameter, they are governed by a probability law P_θ on $(\Omega, \mathscr{F})$. The expectation under P_θ is denoted E_θ. Also, $Y \overset{\mathscr{L}}{\sim} Q$ denotes "Y has distribution Q".

7.2 The Cramér-Rao Bound

Theorem 7.1 (Cramér-Rao Information Inequality). *Suppose* $\mathbf{X}$ *has p.d.f.* $g(\mathbf{x}; \theta)$ *(with respect to a sigma-finite measure* μ*) on a space* $\mathscr{X}$ *satisfying*

(i) $g(\mathbf{x}; \theta) > 0 \;\forall\; \mathbf{x} \in \mathscr{X}, \;\forall\; \theta \in \Theta$—*an open interval,*

(ii) $\int_{\mathscr{X}} \frac{d}{d\theta} g(\mathbf{x}; \theta)\mu(d\mathbf{x}) = \frac{d}{d\theta} \int_{\mathscr{X}} g(\mathbf{x}; \theta)\mu(d\mathbf{x}) \equiv 0.$

Let $T = t(\mathbf{X})$ *be a (real-valued) statistic with* $c(\theta) := E_\theta T$, $E_\theta T^2 < \infty \;\forall\; \theta$, *satisfying*

(iii) $\frac{d}{d\theta} \int_{\mathscr{X}} t(\mathbf{x})g(\mathbf{x}; \theta)\mu(d\mathbf{x}) \equiv c'(\theta) = \int_{\mathscr{X}} t(\mathbf{x}) \frac{dg(\mathbf{x};\theta)}{d\theta}\mu(d\mathbf{x}).$

Then

$$\mathrm{var}_\theta T \equiv E_\theta(T - c(\theta))^2 \geq \frac{c'(\theta)^2}{E_\theta \left(\frac{d \log g(\mathbf{X};\theta)}{d\theta} \right)^2}. \tag{7.1}$$

Proof. Condition (ii) may be restated as

$$0 = E_\theta \frac{d \log g(\mathbf{X}; \theta)}{d\theta} \qquad \left(\equiv \int_{\mathscr{X}} \frac{\frac{d}{d\theta} g(\mathbf{x}; \theta)}{g(\mathbf{x}; \theta)} g(\mathbf{x}; \theta)\mu(d\mathbf{x}) \right) \tag{7.2}$$

Similarly, condition (iii), together with (ii), says

$$\mathrm{cov}_\theta \left(T, \frac{d \log g(\mathbf{X}; \theta)}{d\theta} \right) = c'(\theta), \tag{7.3}$$

The inequality (7.1) now follows from (7.3) by the Cauchy-Schwartz inequality.

$\square$

Remark 7.1. Assumptions (ii), (iii) in the theorem concern the interchangeability of the order of differentiation and integration. If $\mathscr{X}$ depends on θ, then these generally do not hold. For example, let $\mathbf{X} = (X_1, \ldots, X_n)$ with X_j's i.i.d. uniform on $(0, \theta)$. Then $\mathscr{X} = (0, \theta)^n$. Take $T = M_n \equiv \max\{X_1, \ldots, X_n\}$. Note that $g(\mathbf{x}; \theta) = 1/\theta^n$ on $\mathscr{X}$, so that $dg(\mathbf{x}; \theta)/d\theta = -n/\theta^{n+1}$, $\int_{\mathscr{X}} \frac{d}{d\theta} g(\mathbf{x}; \theta)d\mathbf{x} \equiv \int_{(0,\theta)^n} (-n/\theta^{n+1})d\mathbf{x} = -\frac{n}{\theta^{n+1}} \cdot \theta^x = -n/\theta \neq 0$. Also, letting $c(\theta) = E_\theta M_n = \frac{n}{n+1}\theta$ (Exercise 7.1), one has $c'(\theta) = \frac{n}{n+1}$, while $\int_{\mathscr{X}} t(\mathbf{x})(dg(\mathbf{x}; \theta)/d\theta)d\mathbf{x} = E_\theta(Td \log g \,(\mathbf{X}; \theta)/d\theta) = E_\theta(M_n d[-n \log \theta]/d\theta) = E_\theta \left(M_n(-\tfrac{n}{\theta}) \right) = -\frac{n^2}{n+1}.$

Remark 7.2. The most common $\mathbf{X}$ encountered in this course is that of a random vector $\mathbf{X} = (X_1, \ldots, X_n)$ with X_j's i.i.d. and having a (common) p.d.f. or p.m.f. $f(x; \theta)$. In this case

$$g(\mathbf{x}; \theta) = \prod_{j=1}^{n} f(x_j; \theta), \quad \log g(\mathbf{x}; \theta) = \sum_{j=1}^{n} \log f(x_j; \theta), \quad x = (x_1, \ldots, x_n). \tag{7.4}$$

so that (ii) or (7.2) is equivalent to

$$E_\theta \frac{d\log f(X_1;\theta)}{d\theta} \qquad \left(\equiv \int_{\mathscr{X}} \frac{df(x;\theta)}{d\theta}dx, \text{ or } \sum_{x\in\mathscr{X}} \frac{df(x;\theta)}{dx}\right) = 0$$

Also,

$$E_\theta\left(\frac{d\log g(\mathbf{X};\theta)}{d\theta}\right)^2 \equiv \mathrm{var}_\theta\left(\frac{d\log g(\mathbf{X};\theta)}{d\theta}\right)$$

$$= \mathrm{var}_\theta\left(\sum_{j=1}^n \frac{d\log f(X_j;\theta)}{d\theta}\right) = \sum_{j=1}^n \mathrm{var}_\theta\left(\frac{d\log f(X_j;\theta)}{d\theta}\right)$$

$$= n\,\mathrm{var}_\theta\left(\frac{d\log f(X_1;\theta)}{d\theta}\right) = nE_\theta\left(\frac{d\log f(X_1;\theta)}{d\theta}\right)^2.$$

The quantity $I(\theta) := E_\theta\left(\frac{d\log f(X_1;\theta)}{d\theta}\right)^2$ is called, in the case of i.i.d. observations, the *information per observation*, and $E_\theta\left(\frac{d\log g(\mathbf{X};\theta)}{d\theta}\right)^2 = nI(\theta)$ is the *information contained in the whole sample*. We have derived the following corollary of the theorem.

Corollary 7.1. *Suppose $X_1,\ldots,X_n$ are i.i.d. with a common density $f(x;\theta)$ with respect to a sigma-finite measure μ, and let $\mathbf{X} = (X_1,\ldots,X_n)$. Then, under the hypothesis of the theorem above, one has*

$$\mathrm{var}_\theta(T) \geq \frac{(c'(\theta))^2}{nI(\theta)}, \tag{7.5}$$

where $I(\theta) = E_\theta\left(\frac{d\log f(X_1;\theta)}{d\theta}\right)^2$.

Remark 7.3. The hypothesis of Corollary 7.1 holds if the following conditions hold:

(R_0) $f(\mathbf{x};\theta) > 0 \ \forall \ \mathbf{x} \in \mathscr{X}$, $\theta \in \Theta$, where $\mathscr{X}$ does not depend on θ.

(R_1) $\frac{df(\mathbf{x};\theta)}{d\theta}$ is continuous on Θ ($\forall \ \mathbf{x} \in \mathscr{X}$), and for each $\theta_0 \in \Theta$ there exists $h_1 = h_1(\theta_0) > 0$ such that $\sup\left\{|\frac{df(\mathbf{x};\theta)}{d\theta}| : |\theta - \theta_0| \leq h_1\right\} \leq g_1(\mathbf{x})$, where $\int_{\mathscr{X}} g_1(\mathbf{x})\mu(d\mathbf{x}) < \infty$.

(R_2) For each $\theta_0 \in \Theta$ there exists $h_2 = h_2(\theta_0) > 0$ such that $\sup\left\{|\frac{df(\mathbf{x};\theta)}{d\theta}| : |\theta - \theta_0| \leq h_2\right\} \leq g_2(\mathbf{x})$ where $E_{\theta_0} g_2^2(\mathbf{X})$ $(\equiv \int_{\mathscr{X}} g_2^2(\mathbf{x})f(\mathbf{x};\theta_0)\mu(d\mathbf{x}) < \infty)$.

Remark 7.4. If one estimates some function $h(\theta)$ by T, and $E_\theta T = c(\theta)$, then under the hypothesis of Theorem 7.1 one has

$$E_\theta(T - h(\theta))^2 = E_\theta(T - c(\theta))^2 + (c(\theta) - h(\theta))^2$$

$$\geq \frac{(c'(\theta))^2}{E_\theta\left(\frac{d\log g(\mathbf{X};\theta)}{d\theta}\right)} + (c(\theta) - h(\theta))^2 \tag{7.6}$$

Remark 7.5. It may be shown in the case of i.i.d. observations $X_1, \ldots, X_n$ from a one-parameter exponential family (with p.d.f. or p.m.f. $f(x;\theta) = c_1(\theta)h(x)e^{\theta t_1(x)}, \theta \in \Theta$—an open interval) that (R_0)–(R_2) hold. Hence (7.5) is valid in this case. However, the lower bound is attained only by $T = t(X) = \sum_{j=1}^n t_1(X_j)$ (and its linear functions). (Exercise 7.2(b).)

7.3 Maximum Likelihood: The One Parameter Case

As mentioned in Sect. 7.1, one of the main reasons why the Rao–Cramér bound is so important is that the MLE's, under appropriate regularity conditions, attain this bound in an asymptotic sense. In other words, the MLE is asymptotically optimal or, *asymptotically efficient*. The following result is a precise statement of this fact.

Theorem 7.2 (CLT for the MLE). *Let $f(x;\theta)$ be the p.d.f. with respect to a sigma-finite measure μ of the common distribution of i.i.d. random variables $X_1, X_2, \ldots$, for θ belonging to an open interval Θ. Assume that $f(x;\theta) > 0 \ \forall\ x \in \mathcal{X}, \ \forall\ \theta \in \Theta$, where $P_\theta(X_1 \in \mathcal{X}) = 1 \ (\forall\ \theta \in \Theta)$. Assume also the following*

(A_1^*) *$\theta \to f(x;\theta)$ is three times continuously differentiable on Θ, $\forall\ x \in \mathcal{X}$;*

(A_2^*) *$\int_{\mathcal{X}} \frac{d}{d\theta} f(x;\theta)\mu(dx) = 0 \ (\equiv \frac{d}{d\theta} \int_{\mathcal{X}} f(x;\theta)\mu(dx)), \ \int_{\mathcal{X}} \frac{d^2}{d\theta^2} f(x;\theta)\mu(dx) = 0 (\equiv \frac{d^2}{d\theta^2} \int_{\mathcal{X}} f(x;\theta)\mu(dx));$*

(A_3^*) *$0 < I(\theta) := E_\theta \left(\frac{d \log f(X_1;\theta)}{d\theta} \right)^2 < \infty \ \forall\ \theta \in \Theta;$*

(A_4^*) *For each $\theta_0 \in \Theta$ there exists an $\varepsilon \ (= \varepsilon(\theta_0)) > 0$ such that $|\frac{d^3 \log f(x;\theta)}{d\theta^3}|^3 \leq g(x) \ \forall\ \theta \in [\theta_0 - \varepsilon, \theta_0 + \varepsilon]$, where $\int_{\mathcal{X}} g(x)f(x;\theta_0)\mu(dx) < \infty$.*

(A_5^*) *The likelihood equation (writing $\ell(\theta) := \sum_{j=1}^n \log f(X_j;\theta)$)*

$$\frac{d\ell(\theta)}{d\theta} = 0, \qquad i.e., \quad \sum_{j=1}^n \frac{d \log f(X_j;\theta)}{d\theta} = 0, \tag{7.7}$$

has a consistent solution $\hat{\theta}_n$.

Then $\sqrt{n}(\hat{\theta}_n - \theta_0)$ converges in distribution to $N(0, 1/I(\theta_0))$ if θ_0 is the true parameter value (i.e., under P_{θ_0}), as $n \to \infty$.

Proof. Using a Taylor expansion of $d\ell(\theta)/d\theta$ around $\theta = \theta_0$, one has

$$0 = \frac{d\ell(\theta)}{d\theta}\big|_{\theta=\hat{\theta}_n} = \frac{d\ell(\theta)}{d\theta}\big|_{\theta=\theta_0} + (\hat{\theta}_n - \theta_0)\frac{d^2\ell(\theta)}{d\theta^2}\big|_{\theta=\theta_0} + \frac{(\hat{\theta}_n - \theta_0)^2}{2}\frac{d^3\ell(\theta)}{d\theta^3}\big|_{\theta=\theta^*}. \tag{7.8}$$

where θ^* lies in the line segment joining θ_0 and $\hat{\theta}_n$. Thus,

$$\sqrt{n}(\hat{\theta}_n - \theta_0) = \frac{-\frac{1}{\sqrt{n}}\left(\frac{d\ell(\theta)}{d\theta} \right)_{\theta=\theta_0}}{\frac{1}{n}\left(\frac{d^2\ell(\theta)}{d\theta^2} \right)_{\theta=\theta_0} + \frac{1}{n}\frac{\hat{\theta}_n-\theta_0}{2}\frac{d^3\ell(\theta)}{d\theta^3}\big|_{\theta=\theta^*}}. \tag{7.9}$$

First note that

$$
\begin{aligned}
E_\theta \frac{d^2 \log f(X_j;\ \theta)}{d\theta^2} &= \int_{\mathscr{X}} \frac{d}{d\theta} \left(\frac{\frac{d}{d\theta} f(x;\theta)}{f(x;\theta)} \right) f(x;\theta)\,dx \\
&= \int_{\mathscr{X}} \frac{f''(x;\theta)f(x;\theta) - (f'(x;\theta))^2}{f^2(x;\theta)} f(x;\theta)\,dx \\
&= \int_{\mathscr{X}} f''(x;\theta)\mu(dx) - \int_{\mathscr{X}} \left(\frac{d \log f(x;\theta)}{d\theta} \right)^2 f(x;\theta)\mu(dx) \\
&= -I(\theta),
\end{aligned}
\tag{7.10}
$$

using the second relation in (A_2^*) for the last step. Thus by the strong law of large numbers (SLLN),

$$
\frac{1}{n} \left(\frac{d^2 \ell(\theta)}{d\theta^2} \right)_{\theta=\theta_0} \equiv \frac{1}{n} \sum_{j=1}^{n} \left(\frac{d^2 \log f(X_j;\theta)}{d\theta^2} \right)_{\theta=\theta_0} \longrightarrow -I(\theta_0)
\tag{7.11}
$$

with probability one (under P_{θ_0}). By (A_4^*), $\frac{1}{n}|\frac{d^3 \ell(\theta)}{d\theta^3}| \le \frac{1}{n}\sum_{j=1}^{n} g(X_j)\ \forall\ \theta \in [\theta_0 - \varepsilon, \theta_0+\varepsilon]$, but by (A_5^*), $P_{\theta_0}(|\theta_0 - \theta^*| \le \varepsilon) \le P_{\theta_0}(|\hat{\theta}_n - \theta_0| \le \varepsilon) \to 1$ as $n \to \infty$. Using this and the fact that $\frac{1}{n}\sum_{j=1}^{n} g(X_j) \to E_{\theta_0} g(X_1) < \infty$, one finds that $\frac{1}{n}|\frac{d3\ell(\theta)}{d\theta^3}|_{\theta^*}$ remains bounded in probability as $n \to \infty$ (See Definition 7.1 below). Since, also, $\theta_n - \theta_0 \to 0$ in probability, one gets

$$
\frac{1}{n}\frac{\hat{\theta}_n - \theta_0}{2}\frac{d^3 \ell(\theta)}{d\theta^3} \longrightarrow 0 \quad \text{in} \quad P_{\theta_0}\text{-probability as } n \to \infty.
$$

Using this and (7.11) in (7.9), one gets

$$
\sqrt{n}(\hat{\theta}_n - \theta_0) \approx \frac{\frac{1}{\sqrt{n}} \left(\frac{d\ell(\theta)}{d\theta} \right)_{\theta=\theta_0}}{I(\theta_0)} \equiv \frac{\frac{1}{\sqrt{n}} \sum_{j=1}^{n} \left(\frac{d \log f(X_j;\theta)}{d\theta} \right)_{\theta=\theta_0}}{I(\theta_0)},
\tag{7.12}
$$

where $\approx$ indicates that the difference between its two sides goes to zero. But $\left(\frac{d \log f(X_j;\theta)}{d\theta} \right)_{\theta=\theta_0}$ $(j = 1, 2, \ldots)$ is an i.i.d. sequence of random variables having mean zero (by (A_2^*)) and finite variance $I(\theta_0)$ (by (A_3^*)). Therefore, by the classical CLT, writing Z for a Normal random variable $N(0, I(\theta_0))$,

$$
\sqrt{n}(\hat{\theta}_n - \theta_0) \xrightarrow{\mathscr{L}} \frac{Z}{-I(\theta_0)} \stackrel{\mathscr{L}}{=} N\left(0, \frac{I(\theta_0)}{I^2(\theta_0)} \right) = N\left(0, \frac{1}{I(\theta_0)} \right).
$$

Q.E.D.   □

Remark 7.6. In particular, the above theorem says that (under P_{θ_0}) the "asymptotic variance" of $(\hat{\theta}_n - \theta_0)$ is $\frac{1}{nI(\theta_0)}$, which is the Rao–Cramér lower bound for the expected squared error of any estimator of θ (at $\theta = \theta_0$).

Example 7.1. The hypothesis of Theorem 7.2 is satisfied by one-parameter exponential families with p.d.f.

$$
f(x;\theta) = C(\theta)h(x)e^{\theta t(x)}, \qquad x \in \mathscr{X}
\tag{7.13}
$$

where $\mathscr{X}$ does not depend on θ, $h(x) > 0$ for $x \in \mathscr{X}$. This form of the exponential family is said to be in natural parametric form with θ being the natural parameter (belonging to an open interval Θ). The likelihood equation is

$$\frac{d\log}{d\theta}\left[C^n(\theta)\left(\prod_{j=1}^{n}h(X_j)\right)exp\left\{\theta\sum_{j=1}^{n}t(X_j)\right\}\right] = 0, \tag{7.14}$$

i.e.,

$$-\frac{d\log C(\theta)}{d\theta} = \frac{1}{n}\sum_{j=1}^{n}t(X_j). \tag{7.15}$$

But

$$-\frac{d\log C(\theta)}{d\theta} = -\frac{C'(\theta)}{C(\theta)} = -C'(\theta)\int h(x)\exp\{\theta t(x)\}\mu(dx)$$

$$= -\frac{d}{d\theta}\left(\int h(x)\exp\{\theta t(x)\}\mu(dx)\right)^{-1}\int h(x)\exp\{\theta t(x)\}\mu(dx)$$

$$= \frac{\int t(x)h(x)\exp\{\theta t(x)\}\mu(dx)}{\int h(x)\exp\{\theta t(x)\}\mu(dx)} = \int t(x)f(x;\theta)\mu(dx)$$

$$= E_\theta t(X_1). \tag{7.16}$$

Hence, one may rewrite (7.15) as the equation

$$E_\theta t(X_1) = \frac{1}{n}\sum_{j=1}^{n}t(X_j). \tag{7.17}$$

Now, the second derivative of the log likelihood function is (see the left side of the first equation in (7.15) and the relations (7.16))

$$n\frac{d^2\log C(\theta)}{d\theta^2} =$$

$$= \frac{n[(\int t(x)\exp\{\theta t(x)\}\mu(dx))^2 - (\int h(x)\exp\{\theta t(u)\}\mu(dx))(\int t^2(x)h(x)e^{\theta t(x)}\mu(dx))]}{(\int h(x)\exp\{\theta t(x)\}\mu(dx))^2}$$

$$= -n\,\mathrm{var}_\theta\, t(X_1) < 0. \tag{7.18}$$

Thus the log likelihood function is strictly concave, and, therefore, (7.15) cannot have more than one root $\hat{\theta}$. We will show later (see Theorem 7.5) that in this case the likelihood equation has a unique solution $\hat{\theta}$ on a set with P_{θ_0}-probability tending to 1, as $n \to \infty$, under θ_0, and that $\hat{\theta}$ is consistent. Thus Theorem 7.2 applies.

Example 7.2 (Logistic). In this case

$$f(x;\theta) = \frac{e^{-(x-\theta)}}{(1+e^{-(x-\theta)})^2}, \qquad -\infty < x < \infty \quad (\Theta = (-\infty, \infty)),$$

so that the likelihood equation ($d\log f_n(\mathbf{X};\theta)/d\theta = 0$) is

$$\frac{d}{d\theta}\left\{-\sum_{j=1}^{n}(X_j - \theta) - 2\sum_{j=1}^{n}\log\left(1 + e^{-(X_j-\theta)}\right)\right\} = 0, \tag{7.19}$$

or

$$n - 2 \sum_{j=1}^{n} \frac{e^{-(X_j - \theta)}}{1 + e^{-(X_j - \theta)}} = 0$$

or,

$$n - 2 \sum_{j=1}^{n} \left\{ 1 - \frac{1}{1 + e^{-(X_j - \theta)}} \right\} = 0,$$

or,

$$n - 2n + 2 \sum_{j=1}^{n} \frac{1}{1 + e^{-(X_j - \theta)}} = 0 \tag{7.20}$$

or,

$$\sum_{j=1}^{n} \frac{1}{1 + e^{-(X_j - \theta)}} = \frac{n}{2}. \tag{7.21}$$

Since the left hand side is strictly decreasing in θ, and goes to n as $\theta \downarrow -\infty$ and to 0 as $\theta \uparrow \infty$, there is a unique solution $\hat{\theta}$ of (7.20). Also the left hand side of (7.19) (which gives the first derivative of $\log f_n(\mathbf{X}; \theta)$) is positive for $\theta < \hat{\theta}$ and negative for $\theta > \hat{\theta}$. Hence $\hat{\theta}$ is the MLE. The hypothesis of Theorem 7.2 can be easily verified for this case.

Example 7.3 (Double Exponential). Here

$$f(x; \theta) = \frac{1}{2} e^{-|x - \theta|}, \qquad -\infty < x < \infty, \quad \theta \in (-\infty, \infty).$$

$$\log f_n(\mathbf{X}; \theta) = -n \log 2 - \sum_{j=1}^{n} |X_j - \theta|, \tag{7.22}$$

which is maximized by that value of θ for which $\varphi(\theta) := \sum_{j=1}^{n} |X_j - \theta|$ is minimized. The minimizing value is the median of $X_1, X_2, \ldots, X_n$; i.e., $\hat{\theta}$ is the $\frac{n+1}{2}$th observation when $X_1, X_2, \ldots, X_n$ are arranged in increasing order, for the case when n is odd. If n is even, then $\hat{\theta}$ is any number between the $\frac{n}{2}$th and $(\frac{n}{2}+1)$th observation (arranged in increasing order). To see this, let $X_{(1)} < X_{(2)} < \cdots < X_{(n)}$ denote the ordering of the n observations. Let $\theta \in [X_{(r)}, X_{(r+1)})$, and let $\delta > 0$ be such that $\theta + \delta \in [X_{(r)}, X_{(r+1)}]$. Then $\varphi(\theta + \delta) = \varphi(\theta) + r\delta - (n-r)\delta = \varphi(\theta) + (2r - n)\delta$. Thus $\varphi(\theta + \delta) < \varphi(\theta)$ if and only if $r < \frac{n}{2}$, and $\varphi(\theta + \delta) = \varphi(\theta)$ if $r = \frac{n}{2}$. In other words, if n is odd, then $\varphi(\theta)$ is strictly decreasing on $(-\infty, X_{([\frac{n}{2}]+1)}]$, and strictly increasing on $(X_{([\frac{n}{2}]+1)}, \infty)$, attaining its unique minimum at $\theta = X_{([\frac{n}{2}]+1)}$. On the other hand, if n is even, $\varphi(\theta)$ is strictly decreasing on $(-\infty, X_{(\frac{n}{2})}]$, strictly increasing on $(X_{(\frac{n}{2}+1)}, \infty)$, and constant on $[X_{(\frac{n}{2})}, X_{(\frac{n}{2}+1)}]$.

Although the hypothesis of Theorem 7.2 does not hold, the conclusion holds. Note that

$$\frac{d \log f(x; \theta)}{d\theta} = \begin{cases} -1 & \text{for } x < \theta \\ 1 & \text{for } x > \theta. \end{cases} \tag{7.23}$$

$$I(\theta) = E_\theta \left(\frac{d \log f(X_1; \theta)}{d\theta} \right)^2 = E_\theta 1 = 1. \tag{7.24}$$

We know from Theorem 6.3 that, under P_θ,

$$\sqrt{n}(\hat\theta - \theta) \xrightarrow{\mathscr{L}} N\left(0, \frac{1}{4f^2(\theta;\theta)}\right) = N(0,1).$$

Hence $\hat\theta$ is asymptotically efficient. This also explains the poor performance of the mean relative to the median: $e_{\hat\theta,\overline{X}} = 2$. (Note that $\sqrt{n}(\overline{X} - \theta) \to \mathscr{N}(0,\sigma^2)$ where $\sigma^2 = E_\theta(X_1 - \theta)^2 = \frac{1}{2}\int_{-\infty}^{\infty} x^2 e^{-|x|}dx = \int_0^\infty x^2 e^{-x}dx = \Gamma_3 = 2$).

Example 7.4.

$$f(x;\theta) = \left(\frac{1}{\sqrt{2\pi\theta^2}}\right)^n e^{-\frac{1}{2\theta^2}(x-\theta)^2}, \qquad -\infty < x < \infty,$$

$$\Theta = \{\theta \in \mathbb{R}^1, \theta \neq 0\} = \mathbb{R}^1 \setminus \{0\}.$$

The likelihood equation $d\log f_n(\mathbf{X};\theta)/d\theta = 0$ may be expressed as

$$-\frac{n}{\theta} + \frac{\sum_1^n X_i^2}{\theta^3} - \frac{\sum X_i}{\theta^2} = 0,$$

or,

$$\theta^2 + \overline{X}\theta - m_2 = 0, \qquad \left(\overline{X} := \frac{\sum X_i}{n}, \ m_2 := \frac{\sum X_i^2}{n}\right), \tag{7.25}$$

whose roots are

$$\theta_+ = \frac{-\overline{X} + \sqrt{\overline{X}^2 + 4m_2}}{2}, \qquad \theta_- = \frac{-\overline{X} - \sqrt{\overline{X}^2 + 4m_2}}{2}. \tag{7.26}$$

Note that (1) as $\theta \to 0$, $f_n(\mathbf{X};\theta) \to 0$ (except for $\mathbf{X} = (\theta,\theta,\ldots,\theta)$ which has zero probability), and (2) as $\theta \to \pm\infty$, $\max_{\mathbf{x}} f_n(\mathbf{x};\theta) = (2\pi\theta^2)^{-n/2} \to 0$. There are, therefore, at least two extrema: at least one between $-\infty$ and 0, and at least one between 0 and $+\infty$. However, the only critical points are θ_+, θ_-. Hence there are *exactly* two extrema. Note that $\theta_+ > 0$ and $\theta_- < 0$. Now at $\hat\theta = \theta_+, \theta_-$ one has

$$\begin{aligned}
\log f_n(\mathbf{X};\theta) &= -\frac{n}{2}\log 2\pi - \frac{n}{2}\log\hat\theta^2 - \frac{n}{2\hat\theta^2}\left(\hat\theta^2 - 2\overline{X}\hat\theta + m_2\right)\\
&= -\frac{n}{2}\log 2\pi - \frac{n}{2} - \frac{n}{2}\log\hat\theta^2 - \frac{n}{2\hat\theta^2}\left(\hat\theta^2 - \overline{X}\hat\theta\right)\\
&= -\frac{n}{2}\log 2\pi - \frac{n}{2}\log\hat\theta^2 - n + \frac{n}{2\hat\theta}\overline{X}.
\end{aligned} \tag{7.27}$$

If $\overline{X} > 0$, then (7.27) is larger at $\hat\theta = \theta_+$, and if $\overline{X} < 0$, then (7.27) is larger at $\hat\theta = \theta_-$. Therefore, the *MLE* is

$$\hat\theta_n = \begin{cases} \theta_+ & \text{if } \overline{X} > 0 \\ \theta_- & \text{if } \overline{X} \leq 0. \end{cases} \tag{7.28}$$

Theorem 7.2 applies to *consistent* solutions (i.e., an estimator (sequence) $\hat\theta_n$ which is consistent and solves the likelihood Eq. (7.2) on a set whose probability goes to 1 as $n \to \infty$). How does one obtain such a solution? There is no general method that works in all cases, from the computational point of view. However, under the hypothesis of Theorem 7.2 one may show that there exists essentially one consistent sequence of solutions.[1] A general numerical method which often works is the following: find a consistent estimator (sequence) $\tilde{\theta}_n$ (e.g., by the method of moments if that is applicable, or by using the median in case of a symmetric location problem etc.); taking $\tilde{\theta}_n$ as a trial solution, use iteration (or *Newton–Raphson method*):

$$\tilde{\theta}_n^{(i+1)}(\mathbf{x}) = \tilde{\theta}_n^{(i)}(\mathbf{x}) - \left(\frac{d \log f_n(\mathbf{x}; \theta)/d\theta}{d^2 \log f_n(\mathbf{x}; \theta)/d\theta^2} \right)_{\tilde{\theta}_n^{(i)}}$$

$$\tilde{\theta}_n^{(0)} = \tilde{\theta}_n, \qquad (i = 0, 1, 2, \dots). \tag{7.29}$$

As the following result shows, one may use (7.29) with $i = 0$ and use the estimator $\tilde{\theta}_n^{(1)}$ in place of a consistent root, in situations involving multiple roots and computational difficulties.

Definition 7.1. A sequence of random variables (or, vectors) Y_n is said to be *bounded in probability* if for every $\varepsilon > 0$ there exists $A = A(\varepsilon)$ such that

$$P(|Y_n| > A) < \varepsilon.$$

Note that Y_n, $n \geq 1$, is bounded in probability if $E|Y_n|$ is a bounded sequence.

Theorem 7.3. *Suppose that the hypothesis of Theorem 7.2 concerning $f(x; \theta)$ holds. Assume that $\tilde{\theta}_n$ is an estimator (sequence) such that for each $\theta_0 \in \Theta$, $\sqrt{n}(\tilde{\theta}_n - \theta_0)$ is bounded in P_{θ_0}-probability. Then the estimator (sequence)*

$$\delta_n = \tilde{\theta}_n - \left(\frac{d \log f_n(\mathbf{X}; \theta)/d\theta}{d^2 \log f_n(\mathbf{X}; \theta)/d\theta^2} \right)_{\tilde{\theta}_n} \tag{7.30}$$

is asymptotically efficient.

Proof. One has, by a Taylor expansion of $(d \log f_n(\mathbf{X}; \theta)/d\theta)_{\tilde{\theta}}$

$$\sqrt{n}(\delta_n - \theta_0) = \sqrt{n}(\tilde{\theta}_n - \theta_0) - \frac{\frac{1}{\sqrt{n}}(d \log f_n(\mathbf{X}; \theta/d\theta)_{\theta_0}}{\frac{1}{n}(d^2 \log f_n(\mathbf{X}; \theta)/d\theta^2)_{\tilde{\theta}_n}}$$

$$- \frac{\sqrt{n}(\tilde{\theta}_n - \theta_0)\frac{1}{n}(d^2 \log f_n(\mathbf{X}; \theta)/d\theta^2)_{\theta_n^*}}{\frac{1}{n}(d^2 \log f_n(\mathbf{X}; \theta)/d\theta^2)_{\tilde{\theta}_n}} \tag{7.31}$$

where θ_n^* lies between θ_0 and $\tilde{\theta}_n$. Now, under P_{θ_0},

$$\frac{\frac{1}{n}(d^2 \log f_n(\mathbf{X}; \theta)/d\theta^2)_{\theta_n^*}}{\frac{1}{n}(d^2 \log f_n(\mathbf{X}; \theta)/d\theta^2)_{\tilde{\theta}_n}} - 1 \xrightarrow{P_{\theta_0}} 0. \tag{7.32}$$

One may prove (7.32)) by expanding the numerator and the denominator around θ_0 and using A_4^*.

Using (7.32) in (7.31) one gets

$$\sqrt{n}(\delta_n - \theta_0) - \left\{ -\frac{\frac{1}{\sqrt{n}}(d \log f_n(\mathbf{X}; \theta)/d\theta)_{\theta_0}}{\frac{1}{n}(d^2 \log f_n(\mathbf{X}; \theta)/d\theta^2)_{\tilde{\theta}_n}} \right\} \xrightarrow{P_{\theta_0}} 0. \tag{7.33}$$

But the expression within curly brackets converges in distribution to $N\left(0, \frac{1}{I(\theta_0)}\right)$ under P_{θ_0}. $\qquad \square$

---

[1] See Lehmann and Casella (1998, p. 448).

Remark 7.7. Suppose now that T_n is bounded in probability and U_n converges in probability to zero (written $U_n \xrightarrow{P} 0$ or, $U_n = o_p(1)$). Then one can easily show that $U_n T_n$ converges in probability to zero as $n \to \infty$. [Proof: $P(|U_n T_n| > \delta) \leq P(|T_n| > A_\varepsilon) + P(|U_N| > \delta/A_\varepsilon) < \varepsilon + \varepsilon_n$ for all sufficiently large n, with $\varepsilon_n \to 0$]. This fact has been used in getting (7.33) from (7.31) and (7.32).

Remark 7.8. A variant of (7.30) is

$$\delta_n' = \tilde{\theta}_n + \frac{(d \log f_n(\mathbf{X}; \theta)/d\theta)_{\tilde{\theta}_n}}{nI(\tilde{\theta}_n)}. \qquad (7.34)$$

Since $-(d^2 \log F_n(\mathbf{X} : \theta)/d\theta^2)_{\tilde{\theta}_n}/nI(\tilde{\theta}_n)$ converges to 1 in probability, by Slutsky's Lemma, δ_n and δ_n' converge asymptotically to the same distribution.

7.4 The Multi-Parameter Case

We will now prove a multi-parameter analogue of the Cramér–Rao inequality.

Theorem 7.4 (Multi-Parameter Cramér-Rao Information Inequality).
Let $\mathbf{X}$ be a random quantity whose distribution has density $g(\mathbf{x}; \theta)$ with respect to a sigma finite measure ν (on a sigma field $\mathscr{A}$ on the range space $\mathscr{X}$ of $\mathbf{X}$); here $\theta \in \Theta$—a nonempty open subset of $\mathbb{R}^p$. Assume the following: (Notation: $\boldsymbol{\theta} = (\theta_1, \theta_2, \ldots, \theta_p)$)

(B_1)　　$C = \{\mathbf{x}; g(\mathbf{x}; \boldsymbol{\theta}) > 0\}$ *is independent of θ.*
(B_2)　　$\theta \to g(\mathbf{x}; \boldsymbol{\theta})$ *is once differentiable on Θ, $\forall \mathbf{x}$ (outside a ν-null set).*
(B_3)　　$\int_C (\partial/\partial\theta_r) g(\mathbf{x}; \boldsymbol{\theta})\nu(dx) = (\partial/\partial\theta_r) \int_C g(\mathbf{x}; \boldsymbol{\theta})\nu(dx)(= 0) \; \forall \; r = 1, 2, \ldots, p$
　　　　and $\forall \; \boldsymbol{\theta} \in \Theta$.
(B_4)　　*The matrix $\mathscr{I}(\boldsymbol{\theta}) = ((E_\theta[(\partial \log g(\mathbf{X}; \boldsymbol{\theta})/\partial\theta_r) \cdot (\partial \log g(\mathbf{X}; \boldsymbol{\theta})/\partial\theta_{r'})]))$ is non-*
　　　　singular.
　　　　Let now $T = (T_1, T_2, \ldots, T_p)$ be an unbiased estimator of $\boldsymbol{\theta}$ (i.e., $E_{\boldsymbol{\theta}} T_r =$
　　　　$\theta_r \; \forall \; \boldsymbol{\theta} \in \Theta$) *such that $E_{\boldsymbol{\theta}}(T_r)^2 < \infty \; \forall \; r$ and $\forall \; \boldsymbol{\theta}$. Assume further that*
(B_5)　　$\int_C T_r(\mathbf{x})(\partial/\partial\theta_{r'}) g(\mathbf{x}; \boldsymbol{\theta})\nu(dx) = (\partial/\partial\theta_{r'}) \int_C T_r(x) g(\mathbf{x}; \boldsymbol{\theta})\nu(dx) \; (= \; \delta_{rr'},$
　　　　Kronecker's delta) $\forall \; r, r'$ and $\forall \; \boldsymbol{\theta}$.

　　　　Then one has the inequality

$$\Sigma(\boldsymbol{\theta}) \geq \mathscr{I}^{-1}(\boldsymbol{\theta}), \qquad (7.35)$$

where $\Sigma(\theta) = ((\mathrm{cov}_{\boldsymbol{\theta}}(T_r, T_{r'})))$, and the inequality (7.35) means the $\Sigma(\boldsymbol{\theta}) - \mathscr{I}^{-1}(\boldsymbol{\theta})$ is a nonnegative definite matrix.

Proof. It follows from (B_3) that

$$E_\theta(\partial \log g(\mathbf{X}; \theta)/\partial\theta_{r'}) = \int_C \frac{(\partial g(\mathbf{x}; \boldsymbol{\theta})/\partial\theta_{r'})}{g(\mathbf{x}'; \boldsymbol{\theta})} g(\mathbf{x}; \boldsymbol{\theta})\nu(dx) = 0. \qquad (7.36)$$

Hence, using (B_5),

$$\mathrm{cov}_{\boldsymbol{\theta}}(T_r, (\partial/\partial\theta_{r'}) \log g(\mathbf{X}; \theta)) = E_\theta[T_r(\partial \log g(\mathbf{X}; \boldsymbol{\theta})/\partial\theta_{r'})]$$

$$= \int_C T_r(\mathbf{x})(\partial g(\mathbf{x}; \boldsymbol{\theta})/\partial\theta_{r'})\nu(dx) = \delta_{rr'}. \qquad (7.37)$$

Let $\mathbf{a} = (a_1, a_2, \ldots, a_p)$, $\mathbf{b} = (b_1, b_2, \ldots, b_p)$ be arbitrary vectors in $\mathbb{R}^p$. Then, by (7.37),

$$\mathrm{cov}_{\boldsymbol{\theta}}\left(\sum_1^p a_r T_r, \sum_1^p b_r (\partial \log g(\mathbf{X}; \boldsymbol{\theta})/\partial \theta_r)\right) = \sum_1^p a_r b_r. \tag{7.38}$$

Therefore, by Schwartz' inequality, writing $\sigma_{rr'}(\boldsymbol{\theta}) = \mathrm{cov}_{\boldsymbol{\theta}}(T_r, T_{r'})$, $\mathscr{I}(\boldsymbol{\theta}) = ((\mathscr{I}_{rr'}(\boldsymbol{\theta}))$,

$$\left(\sum_1^p a_r b_r\right)^2 \leq \mathrm{var}_{\boldsymbol{\theta}}\left(\sum_1^p a_r T_r\right) \cdot \mathrm{var}_{\boldsymbol{\theta}}\left(\sum_1^p b_r (\partial \log g(\mathbf{X}; \boldsymbol{\theta})/\partial \theta_r)\right)$$

$$= \left(\sum_{r=1}^p \sum_{r'=1}^p a_r a_{r'} \ \sigma_{rr'}(\boldsymbol{\theta})\right)\left(\sum_{r=1}^p \sum_{r'=1}^p b_r b_{r'} \mathscr{I}_{rr'}(\boldsymbol{\theta})\right). \tag{7.39}$$

Writing $\langle \mathbf{a}, \mathbf{b} \rangle = \sum_{r=1}^p a_r b_r$ (the Euclidean inner product), one may rewrite (7.39) as

$$\langle \mathbf{a}, \mathbf{b} \rangle^2 \leq \langle \mathbf{a}, \Sigma(\boldsymbol{\theta})\mathbf{a} \rangle \langle \mathbf{b}, \mathscr{I}(\boldsymbol{\theta})\mathbf{b} \rangle. \tag{7.40}$$

Now choose $\mathbf{b} = \mathscr{I}^{-1}(\boldsymbol{\theta})\mathbf{a}$ to get

$$\langle \mathbf{a}, \mathscr{I}^{-1}(\boldsymbol{\theta})\mathbf{a} \rangle^2 \leq \langle \mathbf{a}, \Sigma(\boldsymbol{\theta})\mathbf{a} \rangle \langle \mathscr{I}^{-1}(\boldsymbol{\theta})\mathbf{a}, \mathbf{a} \rangle,$$

or,

$$\langle \mathbf{a}, \Sigma(\boldsymbol{\theta})\mathbf{a} \rangle \geq \langle \mathbf{a}\mathscr{I}^{-1}(\boldsymbol{\theta})\mathbf{a} \rangle$$

or,

$$\langle \mathbf{a}, (\Sigma(\boldsymbol{\theta}) - \mathscr{I}^{-1}(\boldsymbol{\theta}))\mathbf{a} \rangle \geq 0 \ \forall \ \mathbf{a} \in \mathbb{R}^p. \tag{7.41}$$

$$\square$$

Remark 7.9. As in the one-parameter case, (see proof of Cramér–Rao Theorem 7.1) the following conditions ensure the validity of the inequality (7.35) at $\boldsymbol{\theta} = \boldsymbol{\theta}_0$: $(B_1), (B_2)$ hold, as well as the following conditions

(R_1'): $\left|\frac{\partial g(\mathbf{x};\boldsymbol{\theta})}{\partial \theta_r}\right| \leq g_1(\mathbf{x})$ $(\forall \ r = 1, 2, \ldots, p)$ in a neighborhood of $\boldsymbol{\theta}_0$, where $g_1(\mathbf{x})$ does not depend on values of $\boldsymbol{\theta}$ in this neighborhood, and $\int g_1(\mathbf{x})\nu(d\mathbf{x}) < \infty$.

(R_2'): In a neighborhood of $\boldsymbol{\theta} = \boldsymbol{\theta}_0$ one has $\left|\frac{\partial g(\mathbf{x};\boldsymbol{\theta})}{\partial \theta_r}\right| \leq g_2(\mathbf{x})$ $(\forall \ r = 1, 2, \ldots, p)$, where $g_2(\mathbf{x})$ does not depend on $\boldsymbol{\theta}$ in this neighborhood, and $\int \frac{g_2^2(\mathbf{x})}{g(\mathbf{x};\boldsymbol{\theta}_0)}\nu(d\mathbf{x}) < \infty$.

Remark 7.10. In case $\mathbf{X} = (X_1, X_2, \ldots, X_n)$ where $X_1, X_2, \ldots, X_n$ are i.i.d. with common p.d.f. (w.r.t. μ) $f(x; \boldsymbol{\theta})$, one takes $g(\mathbf{x}; \boldsymbol{\theta}) = f_n(\mathbf{x}; \boldsymbol{\theta}) = \prod_{j=1}^n f(x_j; \boldsymbol{\theta})$. The information inequality then becomes

$$\Sigma(\boldsymbol{\theta}) \geq \tfrac{1}{n}\mathscr{I}_1^{-1}(\boldsymbol{\theta}), \tag{7.42}$$

where $\mathscr{I}_1(\boldsymbol{\theta}) = ((E_{\boldsymbol{\theta}}\frac{\partial \log f(X_1;\boldsymbol{\theta})}{\partial \theta_r} \cdot \frac{\partial \log f(X_1;\boldsymbol{\theta})}{\partial \theta_{r'}}))$.

Remark 7.11. Letting $\mathbf{a}$ in (7.41) be the vector with 1 as the jth component and zeros elsewhere, one gets

$$\mathrm{var}_{\boldsymbol{\theta}}(T_j) \geq \left(\mathscr{I}^{-1}(\boldsymbol{\theta})\right)_{jj}. \tag{7.43}$$

In the context of Remark 7.10 this becomes

$$\operatorname{var}_\theta(T_j) \geq \frac{1}{n}\left(\mathscr{I}_1^{-1}(\boldsymbol{\theta})\right)_{jj}. \tag{7.44}$$

Note also that (7.41) (and, therefore, (7.35)) says that *the variance of an unbiased estimator of a linear parametric function* $g(\theta) = \sum_1^p a_r \theta_r$ *is at least* $\langle \mathbf{a}, \mathscr{I}^{-1}(\boldsymbol{\theta})\mathbf{a}\rangle$ when $\boldsymbol{\theta}$ is the true parameter value.

Definition 7.2. Assume that $(B_1), (B_2)$, hold for $f(x;\theta)$ (in place of $g(\mathbf{x};\boldsymbol{\theta})$), and that $(R'_1), (R'_2)$ hold for $f_n(\mathbf{x};\boldsymbol{\theta}) = \prod_{j=1}^n f(x_j;\boldsymbol{\theta})$ (in place of $g(\mathbf{x};\boldsymbol{\theta})$). Then a sequence of estimators T_n of $\boldsymbol{\theta}$ is said to be *asymptotically efficient* (in Fisher's sense) if, $\forall\, \boldsymbol{\theta} \in \Theta$,

$$\sqrt{n}(T_n - \boldsymbol{\theta}) \xrightarrow{\mathscr{L}} N(\mathbf{o}, \mathscr{I}_1^{-1}(\boldsymbol{\theta})), \qquad \text{under } P_\theta. \tag{7.45}$$

Remark 7.12. Apart from the motivation for the above definition provided in Remark 7.11, one may show that *an asymptotically efficient estimator* T_n *concentrates more probability around the true parameter value, asymptotically, than a competing asymptotically normal estimator* δ_n *which is not asymptotically efficient.* More precisely,

$$\lim_{n\to\infty} P_{\boldsymbol{\theta}}(\sqrt{n}(T_n - \boldsymbol{\theta}) \in C) = \Phi_{\mathscr{I}^{-1}(\boldsymbol{\theta})}(C) \geq \Phi_{V(\boldsymbol{\theta})}(C)$$

$$\equiv \lim_{n\to\infty} P_{\boldsymbol{\theta}}(\sqrt{n}(\delta_n - \boldsymbol{\theta}) \in C) \tag{7.46}$$

for all *symmetrically convex sets* C (i.e., convex C satisfying: $C = -C = \{-\mathbf{x} : \mathbf{x} \in C\}$). Here $V(\boldsymbol{\theta})$ is the dispersion matrix of the limiting normal distribution of $\sqrt{n}(\delta_n - \boldsymbol{\theta})$ under P_θ. The two equalities in (7.46) are consequences of convergence in distribution, while the inequality follows from a theorem of T.W. Anderson (*Proc. Amer. Math. Soc.* **6**, 170–176): If $V_1(\boldsymbol{\theta}) \geq V_2(\boldsymbol{\theta})$, $V_1(\boldsymbol{\theta})$, V_2, $(\boldsymbol{\theta})$ *nonnegative definite, then* $\Phi_{V_1(\boldsymbol{\theta})}(C) \leq \Phi_{V_2(\boldsymbol{\theta})}(C)$ $\forall$ *symmetric convex sets* C.

To provide an alternative justification in terms of risk functions, define a loss function $L(\boldsymbol{\theta} - \boldsymbol{\theta}')$ (loss when $\boldsymbol{\theta}$ is the true parameter value and $\boldsymbol{\theta}'$ is the estimated value) to be *negative unimodal* (or *bowl shaped*) if $L \geq 0$ and $\{\mathbf{z} : \mathbf{z} \in \mathbb{R}^p, L(\mathbf{z}) \leq r\}$ is convex for all $r \geq 0$; a loss function $L(\boldsymbol{\theta} - \boldsymbol{\theta}')$ is *symmetric* if $L(\mathbf{z}) = L(-\mathbf{z})$ $\forall\, \mathbf{z} \in \mathbb{R}^p$. Let $\mathscr{L}_0$ denote the class of all negative unimodal symmetric and bounded loss functions. For each $L \in \mathscr{L}_0$ define the loss function (sequence) $L_n(\boldsymbol{\theta}, \boldsymbol{\theta}') = L(\sqrt{n}(\boldsymbol{\theta} - \boldsymbol{\theta}'))$. Then one can show, using T.W. Anderson's inequality and weak convergence, that

$$\lim_{n\to\infty} E_\theta L_n(\boldsymbol{\theta}, T_n) = \int_{\mathbb{R}^p} L(\mathbf{z})d\Phi_{\mathscr{I}^{-1}(\boldsymbol{\theta})}(\mathbf{z}) \leq \int_{\mathbf{R}^p} L(\mathbf{z})d\Phi_{V(\boldsymbol{\theta})}(\mathbf{z})$$

$$= \lim_{n\to\infty} E_\theta L_n(\boldsymbol{\theta}, \delta_n) \tag{7.47}$$

$\forall\, L \in \mathscr{L}_0$ (provided $V(\boldsymbol{\theta}) \geq I^{-1}(\boldsymbol{\theta})$).

Finally note that the loss function L_n is a "normalization" of L, which is in the class $\mathscr{L}_0$, and such that the limits in (7.47) become *discriminating*. On the other hand,

$$\lim_{n\to\infty} E_\theta L(\boldsymbol{\theta} - T_n) = L(0) = \lim_{n\to\infty} E_\theta L(\boldsymbol{\theta} - \delta_n). \tag{7.48}$$

Theorem 7.5 (Asymptotic Efficiency of the MLE). *Suppose $X_1, X_2, \ldots$ is an i.i.d. sequence with common p.d.f. $f(x; \boldsymbol{\theta})$ (w.r.t. μ) and that the following assumptions hold: Θ is an open subset of $\mathbb{R}^p$ and*

(A_1): $\quad C = \{x : f(x; \boldsymbol{\theta}) > 0\}$ *is independent of $\boldsymbol{\theta}$.*

(A_2): $\quad \boldsymbol{\theta} \to f(x; \boldsymbol{\theta})$ *is thrice differentiable on Θ, $\forall\, x \in C$ (i.e., $\partial^3 f(x; \boldsymbol{\theta}) / \partial\theta_r \partial\theta_{r'} \partial\theta_{r''}$ exist $\forall\, r, r', r''$).*

(A_3): $\quad \int_C (\partial/\partial\theta_r) f(x; \boldsymbol{\theta}) \mu(dx) = (\partial/\partial\theta_r) \int_C f(x; \boldsymbol{\theta}) \mu(dx)\ (= 0)$, $\int_C (\partial^2/\partial\theta_r \partial\theta_{r'})$ $f(x; \boldsymbol{\theta}) \mu(dx) = (\partial^2/\partial\theta_r \partial\theta_{r'}) \int_C f(x; \boldsymbol{\theta}) \mu(dx)\ (= 0)\ \forall\, r, r'$ *and $\forall\, \boldsymbol{\theta} \in \Theta$.*

(A_4): $\quad$ *The matrix $\mathscr{I}(\boldsymbol{\theta}) = \left(\left(E_{\boldsymbol{\theta}} \left(\frac{\partial \log f(X_1; \boldsymbol{\theta})}{\partial\theta_r} \cdot \frac{\partial \log f(X_1; \boldsymbol{\theta})}{\partial\theta_{r'}} \right) \right) \right)$ is finite and nonsingular, $\forall\, \boldsymbol{\theta} \in \Theta$.*

(A_5): $\quad$ *for each $\boldsymbol{\theta}_0 \in \Theta$ there exists an $\varepsilon > 0$ and a $P_{\boldsymbol{\theta}_0}$-integrable function $g(x)$ such that, $\forall\, r, r', r''$,*

$$\left| \frac{\partial^3 \log f(x; \boldsymbol{\theta})}{\partial\theta_r \partial\theta_{r'} \partial\theta_{r''}} \right| \le g(x) \qquad if\ |\boldsymbol{\theta} - \boldsymbol{\theta}_0| \le \varepsilon.$$

$(|\boldsymbol{\theta} - \boldsymbol{\theta}_0|^2 = \sum_{r=1}^{p} (\theta_r - \theta_{0r})^2)$.

Then for each $\boldsymbol{\theta}_0 \in \Theta$, there exists a sequence $\hat{\boldsymbol{\theta}}_n$ with the following properties:

*(1) $\hat{\boldsymbol{\theta}}_n$ is a solution of the **likelihood equations***

$$P_{\boldsymbol{\theta}_0}((\partial f_n(\mathbf{x}; \boldsymbol{\theta})/\partial\theta_r)_{\hat{\boldsymbol{\theta}}_n} = 0 \qquad for\ \ 1 \le r \le p) \longrightarrow 1 \qquad as\ n \to \infty,$$

(2) $\hat{\boldsymbol{\theta}}_n \xrightarrow{P_{\boldsymbol{\theta}_0}} \boldsymbol{\theta}_0$,

(3) $\sqrt{n}(\hat{\boldsymbol{\theta}}_n - \boldsymbol{\theta}_0) \xrightarrow{\mathscr{L}} N(0, \mathscr{I}_1^{-1}(\boldsymbol{\theta}))$.

Proof. To prove (1) and (2), for each $\mathbf{x} \in \mathscr{X}^n$ consider the set

$$S_{\mathbf{x};n} = \left\{ \boldsymbol{\theta} \in \Theta : \frac{\partial f_n(\mathbf{x}; \boldsymbol{\theta})}{\partial\theta_r} = 0 \qquad for\ 1 \le r \le p \right\}. \tag{7.49}$$

$S_{\mathbf{x},n}$ is a closed set. In case $S_{\mathbf{x},n}$ is nonempty choose the (or a) solution of the likelihood equations

$$\frac{\partial f_n(\mathbf{x}; \boldsymbol{\theta})}{\partial\theta_r} = 0 \qquad for\ 1 \le r \le p, \tag{7.50}$$

which is nearest $\boldsymbol{\theta}_0$, and denote it by $\hat{\boldsymbol{\theta}}_n(\mathbf{x})$. If, for some $\mathbf{x}$, $S_{\mathbf{x},n}$ is empty choose $\hat{\boldsymbol{\theta}}_n(\mathbf{x})$ from Θ arbitrarily. (A *measurable* choice of $\hat{\boldsymbol{\theta}}_n$ is possible). Now consider a $\delta > 0$ such that $\{\boldsymbol{\theta} : |\boldsymbol{\theta} - \boldsymbol{\theta}_0| \le \delta\} \subset \Theta$. Under A_2 one has, for $|\boldsymbol{\theta} - \boldsymbol{\theta}_0| \le \delta$,

$$\frac{1}{n} \sum_{j=1}^{n} \log f(X_j; \boldsymbol{\theta})$$

$$= \frac{1}{n} \sum_{j=1}^{n} \log f(X_j; \boldsymbol{\theta}_0) + \sum_{r=1}^{p} (\theta_r - \theta_{0r}) \frac{1}{n} \sum_{j=1}^{n} \left(\frac{\partial \log f(X_j; \boldsymbol{\theta})}{\partial\theta_r} \right)$$

$$+ \frac{1}{2!} \sum_{r,r'=1}^{p} (\theta_r - \theta_{0r})(\theta_{r'} - \theta_{0r'}) \frac{1}{n} \sum_{j=1}^{n} \left(\frac{\partial^2 \log f(X_j; \boldsymbol{\theta})}{\partial\theta_r \partial\theta_{r'}} \right)_{\boldsymbol{\theta}_0}$$

$$+ \frac{1}{3!} \sum_{r,r',r''=1}^{p} (\theta_r - \theta_{0r})(\theta_{r'} - \theta_{0r'})(\theta_{r''} - \theta_{0r''}) \frac{1}{n} \sum_{j=1}^{n} \left(\frac{\partial^3 \log f(X_j; \boldsymbol{\theta})}{\partial\theta_r \partial\theta_{r'} \partial\theta_{r''}} \right)_{\boldsymbol{\theta}^*} \tag{7.51}$$

where θ^* lies on the line segment joining θ and θ_0, and therefore $|\theta^* - \theta_0| \leq \delta$. The second summand on the right side of (7.51) may be expressed, by the strong law of large numbers, as

$$\sum_{r=1}^{p} (\theta_r - \theta_{0r})\eta_{r,n}, \quad \eta_{r,n} \to \text{a.s.} \quad E_{\theta_0}\left(\frac{\partial \log f(X_1; \theta)}{\partial \theta_r}\right)_{\theta_0} = 0. \tag{7.52}$$

$$(\text{under } P_{\theta_0})$$

The third summand converges a.s. (under P_{θ_0}) to

$$-\frac{1}{2!}\sum_{r,r'=1}^{p} (\theta_r - \theta_{0r})(\theta_{r'} - \theta_{0r'})I_{rr'}(\theta_0); \tag{7.53}$$

for one may show, as in (7.10), that

$$E_{\theta_0}\left(\frac{\partial^2 \log f(X_1; \boldsymbol{\theta})}{\partial \theta_r \partial \theta_{r'}}\right)_{\theta_0} = -E_{\theta_0}\left(\frac{\partial \log f(X_1; \boldsymbol{\theta})}{\partial \theta_r} \cdot \frac{\partial \log f(X_1; \boldsymbol{\theta})}{\partial \theta_{r'}}\right)_{\boldsymbol{\theta}_0}. \tag{7.54}$$

The last summand in (7.51) is, by (A5), bounded by

$$\frac{p^{\frac{3}{2}}}{3!}|\boldsymbol{\theta} - \boldsymbol{\theta}_0|^3 \frac{1}{n}\sum_{j=1}^{n} g(X_j). \tag{7.55}$$

(Note: $(\sum_{r=1}^{p} |\theta_r - \theta_{0r}|)^3 \leq p^{\frac{3}{2}}|\boldsymbol{\theta} - \boldsymbol{\theta}_0|^3$).)

By the strong law of large numbers the expression (7.55) converges a.s., under P_{θ_0}, to

$$\frac{p^{\frac{3}{2}}}{3!}|\boldsymbol{\theta} - \boldsymbol{\theta}_0|^3 \cdot E_{\theta_0}g(X_1). \tag{7.56}$$

Let λ be the smallest eigenvalue of $\mathscr{I}_1(\boldsymbol{\theta}_0)$. Then

$$\frac{1}{2!}\sum_{r,r'=1}^{p} (\theta_r - \theta_{0r})(\theta_{r'} - \theta_{0r'})I_{rr'}(\boldsymbol{\theta}_0) \geq \frac{1}{2!}\lambda_1|\boldsymbol{\theta} - \boldsymbol{\theta}_0|^2. \tag{7.57}$$

Choose δ such that

$$\frac{p^{\frac{3}{2}}}{3!}\delta E_{\theta_0}g(X_1) < \frac{\lambda_1}{8}, \tag{7.58}$$

so that the expression (7.56) is less than $\lambda_1|\boldsymbol{\theta} - \boldsymbol{\theta}_0|^2/8$ for $|\boldsymbol{\theta} - \boldsymbol{\theta}_0| \leq \delta$. Consider the set

$$B_{n_1} = \left\{\mathbf{x} : x_j \in C \; \forall j = 1, 2, \ldots, n, \text{ and the right side of (7.51) is less than}\right.$$

$$\tag{7.59}$$

$$\left.\frac{1}{n}\sum_{j=1}^{n} \log f(X_j; \boldsymbol{\theta}_0) + \sum_{r=1}^{p}(\theta_r - \theta_{0r})\eta_{r,r'} - \frac{1}{4}\lambda_1|\boldsymbol{\theta} - \boldsymbol{\theta}_0|^2 \; \forall \, \boldsymbol{\theta} \text{ s.t. } |\boldsymbol{\theta} - \boldsymbol{\theta}_0| \leq \delta\right\}.$$

It follows from (7.41)–(7.58) that

$$P_{\theta_0}(B_{n1}) \to 1 \qquad \text{as } n \to \infty. \tag{7.60}$$

Now on the (*surface* of the) sphere $\{\theta : |\theta_r - \theta_{\theta_0}| = \delta\}$

$$\sum_{r=1}^{p}(\theta_r - \theta_{0r})\eta_{r,n} - \frac{1}{4}\lambda_1|\theta - \theta_0|^2 \le |\theta - \theta_0| \cdot \left(\sum_{r=1}^{p}\eta_{r,n}^2\right)^{\frac{1}{2}} - \frac{1}{4}\lambda_1|\theta - \theta_0|^2$$

$$= \delta\left(\sum_{r=1}^{p}\eta_{r,n}^2\right)^{\frac{1}{2}} - \frac{1}{4}\lambda_1\delta^2. \tag{7.61}$$

Consider the set

$$B_{n2} = \left\{\mathbf{x} : \left(\sum_{r=1}^{p}\eta_{r,n}^2\right)^{\frac{1}{2}} < \frac{\lambda_1\delta}{8}\right\}. \tag{7.62}$$

On B_{n2} the expression (7.61) is strictly negative. Therefore for $\mathbf{x}$ belonging to

$$B_n = B_{n1} \cap B_{n2} \tag{7.63}$$

one has

$$\frac{1}{n}\sum_{j=1}^{n}\log f(X_j; \theta) < \frac{1}{n}\sum_{j=1}^{n}\log f(X_j; \theta_0) \quad \forall\, \theta \in \{|\theta - \theta_0| = \delta\}. \tag{7.64}$$

i.e.,

$$f_n(\mathbf{x}; \theta) < f_n(\mathbf{x}; \theta_0) \quad \forall\, \theta \in \{|\theta - \theta_0| = \delta\}. \tag{7.65}$$

But (7.65) implies that the maximum of the function $f_n(\mathbf{x}; \theta)$ on the closed ball $\{|\theta - \theta_0| \le \delta\}$ is attained in the interior $\{|\theta - \theta_0| < \delta\}$. Therefore, inside this ball there exists a (at least one) point θ satisfying (7.50). On the other hand by (7.60) $P_{\theta_0}(B_{n1}) \to 1$ as $n \to \infty$. Therefore, by (7.52),

$$P_{\theta_0}(B_n) \longrightarrow 1 \qquad \text{as } n \to \infty. \tag{7.66}$$

Since $\mathbf{x} \in B_n$ imples $S_{\mathbf{x},n}$ is nonempty and $|\hat{\theta}_n(\mathbf{x}) - \theta_0| < \delta$, and since the definition of $\hat{\theta}_n$ does not involve δ, the proof of (1), (2) is complete. (Note that (7.66) holds for all sufficiently small δ).

In order to prove part (3) we proceed as in the proof of Theorem 7.2 i.e., write for $\mathbf{x} \in B_n$

$$0 = \left(\frac{\partial \log f_n(\mathbf{x}; \theta)}{\partial \theta_r}\right)_{\hat{\theta}_n(\mathbf{x})}$$

$$= \left(\frac{\partial \log f_n(\mathbf{x}; \theta)}{\partial \theta_r}\right)_{\theta_0} + \sum_{r'=1}^{p}(\hat{\theta}_{n,r'}(\mathbf{x}) - \theta_{0r'})\left(\frac{\partial^2 \log f_n(\mathbf{x}; \theta)}{\partial \theta_r \partial \theta_{r'}}\right)_{\theta_0} \tag{7.67}$$

$$+ \frac{1}{2!}\sum_{r',r''=1}^{p}(\hat{\theta}_{n,r'}(\mathbf{x}) - \theta_{0r'})(\hat{\theta}_{n,r''}(\mathbf{x}) - \theta_{0r''})\left(\frac{\partial^3 \log f_n(\mathbf{x}; \theta)}{\partial \theta_r \partial \theta_{r'} \partial \theta_{r''}}\right)_{\bar{\theta}(\mathbf{x},r))}, \quad 1 \le r \le p,$$

where $\bar{\theta}(\mathbf{x}; r)$ lies on the line segment joining $\hat{\theta}_n(\mathbf{x})$ and θ_0. Express (7.67) as

$$\sqrt{n}(\hat{\theta}_n(\mathbf{x}) - \theta_0) =$$

$$= -\left(\left(\left(\frac{1}{n}\frac{\partial^2 \log f_n(\mathbf{x}; \theta)}{\partial \theta_r \partial \theta_{r'}}\right)_{\theta_0} + \eta_{n,r,r'}(\mathbf{x})\right)\right)^{-1}\left(\frac{1}{\sqrt{n}}\text{grad} \log f_n(\mathbf{x}; \theta)\right)_{\theta_0} \tag{7.68}$$

where

$$\eta_{n,r,r'}(\mathbf{x}) = \frac{1}{2} \sum_{r''=1}^{p} \left(\hat{\theta}_{n,r''}(\mathbf{x}) - \theta_{0r''} \right) \frac{1}{n} \left(\frac{\partial^3 \log f_n(\mathbf{x};\boldsymbol{\theta})}{\partial\theta_r\partial\theta_{r'}\partial\theta_{r''}} \right)_{\overline{\boldsymbol{\theta}}(\mathbf{x};r)}, \tag{7.69}$$

and (7.68) is defined only for those $\mathbf{x}$ for which the matrix $((\quad))$ (whose (r,r') element is given in (7.68)) is nonsingular. Now, by (A5),

$$\left| \frac{1}{n} \left(\frac{\partial^3 \log f_n(\mathbf{x};\boldsymbol{\theta})}{\partial\theta_r\partial\theta_{r'}\partial\theta_{r''}} \right)_{\overline{\boldsymbol{\theta}}(\mathbf{x},r)} \right| \leq \frac{1}{n} \sum_{j=1}^{n} g(x_j) \tag{7.70}$$

which is bounded in $P_{\boldsymbol{\theta}_0}$ probability (since the right side converges a.s. ($P_{\boldsymbol{\theta}_0}$) to a constant). But $\hat{\theta}_{n,r''} - \tilde{\theta}_{0r''} \to 0$ in $P_{\boldsymbol{\theta}_0}$-probability ($\forall\, r'' = 1,2,\ldots,p$) as $n \to \infty$. Hence

$$\eta_{r,r,r'} \xrightarrow{P_{\boldsymbol{\theta}_0}} 0 \quad \text{as } n \to \infty. \tag{7.71}$$

Also, writing $\mathscr{I}_1 = ((I_{rr'}))$, one has

$$\frac{1}{n} \left(\frac{\partial^2 \log f_n(\mathbf{X};\boldsymbol{\theta})}{\partial\theta_r\partial\theta_{r'}} \right)_{\boldsymbol{\theta}_0} \xrightarrow{\text{a.s.}(P_{\boldsymbol{\theta}_0})} -I_{rr'}(\boldsymbol{\theta}_0), \qquad \text{as } \quad n \to \infty. \tag{7.72}$$

Therefore, the matrix $((\quad))$ in (7.68) converges a.s. to $-\mathscr{I}_1(\boldsymbol{\theta}_0)$, and its inverse converges a.s. to $-\mathscr{I}_1^{-1}(\boldsymbol{\theta}_0)$, as $n \to \infty$. By the multidimensional classical central limit theorem,

$$\frac{1}{\sqrt{n}}(\text{grad } \log f_n(\mathbf{X};\boldsymbol{\theta}))_{\boldsymbol{\theta}_0} = \frac{1}{\sqrt{n}} \sum_{j=1}^{n} (\text{grad } \log f(X_j;\boldsymbol{\theta}))_{\boldsymbol{\theta}_0} \xrightarrow{\mathscr{L}} N(\mathbf{0}, \mathscr{I}(\boldsymbol{\theta}_0)). \tag{7.73}$$

Therefore, by (a vector version of) Slutsky's theorem,

$$\sqrt{n}(\hat{\boldsymbol{\theta}}_n - \boldsymbol{\theta}_0) \xrightarrow{\mathscr{L}} \mathscr{I}_1^{-1}(\boldsymbol{\theta}_0)\mathbf{Z}, \quad Z \overset{\mathscr{L}}{\sim} N(0, \mathscr{I}_1(\boldsymbol{\theta}_0)), \tag{7.74}$$

from which part (3) of the theorem follows. To be completely precise one needs to write

$$\sqrt{n}(\hat{\boldsymbol{\theta}}_n - \boldsymbol{\theta}_0) = \sqrt{n}(\hat{\boldsymbol{\theta}}_n - \boldsymbol{\theta}_0) \cdot \mathbf{1}_{D_n} + \sqrt{n}(\hat{\boldsymbol{\theta}}_n - \boldsymbol{\theta}_0)\mathbf{1}_{D_n^c}, \tag{7.75}$$

where D_n is the subset of B_n on which the matrix $((\cdot))$ in (7.68) is nonsingular. Then use the representation (7.68) for the first summand in (7.75). Since $\mathbf{1}_{D_n} \xrightarrow{P_{\boldsymbol{\theta}_0}} 1$, and, therefore, $\sqrt{n}(\hat{\boldsymbol{\theta}}_n - \boldsymbol{\theta}_0)\mathbf{1}_{D_n^c} \xrightarrow{P_{\boldsymbol{\theta}_0}} 0$, the proof is complete. $\qquad\square$

Example 7.5 (Multi-Parameter Exponential Family). Here

$$f(x;\boldsymbol{\theta}) = C(\boldsymbol{\theta})e^{\sum_{r=1}^{P} \theta_r t_r(x)}h(x), \tag{7.76}$$

$$\Theta = \left\{ \boldsymbol{\theta} = (\theta_1, \theta_2, \ldots, \theta_p) \in \mathbb{R}^p : \int_{\mathscr{X}} e^{\sum_{r=1}^{P} \theta_r t_r(x)} h(x)\mu(dx) < \infty \right\}.$$

Assume that Θ is a nonempty open subset of $\mathbb{R}^p$. In this set up $\boldsymbol{\theta}$ is called a natural parameter. Note that

$$C(\boldsymbol{\theta}) = \left(\int_{\mathscr{X}} e^{\sum_{r=1}^{P} \theta_r t_r(x)} h(x)\mu(dx) \right)^{-1}. \tag{7.77}$$

A family of distributions $\{f(x; \boldsymbol{\theta}) : \boldsymbol{\theta} \in \Theta\}$ given by (7.76) is called a multi-parameter exponential family in the natural parameter form. One may easily show, by Hölder's inequality, that Θ is convex (See Sect. 4.2 and Theorem 4.4). Also under the assumption underlined above $\int_{\mathcal{X}} \exp\{\sum_{r=1}^{p} \theta_r t_r(x)\} h(x) \mu(dx)$ is analytic (in particular, infinitely differentiable) in θ. One has the likelihood equations

$$-\frac{\partial \log C(\boldsymbol{\theta})}{\partial \theta_r} = \frac{1}{n} \sum_{j=1}^{n} t_r(X_j), \qquad (1 \le r \le p), \tag{7.78}$$

(See Exercise 7.12). As in Theorem 7.1 (iii), one may easily show that

$$E_{\boldsymbol{\theta}} t_r(X_1) = -\frac{\partial \log C(\boldsymbol{\theta})}{\partial \theta_r}, \qquad (1 \le r \le p), \tag{7.79}$$

and

$$\frac{\partial^2 \log C(\boldsymbol{\theta})}{\partial \theta_r \partial \theta_{r^i}} = -\mathrm{cov}_{\boldsymbol{\theta}}(t_r(X_1), t_{r'}(X_1)). \tag{7.80}$$

Since

$$\frac{\partial^2 \log f_n(\mathbf{x}, \boldsymbol{\theta})}{\partial \theta_r \partial \theta_{r'}} = n \frac{\partial^2 \log C(\boldsymbol{\theta})}{\partial \theta_r \partial \theta_{r'}} = -n \, \mathrm{cov}_{\boldsymbol{\theta}}(t_r(X_1), t_{r'}(X_1)), \tag{7.81}$$

the matrix of second derivatives of $\log f_n(\mathbf{x}; \boldsymbol{\theta})$ is a strictly concave function of θ. Hence there can not be more than one solution to the likelihood Eq. (7.78). If there is a solution it is the MLE. Since the assumptions $(A_1) - (A_5)$ are all satisfied in this case it follows that the MLE (equivalently, the unique solution of the likelihood equations) exists on a set whose probability goes to one. When extended arbitrarily (but measurably) to these $\mathbf{x}$ for which a solution to (7.78) does not exist, the MLE is, therefore, asymptotically efficient.

The following special multi-parameter families are important in applications.

(a) *Univariate Normal* $N(\mu, \sigma^2)$. In this case write $\boldsymbol{\theta} = (\theta_1, \theta_2)$ with $\theta_1 = \mu/\sigma^2$ and $\theta_2 = -\frac{1}{2\sigma^2}$. Then

$$f(x; \mu, \sigma^2) = \frac{1}{\sqrt{2\pi\sigma^2}} e^{-(x-\mu)^2/2\sigma^2} = \frac{1}{\sqrt{2\pi\sigma^2}} e^{-\frac{x^2}{2\sigma^2} + \frac{\mu x}{\sigma^2} - \frac{\mu^2}{2\sigma^2}}$$

$$= C(\boldsymbol{\theta}) e^{\theta_1 t_1(x) + \theta_2 t_2(x)} = f(x; \boldsymbol{\theta}) \qquad \theta_1 = \frac{\mu}{\sigma^2}, \theta_2 = -\frac{1}{2\sigma^2},$$

$$C(\boldsymbol{\theta}) = \frac{1}{\sqrt{2\pi\sigma^2}} e^{-\mu^2/2\sigma^2} = \left(-\frac{\theta_2}{\pi}\right)^{\frac{1}{2}} e^{\frac{1}{4}\theta_1^2/\theta_2}; \quad t_1(x) = x, t_2(x) = x^2.$$

From (7.78) we get the MLE as

$$\hat{\boldsymbol{\theta}} = (\hat{\theta}_1, \hat{\theta}_2) = \left(\frac{\overline{X}}{s^2}, -\frac{1}{2s^2}\right),$$

where $\overline{X} = \frac{1}{n} \sum_{j=1}^{n} X_j$ and $s^2 = \frac{1}{n} \sum_{j=1}^{n} (X_j - \overline{X})^2$. The information matrix is given by $\mathscr{I}_1(\boldsymbol{\theta}) = ((I_{ij}(\boldsymbol{\theta})))$, where

$$I_{11}(\boldsymbol{\theta}) = -\frac{\partial^2 \log C(\boldsymbol{\theta})}{\partial \theta_1^2} = -\frac{1}{2\theta_2} = \sigma^2,$$

$$I_{12}(\boldsymbol{\theta}) = I_{21}(\theta) = -\frac{\partial^2 \log C(\boldsymbol{\theta})}{\partial \theta_1 \partial \theta_2} = -\frac{\theta_1}{2\theta_2^2} = -2\mu\sigma^2,$$

$$I_{22}(\boldsymbol{\theta}) = -\frac{\partial^2 \log C(\boldsymbol{\theta})}{\partial \theta_2^2} = \frac{1}{2\theta_2^2} + \frac{\theta_1^2}{2\theta_2^3} = 2\sigma^4 + 4\mu^2\sigma^4. \qquad (7.82)$$

Since $(\theta_1, \theta_2) \to (\mu, \sigma^2)$ is one-to-one (and differentiable), the MLE of (μ, σ^2) is (Note: $\mu = -\theta_1/2\theta_2, \sigma^2 = -\frac{1}{2\theta_2}$.)

$$\hat{\mu} = -\frac{\hat{\theta}_1}{2\hat{\theta}_2} = \overline{X}, \qquad \hat{\sigma}^2 = s^2. \qquad (7.83)$$

The information matrix in terms of μ, σ^2 is then given by

$$I_{11}(\mu, \sigma^2) = E_{\mu,\sigma^2} \left(\frac{\partial \log f(X_1; \mu, \sigma^2)}{\partial \mu} \right)^2$$

$$= E_{\mu,\sigma^2} \left(\frac{X_1}{\sigma^2} - \frac{\mu}{\sigma^2} \right)^2 = \frac{1}{\sigma^4} E_{\mu,\sigma^2}(X_1 - \mu) = \frac{1}{\sigma^2},$$

$$I_{12}(\mu, \sigma^2) = E_{\mu,\sigma^2} \left(\frac{\partial \log f(X_1; \mu, \sigma^2)}{\partial \mu} \cdot \frac{\partial \log f(X_1; \mu, \sigma^2)}{\partial \sigma^2} \right)$$

$$= E_{\mu,\sigma^2} \left(\frac{X_1 - \mu}{\sigma^2} \cdot \left\{ -\frac{1}{2\sigma^2} + \frac{X_1^2}{2\sigma^4} - \frac{\mu X_1}{\sigma^4} \right\} \right)$$

$$= E_{\mu,\sigma^2} \left(\frac{X_1 - \mu}{\sigma^2} \left\{ \frac{(X_1 - \mu)^2}{2\sigma^4} - \frac{1}{2\sigma^2} - \frac{\mu^2}{2\sigma^4} \right\} \right)$$

$$= E_{\mu,\sigma^2} \frac{(X_1 - \mu)^3}{2\sigma^6} = 0,$$

$$I_{22}(\mu, \sigma^2) = E_{\mu,\sigma^2} \left(\frac{\partial \log f(X_1; \mu, \sigma^2)}{\partial \sigma^2} \right)^2 = \text{var} \left(\frac{\partial \log f(X_1; \mu, \sigma^2)}{\partial \sigma^2} \right)$$

$$= \text{var} \left\{ \frac{(X_1 - \mu)^2}{2\sigma^4} \right\} = \frac{2\sigma^4}{4\sigma^8} = \frac{1}{2\sigma^4}.$$

$$I_{11}^{-1}(\mu, \sigma^2) = \sigma^2, \qquad I_{12}^{-1} = I_{21}^{-1}(\mu, \sigma^2) = 0, \qquad I_{22}^{-1}(\mu, \sigma^2) = 2\sigma^4.$$

Therefore, $(\sqrt{n}(\overline{X} - \mu), \sqrt{n}(s^2 - \sigma^2)) \xrightarrow{\mathcal{L}} N\left(0, \begin{pmatrix} \sigma^2 & 0 \\ 0 & 2\sigma^4 \end{pmatrix}\right).$

(b) *Multivariate Normal* $N\left(\boldsymbol{\mu} = \begin{pmatrix} \mu_1 \\ \vdots \\ \mu_k \end{pmatrix}, \Sigma = ((\sigma_{ij})) \right).$

Let $X_1 = (X_{11}, X_{12}, \ldots, X_{1k}), \ldots, X_n = (X_{n1}, X_{n2}, \ldots, X_{nk})$ be i.i.d. $N(\boldsymbol{\mu}, \Sigma)$, $\boldsymbol{\mu} \in \mathbb{R}^k$, $\Sigma \in$ set of all positive definite symmetric $k \times k$ matrices. One has, with $\boldsymbol{\theta} = (\boldsymbol{\mu}, \Sigma)$,

$$f_n(\mathbf{x};\theta) = (2\pi)^{-nk/2}\,(\det \Sigma)^{-n/2}\exp\left\{-\frac{1}{2}\sum_{j=1}^{n}\left[\sum_{i,i'=1}^{k}\sigma^{ii'}(x_{ji}-\mu_i)\cdot(x_{ji'}-\mu_{i'})\right]\right\}$$

$$(7.84)$$

where $((\sigma^{ii'})) = \Sigma^{-1}$. If μ is known then (7.84) is a $p(=\ k(k+1)/2)$-parameter exponential family with *natural parameters* $\{\sigma^{ii'}:1\leq i\leq i'\leq p\}$. By (7.78), (7.79) one has the likelihood equations (with $t_{ii'}(x_j) = -(x_{ji}-\mu_i)(x_{ji'}-\mu_{i'})$ for $i<i'$, and $t_{ii}(x_j) = -\frac{1}{2}(x_{ji}-\mu_i)^2$).

$$E_{\boldsymbol\theta}t_{ii'}(X_1) = \frac{1}{n}\sum_{j=1}^{n}t_{ii'}(X_j),\qquad (7.85)$$

or, *the MLE* $\hat{\boldsymbol\theta}_n = \{\hat\sigma_{ii'}:1\leq i\leq i'\leq k\}$ is given by

$$\hat\sigma_{ii'} = s_{ii'}\qquad (1\leq i\leq i'\leq k),\qquad (7.86)$$

where

$$s_{ii'} = \frac{1}{n}\sum_{j=1}^{n}(X_{ji}-\mu_i)(X_{ji'}-\mu_{i'}).\qquad (7.87)$$

The $p = (\frac{k(k+1)}{2})$-dimensional random vectors $\sqrt{n}(\hat{\boldsymbol\theta}_n - \boldsymbol\theta)$ converges in distribution to a p-variate Normal distribution $N(\mathbf{0},V(\boldsymbol\theta))$, where $V(\boldsymbol\theta) = ((v_{(ii'),(\ell,\ell')}(\theta)))$

$$
\begin{aligned}
v_{(i,i'),(\ell,\ell')}(\boldsymbol\theta) &= \mathrm{cov}_\theta((X_{1i}-\mu_i)(X_{1i'}-\mu_{i'}),(X_{1\ell}-\mu_\ell)(X_{1\ell'}-\mu_{\ell'}))\\
&= E_{\boldsymbol\theta}(X_{1i}-\mu_i)(X_{1i'}-\mu_{i'})(X_{1\ell}-\mu_\ell)(X_{1\ell'}-\mu_{\ell'})-\sigma_{ii'}\sigma_{\ell\ell'}\\
&= \left(\frac{\partial^4}{\partial\xi_i\partial\xi_{i'}\partial\xi_\ell\partial\xi_{\ell'}}\exp\left\{\frac{1}{2}\sum_{r,r'=1}^{k}\sigma_{rr'}\xi_r\xi_{r'}\right\}\right)_{\xi=0} -\sigma_{ii'}\sigma_{\ell\ell'}.\quad(7.88)
\end{aligned}
$$

In particular,

$$v_{(i,i'),(i,i')} = \sigma_{ii}\sigma_{i'i'} + \sigma_{ii'}^2.\qquad (7.89)$$

In case μ_i's are also unknown, one has

$$f(x;\boldsymbol\theta) = (2\pi)^{-k/2}\,(\det \Sigma)^{-\frac{1}{2}}\exp\left\{-\frac{1}{2}\sum_{i,i'=1}^{k}\sigma^{ii'}(x_i-\mu_i)(x_{i'}-\mu_{i'})\right\}$$

$$= C(\boldsymbol\theta)\exp\left\{-\frac{1}{2}\sum_{i,i'=1}^{k}x_i x_{i'}\sigma^{ii'} + \sum_{i=1}^{k}x_i\left(\sum_{i'=1}^{k}\mu_{i'}\sigma^{ii'}\right)\right\},$$

with $\boldsymbol\theta = \{\sigma^{ii'}, 1\leq i\leq i'\leq k; \sum_{i'=1}^{k}\mu_{i'}\sigma^{ii'}, 1\leq i\leq k\}$. Hence $f(x;\boldsymbol\theta)$ belongs to the p-parameter exponential family with $p = \frac{k(k+1)}{2} + k = k(k+3)/2$. The *natural parameters* are the coordinates of $\boldsymbol\theta$. The likelihood equations are (see (7.78), (7.79)).

$$E_{\boldsymbol\theta}\tilde{t}_{ii'}(X_1) = \frac{1}{n}\sum_{j=1}^{n}\tilde{t}_{ii'}(X_j),$$

$$E_{\boldsymbol{\theta}} T_i(X_1) = \frac{1}{n} \sum_{j=1}^{n} T_i(X_j), \tag{7.90}$$

with $\tilde{t}_{ii'}(x) = -x_i x_{i'}$ (if $i < i'$), $\tilde{t}_{ii}(x) = -\frac{1}{2} x_i^2$, $T_i(x) = x_i$. Hence one has the equations

$$\sigma_{ii'} + \mu_i \mu_{i'} = \frac{1}{n} \sum_{j=1}^{n} X_{ji} X_{ji'} = s'_{ii'},$$

$$\mu_i = \frac{1}{n} \sum_{j=1}^{n} X_{ji} = \overline{X}_i \quad (1 \le i \le i' \le k), \tag{7.91}$$

whose solution is

$$\hat{\mu}_i = \overline{X}_i, \qquad \tilde{\sigma}_{ii'} = s'_{ii'} - \overline{X}_i \overline{X}_{i'} = \frac{1}{n} \sum_{j=1}^{n} (X_{ji} - \overline{X}_i)(X_{ji'} - \overline{X}_i) \tag{7.92}$$

It may be shown (See Problem 5.6, Lehmann, *Theory of Point Estimation*, Chap. 6) that $\{\tilde{\sigma}_{ii'}; 1 \le i \le i' \le k\}$ are independent of $\{\hat{\mu}_i; 1 \le i \le k\}$ and that the joint distribution of $\{\tilde{\sigma}_{ii'}; 1 \le i \le i' \le k\}$ is the same as that of $\{s_{ii'}; 1 \le i \le i' \le k\}$ based on $n-1$ observations. Thus the asymptotic distribution of $\{\sqrt{n}(\tilde{\sigma}_{ii'} - \sigma_{ii'}); 1 \le i \le i' \le k\}$ is the same as that of $\{\sqrt{n}(s_{ii'} - \sigma_{ii'}); 1 \le i \le i \le k\}$. The (asymptotic) distribution of $\{\sqrt{n}(\hat{\mu}_i - \mu_i); 1 \le i \le k\}$ is $N(\mathbf{0}, \Sigma)$. In view of the independence mentioned above the limiting distribution of $\sqrt{n}(\tilde{\boldsymbol{\theta}} - \boldsymbol{\theta})$ is now completely specified ($\tilde{\boldsymbol{\theta}} = \{\tilde{\sigma}_{ii'}, 1 \le i \le i' \le k; \hat{\mu}_i; 1 \le i \le k\}$). (See Exercise 7.16).

(c) *The multinomial distribution.* A population is divided into k categories, with proportion p_i belonging to the ith category, $0 < p_i < 1$ ($1 \le i \le k$): $\sum_{i=1}^{k} p_i = 1$. Consider n observations taken at random (with replacement) from this population. Let us code an observation as the unit k-dimensional vector e_i with 1 as the ith coordinate and zeros elsewhere, if the observation belongs to the ith category. Let $X_1, X_2, \ldots, X_n$ ($X_j = (X_{j1}, \ldots, X_{jk})$) be the random observations. Then $X_1, X_2, \ldots, X_n$ are i.i.d. with common p.d.f.

$$f(x; \boldsymbol{\theta}) = \prod_{i=1}^{k} p_i^{x_i} = p_1^{x_1} p_2^{x_2} \cdots p_{k-1}^{x_{k-1}} (1 - p_1 - \cdots - p_{k-1})^{x_k}$$

$$\text{for } x = (x_1, \ldots, x_k) \in \{e_i; 1 \le i \le k\} = \mathscr{X}. \tag{7.93}$$

Write

$$f(x; \boldsymbol{\theta}) = e^{\sum_{i=1}^{k-1} x_i \log p_i + (1 - \sum_{1}^{k-1} x_i) \log(1 - \sum_{1}^{k-1} p_i)}$$

$$= c(\boldsymbol{\theta}) e^{\sum_{i=1}^{k-1} \theta_i x_i}, \tag{7.94}$$

where $\theta_i = \log(p_i / (1 - \sum_{1}^{k-1} p_i))$, $1 \le i \le k - 1$, are natural parameters. The likelihood equations are

$$p_i = E_{\boldsymbol{\theta}} X_{1i} = \frac{1}{n} \sum_{j=1}^{n} X_{ji}, \qquad (1 \le i \le k - 1), \tag{7.95}$$

or, the MLE of $(p_1, \ldots, p_{k-1})$ is given by

$$\hat{p}_i = \frac{1}{n} \sum_{j=1}^{n} X_{ji} = \text{ proportion of the sample belonging to } ith \text{ category } (1 \leq i \leq k-1),$$

$$\hat{\mathbf{p}} = (\hat{p}_1, \ldots \hat{p}_{k-1}).$$

Of course, in this case one may represent

$$\sqrt{n}(\hat{\mathbf{p}} - \mathbf{p}) = \sqrt{n} \sum_{j=1}^{n} (Y_j - \mathbf{p}), \qquad \mathbf{p} = \begin{pmatrix} p_1 \\ \vdots \\ p_{k-1} \end{pmatrix} \tag{7.96}$$

where $Y_1, \ldots, Y_n$ are i.i.d. $(k-1)$-dimensional random vectors with

$$\text{Prob}\left[Y_1 = \begin{pmatrix} 0 \\ \vdots \\ 0 \\ 1 \\ 0 \\ \vdots \\ 0 \end{pmatrix} \}i \right] = p_i \quad (1 \leq i \leq k-1), \quad P(Y_1 = \mathbf{0}) = 1 - \sum_{1}^{k-1} p_i \tag{7.97}$$

The classical multidimensional CLT then yields (Exercise 7.12)

$$\sqrt{n}(\hat{\mathbf{p}} - \mathbf{p}) \xrightarrow{\mathscr{L}} N(\mathbf{0}, V), \quad V = ((v_{ij'})),$$

$$v_{ii} = p_i(1 - p_i)(1 \leq i \leq k-1), \quad v_{ii'} = -p_i p_{i'} \ (i \neq i'), \tag{7.98}$$

although one could also use Theorem 7.5.

7.5 Method of Moments

Let $X_1, X_2, \ldots, X_n$ be i.i.d. real-valued with common p.d.f. $f(x; \boldsymbol{\theta})$ where $\boldsymbol{\theta} = (\theta_1, \theta_2, \ldots, \theta_p) \in \Theta$ which is an open subset of $\mathbb{R}^p$. Assume that

$$E|X_1|^p < \infty. \tag{7.99}$$

The method of moments consists in solving the equations

$$E_{\boldsymbol{\theta}} X_1^r = \frac{1}{n} \sum_{j=1}^{n} X_j^r \quad (1 \leq r \leq p) \tag{7.100}$$

for $\theta_1, \theta_2, \ldots, \theta_p$. In case (7.99) does not hold, or X_j's are not real-valued, or (7.100) are difficult to solve, one may take some suitable real-valued functions $g_1(x), g_2(x), \ldots, g_p(x)$ such that

$$E_{\boldsymbol{\theta}}|g_r(X_1)| < \infty, \qquad 1 \leq r \leq p, \tag{7.101}$$

and solve the equations

$$M_r(\boldsymbol{\theta}) := E_\theta g_r(X_1) = \frac{1}{n}\sum_{j=1}^{n} g_r(X_j), \qquad 1 \le r \le p, \tag{7.102}$$

for $\theta_1, \theta_2, \ldots, \theta_p$.

Proposition 7.1. *Let Θ be an open subset of $\mathbb{R}^p$. Let g_r, $1 \le r \le p$, be real-valued functions on $\mathscr{X}$ such that*

$$E_\theta g_r^2(X_1) < \infty \quad \text{for} \quad 1 \le r \le p, \quad \forall\, \boldsymbol{\theta} \in \Theta. \tag{7.103}$$

Assume (i) (7.102) has a unique solution $\tilde{\boldsymbol{\theta}} = (\tilde{\theta}_1, \tilde{\theta}_2, \ldots, \tilde{\theta}_p)$ (a.s. $P_{\boldsymbol{\theta}_0}\ \forall\, \boldsymbol{\theta}_0 \in \Theta$), (ii) the map $\theta \to M(\theta) = (M_1(\theta), \ldots, M_p(\theta))$ is a diffeomorphism, i.e., M and its inverse H (say) are both continuously differentiable. Then, for each $\theta_0 \in \Theta$,

$$\sqrt{n}(\tilde{\boldsymbol{\theta}} - \boldsymbol{\theta}_0) \xrightarrow{\mathscr{L}} N(0, V(\boldsymbol{\theta}_0)) \qquad \text{under } P_{\boldsymbol{\theta}_0}, \tag{7.104}$$

where

$$V(\boldsymbol{\theta}_0) = ((v_{ii'}(\boldsymbol{\theta}_0))),$$

$$v_{ii'}(\boldsymbol{\theta}_0) = \sum_{r,r'=1}^{p} \sigma_{rr'}(\boldsymbol{\theta}_0)\ell_{ir}(\boldsymbol{\theta}_0)\ell_{i'r'}(\boldsymbol{\theta}_0),$$

$$\sigma_{rr'}(\boldsymbol{\theta}_0) = E_{\boldsymbol{\theta}_0}[(g_r(X_1) - M_r(\boldsymbol{\theta}_0))(g_{r'}(X_1) - M_{r'}(\boldsymbol{\theta}_0))],$$

$$\ell_{ir}(\boldsymbol{\theta}_0) = \left(\frac{\partial H_i(M)}{\partial M_r}\right)_{M = M(\boldsymbol{\theta}_0)} \tag{7.105}$$

Proof.

$$\sqrt{n}(\tilde{\boldsymbol{\theta}} - \boldsymbol{\theta}_0)' = \sqrt{n}\left[H\left(\frac{1}{n}\sum_{j=1}^{n} g_1(X_j), \ldots, \frac{1}{n}\sum_{j=1}^{n} g_p(X_j)\right) - H(M(\boldsymbol{\theta}_0))\right]'$$

$$= \sqrt{n}\,(\mathrm{Grad}\, H(M(\boldsymbol{\theta}_0)) + \varepsilon_n)\begin{pmatrix} \frac{1}{n}\sum_{j=1}^{n} g_1(X_j) - M_1(\theta_0) \\ \vdots \\ \frac{1}{n}\sum_{j=1}^{n} g_p(X_j) - M_p(\theta_0) \end{pmatrix}, \tag{7.106}$$

where $\mathrm{Grad}\, H(M(\boldsymbol{\theta}_0))$ is the $p \times p$ matrix who (i, r) element is $\ell_{ir}(\boldsymbol{\theta}_0)$, and ε_n is a $p \times p$ matrix whose elements converge in probability $(P_{\boldsymbol{\theta}_0})$ to zero. $\qquad\square$

Example 7.6. Let $X_1, X_2, \ldots, X_n$ be i.i.d. with common p.d.f. $f(x; \mu, \alpha) = \frac{1}{\mu^\alpha \Gamma(\alpha)} e^{-x/\mu} x^{\alpha-1}$, $0 < x < \infty$; $\mu \in (0, \infty)$, $\alpha \in (0, \infty)$. Note that $EX_1 = \alpha\mu$, $EX_1^2 = (\alpha+1)\alpha\mu^2$. Therefore, solve

$$\alpha\mu = \frac{1}{n}\sum_{j=1}^{n} X_j = \overline{X}, \qquad (\alpha+1)\alpha\mu^2 = \frac{1}{n}\sum_{j=1}^{n} X_j^2 = m_2', \tag{7.107}$$

for α, μ to get

$$\tilde{\alpha} = \frac{\overline{X}^2}{m_2' - \overline{X}^2} = \frac{\overline{X}^2}{s^2}, \qquad \left(s^2 = \frac{1}{n} \sum_{j=1}^{n} (X_j - \overline{X})^2 \right),$$

$$\tilde{\mu} = \frac{\overline{X}}{\tilde{\alpha}} = \frac{s^2}{\overline{X}}. \tag{7.108}$$

Here (check)

$$M_1(\mu, \alpha) = \alpha\mu, \quad M_2(\mu, \alpha) = \alpha^2 \mu^2 + \alpha\mu^2;$$

$$H_1(z_1, z_2) = \frac{z_2 - z_1^2}{z_1}, \quad H_2(z_1, z_2) = \frac{z_1^2}{z_2 - z_1^2}. \tag{7.109}$$

Therefore, with $\boldsymbol{\theta} = (\mu, \alpha)$,

$$\ell_{11}(\boldsymbol{\theta}_0) = -\left(\frac{z_2 + z_1^2}{z_1^2} \right)_{z_1 = \alpha_0\mu_0, z_2 = (\alpha_0+1)\alpha_0\mu_0^2} = -\frac{2\alpha_0^2\mu_0^2 + \alpha_0\mu_0^2}{\alpha_0^2\mu_0^2}$$

$$= -\left(2 + \frac{1}{\alpha_0} \right),$$

$$\ell_{12}(\boldsymbol{\theta}_0) = \frac{1}{\alpha_0\mu_0}, \quad \ell_{22}(\boldsymbol{\theta}_0) = -\frac{\alpha_0^2\mu_0^2}{\alpha_0^2\mu_0^4} = -\frac{1}{\mu_0^2},$$

$$\ell_{21}(\boldsymbol{\theta}_0) = \frac{2\alpha_0\mu_0(\alpha_0\mu_0^2) - \alpha_0^2\mu_0^2(-2\alpha_0\mu_0)}{\alpha_0^2\mu_0^4} = \frac{2(1 + \alpha_0)}{\mu_0};$$

$$\sigma_{11}(\boldsymbol{\theta}_0) = \mathrm{var}_{\theta_0}(X_1) = E_{\theta_0} X_1^2 - (E_{\theta_0} X_1)^2 = \alpha_0\mu_0^2,$$

$$\sigma_{12}(\boldsymbol{\theta}_0) = E_{\theta_0} X_1^3 - (E_{\theta_0} X_1)(E_{\theta_0} X_1^2) = (\alpha_0 + 2)(\alpha_0 + 1)\alpha_0\mu_0^3 - (\alpha_0 + 1)\alpha_0^2\mu_0^3$$
$$= 2(\alpha_0 + 1)\alpha_0\mu_0^3 = \sigma_{21}(\boldsymbol{\theta}_0),$$

$$\sigma_{22}(\boldsymbol{\theta}_0) = E_{\theta_0} X_1^4 - (E_{\theta_0} X_1^2)^2 = (\alpha_0 + 3)(\alpha_0 + 2)(\alpha_0 + 1)\alpha_0\mu_0^4 - (\alpha_0 + 1)^2\alpha_0^2\mu_0^4$$
$$= 2(2\alpha_0 + 3)(\alpha_0 + 1)\alpha_0\mu_0^4. \tag{7.110}$$

$V(\boldsymbol{\theta})$ can be computed from (7.110). For example,

$$v_{11}(\boldsymbol{\theta}_0) = \sigma_{11}(\theta_0)\ell_{11}^2(\theta_0) + \sigma_{12}(\theta_0)(\ell_{11}(\theta_0)\ell_{12}(\theta_0)$$
$$+ \ell_{12}(\theta_0) + \ell_{11}(\theta_0)) + \sigma_{22}(\theta_0)\ell_{12}^2(\theta_0). \tag{7.111}$$

One may show that $\tilde{\boldsymbol{\theta}}$ is not asymptotically efficient in the above example. However, since $\sqrt{n}(\tilde{\boldsymbol{\theta}} - \boldsymbol{\theta}_0)$ is bounded in probability (under $P_{\boldsymbol{\theta}_0}$) one may obtain an asymptotically efficient estimator by adding a "correction" term, as in the one-parameter case (See Theorem 7.3). You may also check that the likelihood equations are intractable for this example.

Theorem 7.6. *Suppose the hypothesis of Theorem 7.1 holds, and that $\tilde{\boldsymbol{\theta}}_n$ is an estimator (sequence) of $\boldsymbol{\theta}$ such that $\sqrt{n}(\tilde{\boldsymbol{\theta}}_n - \boldsymbol{\theta}_0)$ is bounded in probability under $P_{\boldsymbol{\theta}_0}$ (for each $\boldsymbol{\theta}_0 \in \Theta$). Then*

$$\delta_n = \tilde{\boldsymbol{\theta}}_n - \left(\left(\frac{\partial^2 \log f_n(\mathbf{X}; \boldsymbol{\theta})}{\partial \theta_r \partial \theta_{r'}}\right)\right)^{-1}_{\tilde{\theta}_n} (\text{Grad } \log f_n(\mathbf{X}; \boldsymbol{\theta}))_{\tilde{\theta}_n} \qquad (7.112)$$

is an asymptotically efficient estimator of θ. Here

$$\text{Grad } \log f_n(\mathbf{x}; \boldsymbol{\theta}) = \begin{pmatrix} \partial \log f_n(\mathbf{x}; \boldsymbol{\theta})/\partial \theta_1 \\ \vdots \\ \partial \log f_n(\mathbf{x}; \boldsymbol{\theta})/\partial \theta_p \end{pmatrix}. \qquad (7.113)$$

Proof. Entirely analogous to the proof of Theorem 7.3. $\qquad\qquad\qquad\qquad$ $\square$

Example 7.7. To find an asymptotically efficient estimator of $\theta = (\mu, \alpha)$ in the above example, write $f(x; \mu, \alpha) = c(\mu, \alpha)e^{-x(1/\mu)+(\log x)\alpha}x^{-1}$, $0 < x < \infty$. Thus this is a two-parameter exponential family with $\theta_1 = 1/\mu$, $\theta_2 = \alpha$, $t_1(x) = -x$, $t_2(x) = \log x$. The likelihood equations, therefore, may be expressed as (See (7.78), (7.79))

$$EX_1 = \frac{1}{n}\sum_{j=1}^{n} X_j, \quad E\log X_1 = \frac{1}{n}\sum_{j=1}^{n}\log X_j,$$

or,

$$\alpha\mu = \frac{1}{n}\sum_{j=1}^{n} X_j, \quad E\log X_1 = \frac{1}{n}\sum_{j=1}^{n}\log X_j. \qquad (7.114)$$

But $E\log X_1$ cannot be computed in a tractable form. Hence the likelihood equations are impossible to solve explicitly. One may, therefore, take recourse to Theorem 7.6 with $\tilde{\theta}_n$ as the method-of-moments estimator given by (7.114): $\tilde{\theta}_{n1} = \frac{1}{\tilde{\mu}} = \frac{\overline{X}}{s^2}, \tilde{\theta}_{n2} = \tilde{\alpha} = \frac{\overline{X}}{s^2}$. One has

$$\partial \log f_n(\mathbf{x}; \boldsymbol{\theta})/\partial \theta_1 = n\frac{\partial \log C(\boldsymbol{\theta})}{\partial \theta_1} - \sum_{j=1}^{n} X_j = n\alpha\mu - \sum_{j=1}^{n} X_j,$$

$$\partial \log f_n(\mathbf{x}; \boldsymbol{\theta})/\partial \theta_2 = n\frac{\partial \log C(\boldsymbol{\theta})}{\partial \theta_2} + \sum_{j=1}^{n}\log X_j, \qquad (7.115)$$

with $\log C(\boldsymbol{\theta}) = \theta_2 \log \theta_1 - \log \Gamma(\theta_2)$. Hence $\partial \log C(\boldsymbol{\theta})/\partial \theta_1 = \theta_2/\theta_1$,

$$\partial \log C(\boldsymbol{\theta})/\partial \theta_2 = \log \theta_1 - \frac{1}{\Gamma(\theta_2)}\int_0^{\infty} e^{-x}x^{\theta_2-1}\log x \, dx,$$

$$\partial^2 \log C(\boldsymbol{\theta})/\partial \theta_1^2 = -\theta_2/\theta_1^2,$$

$$\partial^2 \log C(\boldsymbol{\theta})/\partial \theta_1 \partial \theta_2 = \frac{1}{\theta_1},$$

$$\partial^2 \log C(\boldsymbol{\theta})/\partial \theta_2^2 = \left(\frac{1}{\Gamma(\theta_2)}\right)^2 \left(\int_0^{\infty} e^{-x}x^{\theta_2-1}\log x \, dx\right)^2$$
$$- \frac{1}{\Gamma(\theta_2)}\left(\int_0^{\infty} e^{-x}x^{\theta_2-1}(\log x)^2 dx\right).$$

Since

$$\frac{\partial^2 \log f_n(\mathbf{x};\boldsymbol{\theta})}{\partial \theta_i \partial \theta_{i'}} = n\,\frac{\partial^2 \log C(\boldsymbol{\theta})}{\partial \theta_i \partial \theta_{i'}}\,. \tag{7.116}$$

The estimator δ_n in (7.115) may now be computed numerically. Note that δ_n is a first approximation to the solution of the likelihood Eq. (7.114).

7.6 Asymptotic Efficiency of Bayes Estimators

Suppose that the observation vector $\mathbf{X}$ in a statistical experiment has a p.d.f. $g(\mathbf{x};\boldsymbol{\theta})$ (w.r.t. a sigma-finite measure $\nu(d\mathbf{x})$), and $\boldsymbol{\theta} \in \Theta$—an open subset of $\mathbb{R}^p$. Let $\pi(d\boldsymbol{\theta})$ be a probability measure (the *prior*) on the Borel sigma-field of Θ. The Bayesian thinks of $\boldsymbol{\theta}$ as a random variable with distribution $\pi(d\boldsymbol{\theta})$, and $g(\mathbf{x};\boldsymbol{\theta})$ is regarded as the *conditional p.d.f. of* $\mathbf{X}$ *given* $\boldsymbol{\theta}$. Hence the joint distribution of $\mathbf{X}$ and $\boldsymbol{\theta}$ is

$$g(\mathbf{x};\boldsymbol{\theta})\nu(d\mathbf{x})\pi(d\boldsymbol{\theta}). \tag{7.117}$$

Let $\pi(d\theta|\mathbf{x})$ denote the *conditional distribution of* θ *given* $\mathbf{X} = \mathbf{x}$, i.e.,

$$\pi(d\theta|\mathbf{x}) = \frac{g(\mathbf{x};\theta)\pi(d\theta)}{\int_\Theta g(\mathbf{x};\theta')\pi(d\theta')} = \frac{g(\mathbf{x};\theta)\pi(d\theta)}{\overline{g}(\mathbf{x})}. \tag{7.118}$$

where $\overline{g}(\mathbf{x})$ is the *marginal p.d.f. of* $\mathbf{X}$:

$$\overline{g}(\mathbf{x}) = \int_\Theta g(\mathbf{x};\boldsymbol{\theta})\pi(d\boldsymbol{\theta}). \tag{7.119}$$

Then $\pi(d\theta|\mathbf{x})$ is called the *posterior distribution* of θ given $\mathbf{X} = \mathbf{x}$.

In this section it will be shown that, under the regularity conditions of Theorem 7.5, and with respect to a prior $\pi(d\theta)$ having a positive density on Θ (with respect to Lebesgue measure on Θ) the posterior distribution of θ is asymptotically normal with mean $\hat{\theta}(\mathbf{x})$ (the *maximum likelihood estimator*, i.e., the consistent solution of the likelihood equation) and covariance matrix $g'^{-1}(\boldsymbol{\theta}')/n$, a.s. $(P_{\boldsymbol{\theta}'})$ for every $\boldsymbol{\theta}'$. Thus, irrespective of the prior density, the Bayes estimator is asymptotically the same as the MLE. We will outline the proof whose details may be found in Bickel and Doksum (2001, pp. 337–345).

Assume for simplicity that $p = 1$ and $\Theta = (a,b)$, $-\infty \le a < b \le \infty$, and $\pi(d\theta)$ has a positive and continuous density on (a,b). Let $X_1,\ldots,X_n$ be i.i.d. with common p.d.f. $f(x;\theta)$, $\mathbf{X} = (X_1,\ldots,X_n)$ and $f_n(\mathbf{x};\theta) = \prod_{i=1}^n f(x_i;\theta)\ \forall\ \mathbf{x} = (x_1,\ldots,x_n)$. The posterior density of $\sqrt{n}(\theta - \hat{\theta}_n)$ at t is

$$\pi\!\left(\hat{\theta}_n + \frac{t}{\sqrt{n}} \mid \mathbf{X}\right) = \frac{\pi(\hat{\theta}_n + \frac{t}{\sqrt{n}})f_n(\mathbf{X};\hat{\theta}_n j + \frac{t}{\sqrt{n}})}{\int \pi(\hat{\theta}_n + \frac{s}{\sqrt{n}})f_n(\mathbf{X};\hat{\theta}_n + \frac{s}{\sqrt{n}})ds}, \tag{7.120}$$

making the change of variables $\theta \to s = (\theta - \hat{\theta}_n)/\sqrt{n}$. Now, given $\theta = \theta'$,

$$\log f_n\!\left(\mathbf{X};\hat{\theta}_n + \frac{s}{\sqrt{n}}\right) = \log f_n(\mathbf{X};\hat{\theta}_n) + \frac{s^2}{2n}\sum_{i=1}^n \frac{d^2 \log f(X_i;\theta)}{d\theta^2}\bigg|_{\hat{\theta}_n} + R_n$$

$$= \log f_n(\mathbf{X};\hat{\theta}_n) + \frac{s^2}{2}\,\mathscr{I}_1(\hat{\theta}_n) + R'_n, \tag{7.121}$$

where R_n, R'_n go to zero in $P_{\theta'}$-probability as $n \to \infty$, uniformly for $|s| \le M\sqrt{n}$, for every $M > 0$. For simplicity assume this convergence is a.s. $(P_{\theta'})$. In particular, $R'_n \to 0$ for every s a.s. $(P_{\theta'})$. One may now express (7.121) as

$$f_n\left(\mathbf{X}; \hat{\theta}_n + \frac{s}{\sqrt{n}}\right) = f_n(\mathbf{X}; \hat{\theta}_n) \exp\left\{\frac{s^2}{2}\mathscr{I}_1(\hat{\theta}_n)\right\}(1 + o(1)) \ \forall\, s, \qquad (7.122)$$

a.s. $(P_{\theta'})$. Applying this to (7.120) one arrives at

$$
\begin{aligned}
\pi\left(\hat{\theta}_n + \frac{t}{\sqrt{n}}\,\middle|\,\mathbf{X}\right) &= \frac{\pi(\hat{\theta}_n + \frac{t}{\sqrt{n}})f_n(\mathbf{X}; \hat{\theta}_n)\exp\{-\frac{t^2}{2}\mathscr{I}_1(\hat{\theta}_n)\}(1 + o(1))}{\int \pi(\hat{\theta}_n + \frac{s}{\sqrt{n}})f_n(\mathbf{X}; \hat{\theta}_n)\exp\{-\frac{s^2}{2}\mathscr{I}_1(\hat{\theta}_n)\}(1 + o(1))ds} \\[2mm]
&= \frac{\pi(\hat{\theta}_n + \frac{t}{\sqrt{n}})\exp\{-\frac{t^2}{2}\mathscr{I}_1(\hat{\theta}_n)\}(1 + o(1))}{\int \pi(\hat{\theta}_n + \frac{s}{\sqrt{n}})\exp\{-\frac{s^2}{2}\mathscr{I}_1(\hat{\theta}_n)\}(1 + o(1))ds} \\[2mm]
&\longrightarrow \frac{\exp\{-\frac{t^2}{2}\mathscr{I}_1(\theta')\}}{\int \exp\{-\frac{s^2}{2}\mathscr{I}_1(\theta')\}ds} = \frac{\exp\{-\frac{t^2}{2}\mathscr{I}_1(\theta')\}}{\sqrt{2\pi/\mathscr{I}_1(\theta')}} \\[2mm]
&\approx \sqrt{\frac{\mathscr{I}_1(\hat{\theta}_n)}{2\pi}}\exp\left\{-\frac{t^2}{2}\mathscr{I}_1(\hat{\theta}_n)\right\},
\end{aligned}
\qquad (7.123)
$$

where the difference between the two sides of $\approx$ goes to zero a.s. $(P_{\theta'})$.

Thus the total variation distance between the posterior distribution of θ and the Normal distribution $N(\hat{\theta}_n, \frac{1}{n}\mathscr{I}_1(\hat{\theta}_n))$, or $N(\hat{\theta}_n, \frac{1}{n}\mathscr{I}_1(\theta'))$, goes to zero as $n \to \infty$, a.s. $P_{\theta'}$, for every (true) parameter value θ'. The posterior distribution concentrates most of its mass near the maximum likelihood estimator $\hat{\theta}_n$. In particular, the asymptotic mean and the asymptotic median of the posterior are both $\hat{\theta}_n$.

We state the result for the general multi-parameter case for which the proof is not substantially different from that outlined above.

Theorem 7.7 (Bernstein–von Mises Theorem). *If the assumptions (A_1)– (A_5) of Theorem 7.5 hold and the prior π of $\boldsymbol{\theta}$ has a continuous and positive density on Θ then, under $P_{\boldsymbol{\theta}'}$, the total variation distance between the posterior distribution of $\boldsymbol{\theta}$ and the Normal distribution $N(\hat{\boldsymbol{\theta}}_n, \mathscr{I}^{-1}(\boldsymbol{\theta}')/n)$ converges to zero as $n \to \infty$, a.s. $(P_{\boldsymbol{\theta}'})$.*

Example 7.8. Let $X_1, \ldots, X_n$ be i.i.d. Bernoulli with $P_\theta(X_i = 1) = \theta$, $P_\theta(X_i = 0) = 1 - \theta$, $\theta \in \Theta = (0,1)$. Let $\pi(\theta) =$ Beta density with parameters $\alpha, \beta : \pi(\theta) = \frac{\Gamma(\alpha+\beta)}{\Gamma(\alpha)\Gamma(\beta)}\theta^{\alpha-1}(1 - \theta)^{\beta-1}$ $(\alpha > 0, \beta > 0)$. Then

$$f_n(\mathbf{x}; \theta) = \theta^{\sum_1^n x_i}(1 - \theta)^{n - \sum_1^n x_i},$$

$$\pi_n(\theta|\mathbf{x}) = \frac{\theta^{\sum_{i=1}^n x_i + \alpha - 1}(1 - \theta)^{n - \sum_1^n x_i + \beta - 1}}{\Gamma(\sum_1^n x_i + \alpha - 1)\Gamma(n - \sum_1^n x_i + \beta - 1)/\Gamma(n + \alpha + \beta - 2)},$$

i.e., the posterior distribution is a beta distribution with parameters $\sum x_i + \alpha$ and $n - \sum x_i + \beta$, so that one can directly show that a beta random variable Y with this distribution is asymptotically Normal $N(\hat{\theta}_n, 1/n\, I_1(\theta))$ under P_θ, a.s. as $n \to \infty$ (Exercise 7.12).

7.7 Asymptotic Normality of M-estimators

Let X_i $(i = 1, 2, \ldots, n)$ be i.i.d. with values in $(\mathscr{X}, \mathscr{S})$ defined on $(\Omega, \mathscr{F}, P_\theta)$.

An M-estimator (or a maximum likelihood type estimator), a term coined by Huber (1981), may be defined to be an estimator $\hat{\theta}_n = \hat{\theta}_n(X_1, \ldots, X_n)$ which maximizes (on Θ) a real-valued function of the form

$$g_n(\theta) := \frac{1}{n}\sum_{i=1}^{n} g(X_i, \theta), \qquad (\theta \in \Theta \text{ open } \subset \mathbb{R}^k) \tag{7.124}$$

where $g(x, \theta) : \mathscr{X} \times \Theta \to \mathbb{R}$ is measurable (w.r.t. the given sigma-field $\mathscr{S}$ on $\mathscr{X}$) for each $\theta \in \Theta$. Just as in the case of the MLE, it is more common to define an M-estimator as a critical point of $g_n(\theta)$, i.e., an estimator which satisfies

$$\frac{\partial g_n(\theta)}{\partial \theta_r} = 0, \qquad 1 \le r \le k. \tag{7.125}$$

Writing $h(x, \theta) = \operatorname{grad} g(x, \theta) \equiv (\partial g(x, \theta)/\partial \theta_1, \ldots, \partial g(x, \theta)/\partial \theta_k)$, one may rewrite (7.125) as

$$h_n^{(r)}(\theta) = 0 \qquad (1 \le r \le k), \tag{7.126}$$

where

$$h_n^{(r)}(\theta) = \frac{1}{n}\sum_{i=1}^{n} h^{(r)}(X_i, \theta), \qquad 1 \le r \le k. \tag{7.127}$$

More generally, we define an M-estimator $\hat{\theta}_n$ to be a solution of a vector equation

$$h_n(\theta) = 0 \qquad [\text{i.e., } h_n^{(r)}(\theta) = 0 \quad (1 \le r \le k)] \tag{7.128}$$

where $h_n(\theta) \equiv (h_n^{(1)}(\theta, \ldots, h_n^{(k)}(\theta))'$ is defined by (7.127), $h^{(r)}(x, \theta)$ being, for each $r\ (= 1, 2, \ldots, k)$, a real-valued (measurable) function on $\mathscr{X} \times \Theta$.

Theorem 7.8. *Let Θ be an open subset of $\mathbb{R}^k$, and let $\hat{\theta}_n$ be a consistent solution of (7.128). Assume (i) $E_\theta h(X_1, \theta) = 0 \ \forall\ \theta \in \Theta$, (ii) $\theta \to h(x, \theta)$ is twice continuously differentiable on Θ, for every $x \in \mathscr{X}$, (iii) $\Gamma(\theta) \equiv E_\theta \operatorname{Grad} h(X_1, \theta)$ is a nonsingular $k \times k$ matrix $(\forall\ \theta \in \Theta)$, whose r-th row is given by*

$$E_\theta(\operatorname{grad} h^{(r)}(X_1, \theta))' \equiv (E_\theta(\partial h^{(r)}(X_1, \theta)/\partial \theta_1), \ldots, E_\theta(\partial h^{(r)}(X_1, \theta)/\partial \theta_k)),$$

(iv) $V(\theta) \equiv \operatorname{Cov}_\theta h(X_1, \theta)$ is finite and nonsingular $\forall\ \theta \in \Theta$, and (v) for each $\theta \in \Theta$, there exists $\delta(\theta) > 0$ such that $\sup_{\{\theta' : |\theta' - \theta| \le \delta(\theta)\}} \left| \dfrac{\partial^2 h^{(r)}(x, \theta)}{\partial \theta_i \partial \theta_j} \right|_{\theta'} \le b_{ij}^{(r)}(x)$ where

$$E_\theta b_{ij}^{(r)}(X_1) < \infty \qquad \forall\ 1 \le r, i, j \le k. \tag{7.129}$$

Then

$$\sqrt{n}(\hat{\theta}_n - \theta_0) \xrightarrow{\mathscr{L}} N(0, \Gamma^{-1}(\theta_0)V(\theta_0)(\Gamma^{-1}(\theta_0))')$$

under P_{θ_0} $(\forall\ \theta_0 \in \Theta)$.

Proof. For simplicity of notation, let us first consider the case $k = 1$. Then h is real-valued and θ is one-dimensional. One has, by a Taylor expansion,

$$0 = h_n(\hat{\theta}_n) = h_n(\theta_0) + (\hat{\theta}_n - \theta_0)\left(\frac{dh_n(\theta)}{d\theta}\right)_{\theta=\theta_0} + \frac{1}{2}(\hat{\theta}_n - \theta_0)^2\left(\frac{d^2h_n(\theta)}{d\theta^2}\right)_{\theta=\theta_n^*}$$

$$(7.130)$$

on the set $A_n = \{|\hat{\theta}_n - \theta_0| < \delta(\theta_0)\}$, where $|\theta_n^* - \theta_0| \le |\hat{\theta}_n - \theta_0| < \delta(\theta_0)$. Let $B_n = \left\{\left(\frac{dh_n(\theta)}{d\theta}\right)_{\theta_0} + \frac{1}{2}(\hat{\theta}_n - \theta_0)\left(\frac{d^2h_n(\theta)}{d\theta^2}\right)_{\theta_n^*} \ne 0\right\}$. Then $P_{\theta_0}(A_n) \to 1$, by consistency of $\hat{\theta}_n$. Also, under P_{θ_0},

$$\left(\frac{dh_n(\theta)}{d\theta}\right)_{\theta_0} \equiv \frac{1}{n}\sum_{i=1}^{n}\left(\frac{dh(X_i,\theta)}{d\theta}\right)_{\theta_0} \longrightarrow E_{\theta_0}\left(\frac{dh(X_1,\theta)}{d\theta}\right)_{\theta_0} = \Gamma(\theta_0) \quad \text{a.s.}$$

$$(7.131)$$

Moreover, on A_n (using (v) and writing b for b_{11}),

$$\left|\left(\frac{d^2h_n(\theta)}{d\theta^2}\right)_{\theta_n^*}\right| \le \frac{1}{n}\sum_{i=1}^{n}b(X_i),$$

so that, under P_{θ_0},

$$\overline{\lim_{n\to\infty}}\left|\left(\frac{d^2h_n(\theta)}{d\theta^2}\right)_{\theta_n^*}\right| \le \overline{\lim_{n\to\infty}}\frac{1}{n}\sum_{i=1}^{n}b(X_i) = E_{\theta_0}b(X_1) < \infty.$$

It follows that, under P_{θ_0},

$$\left(\frac{dh_n(\theta)}{d\theta}\right)_{\theta_0} + \frac{1}{2}(\hat{\theta}_n - \theta_0)\left(\frac{d^2h_n(\theta)}{d\theta^2}\right)_{\theta_n^*} \quad \text{converges in } (P_{\theta_0^-}) \text{ probability to } \Gamma(\theta_0).$$

$$(7.132)$$

In particular, $P_{\theta_0}(B_n) \to 1$, and (7.130) yields

$$\sqrt{n}(\hat{\theta}_n - \theta_0) = \frac{-\sqrt{n}\,h_n(\theta_0)}{\Gamma(\theta_0) + o_p(1)} \xrightarrow{\mathscr{L}} N(0, \Gamma^{-2}(\theta_0)\,V(\theta_0)), \tag{7.133}$$

since $-\sqrt{n}\,h_n(\theta_0) = \frac{1}{\sqrt{n}}\sum_{i=1}^{n}(-h(X_i,\theta_0))$ and $-h(X_i,\theta_0)$ has mean zero and variance $V(\theta_0)$, under P_{θ_0}.

For $k > 1$ one may write, in place of (7.130)

$$0 = h_n^{(r)}(\hat{\boldsymbol{\theta}}_n) \tag{7.134}$$
$$= h_n^{(r)}(\boldsymbol{\theta}_0) + (\operatorname{grad} h_n^{(r)}(\boldsymbol{\theta}_0))'(\hat{\boldsymbol{\theta}}_n - \boldsymbol{\theta}_0)$$
$$+ \frac{1}{2}(\hat{\boldsymbol{\theta}}_n - \boldsymbol{\theta}_0)'H_n^{(r)}(\boldsymbol{\zeta}_n^{(r)})(\hat{\boldsymbol{\theta}}_n - \boldsymbol{\theta}_0) \qquad (1 \le r \le k),$$

where $|\boldsymbol{\zeta}_n^{(r)} - \boldsymbol{\theta}_0| \le \delta(\boldsymbol{\theta}_0)$, and $H_n^{(r)}(\boldsymbol{\theta}) = \left(\left(\frac{\partial^2}{\partial\theta_i\partial\theta_j}h_n^{(r)}(\boldsymbol{\theta})\right)\right)_{1\le i,j\le k}$. In matrix notation one may express (7.134) as

$$0 = h_n(\boldsymbol{\theta}_0) + \operatorname{Grad} h_n(\boldsymbol{\theta}_0)(\hat{\boldsymbol{\theta}}_n - \boldsymbol{\theta}_0) + \left[\frac{1}{2}(\hat{\boldsymbol{\theta}}_n - \boldsymbol{\theta}_0)'H_n^{(r)}(\boldsymbol{\zeta}_n^{(r)})\right]_{1\le r\le k}(\hat{\boldsymbol{\theta}}_n - \boldsymbol{\theta}_0)$$

$$(7.135)$$

where $[\ \ ]_{1\le r\le k}$ is a $k \times k$ matrix whose r-th column is $\frac{1}{2}(\hat{\boldsymbol{\theta}}_n - \boldsymbol{\theta}_0)'H_n^{(r)}(\boldsymbol{\zeta}_n^{(r)})$. By assumption (v) and the consistency of $\hat{\boldsymbol{\theta}}_n$, all elements of this matrix converge to zero in probability, by the same argument as in the case $k = 1$. Hence one may rewrite (7.135) as

$$-\sqrt{n}\,h_n(\boldsymbol{\theta}_0) = \left\{\mathrm{Grad}\,h_n(\boldsymbol{\theta}_0) + \left[\frac{1}{2}(\hat{\boldsymbol{\theta}}_n - \boldsymbol{\theta}_0)' H_n^{(r)}(\boldsymbol{\zeta}_n^{(r)})\right]_{1\le r\le k}\right\}\sqrt{n}(\hat{\boldsymbol{\theta}}_n - \boldsymbol{\theta}_0),$$
$$(7.136)$$

or, on the set $A_n \cap B_n$, with $B_n =$ the expression $\{\ \}$ within curly brackets in (7.136) is nonsingular,

$$\sqrt{n}(\hat{\boldsymbol{\theta}}_n - \boldsymbol{\theta}_0) = \left\{\qquad\right\}^{-1}(-\sqrt{n}\,h_n(\boldsymbol{\theta}_0)). \qquad (7.137)$$

Since (7.124) $-\sqrt{n}\,h_n(\boldsymbol{\theta}_0) \to N(0, V(\boldsymbol{\theta}_0))$, under $P_{\boldsymbol{\theta}_0}$ (by CLT), and $\{\ \}$ converges to $\Gamma(\boldsymbol{\theta}_0)$ in $P_{\boldsymbol{\theta}_0}$-probability, it follows that

$$\sqrt{n}(\hat{\boldsymbol{\theta}}_n - \boldsymbol{\theta}_0) \xrightarrow{\mathscr{L}} \Gamma^{-1}(\boldsymbol{\theta}_0)\mathbf{Z}, \qquad \mathbf{Z} \overset{\mathscr{L}}{\sim} N(\mathbf{0}, V(\boldsymbol{\theta}_0)).$$

$\square$

Remark 7.13. The hypothesis of Theorem 7.8 may be relaxed. Instead of assuming $\theta \to h(x, \theta)$ is twice continuously differentiable (and (7.129) holds), it is enough to assume that $\theta \to h(x; \theta)$ is (once) continuously differentiable, and that for each $\theta_0 \in \Theta$, and every $\varepsilon > 0$, there exists $\delta(\varepsilon, \theta_0) > 0$ such that

$$\sup\left\{\left|\left(\frac{\partial h^{(r)}(x, \theta^n)}{\partial\theta_i}\right)_{\theta'} - \left(\frac{\partial h^{(r)}(x, \theta)}{\partial\theta_i}\right)_{\theta_0}\right| : |\theta' - \theta_0| \le \delta(\varepsilon, \theta_0)\right\} \le \varepsilon b_i(x),$$

with $E_{\theta_0} b_i(X_1) < \infty \ \forall\ i$.

Example 7.9 (MLE). Let $f(x; \boldsymbol{\theta})$ be the p.d.f. of X_i (under $P_{\boldsymbol{\theta}}$) w.r.t. a σ-finite measure $\mu(dx)$. The log likelihood equations for the MLE $\hat{\boldsymbol{\theta}}_n$ are

$$\sum_{i=1}^{n} \frac{\partial \log f(X_i; \boldsymbol{\theta})}{\partial\theta_r} = 0 \qquad (1 \le r \le k), \qquad (7.138)$$

which one may write as (7.126) or (7.128), with $h^{(r)}(x, \boldsymbol{\theta}) = \partial \log f(x; \boldsymbol{\theta})/\partial\theta_r$ $(1 \le r \le k)$.

Example 7.10 (Method of Moments). Let X_i be real-valued, and

$$m_r(\boldsymbol{\theta}) = EX_i^r \qquad (r = 1, 2, \ldots, k) \qquad (7.139)$$

finite. The *method of moments* provides an estimator $\hat{\boldsymbol{\theta}}_n$ which solves the equations

$$h_n^{(r)}(\boldsymbol{\theta}) = \frac{1}{n}\sum_{i=1}^{n} X_i^r - m_r(\boldsymbol{\theta}) = 0 \qquad (1 \le r \le k). \qquad (7.140)$$

Here $h^{(r)}(x, \boldsymbol{\theta}) = X^r - m_r(\boldsymbol{\theta})$, $1 \le r \le k$. More generally, one may choose k functions $\psi_r(x)$, $1 \le r \le k$, such that $E_{\boldsymbol{\theta}}|\psi_r(X_1)| < \infty$, and solve (for θ)

$$h_n^{(r)}(\boldsymbol{\theta}) \equiv \frac{1}{n}\sum_{i=1}^{n} \psi_r(X_i) - E_{\boldsymbol{\theta}}\psi_r(X_1) = 0 \qquad (1 \le r \le k), \qquad (7.141)$$

so that $h^{(n)}(x, \boldsymbol{\theta}) = \psi_r(x) - E_{\boldsymbol{\theta}}\psi_r(X_1)$.

7.8 Asymptotic Efficiency and Super Efficiency

When does the variance of an unbiased estimator T_n attain the lower bound given by the information inequality? Under (R_0)–(R_2), this happens if and only if the coefficient of correlation between $T_n - E_\theta T_n$ and $\frac{d \log f_n(\mathbf{X};\theta)}{d\theta}$ is $+1$ or -1, i.e.,

$$T_n(\mathbf{x}) - E_\theta T_n = \lambda_n(\theta) \frac{d \log f_n(\mathbf{x};\theta)}{d\theta}$$

with P_θ-probability one for some λ_n. Thus

$$\log f_n(\mathbf{x};\theta) = \int \frac{(T_n(\mathbf{x}) - g_n(\theta))}{\lambda_n(\theta)}\, d\theta = \pi_n(\theta) T_n(\mathbf{x}) - \gamma_n(\theta) + \varphi_n(\mathbf{x}).$$

In view of the information inequality (7.1) or (7.5), the following definition is reasonable.

Definition 7.3. Under (R_0)–(R_2) a sequence of estimators T_n $(n = 1, 2, \ldots)$ of $g(\theta)$ is *asymptotically efficient (in Fisher's sense)* if

$$\sqrt{n}(T_n - g(\theta)) \xrightarrow{\mathscr{L}} N(0, \sigma^2(\theta)) \tag{7.142}$$

with

$$\sigma^2(\theta) = \frac{1}{E_\theta \left(\frac{d \log f(X_1;\theta)}{d\theta} \right)^2}. \tag{7.143}$$

Remark 7.14. Note that an asymptotically efficient T_n in the above sense is asymptotically normal with mean $g(\theta)$ and variance equal to the information lower bound. This *does not* necessarily *imply that $g(\theta)$ is the mean of T_n*, i.e., T_n *is unbiased*. But it implies that T_n *is consistent.* Suppose $E_\theta T_n = g_n(\theta)$ and var $E_\theta T_n^2 \to 0$. Then one may show by Chebyshev's inequality that $T_n - g_n(\theta) \xrightarrow{P} 0$. But by consistency of T_n as an estimator of $g(\theta)$, $T_n - g(\theta) \xrightarrow{P} 0$. Therefore, the *bias* $g_n(\theta) - g(\theta) \to 0$ as $n \to \infty$. Note that (7.5) also implies for biased estimators T_n

$$E_\theta(T_n - g(\theta))^2 \geq \operatorname{var}_\theta T_n \geq \frac{\left(\frac{d}{d\theta} E_\theta T_n \right)^2}{n E_\theta \left(\frac{d \log f(X_1;\theta)}{d\theta} \right)^2}$$

$$= \frac{\left[\frac{d}{d\theta}(g(\theta) + E_\theta T_n - g(\theta)) \right]^2}{n E_\theta \left(\frac{d \log f(X_1;\theta)}{d\theta} \right)^2} = \frac{(g'(\theta) + b_n'(\theta))^2}{n E_\theta \left(\frac{d \log f(X_1;\theta)}{d\theta} \right)^2}, \tag{7.144}$$

where $b_n(\theta) = E_\theta T_n - g(\theta)$ is the *bias*. Since in the present case $b_n(\theta) \to 0$ as $n \to \infty$, it is usually true that $b_n'(\theta) \to 0$ as $n \to \infty$. Hence, from that asymptotic point of view, one need not restrict attention only to unbiased estimators in order to compare efficiencies in Fisher's sense.

Remark 7.15. Relation (7.142) *does not imply*

$$E_\theta \left(\sqrt{n}(T_n - g(\theta)) \right)^2 \longrightarrow \sigma^2(\theta) \quad \text{as } n \to \infty. \tag{7.145}$$

For the function $f(x) = x^2$ is *not bounded*. Thus $\frac{\sigma^2(\theta)}{n}$ is *not*, in general, *the asymptotic variance of T_n; it is the variance of the asymptotic distribution of T_n.* In general one can only prove

$$\lim_{n\to\infty} E_\theta \left(\sqrt{n}(T_n - g(\theta))\right)^2 \geq \sigma^2(\theta), \tag{7.146}$$

if (7.142) holds. (See Lehmann, *Theory of Point Estimation*, Lemma 5.1.2).

Remark 7.16. For statistical experiments satisfying (R_0)–(R_2) the information inequality (7.5) holds. However, the equality in (7.5) rarely holds. We have seen that the equality in (7.5), under mild additional conditions, implies that $f(x;\theta)$ belongs to the exponential family. Even in the exponential case, however, the equality holds only for the estimation of $\frac{1}{n}E_\theta T_n(\mathbf{x}) = (\frac{d}{d\pi}) \log \int h(x)e^{\pi T_1(x)}\mu(dx)$ or some linear function of it, and *not* for other parametric functions. On the other hand, it will be shown that there are asymptotically efficient estimators of every 'smooth' parametric function in the exponential family case and, more generally, when (R_0)–(R_2) and some additional conditions are satisfied.

Remark 7.17 (Superefficiency). Let $X_1, X_2, \ldots, X_n$ be i.i.d. $N(\theta,1)$. Hodges pointed out that the estimator

$$T_n = \begin{cases} \overline{X} = \frac{1}{n}\sum_{j=1}^{n} X_j & \text{for } |\overline{X}| > n^{-\frac{1}{4}} \\ \alpha\overline{X} & \text{for } |\overline{X}| \leq n^{-\frac{1}{4}}, \end{cases} \tag{7.147}$$

where α is a constant, $0 \leq \alpha < 1$, satisfies (where θ is the true value)

$$\begin{aligned} \sqrt{n}(T_n - \theta) &\xrightarrow{\mathscr{L}} N(0,1) \quad \text{if } \theta \neq 0, \\ \sqrt{n}(T_n - 0) &\longrightarrow N(0,\alpha^2) \quad \text{if } \theta = 0. \end{aligned} \tag{7.148}$$

In other words, at all $\theta \neq 0$ the estimator T_n is asymptotically efficient; and at $\theta = 0$ the variance of the asymptotic distribution of $\sqrt{n}(T_n - 0)$ is α^2, which is *smaller than* the information lower bound 1 (note that $\sqrt{n}(\overline{X} - \theta) \to N(0,1)$ for all θ). Such an estimator has come to be known as *superefficient*, and the point $\theta = 0$ as a *point of superefficient* of T_n. To check (7.148) note that, for $\theta \neq 0$,

$$P_\theta \left(|\sqrt{n}(T_n - \theta) - \sqrt{n}(\overline{X} - \theta)| > 0\right)$$

$$= P_\theta \left(|\overline{X}| \leq n^{-\frac{1}{4}}\right) = \int_{-n^{-\frac{1}{4}}}^{n^{-\frac{1}{4}}} \frac{\sqrt{n}}{\sqrt{2\pi}} e^{-n(x-\theta)^2/2}dx$$

$$= \frac{1}{\sqrt{2\pi}} \int_{-\sqrt{n}(-n^{-\frac{1}{4}}-\theta)}^{\sqrt{n}(n^{-\frac{1}{4}}-\theta)} e^{-y^2/2}dy \leq \frac{1}{\sqrt{2\pi}} \left(2\sqrt{n}\,n^{\frac{1}{4}}e^{-(|\theta\sqrt{n}|-n^{\frac{1}{4}})^2/2}\right)$$

$$= \frac{2n^{\frac{1}{2}}}{\sqrt{2\pi}} e^{\frac{n^{\frac{1}{2}}}{2}(|n^{\frac{1}{4}}\theta|-1)^2} = n^{\frac{1}{4}} \cdot o\left(e^{-n^{\frac{1}{2}}/2}\right) \longrightarrow 0 \text{ as } n \to \infty. \tag{7.149}$$

This means that

$$\sqrt{n}(T_n - \theta) = \sqrt{n}(\overline{X} - \theta) + R_n \quad \text{with } R_n \xrightarrow{P} 0. \tag{7.150}$$

Since $\sqrt{n}(\overline{X} - \theta)$ is $N(0,1)$, the first relation in (7.148) follows (by Theorem 6.1). Now, if $\theta = 0$, then

$$\sqrt{n}\,T_n = \alpha\sqrt{n}\,\overline{X} \cdot I_{\{|\overline{X}|\leq n^{-\frac{1}{4}}\}} + R'_n, \quad R'_n = \sqrt{n}\,\overline{X} \cdot I_{\{|\overline{X}|>n^{-\frac{1}{4}}\}}. \tag{7.151}$$

Since $\theta = 0$, $\sqrt{n}\,\overline{X}$ is $N(0,1)$. Also,

$$\text{Prob}\left(I_{\{|\overline{X}|\leq n^{-\frac{1}{4}}\}} = 1\right) = \text{Prob}\left(|\overline{X}| \leq n^{-\frac{1}{4}}\right) = \frac{1}{\sqrt{2\pi}}\int_{-n^{\frac{1}{4}}}^{n^{\frac{1}{4}}} e^{-y^2/2}dy \longrightarrow 1$$
$$\text{as } n \to \infty;$$
$$\text{Prob}(R'_n = 0) = \text{Prob}\left(|\overline{X}| \leq n^{-\frac{1}{4}}\right) \longrightarrow 1 \text{ as } n \to \infty. \tag{7.152}$$

Therefore, (by Theorem 6.1) $\sqrt{n}\,T_n \to \alpha Z \cdot 1 + 0 = \alpha Z$ where Z is $N(0,1)$, proving (7.148). Hence the first term on the right side in (7.151) converges in distribution to αZ, where Z is $N(0,1)$, while the second term goes to $Z \cdot 0 = 0$ in probability. Thus, under $\theta = 0$, $\sqrt{n}\,T_n \xrightarrow{\mathscr{L}} N(0,\alpha^2)$.

A calculation of the risk function $\mathscr{R}_n(\theta, T_n) \equiv E_\theta(T_n - \theta)^2$ shows, however, that (Exercise 7.22)

$$\mathscr{R}_n(\theta, T_n) = \frac{A_n(\theta)}{n} \quad \text{with} \quad \sup_{|\theta|\leq n^{-\frac{1}{4}}} A_n(\theta) = \infty, \tag{7.153}$$

whereas

$$\mathscr{R}_n(\theta, \overline{X}) = \frac{1}{n} \quad \forall\, \theta. \tag{7.154}$$

Thus an event near the point of superefficiency T_n behaves poorly compared to $\overline{X}$, if one would like to control the risk uniformly for all θ in at least a small neighborhood of each parameter point.

Remark 7.18. Under slightly stronger conditions than (R_0)–(R_2) it has been proved by Le Cam (1953), and by Bahadur (1958), that if an estimator T_n of θ is asymptotically $N(\theta, \frac{\sigma^2(\theta)}{n})$ (i.e., $\sqrt{n}(T_n - \theta) \xrightarrow{\mathscr{L}} N(0,\sigma^2(\theta))$), then the points of superefficiency (i.e., θ such that $\sigma^2(\theta) < 1/E_\theta((d\log f(X_1;\theta)/d\theta)^2)$ constitute at most a set of Lebesgue measure zero). If one requires that the normalized risk function $n\mathscr{R}_n(\theta, T_n)$ perform *uniformly* well in an interval of width $O(n^{-\frac{1}{2}})$ around each parameter point θ_0, then under the same regularity conditions LeCam and Hajek proved that there do not exist estimators which are superior to asymptotically normal estimators.

Exercises for Chap. 7

Exercises for 7.2, 7.3

Ex. 7.1. Let $\mathbf{X} = (X_1,\ldots,X_n)$ where X_i's are i.i.d. uniform on $(0,\theta)$ ($\theta \in \Theta = (0,\infty)$). Show that

(a) $M_n \equiv \max\{X_1,\ldots,X_n\}$ is the MLE of θ, and
(b) $E_\theta M_n = \frac{n}{n+1}\theta$, $\text{var}_\theta M_n = \frac{n}{(n+2)(n+1)^2}\theta^2$, $E_\theta\left(\frac{n+1}{n}M - \theta\right)^2 = \frac{1}{n(n+2)}\theta^2$.

Ex. 7.2. (a) Show that the hypothesis of Theorem 7.1 holds if (R_0), (R_1), (R_2) stated in Remark 7.3 hold.
(b) Show that the hypothesis of Theorem 7.1 holds for one-parameter exponential families as stated in Remark 7.5.

(c) In the context of Remark 7.5 show that the Cramér-Rao lower bound is only attained by linear functions of T.

Ex. 7.3. Show that *truncated exponential families* are also exponential.

Ex. 7.4 (Dasgupta 2008). *(Missing data)*. In a city the number of traffic accidents per day is assumed to follow the Poisson distribution $\mathscr{P}(\lambda)$.

(a) Suppose records are available only for days with more than r accidents, and the number of days with r or less accidents is not known. Show that Theorem 7.2 applies for the estimation of λ, and find the asymptotic distribution of the MLE $\hat{\lambda}$. Do not attempt to find closed form expressions. [Hint: Use the truncated $\mathscr{P}(\lambda)$.]

(b) Suppose the over a period of n days $n - m$ days have more than r accidents and the numbers of accidents on these days are recorded. For the remaining m days with r or less accidents the actual numbers of accidents are not recorded. Regarding m as random write down the likelihood equation. Show that Theorem 7.2 applies for the MLE.

(c) One way of dealing with the missing observations in (b) is to compute $\hat{\lambda}$ as in (a) based on the $n - m$ observations fully recorded, and then replace the missing data by a random sample of size m drawn from $\mathscr{P}(\hat{\lambda})$. Considering the n observations so obtained as a random sample from $\mathscr{P}(\lambda)$, the "MLE" λ^* is computed. As an example, compute numerically the estimates of λ in (a) (with $n - m$ observations), (b), (c), from a random sample of size $n = 100$ from $\mathscr{P}(10)$, letting $r = 2$. Do the same with $r = 1$.

Ex. 7.5 (Bhattacharyya Inequality). Suppose the pd.f. $g(\mathbf{x}; \theta)$ of $\mathbf{X}$ belongs to a one-parameter exponential family. Let $T = t(\mathbf{X})$ be as in Theorem 7.1, and write $u_r = c^{(r)}(\theta)$, the r-th derivative of $c(\theta) = E_\theta T$, $\mathbf{u} = (u_1, \ldots, u_k)'$, $k \geq 1$. Also write $a_{rs} = E_\theta((g^{(r)}(\mathbf{X}; \theta)/g)(g^{(s)}(\mathbf{X}; \theta)/g))$, $A = ((a_{rs}))_{1 \leq r,s \leq k}$.

(a) Derive the inequality (Bhattacharyya 1946)

$$\mathrm{var}_\theta T \geq \mathbf{u}' A^{-1} \mathbf{u}.$$

[Hint: $\mathrm{cov}_\theta(T, \sum_{r=1}^{k} b_r g^{(r)}/g) = \sum_r b_r \mathrm{cov}_\theta(T, g^{(r)}/g) = \sum_r b_r u_r$, so that $\mathrm{var}_\theta T \geq (\mathbf{b}'\mathbf{u})^2 / \sum_{r,s} b_r b_s a_{rs} = (\mathbf{b}'\mathbf{u})^2/\mathbf{b}'A\mathbf{b} = (\gamma' A^{-\frac{1}{2}})^2)/\|\gamma\|^2$ ($\gamma := A^{-\frac{1}{2}}\mathbf{b}$). The supremum of the last quantity over all $\mathbf{b} \neq \mathbf{0}$ equals $\sup_{\{\|\gamma\|=1\}} (\gamma' A^{-\frac{1}{2}})^2 = ((\mathbf{u}'/\|\mathbf{u}\|)A^{-\frac{1}{2}}\mathbf{u})^2 = \mathbf{u}' A^{-1} \mathbf{u}$.]

(b) Let $\mathbf{X} = (X_1, \ldots, X_n)$, X_i's i.i.d. $N(\mu, 1)$, $\mu \in \mathbb{R}$. Let the MLE $T = \overline{X}^2$ estimate μ^2. Find the lower bound of $\mathrm{var}_\theta T$ using $k = 1$ (Cramér-Rao) and $k = 2$ (Bhattacharyya), and compare these with the true variance of $\overline{X}^2$. Note that the UMVU estimator of μ^2 is $\overline{X}^2 - \frac{1}{n}$, having the same asymptotic distribution around μ^2 as the MLE.

Ex. 7.6 (Hardy-Weinberg Model). The Hardy-Weinberg formula for probabilities of the three genotypes of a single gene with two alleles is $p_1 = \theta^2$, $p_2 = 2\theta(1 - \theta)$, $p_3 = (1 - \theta)^2$, $0 < \theta < 1$. A random sample of size n from the populations yields frequencies n_1, n_2, n_3 for the three genotypes, $n_1 + n_2 + n_3 = n$.

(a) Show that this is a one-parameter exponential family. Find the UMVU estimator of θ and show that it coincides with the MLE $\hat{\theta}$.
(c) Compare $\hat{\theta}$ with the naive estimator $\sqrt{n_1/n}$.

Ex. 7.7. Instead of the Hardy-Weinberg formula in Exercise 7.6, assume $p_1 = \pi$, $p_2 = \pi^2$, $P_3 = 1 - \pi - \pi^2$ $(0 < \pi < (\sqrt{5} - 1)/2)$.

(a) Show that the (log-) likelihood equation has a consistent solution $\hat{\pi}$.
(b) Find the asymptotic distribution of $\hat{\pi}$.
(c) Find the asymptotic distribution of $\hat{p}_1$ and compute the asymptotic relative efficiency $e_{\hat{p}_1,\hat{\pi}}$.

Ex. 7.8. In Example 7.4 let the parameter space Θ be $(0, \infty)$.

(a) Find the asymptotic distribution of the MLE.
(b) Compare the estimate in (a) with $\overline{X}^+ \equiv \max\{\overline{X}, 0\}$.

Ex. 7.9. To study if animals are free of a toxin after being fed a contaminated feed, a veterinarian combines blood samples from m different animals to determine if the combined specimen is free of the toxin. Let $X = 0$ if the combined specimen indicates presence of the toxin and $X = 1$ otherwise. Let π be the probability that an animal is free of the toxin. What is the distribution of X? Suppose a random sample $\{X_1, \ldots, X_n\}$ of such combined samples is drawn. Find the MLE $\hat{\pi}$ of π and the asymptotic distribution of $\hat{\pi}$.

Ex. 7.10. Let $X_1, \ldots, X_n$ be i.i.d. binomial $\mathscr{B}(m, p)$, $0 < p < 1$.

(a) Find the UMVU estimate $\tilde{\pi}$ of $\pi(p) = (1-p)^m$ (= probability of "no success").
(b) Find the MLE $\hat{\pi}$ of $\pi(p)$ and show that $\sqrt{n}(\tilde{\pi} - \hat{\pi}) \to 0$ in probability as $n \to \infty$.
(c) Find the asymptotic distributions of $\tilde{\pi}$, $\hat{\pi}$.

Ex. 7.11. Suppose $X_1, \ldots, X_n$ are independent random variables with X_i having the Poisson distribution $\mathscr{P}(c_i\theta)$ where c_i's are known positive constants $(i = 1, 2, \ldots)$, and $\theta > 0$ is an unknown parameter.

(a) Find the MLE $\hat{\theta}$ of θ and prove that it is also UMVU.
(b) Find a reasonably broad criterion (i.e., condition on the sequence of constants c_i, $i \geq 1$) for the consistency of $\hat{\theta}$. [Hint: Calculate $\mathrm{var}_\theta(\hat{\theta})$.]
(c) Prove that $\hat{\theta}$ is inconsistent if $\sum c_i$ converges to a finite limit δ. [Hint: In this case $\hat{\theta} - \theta$ is approximately of the form $\sum_1^n Y_i - \delta$ where $Y_1, Y_2, \ldots,$ are independent positive random variables.]
(d) Use the Lindeberg CLT to find a broad condition for the asymptotic Normality of $\hat{\theta}$ around θ.

Ex. 7.12 (Neyman-Scott). Let X_{ij} be independent $N(\mu_i, \sigma^2)$ random variables $(i = 1, \ldots, n; j = 1, \ldots, k)$.

(a) Show that the MLE of the parameter $\boldsymbol{\theta} = (\mu_1, \ldots, \mu_n, \sigma^2)$ is given by $\hat{\mu}_i = \overline{X}_{i\cdot} = \frac{1}{k}\sum_{j=1}^k X_{ij}$, $1 \leq i \leq n$, and $\hat{\sigma}^2 = \frac{1}{nk}\sum_{i,j}(X_{ij} - \overline{X}_{i\cdot})^2$.

(b) Show that $\hat{\sigma}^2$ is not consistent. [Hint: $nk\hat{\sigma}^2/\sigma^2$ has the chi-square distribution with $n(k-1)$ degrees of freedom.]

(c) Show that the *bias-corrected MLE* $\tilde{\sigma}^2 = nk\hat{\sigma}^2/n(k-1)$ is a consistent estimator of σ^2.

Ex. 7.13. Suppose the hypothesis of Theorem 7.5 holds. Let $\gamma : \Theta \to \mathbb{R}^s$ be a continuously differentiable function in a neighborhood of $\boldsymbol{\theta}_0$ contained in Θ. Find the asymptotic distribution of the MLE of $\gamma(\boldsymbol{\theta}_0)$ under P_{θ_0}.

Ex. 7.14. Consider the multivariate Normal $N(\boldsymbol{\mu}, \Sigma)$ distribution of Example 7.5(b).

(a) Prove that $\{\hat{\sigma}_{ii'} : 1 \le i \le i' \le k\}$ and $\{\hat{\mu}_i : 1 \le i \le k\}$ defined by (7.92) are independent.

(b) Prove that the joint distribution of $\{n\hat{\sigma}_{ii'} : 1 \le i \le i' \le k\}$ based on n independent observations from $N(\boldsymbol{\mu}, \Sigma)$ is the same as that of $\{(n-1)\sigma_{ii'} \equiv \sum_{j=1}^{n-1} X_{ji}X_{ji'} : 1 \le i \le i' \le k\}$ based on $(n-1)$ observations from $N(\mathbf{0}, \Sigma)$. [Hint: Proceed as in the case of $N(\mu, \sigma^2)$ considered in Proposition 2.1, Chap. 2, Part I.]

Exercises for Sect. 7.4

Ex. 7.15. Let $X_j = (X_{j1}, \ldots, X_{jk})$, $j \ge 1$, be i.i.d. with $P(X_j = e_i) = p_i$, $0 < p_i < 1$ for $1 \le i \le k$, $\sum_{i=1}^{k} p_i = 1$, where e_i's $(i = 1, \ldots, k)$ are the standard unit vectors in $\mathbb{R}^k$ as defined in Example 7.5(c).

(a) Show that $\sqrt{n}(\hat{\mathbf{p}} - \mathbf{p}) \xrightarrow{\mathcal{L}} N(0, \Sigma)$, where $\hat{p}_i = \frac{1}{n}\sum_{j=1}^{n} X_{ji}$, $\hat{\mathbf{p}} = (\hat{p}_1, \ldots, \hat{p}_k)'$, $\mathbf{p} = (p_1, \ldots, p_k)'$, and $\sigma_{ii} = p_i(1 - p_i)$, $\sigma_{ii'} = -p_i p_{i'}$ $(i \ne i')$.

(b) Restricting the result (a) to the first $k-1$ coordinates of $\hat{\mathbf{p}}$, derive (7.98).

Ex. 7.16. Suppose $\sqrt{n}(\tilde{\theta}_n - \theta_0)$ converges in distribution, under P_{θ_0}. Check that $\sqrt{n}(\tilde{\theta}_n - \theta_0)$ is bounded in probability.

Ex. 7.17. Assume Θ in Example 7.5, defined by (7.76) is a nonempty open subset of R^p.

(a) Show that $C(\boldsymbol{\theta})$, $1/C(\boldsymbol{\theta})$ and $f(\mathbf{x}, \boldsymbol{\theta})$ are analytic functions of $\boldsymbol{\theta}$ in Θ. [Hint: Fix $\boldsymbol{\theta}_0 \in \Theta$ and let $\delta > 0$ be such that $B(\boldsymbol{\theta}_0 : \delta) \equiv \{\boldsymbol{\theta} : |\boldsymbol{\theta} - \boldsymbol{\theta}_0| < \delta\} \subset \Theta$. Express $1/C(\boldsymbol{\theta})$ as $(1/C(\boldsymbol{\theta}_0))$-times the m.g.f. of $\mathbf{t}(\mathbf{X}) = (t_1(\mathbf{X}), \ldots, t_p(X))$ in a neighborhood of the origin $\mathbf{0} \in \mathbb{R}^p$, under the distribution of X given by $f(x; \boldsymbol{\theta}_0)\mu(dx)$.]

(b) Check that the hypothesis of Theorem 7.5 is satisfied by the family (7.76).

Exercises for 7.5

Ex. 7.18. Let $X_1, \ldots, X_n$ be i.i.d. observations from the gamma distribution $\mathscr{G}(\mu, \alpha)$, $\mu > 0$, $\alpha > 0$.

(a) Suppose μ is known. Show that both $T_1 = \overline{X}/\mu$ and $T_2 = (\sum X_i^2/n - \overline{X}^2)/(1 - \frac{1}{n})\mu^2$ are unbiased estimates of α, and find the asymptotic relative efficiency e_{T_1, T_2}.

(b) Suppose α is known. Show that both $T_3 = \overline{X}/\alpha$ and $T_4 = [\sum X_i^2/(n\alpha(\alpha+1))]^{\frac{1}{2}}$ are consistent estimators of μ. Find the ARE e_{T_3,T_4}.

Exercises for 7.6

Ex. 7.19. Directly check the conclusion of the Bernstein–von Mises Theorem for the following examples: (i) Example 7.8, (ii) $X_1, \ldots, X_n$ i.i.d. Poisson $\mathscr{P}(\theta)$ with $\theta \in (0, \infty)$ and prior given by Gamma $\mathscr{G}(\alpha, \beta)$, (iii) Example 3.6, Chap. 3, Part I, where $X_1, \ldots, X_n$ are i.i.d. $N(\theta, \sigma^2)$, $\sigma^2 > 0$ known, $\theta \in \mathbb{R}$, and the prior is $N(0, \beta^2)$.

Ex. 7.20. Show that in Examples (i)–(iii) in the preceding Exercise (7.19) the Bayes estimator under squared error, or absolute error, loss is asymptotically efficient.

Exercises for 7.7, 7.8

Ex. 7.21. Let $X_1, \ldots, X_n$, be independent observations with the common distribution Q on $\mathbb{R}$, having a finite sixth moment. Define $g(\theta) = E(X_1 - \theta)^4$ and $g_n(\theta) = n^{-1}\sum_{j=1}^{n}(X_j - \theta)^4$, $\theta \in \mathbb{R}$.

(a) Prove that $g(\theta)$ has a unique minimizer θ_0, and $g_n(\theta)$ has a unique minimizer $\hat{\theta}$. [Hint: g, g_n are strictly convex.]
(b) First assume $\hat{\theta} \to \theta_0$ a.s. as $n \to \infty$, and obtain a Taylor expansion of $\theta \equiv g_n'(\hat{\theta})$ around θ_0 to prove that $\sqrt{n}(\hat{\theta} - \theta_0) \xrightarrow{\mathscr{L}} N(0, \gamma^2)$, where $\gamma^2 = E(X_1 - \theta_0)^6/\{(6(E(X_1 - \theta_0)^2)^2\}$.
(c*) Prove $\hat{\theta} \to \theta_0$ a.s. [Hint: On a compact interval $g_n(\theta) \to g(\theta)$ uniformly, a.s.]

Ex. 7.22. Prove (7.153), (7.154).

References

Bahadur, R.R. (1958). Examples of inconsistency of maximum likelihood estimates. *Sankhya, 20*, 207–210.

Bhattacharyya, A. (1946). On some analogues of the amount of information and their use in statistical estimation. *Sankhya, 8*, 1–14.

Bickel, P. J., & Doksum, K. (2001). *Mathematical statistics* (2nd ed.). Englewood Cliffs, NJ: Prentice Hall.

Cramér, H. (1946). *Mathematical methods of statistics.* Princeton: Princeton University Press.

Dasgupta, A. (2008). *Asymptotic theory of statistics and probability.* New York: Springer.

Ferguson, T. S. (1996). *A course in large sample theory.* London: Taylor and Francis.

Huber, P. J. (1981). *Robust statistics.* New York, Wiley.

Le Cam, L. (1953). On some asymptotic properties of maximum likelihood estimates and related Bayes estimates. *University of California Publications in Statistics, 1*, 277–320.

Lehmann, E., & Casella, G. (1998). *Theory of point estimation.* New York: Springer.

Prakasa Rao, B. L. S. (1987). *Asymptotic theory of statistical inference.* New York: Wiley.

Rao, C. R. (1945). Information and accuracy attainable in the estimation of statistical parameters. *Bulletin of the Calcutta Mathematical Society, 37*(3), 81–91.

Rao, C. R. (1973). *Linear statistical inference and its applications* (2nd ed.). New York: Wiley.

Serfling, R. (1980). *Approximation theorems of mathematical statistics.* New York: Wiley.

Stigler, S. M., Wong, W. H., & Xu, D. (2002). *R. R. Bahadur's lectures on the theory of estimation.* Institute of Mathematical Statistics. Beachwood.

Chapter 8
Tests in Parametric and Nonparametric Models

Abstract The asymptotic theory of tests in parametric and nonparametric models and their relative efficiency is presented here. In particular, livelihood ratio, Wald's test and chisquare tests are derived in parametric models. The nonparametric tests discussed include two-sample rank tests and the Kolmogorov–Smirnov tests. Also presented are goodness-of-fit tests and inference for linear time series models.

8.1 Pitman ARE (Asymptotic Relative Efficiency)

Example 8.1. Let $X_1, X_2, \ldots$ be i.i.d. real valued with common distribution function $F(x - \theta)$, where $F'(x) = f(x)$ is symmetric about $x = 0$ and is continuous at $x = 0$. Consider the problem of testing $H_0 : \theta = 0$ against $H_1 : \theta > 0$, by the following procedures:

(1) **Mean Test:** Assume

$$\sigma_f^2 \equiv \int_{-\infty}^{\infty} x^2 f(x) dx < \infty.$$

Reject H_0 iff $\overline{X} > a_n$, where $\overline{X} = \frac{1}{n} \sum_{j=1}^{n} X_j$ and (the sequence) a_n is so chosen that the test has (asymptotic) size α $(0 < \alpha < 1) : P_0(\overline{X} > a_n) \to \alpha$. This may be expressed as

$$P_0(\overline{X} > a_n) = P_0\left(\sqrt{n}\,\overline{X} > \sqrt{n}\,a_n\right) \longrightarrow \alpha, \tag{8.1}$$

which implies by the CLT that

$$\frac{\sqrt{n}\,a_n}{\sigma_f} = \Phi_{1-\alpha} + o(1), \tag{8.2}$$

$\Phi(x)$ denoting the distribution function of $N(0, 1)$, and Φ_β its β-th quantile.

© Springer-Verlag New York 2016
R. Bhattacharya et al., *A Course in Mathematical Statistics and Large Sample Theory*, Springer Texts in Statistics, DOI 10.1007/978-1-4939-4032-5_8

(2) **Sign Test:** Reject H_0 iff $T_n > b_n$, when $T_n = \frac{1}{n} \sum_{j=1}^{n} \mathbf{1}_{\{X_j > 0\}}$ and (the sequence) b_n is so chosen that the test has (asymptotic) size α : $P_0(T_n > b_n) \to \alpha$. This yields, by the CLT (and the fact that $\mathbf{1}_{\{X_j > 0\}}$ are i.i.d. Bernoulli which have, under H_0, common mean $\frac{1}{2}$ and common variance $\frac{1}{2}(1 - \frac{1}{2}) = \frac{1}{4}$),

$$P_0 \left(T_n - \frac{1}{2} > b_n - \frac{1}{2} \right) = P_0 \left(2\sqrt{n} \left(T_n - \frac{1}{2} \right) > 2\sqrt{n} \left(b_n - \frac{1}{2} \right) \right) \longrightarrow \alpha,$$

or,

$$2\sqrt{n} \left(b_n - \frac{1}{2} \right) = \Phi^{-1}(1 - \alpha) + o(1). \tag{8.3}$$

Now if one fixes an alternative $\theta_1 > 0$, then it is simple to check that the *probability of the type* II *error* of each of the above tests converges to zero, i.e., the *power* goes to one. (This property is referred to as *consistency* of a test.) In order to make a comparison among such tests one may choose (a sequence of) alternatives $\theta_n \downarrow \theta_0 = 0$ as $n \uparrow \infty$. Suppose that $\theta_n \downarrow 0$ are so chosen that the probability of type II error $\beta_n^{(1)}$ of test (1) converges to a desired level β, $0 < \beta < 1 - \alpha$ ($\beta < 1 - \alpha$ means *unbiasedness* for a test; hence we are requiring *asymptotic unbiasedness*). Thus

$$\beta_n^{(1)} \equiv P_{\theta_n} \left(\overline{X} \leq a_n \right) \longrightarrow \beta. \tag{8.4}$$

This means

$$P_{\theta_n}(\overline{X} - \theta_n \leq a_n - \theta_n) = P_0(\overline{X} \leq a_n - \theta_n)$$
$$= P_0 \left(\sqrt{n}\, \overline{X}/\sigma_f \leq \sqrt{n}(a_n - \theta_n)/\sigma_f \right) \longrightarrow \beta, \tag{8.5}$$

or, by the CLT,

$$\sqrt{n}(a_n - \theta_n)/\sigma_f = \Phi^{-1}(\beta) + o(1)$$

$$\theta_n = a_n - \sigma_f \Phi^{-1}(\beta) n^{-\frac{1}{2}} + o \left(n^{-\frac{1}{2}} \right)$$
$$= \sigma_f \left(\Phi^{-1}(1 - \alpha) - \Phi^{-1}(\beta) \right) n^{-\frac{1}{2}} + o \left(n^{-\frac{1}{2}} \right). \tag{8.6}$$

A fruitful way of comparing test (2) with test (1) is to find the sample size $h(n)$ required for test (2) (which has asymptotic size α—the same as that of test (1)) to have the same limiting power $1 - \beta$ or probability of the type II error β under the alternative θ_n. One may then define the *Pitman Asymptotic Relative Efficiency of test* (2) *relative to test* (1) as

$$\lim_{n \to \infty} \frac{n}{h(n)} = e_p(T_n, \overline{X}) \tag{8.7}$$

provided this limit exists (and is independent of α, β for $0 < \beta < 1 - \alpha$). One must then find $h(n)$ such that

$$\beta_{h(n)}^{(2)} \equiv P_{\theta_n} \left(T_{h(n)} \leq b_{h(n)} \right) \longrightarrow \beta. \tag{8.8}$$

Now the distribution of (the finite sequence) $\mathbf{1}_{\{X_j > 0\}} \equiv \mathbf{1}_{\{X_j - \theta_n > -\theta_n\}}$ ($j = 1, 2, \ldots, h(n)$) under $\theta = \theta_n$ is the same as that of $\mathbf{1}_{\{X_j > -\theta_n\}}$ under $H_0 : \theta = 0$. Hence (by an application of Liapounov's CLT to triangular arrays)

$$\beta_{h(n)}^{(2)} = P_{\theta_n}\left(\frac{1}{h(n)}\sum_{j=1}^{h(n)}\mathbf{1}_{\{X_j>0\}} \le b_{h(n)}\right) = P_0\left(\frac{1}{h(n)}\sum_{j=1}^{h(n)}\mathbf{1}_{\{X_j>-\theta_n\}} \le b_{h(n)}\right)$$

$$= P_0\left(2\sqrt{h(n)}\left(\frac{1}{h(n)}\sum_{j=1}^{h(n)}\left(\mathbf{1}_{\{X_j>-\theta_n\}} - F(\theta_n)\right)\right) \le 2\sqrt{h(n)}(b_{h(n)} - F(\theta_n))\right)$$

$$= \Phi\left(2\sqrt{h(n)}\left(b_{h(n)} - F(\theta_n)\right)\right) + o(1) \longrightarrow \beta. \tag{8.9}$$

Here we have made use of the facts

$$P_0(X_j > -\theta_n) = P_0(X_j < \theta_n) = F(\theta_n),$$

$$\mathrm{var}_0\,\mathbf{1}_{\{X_j>-\theta_n\}} = F(\theta_n)(1 - F(\theta_n)) = \frac{1}{4} + o(1). \tag{8.10}$$

Thus, by (8.9), (8.3), and (8.6)

$$\Phi^{-1}(\beta) + o(1) = 2\sqrt{h(n)}\left(b_{h(n)} - F(\theta_n)\right) = 2\sqrt{h(n)}\left(b_{h(n)} - \frac{1}{2} + \frac{1}{2} - F(\theta_n)\right)$$

$$= \Phi^{-1}(1-\alpha) - 2\sqrt{h(n)}\left(F(\theta_n) - \frac{1}{2}\right) + o(1)$$

$$= \Phi^{-1}(1-\alpha) - 2\sqrt{h(n)}\left(\theta_n(f(0) + o(1))\right) + o(1)$$

$$= \Phi^{-1}(1-\alpha) - 2\sqrt{\frac{h(n)}{n}}\,\sigma_f\left(\Phi^{-1}(1-\alpha) - \Phi^{-1}(\beta)\right)f(0)$$

$$+ o\left(\sqrt{\frac{h(n)}{n}}\right) + o(1),$$

which yields

$$2\sigma_f f(0) \sim \sqrt{\frac{n}{h(n)}}. \tag{8.11}$$

The symbol "$\sim$" indicates that the ratio of the two sides (of $\sim$) goes to one.
It follows from (8.11) (and (8.7)) that

$$e_P(T_n, \overline{X}) = 4\sigma_f^2 f^2(0). \tag{8.12}$$

Consider now the following special cases:

F	$e_P(T_n, \overline{X})$
$N(0, \sigma^2)$	$2/\pi$
Double exponential	2
Uniform on $[-\frac{1}{2}, \frac{1}{2}]$	$1/3$

A more realistic version of the mean test is the *t-test:*

$$\text{Reject } H_0 \text{ iff } t \equiv \frac{\overline{X}}{s} > a'_n, \tag{8.13}$$

where $s = [\frac{1}{n-1}\sum_{i=1}^{n}(X_i - \overline{X})^2]^{\frac{1}{2}}$. Since, under H_0, $\sqrt{n}\,\overline{X}/s \to N(0,1)$ (provided $\sigma_f^2 < \infty$), one has

$$\sqrt{n}\,a_n' = \Phi^{-1}(1-\alpha) + o(1),$$
$$a_n' = n^{-\frac{1}{2}}\left[\Phi^{-1}(1-\alpha) + o(1)\right]. \qquad (8.14)$$

Usually, the p.d.f. f may not be known, so that σ_f^2 is unknown. Proceeding as above one shows that $e_P(T_n, t) = 4\sigma_f^2 f^2(0)$, the same as in (8.12). (Exercise 8.1.)

Remark 8.1. The reason why one may use the sign test and not the t-test is that one may not be sure of the underlying distribution. The t-test is not nearly as "robust" (against model variation) as the sign test. In particular, $e_p(T_n, \overline{X}) = \infty$ if $\sigma_f^2 = \infty$. Although one may look at the asymptotic relative efficiency *(ARE)* of two tests in a parametric model also, there do exist under broad assumptions asymptotically optimal tests in parametric models—for example, the likelihood ratio test is asymptotically optimal under appropriate regularity conditions. We shall consider such tests in the next section.

The theorem below allows one to compute Pitman ARE of tests of the type in Example 8.1. Consider two tests $\delta_{i,n}$ $(i = 1, 2)$ which may be expressed as

$$\delta_{i,n} : \text{Reject } H_0 \text{ if } T_{i,n} > a_{i,n} \qquad (i = 1, 2), \qquad (8.15)$$

where $a_{i,n}$ are such that

$$\lim_{n\to\infty} P_{\theta_0}\left(T_{i,n} > a_{i,n}\right) = \alpha \qquad (i = 1, 2), \qquad (8.16)$$

for a given $\alpha \in (0,1)$. Fix $\beta \in (0,1)$ such that $\beta < 1 - \alpha$.

Theorem 8.1. (A_1) *Assume that there exists functions* $\mu_{i,n}(\theta)$, $\sigma_{i,n}(\theta) > 0$ $(i = 1, 2)$ *and* $\delta > 0$ *such that*

$$\sup_{\theta\in[\theta_0-\delta,\theta_0+\delta]}\ \sup_{x\in(a,b)}\left|P_\theta\left(\frac{T_{i,n}-\mu_{i,n}(\theta)}{\sigma_{i,n}(\theta)} \le x\right) - G(x)\right| \longrightarrow 0 \qquad (8.17)$$

as $n \to \infty$, *where G is a continuous distribution function which is strictly increasing on an interval (a, b) with $G^{-1}(a) = 0$, $G^{-1}(b) = 1$ $(-\infty \le a < b \le \infty)$. (A_2) Assume that $\theta \to \mu_{i,n}(\theta)$ is k times continuously differentiable in a neighborhood of θ_0, where k is the smallest positive integer such that $\mu_{i,n}^{(k)}(\theta_0) \neq 0$ $(i = 1, 2)$. (A_3) Assume that*

$$\lim_{n\to\infty}\frac{\sigma_{i,n}(\theta_n')}{\sigma_{i,n}(\theta_0)} = 1, \qquad \lim_{n\to\infty}\frac{\mu_{i,n}^{(k)}(\theta_n')}{\mu_{i,n}^{(k)}(\theta_0)} = 1 \qquad (i = 1, 2) \qquad (8.18)$$

for every sequence $\{\theta_n'\}$ converging to θ_0. Finally assume (A_4) that there exist positive constants c_i $(i = 1, 2)$, γ such that

$$\lim_{n\to\infty} n^{-\gamma}\frac{\mu_{i,n}^{(k)}(\theta_0)}{\sigma_{i,n}(\theta_0)} = c_i, \qquad (i = 1, 2). \qquad (8.19)$$

Under assumptions (A_1)–(A_4), for the test $H_0 : \theta = \theta_0$, $H_1 : \theta > \theta_0$, one has

$$e_P(\delta_{2,n}, \delta_{1,n}) = \left(\frac{c_2}{c_1}\right)^{\frac{1}{\gamma}}. \tag{8.20}$$

Proof. Let

$$\theta_n = \theta_0 + \left(\frac{k!}{c_1}\right)^{\frac{1}{k}} \left(G^{-1}(1-\alpha) - G^{-1}(\beta)\right)^{\frac{1}{k}} n^{-\frac{\gamma}{k}}(1 + o(1)),$$

$$h(n) = \left[n\left(\frac{c_1}{c_2}\right)^{\frac{1}{\gamma}}\right] + 1 \qquad ([x] := \text{integer part of } x). \tag{8.21}$$

By assumptions (A_2), (A_3), (A_4), there exists $\theta_{n,i}^*$ lying between θ_n and θ_0 such that

$$
\begin{aligned}
\mu_{i,n}(\theta_n) - \mu_{i,n}(\theta_0) &= \frac{(\theta_n - \theta_0)^k}{k!}\, \mu_{i,n}^{(k)}(\theta_{n,i}^*) \\
&= \frac{1}{c_1}\left\{G^{-1}(1-\alpha) - G^{-1}(\beta)\right\} n^{-\gamma}\mu_{i,n}^{(k)}(\theta_0)(1 + o(1)) \\
&= \frac{c_i}{c_1}\left\{G^{-1}(1-\alpha) - G^{-1}(\beta)\right\} \sigma_{i,n}(\theta_0)(1 + o(1)) \ (i = 1, 2).
\end{aligned}
$$
$$\tag{8.22}$$

Now (8.16) may be expressed as

$$P_{\theta_0}\left(\frac{T_{i,n} - \mu_{i,n}(\theta_0)}{\sigma_{i,n}(\theta_0)} > \frac{a_{i,n} - \mu_{i,n}(\theta_0)}{\sigma_{i,n}(\theta_0)}\right) \longrightarrow \alpha \qquad (i = 1, 2). \tag{8.23}$$

Therefore, by (A_1),

$$\lim_{n\to\infty} \frac{a_{i,n} - \mu_{i,n}(\theta_0)}{\sigma_{i,n}(\theta_0)} == G^{-1}(1-\alpha) \qquad (i = 1, 2). \tag{8.24}$$

Now

$$P_{\theta_0}\left(T_{1,n} \le a_{1,n}\right) = P_{\theta_0}\left(\frac{T_{1,n} - \mu_{1,n}(\theta_n)}{\sigma_{1,n}(\theta_n)} \le \frac{a_{1,n} - \mu_{1,n}(\theta_n)}{\sigma_{1,n}(\theta_n)}\right), \tag{8.25}$$

and, by (A_3) and (8.22), (8.24),

$$
\begin{aligned}
&\frac{a_{1,n} - \mu_{1,n}(\theta_n)}{\sigma_{1,n}(\theta_n)} \\
&= \frac{a_{1,n} - \mu_{1,n}(\theta_0) + \mu_{1,n}(\theta_0) - \mu_{1,n}(\theta_n)}{\sigma_{1,n}(\theta_0)}(1 + o(1)) \\
&= \left[G^{-1}(1-\alpha) - \left\{G^{-1}(1-\alpha) - G^{-1}(\beta)\right\}\right](1 + o(1)) \longrightarrow G^{-1}(\beta). \tag{8.26}
\end{aligned}
$$

Similarly,

$$P_{\theta_n}\left(T_{2,h(n)} \le a_{2,h(n)}\right) = P_{\theta_n}\left(\frac{T_{2,h(n)} - \mu_{2,h(n)}(\theta_n)}{\sigma_{2,h(n)}(\theta_n)} \le \frac{a_{2,h(n)} - \mu_{2,h(n)}(\theta_n)}{\sigma_{2,h(n)}(\theta_n)}\right)$$

and $\mu_{2,h(n)}(\theta_0) - \mu_{2,h(n)}(\theta_n) = \{-(\theta_n - \theta_0)^k/k!\}[\mu_{2,h(n)}^{(k)}(\theta_0) + o(1)]$, so that
by (8.19), (8.21), one has

$$
\begin{aligned}
\frac{a_{2,h(n)} - \mu_{2,h(n)}(\theta_n)}{\sigma_{2,h(n)}(\theta_n)} &= \left\{ \frac{a_{2,h(n)} - \mu_{2,h(n)}(\theta_0)}{\sigma_{2,h(n)}(\theta_0)} + \frac{\mu_{2,h(n)}(\theta_0) - \mu_{2,h(n)}(\theta_n)}{\sigma_{2,h(n)}(\theta_0)} \right\} (1 + o(1)) \\
&= \left[G^{-1}(1 - \alpha) - \frac{1}{c_1} \left\{ G^{-1}(1 - \alpha) - G^{-1}(\beta) \right\} \frac{(1/n)^\gamma}{\sigma_{2,h(n)}(\theta_0)} \right. \\
&\qquad \left. \cdot \mu_{2,h(n)}^{(k)}(\theta_0) \right] (1 + o(1)) \\
&\longrightarrow G^{-1}(1 - \alpha) - \left\{ G^{-1}(1 - \alpha) - G^{-1}(\beta) \right\} = G^{-1}(\beta). \quad \square
\end{aligned}
$$

8.2 CLT for U-Statistics and Some Two-Sample Rank Tests

Let $X_1, X_2, \ldots, X_m$ be a random sample drawn from a population with p.d.f.
$f\left(\frac{x}{\sigma}\right)$, and let $Y_1, Y_2, \ldots, Y_n$ be a random sample (independent of X_j's) from another population with p.d.f. $f\left(\frac{x-\theta}{\sigma}\right)$. We will suppress σ. Consider

$$
H_0 : \theta = 0, \qquad H_1 : \theta > 0. \tag{8.27}
$$

Here σ is an unknown scale parameter.

The *Wilcoxon* (or *Mann–Whitney*) test rejects H_0 iff

$$
\sum_{j=1}^{n} R_j > c_{m,n}, \tag{8.28}
$$

where R_j is the rank of Y_j among the $m + n$ observations $X_1, X_2, \ldots, X_m$,
$Y_1, Y_2, \ldots, Y_n$. Assume that $m \to \infty$, $n \to \infty$ in such a way that

$$
\frac{m}{m + n} \longrightarrow \lambda \quad \text{for some } \lambda \in (0, 1). \tag{8.29}
$$

One chooses $c_{m,n}$ in such a way that (P_θ denotes probability under $f(\frac{x-\theta}{\sigma})$),

$$
\lim P_0 \left(\sum_{i=1}^{n} R_j > c_{m,n} \right) = \alpha, \tag{8.30}
$$

where the limit is through sequences of values of m, n satisfying (8.29), and $\alpha \in (0, 1)$. Now, under H_0, the distribution of $(R_1, R_2, \ldots, R_n)$ is

$$
P_0 (R_1 = r_1, R_2 = r_2, \ldots, R_n = r_n) = \frac{m!}{(m + n)!} \tag{8.31}
$$

for every m-tuple of distinct integers $(r_1, r_2, \ldots, r_n)$ from $\{1, 2, \ldots, m + n\}$. Hence
the null distribution of the Wilcoxon statistic $\sum_1^n R_j$ does not depend on the
underlying f.

The *Fisher–Yates test* rejects H_0 iff

$$\sum_{i=1}^{n} Z_{(R_i)} > d_{m,n} \tag{8.32}$$

where $Z_{(r)}$ is the expected value of the rth *order statistic* of a sample of size $m+n$ from a standard normal distribution. This test is also called the *normal scores test.* Once again, since the distribution of the statistic on the left (8.32) depends only the distribution of $(R_1, \ldots, R_m)$, the null distribution of the normal scores statistic is independent of the underlying p.d.f. f, and $d_{m,n}$ depends only on m, n and α.

In order to compute the ARE's of the above tests (with respect to each other, or with respect to their parametric competitors) we will use the following central limit theorem for U-statistics.

Theorem 8.2. *Let $\varphi(x,y)$ be a real-valued measurable function of (x,y). Assume*

$$E\varphi(X_i, Y_j) = 0, \quad 0 < \sigma_1^2 := Eg^2(X_1) < \infty,$$
$$0 < \sigma_2^2 := Eh^2(Y_1) < \infty, \tag{8.33}$$

where $g(x) := E\varphi(x, Y_1)$, $h(y) := E\varphi(X_1, y)$. Then under the assumption (8.29),

$$\frac{\sum_{i=1}^{m} \sum_{j=1}^{n} \varphi(X_i, Y_j)}{\sqrt{\operatorname{var} \sum_{i=1}^{m} \sum_{j=1}^{n} \varphi(X_i, Y_j)}} \xrightarrow{\mathscr{L}} N(0, 1). \tag{8.34}$$

Proof. Write $U = \sum \sum \varphi(X_i, Y_j)$. Then, writing $\sigma_0^2 = E\varphi^2(X_1, Y_1) = \operatorname{var}\varphi(X_1, Y_1)$,

$$EU = mnE\varphi(X_1, Y_1) = 0,$$
$$\operatorname{var} U = mn \operatorname{var} \varphi(X_1, Y_1) + mn(n-1)\sigma_1^2 + mn(m-1)\sigma_2^2$$
$$= mn\sigma_0^2 + mn(n-1)\sigma_1^2 + mn(m-1)\sigma_2^2, \tag{8.35}$$

since (taking conditional expectation given X_1 first)

$$\operatorname{cov}(\varphi(X_1, Y_1), \varphi(X_1, Y_2)) = Eg^2(X_1) = \sigma_1^2,$$

and (taking conditional expectation given Y_1 first)

$$\operatorname{cov}(\varphi(X_1, Y_1), \varphi(X_2, Y_1)) = Eh^2(Y_1) = \sigma_2^2.$$

In the computation of variance (U) in (8.35) we have also used the fact that $\operatorname{cov}(\varphi(X_i, Y_j), \varphi(X_{i'}, Y_{j'})) = 0$ if $i \neq i'$ and $j \neq j'$.

Now consider the following approximation of U:

$$S := \sum_{i=1}^{m} E(U|X_i) + \sum_{j=1}^{n} E(U|Y_j)$$
$$= n\sum_{i=1}^{m} g(X_i) + m\sum_{j=1}^{n} h(Y_j). \tag{8.36}$$

Note that S is the projection of U onto the subspace (of L^2-functions on the probability space) $(\oplus \mathscr{L}_i) \oplus (\oplus \mathscr{L}_j')$, where $\mathscr{L}_i$ is the space of $(L^2\text{-})$ functions of X_i, $\mathscr{L}_j'$ is the space of $(\mathscr{L}^2\text{-})$ functions of Y_j. *In any case*

$$\operatorname{var} S = mn^2 \sigma_1^2 + m^2 n \sigma_2^2, \tag{8.37}$$

$$\operatorname{cov}(S, U) = n \sum_{i=1}^{m} \operatorname{cov}(g(X_i), U) + m \sum_{j=1}^{n} \operatorname{cov}(h(Y_j), U)$$

$$= n \sum_{i=1}^{m} n \operatorname{cov}(g(X_i), \varphi(X_i, Y_1)) + m \sum_{j=1}^{n} m \operatorname{cov}(h(Y_j), \varphi(X_1, Y_j))$$

$$= mn^2 \sigma_1^2 + m^2 n \sigma_2^2 = \operatorname{var} S,$$

$$\operatorname{cov}(S, U - S) = \operatorname{cov}(S, U) - \operatorname{var} S = 0. \tag{8.38}$$

Also,

$$\operatorname{var}(U - S) = \operatorname{var}(U) - \operatorname{var} S = mn(\sigma_0^2 - \sigma_1^2 - \sigma_2^2). \tag{8.39}$$

Therefore,

$$\operatorname{var}\left(\frac{U - S}{\sqrt{\operatorname{var} U}}\right) \longrightarrow 0, \qquad \frac{\operatorname{var} U}{\operatorname{var} S} \longrightarrow 1. \tag{8.40}$$

Hence it is enough to prove that

$$\frac{S}{\sqrt{\operatorname{var} S}} \xrightarrow{\mathscr{L}} N(0, 1). \tag{8.41}$$

But (8.41) is an immediate consequence of the classical CLT applied separately to the two components of (8.36), using their independence and (6.29). $\qquad \square$

Remark 8.2. Suppose, under P_θ, $\varphi(X_i, Y_j)$ is of the form

$$\varphi(X_i, Y_j) = \psi(X_i, Y_j) - E_\theta \psi(X_1, Y_1),$$

Also, write $\sigma_0^2(\theta)$, $\sigma_1^2(\theta)$, $\sigma_2^2(\theta)$ to indicate the dependence of the variances on θ. Write

$$\ell_{m,n}(\theta) := \frac{n^3 m E_\theta |g(X_1)|^3 + nm^3 E_\theta |h(Y_1)|^3}{(n^2 m \sigma_1^2(\theta) + nm^2 \sigma_2^2(\theta))^{\frac{3}{2}}}. \tag{8.42}$$

By the *Berry–Esséen Theorem*[1]

$$\sup_x \left| P_\theta\left(\frac{S}{\sqrt{mn^2 \sigma_1^2(\theta) + m^2 n \sigma_2^2(\theta)}} \leq x\right) - \Phi(x) \right| \leq c \ell_{m,n}(\theta) \tag{8.43}$$

where Φ is the standard normal distribution function, and c is an absolute constant ($c = 1$ will do). Also,

$$\Delta_{m,n}(\theta) := E_\theta \left(\frac{U}{\sqrt{\operatorname{var}_\theta(U)}} - \frac{S}{\sqrt{\operatorname{var}_\theta(S)}}\right)^2$$

$$= \frac{E_\theta(U - S)^2}{\operatorname{var}_\theta(U)} + \operatorname{var}_\theta(S)\left(\frac{1}{\operatorname{var}_\theta(S)} - \frac{1}{\operatorname{var}_\theta(U)}\right)^2. \tag{8.44}$$

---

[1] See Bhattacharya and Rao Ranga (2010), pp. 104, 186.

Now assume

$$(B): \begin{cases} \text{(i)} & E_\theta|\varphi(X_1,Y_1)|^3 \text{ is bounded away from infinity in a neighborhood of } \theta_0, \\ \text{(ii)} & \sigma_1^2(\theta),\ \sigma_2^2(\theta) \text{ are bounded away from zero in a neighborhood of } \theta_0. \end{cases}$$

Then both $\ell_{m,n}(\theta)$ and $\Delta_{m,n}(\theta)$ (see (8.42), (8.44)) go to zero uniformly for θ in a neighborhood $[\theta_0 - \delta, \theta_0 + \delta]$, as $m, n \to \infty$ (still assuming (8.29)). It may be shown from this that (Exercise 8.2)

$$\sup_{\theta\in[\theta_0-\delta,\theta_0+\delta]} \sup_{x\in\mathbb{R}^1} \left| P_\theta\left(\frac{\sum_{i,j} \psi(X_i,Y_j) - mnE_\theta\psi(X_1,Y_1)}{\sqrt{\mathrm{var}_\theta U}} \le x \right) - \Phi(x) \right| \to 0, \quad (8.45)$$
$$\text{as } m, n \to \infty.$$

Example 8.2. Let us now compute the ARE of the Wilcoxon test (8.28) with respect to the two-sample t-test. First note that, writing $Y_{(1)} < Y_{(2)} < \cdots < Y_{(n)}$ as the *order statistics* of Y_j's,

$$\sum_{i=1}^{m}\sum_{j=1}^{n} \mathbf{1}_{\{X_i<Y_j\}} = \sum_{j=1}^{n} (\#\{i : X_i < Y_j\}) = \sum_{j=1}^{n} (\#\{i : X_i < Y_{(j)}\}),$$

since each Y_j is a unique $Y_{(k)}$. But $\#\{i : X_i < Y_{(j)}\}$ equals the rank of $Y_{(j)}$ among $\{X_1,\ldots,X_m,Y_1,\ldots,Y_n\}$ minus j (since there are j Y_k's $\le Y_{(j)}$). Hence

$$\sum_{i=1}^{m}\sum_{j=1}^{n} \mathbf{1}_{\{X_i<Y_j\}} = \sum_{j=1}^{n} R_{(j)} - \sum_{j=1}^{n} j = \sum_{j=1}^{n} R_j - \frac{n(n+1)}{2}. \quad (8.46)$$

Here $R_{(j)}$ is the j-th order statistic among $R_1,\ldots,R_n$ (which is the same as the rank of $Y_{(j)}$ among $\{X_1,\ldots,X_m,Y_1,\ldots,Y_n\}$). Hence the Wilcoxon test (8.28) may be expressed as

$$\text{Reject } H_0 \text{ iff } \sum_{i=1}^{m}\sum_{j=1}^{n} \mathbf{1}_{\{X_i<Y_j\}} > a_{m,n} \quad (8.47)$$

where $a_{m,n}$ is so chosen that

$$\lim_{m,n\to\infty} P_0\left(\sum_{i=1}^{m}\sum_{j=1}^{n} \mathbf{1}_{\{X_i<Y_j\}} > a_{m,n} \right) = \alpha. \quad (8.48)$$

Now write $\psi(x,y) = \mathbf{1}_{\{x<y\}}$, $\varphi(x,y) = \psi(x,y) - E_\theta(\mathbf{1}_{\{X_i<Y_j\}})$, to get, from Theorem 8.2,

$$\lim_{m,n\to\infty} P_0\left(\frac{U}{\sqrt{\frac{mn}{12}(m+n+1)}} > \frac{a_{m,n} - \frac{1}{2}mn}{\sqrt{\frac{mn}{12}(m+n+1)}} \right) = \alpha. \quad (8.49)$$

Note that, $\forall\, \theta > 0$,

$$p_\theta := E_\theta\left(\mathbf{1}_{\{X_i<Y_j\}}\right) = P_\theta(X_i < Y_j) = \int_{-\infty}^{\infty} F(y)f(y-\theta)dy$$

$$= \int_{-\infty}^{\infty} F(z+\theta)f(z)dz \ge \int_{-\infty}^{\infty} F(z)f(z)dz$$

$$= \frac{1}{2} \int_{-\infty}^{\infty} dF^2(z) = \frac{1}{2} = E_0\left(\mathbf{1}_{\{X_i < Y_j\}}\right);$$

$$\sigma_0^2 = \mathrm{var}_\theta \mathbf{1}_{\{X_i < Y_j\}} = p_\theta(1 - p_\theta);$$

$$g(x) = 1 - F(x - \theta) - p_\theta, \quad h(y) = F(y) - p_\theta;$$

$$\sigma_1^2 = E_\theta g^2(X_1) = \mathrm{var}_\theta F(X_1 - \theta) = E_\theta F^2(X_1 - \theta) - (E_\theta F(X_1 - \theta))^2$$

$$= \int_{-\infty}^{\infty} F^2(x - \theta) f(x) dx - \left(\int_{-\infty}^{\infty} F(x - \theta) f(x) dx\right)^2;$$

$$\sigma_2^2 = \mathrm{var}_\theta F(Y_1) = \int_{-\infty}^{\infty} F^2(y) f(y - \theta) dy - \left(\int_{-\infty}^{\infty} F(y) f(y - \theta) dy\right)^2$$

$$= \int_{-\infty}^{\infty} F^2(x + \theta) f(x) dx - p_\theta^2;$$

$$\sigma_0^2 - \sigma_1^2 - \sigma_2^2 = p_\theta - \int_{-\infty}^{\infty} \left(F^2(x - \theta) + F^2(x + \theta)\right) f(x) dx$$

$$+ \left(\int_{-\infty}^{\infty} F(x - \theta) f(x) dx\right)^2 \tag{8.50}$$

Under $\theta = 0$,

$$\sigma_0^2 = \frac{1}{4}, \qquad \sigma_1^2 = \frac{1}{3} - \frac{1}{4} = \frac{1}{12} = \sigma_2^2,$$

$$\mathrm{var}_0(U) = mn\left(\sigma_0^2 - \sigma_1^2 - \sigma_2^2\right) + \frac{mn}{12}(n + m)$$

$$= \frac{mn}{12} + \frac{mn}{12}(m + n) = \frac{mn}{12}(m + n + 1). \tag{8.51}$$

In particular, $a_{m,n}$ is determined by the relation

$$\frac{a_{m,n} - \frac{1}{2}mn}{\sqrt{\frac{mn}{12}(m + n + 1)}} = z_{1-\alpha},$$

where $z_\beta = \Phi^{-1}(\beta)$. In general

$$\mathrm{var}_\theta(U) = mn\left(\sigma_0^2 - \sigma_1^2 - \sigma_2^2\right) + mn\left(n\sigma_1^2 + m\sigma_2^2\right) = \sigma_{m,n}^2(\theta),$$

$$E_\theta(U) = mnp_\theta = \mu_{2,m,n}(\theta). \tag{8.52}$$

Now, $\frac{d}{d\theta} p_\theta = \int_{-\infty}^{\infty} f(z + \theta) f(z) dz$ (if, e.g., f is bounded and continuous), so that

$$\left(\mu_{2,m,n}^{(1)}(\theta)\right)_{\theta=0} = mn \int_{-\infty}^{\infty} f(z) f(z) dz = mn \int_{-\infty}^{\infty} f^2(z) dz > 0, \tag{8.53}$$

Also,

$$\sigma_{m,n}(0) = \sqrt{\mathrm{var}_0(U)} = \sqrt{\frac{mn}{12}(m + n + 1)}, \tag{8.54}$$

so that $k = 1$ (in the application of Theorem 8.1) and

$$n^{-\frac{1}{2}} \frac{\mu_{2,m,n}^{(1)}(0)}{\sigma_{m,n}(0)} = n^{-\frac{1}{2}} \frac{(n^2\sqrt{\lambda})(1 + o(1)) \int_{-\infty}^{\infty} f^2(z)dz}{n^{\frac{3}{2}} \left(\frac{1}{12}\right)^{\frac{1}{2}} (1 + o(1))}$$

$$\longrightarrow \sqrt{\lambda}\sqrt{12} \int_{-\infty}^{\infty} f^2(z)dz = c_2. \tag{8.55}$$

Thus $\gamma = \frac{1}{2}$. (See (8.18).)

On the other hand for the Student's t,

$$t = \frac{\overline{Y} - \overline{X}}{s\sqrt{\frac{1}{m} + \frac{1}{n}}} \qquad \left[s^2 := \frac{\sum_{i=1}^{m}(X_i - \overline{X})^2 + \sum_{j=1}^{n}(Y_j - \overline{Y})^2}{m + n - 2} \right], \tag{8.56}$$

the test of asymptotic size α is

$$\text{Reject } H_0 \text{ iff } \overline{Y} - \overline{X} > z_\alpha s\sqrt{\frac{1}{m} + \frac{1}{n}} = b_{m,n}. \tag{8.57}$$

Writing $\sigma_f^2 = \text{var}_\theta X = \text{var}_\theta Y = \int(x - \mu_x)^2 f(x)dx$ $(\mu_x = \int xf(x)d\mu)$, one has

$$\mu_{1,m,n}(\theta) = E_\theta Y_1 - EX_1 = \int_{-\infty}^{\infty} xf(x - \theta)dx - \int_{-\infty}^{\infty} xf(x)dx$$

$$= \int_{-\infty}^{\infty} (y + \theta)f(y)dy - \int_{-\infty}^{\infty} xf(x)dx = \theta,$$

$$\sigma_{m,n}(\theta) = \sqrt{\text{var}(\overline{Y}) + \text{var}(\overline{X})} = \sqrt{\sigma_f^2/n + \sigma_f^2/m}$$

$$= \sigma_f\sqrt{1 + \frac{1 - \lambda}{\lambda}} \, n^{-\frac{1}{2}}(1 + o(1)) = \left(\frac{\sigma_f}{\sqrt{\lambda}}\right) n^{-\frac{1}{2}}(1 + o(1)). \tag{8.58}$$

Thus

$$\mu_{1,m,n}^{(1)}(\theta) = \frac{d}{d\theta}\theta = 1, \tag{8.59}$$

$$\sigma_{m,n}(0) = n^{-\frac{1}{2}} \frac{\sigma_f}{\sqrt{\lambda}}(1 + o(1)). \tag{8.60}$$

Thus, as before, $\gamma = \frac{1}{2}$, $k = 1$, and

$$c_1 = \lim_{n\to\infty} n^{-\frac{1}{2}} \frac{\mu_{1,m,n}^{(1)}(0)}{\sigma_{m,n}(0)} = \frac{1}{\sigma_f\sqrt{\frac{1}{\lambda}}}$$

$$= \frac{\sqrt{\lambda}}{\sigma_f}. \tag{8.61}$$

From (8.55) and (8.61), we get

$$e_P(W, t) \equiv e_P(\text{Wilcoxon}, t) = \left(\frac{c_2}{c_1}\right)^{\frac{1}{1/2}} = \left(\frac{c_2}{c_1}\right)^2$$

$$= 12\sigma_f^2 \left(\int_{-\infty}^{\infty} f^2(x)dx \right)^2 \tag{8.62}$$

$$= \begin{cases} \frac{3}{\pi} \approx 0.95 & \text{if } f \text{ is normal density,} \\[2mm] \frac{3}{2} & \text{if } f \text{ is double exponential,} \\[2mm] 1 & \text{if } f \text{ is uniform on } [-1,1]. \end{cases}$$

It has been shown by Hodges and Lehmann (1956) that $e_P(W,t) \geq 108/125 = 0.864$, whatever be f. On the other hand, $e_P(W,t) = \infty$ if $\sigma_f^2 = \infty$. Even in the class of f with $\sigma_f^2 < \infty$, the supremum of $e_P(W,t)$ is infinity. Thus the t-test can be very bad for certain f's.

Similarly, one may compute the ARE of the Fisher–Yates test NS with respect to the two-sample t-test. Indeed,

$$e_P(NS,t) = \sigma_f^2 \left(\int_{-\infty}^{\infty} \frac{f^2(x)}{\varphi\{\Phi^{-1}[F(x)]\}} \, dx \right)^2 .$$

One may show that $e_P(NS,t) > 1$ for all $f \neq \varphi$, and equals 1 when $f = \varphi$. It also follows that

$$e_P(W,NS) = \frac{3}{\pi} \quad \text{if } f \text{ is normal.}$$

See Hodges and Lehmann (1960).

Remark 8.3. The two-sample rank tests are often used to test $H_0 : F(x) = G(x) \,\forall\, x$ against $H_1 : F(x) \geq G(x) \,\forall\, x$ with strict inequality for some $x \in \mathbb{R}^1$. Here F, G are the (common) distribution functions of the X's and the Y's, respectively, and it is assumed that F, G are continuous. (Note that H_1 says that Y's are *stochastically larger* than X's: $P(Y > x) \geq P(X > x) \,\forall\, x$, with strict inequality for some x.) Consider now the group $\mathscr{G}$ (under composition of maps) of all continuous homeomorphisms of $\mathbb{R}^1$. An element φ of $\mathscr{G}$ transforms an observation vector $(x_1, \ldots, x_m, y_1, \ldots, y_n) \in \mathbb{R}^{m+n}$ into the vector $(\varphi(x_1), \ldots, \varphi(x_m), \varphi(y_1), \ldots, \varphi(y_n))$. Thus φ induces a transformation $\tilde{\varphi}$ on $\mathbb{R}^{m+n}$ (onto $\mathbb{R}^{m+n} \equiv$ observation space $\mathscr{X}$). Let $\tilde{\mathscr{G}}$ denote the group of these transformations on $\mathbb{R}^{m+n}$. Because $\tilde{\varphi}$ does not change the orders (or *ranks*) among the observations, it is reasonable to require that the test based on $\tilde{\varphi}\mathbf{x}$ rejects H_0 *iff* the same test based on $\mathbf{x} \equiv (x_1, \ldots, x_m, y_1, \ldots, y_n)$ does so. For the form of H_0 and H_1 remain unchanged if F, G are replaced by the distributions of $\varphi(X_1), \varphi(Y_1)$, respectively. (Also, F, G are completely unknown, except for the properties assumed.) The only tests which are *invariant* under every $\tilde{\varphi}$ in $\tilde{\mathscr{G}}$ are the *rank tests*, i.e., tests based on the ranks of $\{X_1, \ldots, X_m, Y_1, \ldots, Y_n\}$. (See Proposition 5.4.)

The *power* of a rank test of course depends on specific pairs (F, G). For example, if F, G have densities $f(x)$, $f(x - \theta)$, $\theta > 0$, and $f(x)$ *is the normal p.d.f.* with mean 0, then the most powerful rank test is the Fisher–Yates test. If, on the other hand $f(x)$ *is the logistic with mean zero*, then the most powerful rank test is the Wilcoxon test. For a simple derivation of these facts, see T. Ferguson (1967), pp. 250–257.

Remark 8.4. There is a one-sample version of Theorem 8.2, which is even simpler to prove (Exercise 8.4):

Proposition 8.1 (CLT for One-Sample U-Statistics).

$$\sum_{1 \leq i \neq j \leq n} \frac{\varphi(X_i, X_j)}{\sqrt{\operatorname{var} \sum_{i \neq j} \varphi(X_i, X_j)}} \xrightarrow{\mathscr{L}} N(0, 1) \quad if$$

(i) $E\varphi(X_1, X_2) = 0,$
(ii) $0 < \sigma_1^2 := Eg^2(X_1) < \infty.$

Example 8.3. One may write

$$\sum_1^n (X_i - E(X_i)) = \frac{1}{n-1} \sum_{i \neq j} \left[(X_i - E(X_i)) + (X_j - EX_j) \right], \tag{8.63}$$

$$\sum_{i=1}^n (X_i - \overline{X})^2 = \frac{1}{2n} \sum_{1 \leq i \neq j \leq n} (X_i - X_j)^2. \tag{8.64}$$

Remark 8.5. Let $T = T(\mathbf{X})$ be an unbiased estimator of some parameter $\theta = ET(\mathbf{X})$, where $\mathbf{X} = (X_1, \ldots, X_n)$ and $X_1, X_2, \ldots, X_n$ are i.i.d. The U-statistic $\varphi(\mathbf{X}) = \frac{1}{n!} \sum \varphi(X_{i_1}, X_{i_2}, \ldots, X_{i_n})$ is also an unbiased estimator of θ; here the sum is over all $n!$ permutations $(i_1, i_2, \ldots, i_n)$ of the indices $(1, 2, \ldots, n)$. If $E|T|^p < \infty$ for some $p \geq 1$, then $E|\varphi(\mathbf{X}) - \theta|^p \leq E|T - \theta|^p$ (See Exercise 8.1).

8.3 Asymptotic Distribution Theory of Parametric Large Sample Tests

Let $X_1, X_2, \ldots, X_n$ be i.i.d with common p.d.f. $f(x; \theta)$ (w.r.t. a σ-finite measure μ), with the parameter space Θ an open subset of $\mathbb{R}^p$. Assume that the hypothesis of Theorem 7.2, Chap. 7, holds. Suppose that the *null hypothesis* may be expressed in the form

$$H_0 : \pi_i(\boldsymbol{\theta}) = 0, \qquad 1 \leq i \leq k, \tag{8.65}$$

where $1 \leq k \leq p$. To test this, against the alternative that H_0 is not true (i.e., $H_1 : \theta \notin \Theta_0 \equiv \{\boldsymbol{\theta} \in \Theta : \pi_i(\boldsymbol{\theta}) = 0 \ \forall \ i\}$), a natural procedure would be to see if $\pi_i(\tilde{\boldsymbol{\theta}}_n)$ are close to zero (for $1 \leq i \leq k$) or not; here $\tilde{\boldsymbol{\theta}}_n$ is an asymptotically efficient estimator of $\boldsymbol{\theta}$. If $\boldsymbol{\theta}_0$ is the true parameter value, then under H_0 (i.e., for $\boldsymbol{\theta}_0 \in \Theta_0$) one has

$$\pi_i(\tilde{\boldsymbol{\theta}}_n) = \pi_i(\boldsymbol{\theta}_0) + \sum_{r=1}^p \left(\tilde{\theta}_n^{(r)} - \theta_0^{(r)} \right) \left(\frac{\partial \pi_i}{\partial \theta_r} \right)_{\boldsymbol{\theta}^*}$$

$$= \sum_{r=1}^p \left(\tilde{\theta}_n^{(r)} - \theta_0^{(r)} \right) \left[\left(\frac{\partial \pi_i}{\partial \theta^{(r)}} \right)_{\boldsymbol{\theta}_0} + o_p(1) \right], \tag{8.66}$$

assuming $\pi_i(\boldsymbol{\theta})$ is continuously differentiable (in $\boldsymbol{\theta}$) on Θ. One may express (8.66) for all i compactly as

$$\sqrt{n}\, \pi(\tilde{\boldsymbol{\theta}}_n) = \left[(\operatorname{Grad} \pi)_{\boldsymbol{\theta}_0} + o_p(1) \right] \sqrt{n}(\tilde{\boldsymbol{\theta}}_n - \boldsymbol{\theta}_0) \tag{8.67}$$

where $\pi(\boldsymbol{\theta})$ is a $k \times 1$ (column) vector with i-th element $\pi_i(\boldsymbol{\theta})$, $(\tilde{\boldsymbol{\theta}}_n - \boldsymbol{\theta}_0)$ is a $p \times 1$ vector, and $(\mathrm{Grad}\,\pi)$ is a $k \times p$ matrix with (i,r) element $\partial\pi_i(\boldsymbol{\theta})/\partial\theta^{(r)}$. Assume now that $\mathrm{Grad}\,\pi$ is of rank k on Θ_0. Then, under H_0,

$$\sqrt{n}\,\pi(\tilde{\boldsymbol{\theta}}_n) \xrightarrow{\mathscr{L}} N_k(0, V(\boldsymbol{\theta}_0)) \tag{8.68}$$

where $V(\boldsymbol{\theta}_0)$ is the $k \times k$ matrix

$$V(\boldsymbol{\theta}_0) = (\mathrm{Grad}\,\pi)_{\boldsymbol{\theta}_0} I^{-1}(\boldsymbol{\theta}_0)(\mathrm{Grad}\,\pi)'_{\boldsymbol{\theta}_0} \tag{8.69}$$

with $I(\boldsymbol{\theta}_0)$ as the $p \times p$ information matrix. It follows that

$$\sqrt{n}\,\pi(\tilde{\boldsymbol{\theta}}_n)'V^{-1}(\boldsymbol{\theta}_0)\sqrt{n}\,\pi(\tilde{\boldsymbol{\theta}}_0) \xrightarrow{\mathscr{L}} Z_k^0 \tag{8.70}$$

where Z_k^0 denotes the *chi-square distribution with k degrees of freedom*. Since $V(\tilde{\boldsymbol{\theta}}_n) \to V(\boldsymbol{\theta}_0)$ almost surely $(P_{\boldsymbol{\theta}_0})$ as $n \to \infty$, $V(\tilde{\boldsymbol{\theta}}_n)^{-1}$ exists on a set whose probability converges to 1 as $n \to \infty$, and $V(\tilde{\boldsymbol{\theta}}_n)^{-1} \to V(\theta_0)^{-1}$ a.s. as $n \to \infty$. Therefore,

$$W_n \equiv \sqrt{n}\,\pi(\tilde{\boldsymbol{\theta}}_n)'V^{-1}(\tilde{\boldsymbol{\theta}}_n)\sqrt{n}\,\pi(\tilde{\boldsymbol{\theta}}_n) \xrightarrow{\mathscr{L}} Z_k^0. \tag{8.71}$$

Therefore, a reasonable test of (asymptotic) size α is given by

$$\text{Reject } H_0 \text{ iff } W_n > \chi^2_{1-\alpha}(k), \tag{8.72}$$

where

$$\frac{1}{2^{k/2}\Gamma\left(\frac{k}{2}\right)} \int_{\chi^2_{1-\alpha}(k)}^{\infty} e^{-u/2} u^{\frac{k}{u}-1} du = \alpha. \tag{8.73}$$

In case H_0 is simple, i.e., $k = p$ and $\Theta_0 = \{\boldsymbol{\theta}_0\}$, one may use the statistic in (8.70) rather than (8.71) (See Example 5.15 for a motivation). We have thus arrived at the following result.

Theorem 8.3. *Assume that the hypothesis of Theorem 7.3, Chap. 7, holds, and that $\mathrm{Grad}\,\pi$ is continuous and of full rank on Θ. If $\tilde{\boldsymbol{\theta}}_n$ is an asymptotically efficient estimator of $\boldsymbol{\theta}$, then the test statistic $W_n \xrightarrow{\mathscr{L}} Z_k^0$, where Z_k^0 has a χ_k^2 distribution under the null hypothesis H_0, and the test (8.72) is of asymptotic size α.*

Definition 8.1. A test (sequence) is said to be *consistent* if its power ($= 1-$ probability of type II error) goes to 1 as $n \to \infty$.

The test (8.72) may be shown to be consistent. For this let $\boldsymbol{\theta} \notin \Theta_0$, i.e., $\pi_i(\boldsymbol{\theta}) \neq 0$ for some i, say $i = 1$. Then, under $P_{\boldsymbol{\theta}}$, $n(\pi(\tilde{\boldsymbol{\theta}}_n) - \pi(\boldsymbol{\theta}))'V^{-1}(\tilde{\boldsymbol{\theta}}_n)(\pi(\tilde{\boldsymbol{\theta}}_n) - \pi(\boldsymbol{\theta})) \xrightarrow{\mathscr{L}} Z_k^0$, where Z_k^0 has a χ_k^2 distribution as $n \to \infty$. The probability of a type II error under $P_{\boldsymbol{\theta}}$ is

$$\beta_n(\boldsymbol{\theta}) = P_{\boldsymbol{\theta}}\left(W_n \le \chi^2_{1-\alpha}(k)\right) \tag{8.74}$$

$$= P_{\boldsymbol{\theta}}\Big(n\left(\pi(\tilde{\boldsymbol{\theta}}_n) - \pi(\boldsymbol{\theta})\right)' V^{-1}(\tilde{\boldsymbol{\theta}}_n)\left(\pi(\tilde{\boldsymbol{\theta}}_n) - \pi(\boldsymbol{\theta})\right)$$

$$\le \chi^2_{1-\alpha}(k)$$

$$+ n\left\{\pi(\boldsymbol{\theta})'V^{-1}(\tilde{\boldsymbol{\theta}}_n)\pi(\boldsymbol{\theta}) - \pi(\tilde{\boldsymbol{\theta}}_n)'V^{-1}(\tilde{\boldsymbol{\theta}}_n)\pi(\boldsymbol{\theta}) - \pi(\boldsymbol{\theta})'V^{-1}(\tilde{\boldsymbol{\theta}}_n)\pi(\tilde{\boldsymbol{\theta}}_n)\right\}\Big).$$

The expression within curly brackets in (8.74) converges to $-\pi(\boldsymbol{\theta})'V^{-1}(\boldsymbol{\theta})\pi(\boldsymbol{\theta})$ almost surely $(P_{\boldsymbol{\theta}})$. Since this last quantity is strictly negative, $\chi^2_{1-\alpha}(k) + n\{\ldots\}$ converges to $-\infty$ almost surely as $n \to \infty$. Hence $\beta_n \to 0$.

Since all reasonable tests are consistent, to discriminate among them one must consider a sequence of alternatives $\boldsymbol{\theta}_n$ such that $\boldsymbol{\theta}_n \notin \Theta_0$, $\boldsymbol{\theta}_n \to \boldsymbol{\theta}_0 \in \Theta_0$, and

$$\beta_n(\boldsymbol{\theta}_n) \longrightarrow \beta. \tag{8.75}$$

for some $\beta < 1 - \alpha$. This requires that one takes

$$\boldsymbol{\theta}_n = \boldsymbol{\theta}_0 + n^{-\frac{1}{2}}\boldsymbol{\delta} \qquad (\boldsymbol{\delta} = (\delta_1,\ldots,\delta_p)') \tag{8.76}$$

such that θ_0 is an element of Θ_0 and $\boldsymbol{\delta} \neq \mathbf{0}$ is an element of $\mathbb{R}^p$, $\boldsymbol{\theta}_0 + n^{-\frac{1}{2}}\boldsymbol{\delta} \notin \Theta_0$. Since

$$\pi_i(\boldsymbol{\theta}_n) = \pi_i(\boldsymbol{\theta}_0) + n^{-\frac{1}{2}} \sum_{r=1}^{p} \delta_r \left(\frac{\partial \pi_i(\boldsymbol{\theta})}{\partial \theta^{(r)}}\right)_{\theta_0} + o\left(n^{-\frac{1}{2}}\right), \tag{8.77}$$

$$\pi(\boldsymbol{\theta}_n) = o + n^{-\frac{1}{2}} \cdot \operatorname{Grad} \pi(\boldsymbol{\theta}_0)\boldsymbol{\delta} + o\left(n^{-\frac{1}{2}}\right),$$

one ought to choose $\boldsymbol{\delta}$ so that it is not orthogonal to all the vectors $\operatorname{Grad} \pi_i(\boldsymbol{\theta}_0)$, $1 \le i \le k$. Then under (8.74)–(8.77) we have

$$\begin{aligned}
\beta_n(\boldsymbol{\theta}_n) &= P_{\boldsymbol{\theta}_n}(W_n \le \chi^2_{1-\alpha}(k)) \\
&= P_{\boldsymbol{\theta}_n}(n(\pi(\tilde{\boldsymbol{\theta}}_n) - \pi(\boldsymbol{\theta}_n))'V^{-1}(\tilde{\boldsymbol{\theta}}_n)(\pi(\tilde{\boldsymbol{\theta}}_n) - \pi(\boldsymbol{\theta}_n)) \\
&\le \chi^2_{1-\alpha}(k) - \boldsymbol{\delta}' \operatorname{Grad} \pi(\boldsymbol{\theta}_0)V^{-1}(\boldsymbol{\theta}_0)(\operatorname{Grad} \pi(\boldsymbol{\theta}_0))'\boldsymbol{\delta} + o_p(1)).
\end{aligned}$$

Assume now that there exists $\varepsilon > 0$ such that for all $x > 0$

$$\sup_{|\boldsymbol{\theta}-\boldsymbol{\theta}_0|\le\varepsilon} \left| P_{\boldsymbol{\theta}}(n(\pi(\tilde{\boldsymbol{\theta}}_n)-\pi(\boldsymbol{\theta}))'V^{-1}(\boldsymbol{\theta})(\pi(\tilde{\boldsymbol{\theta}}_n)-\pi(\boldsymbol{\theta})) \le x) - G_k(x) \right| \to 0 \text{ as } n \to \infty,$$

$$\tag{A}$$

where G_k is the distribution function of a chi-square random variable with k degrees of freedom. Then

$$\lim_{n\to\infty} \beta_n(\boldsymbol{\theta}_n) = G_k\left(\chi^2_{1-\alpha}(k) - \boldsymbol{\delta}'(\operatorname{Grad} \pi(\boldsymbol{\theta}_0))'V^{-1}(\boldsymbol{\theta}_0)\operatorname{Grad} \pi(\boldsymbol{\theta}_0)\boldsymbol{\delta}\right). \tag{8.78}$$

In order that this limit be β, one must choose δ so that $G_k(\chi^2_{1-\alpha}(k) - \gamma(\boldsymbol{\delta})) = \beta$, where

$$\gamma(\boldsymbol{\delta}) := \boldsymbol{\delta}'(\operatorname{Grad} \pi(\boldsymbol{\theta}_0))'V^{-1}(\boldsymbol{\theta}_0)\operatorname{Grad} \pi(\boldsymbol{\theta}_0)\boldsymbol{\delta}. \tag{8.79}$$

This is possible since $G_k(\chi^2_{1-\alpha}(k)) = 1 - \alpha$ and $\beta < 1 - \alpha$.

Remark 8.6. Theorem 8.3 is due to Wald (1943). To motivate the test (8.72), suppose (as is often the case) that *the MLE $\hat{\boldsymbol{\theta}}_n$ is a sufficient statistic for θ.* It is then enough to confine one's attention to tests based on $\hat{\boldsymbol{\theta}}_n$. By Theorem 7.5, Chap. 7, the asymptotic distribution of $\hat{\boldsymbol{\theta}}_n$ is the k-dim. Normal distribution $N(\theta, \frac{1}{n}I^{-1}(\boldsymbol{\theta}))$. If the latter was the exact distribution of $\hat{\boldsymbol{\theta}}_n$, and if $I(\boldsymbol{\theta})$ was known, then an optimal test for $H_0 : \theta_1 = \theta_2 = \cdots = \theta_k = 0$ would be given by (See Lehmann, *Testing Statistical Hypotheses*)

$$\text{Reject } H_0 \text{ if } n\left(\hat{\boldsymbol{\theta}}_n^{(1)}\right)' V^{-1}\hat{\boldsymbol{\theta}}_n^{(1)} > \chi^2_{1-\alpha}(k), \tag{8.80}$$

where $\hat{\boldsymbol{\theta}}_n^{(1)} = (\hat{\theta}_{n1}, \ldots, \hat{\theta}_{nk})'$ and $V =$ the matrix formed by the elements in the first k rows and the first k columns of the information matrix. In case $I = I(\boldsymbol{\theta})$ depends on $\boldsymbol{\theta}$, one may replace $\boldsymbol{\theta}$ by $\hat{\boldsymbol{\theta}}_n$. This yields test (8.72). The case (8.65) simply needs a reparametrization.

An alternative test of (8.65) is the *likelihood ratio test* originally proposed by Neyman and Pearson (1928) and explored by Wilks (1938). According to this procedure one calculates the *likelihood ratio statistic*

$$\Lambda_n = \frac{\max_{\boldsymbol{\theta} \in \Theta_0} f_n(\mathbf{X}; \boldsymbol{\theta})}{\max_{\boldsymbol{\theta} \in \Theta} f_n(\mathbf{X}; \boldsymbol{\theta})} = \frac{f_n(\mathbf{X}; \hat{\hat{\boldsymbol{\theta}}}_n)}{f_n(\mathbf{X}; \hat{\boldsymbol{\theta}}_n)} . \tag{8.81}$$

The test statistic is then

$$\lambda_n = -2 \log \Lambda_n = 2 \log f_n(\mathbf{X}; \hat{\boldsymbol{\theta}}_n) - 2 \log f_n(\mathbf{X}; \hat{\hat{\boldsymbol{\theta}}}_n) \tag{8.82}$$

and the *likelihood ratio test* is

$$\text{Reject } H_0 \text{ iff } \lambda_n > \chi^2_{1-\alpha}(k). \tag{8.83}$$

Theorem 8.4. *Assume the hypothesis of Theorem 8.3.*

(a) If H_0 holds, then
$$\lambda_n = W_n + o_p(1), \tag{8.84}$$

so that λ_n converges in law to a chi-square distribution with k d.f., and the test (8.83) has asymptotic size α.

(b) The likelihood ratio test (8.83) is consistent. Also, for alternatives $\boldsymbol{\theta}_n$ given by (8.76) (with $\boldsymbol{\delta}$ not orthogonal to the linear span of $\mathrm{Grad}\, \pi_i(\boldsymbol{\theta})$ $(1 \leq i \leq k)$ on Θ_0),
$$\lim_{n \to \infty} P_{\boldsymbol{\theta}_n}\left(\lambda_n \leq \chi^2_{1-\alpha}(k)\right) = \beta, \tag{8.85}$$

provided $\beta < 1 - \alpha$, $\gamma(\boldsymbol{\delta})$ in (8.79) satisfies $G_k(\chi^2_{1-\alpha}(k) = \gamma(\boldsymbol{\delta})) = \beta$, and an analogue of (A) holds for λ_n.

Proof. (a) Fix $\boldsymbol{\theta}_0 \in \Theta_0$. All $o_p(1)$ errors below are under $P_{\boldsymbol{\theta}_0}$. Assume for the sake of simplicity that $\pi_i(\boldsymbol{\theta}) = \theta^{(i)}$ $(\boldsymbol{\theta} := (\theta^{(1)}, \ldots, \theta^{(p)})' \in \Theta)$, $1 \leq i \leq k$. This may be achieved, at least in a neighborhood of $\boldsymbol{\theta}_0$, by reparametrization. Write

$$I(\boldsymbol{\theta}_0) = \begin{bmatrix} I_{11} & I_{12} \\ I_{21} & I_{22} \end{bmatrix}, \qquad I^{-1}(\boldsymbol{\theta}_0) = \begin{bmatrix} I^{11} & I^{12} \\ I^{21} & I^{22} \end{bmatrix}, \tag{8.86}$$

where I_{11} comprises the elements of $I(\boldsymbol{\theta}_0)$ belonging to the first k rows and the first k columns, etc. Also write $(x)^r_{r'} = (x^{(r)}, \ldots, x^{(r')})'$ for $x = (x^{(1)}, \ldots, x^{(p)})' \in \mathbb{R}^p$. With this notation, Wald's statistic (see (8.70), (8.71)) becomes

$$W_n = n(\hat{\boldsymbol{\theta}}_n)^{1'}_k (I^{11})^{-1}(\hat{\boldsymbol{\theta}}_n)^1_k. \tag{8.87}$$

Since $(\hat{\boldsymbol{\theta}}_n)^1_k = 0 = (\hat{\hat{\boldsymbol{\theta}}}_n)^1_k$, λ_n may be expressed as

$$\lambda_n = n(\hat{\boldsymbol{\theta}}_n - \hat{\hat{\boldsymbol{\theta}}}_n)' I(\boldsymbol{\theta}_0)(\hat{\boldsymbol{\theta}}_n - \hat{\hat{\boldsymbol{\theta}}}_n) + o_p(1)$$
$$= (I(\boldsymbol{\theta}_0)\sqrt{n}(\hat{\boldsymbol{\theta}}_n - \hat{\hat{\boldsymbol{\theta}}}_n))' \sqrt{n}(\hat{\boldsymbol{\theta}}_n - \hat{\hat{\boldsymbol{\theta}}}_n) + o_p(1)$$

$$= \left[\begin{matrix} I_{11}\sqrt{n}(\hat{\boldsymbol{\theta}}_n)^1_k + I_{12}\sqrt{n}(\hat{\boldsymbol{\theta}}_n - \hat{\hat{\boldsymbol{\theta}}}_n)^{k+1}_p \\ I_{21}\sqrt{n}(\hat{\boldsymbol{\theta}}_n)^1_k + I_{22}\sqrt{n}(\hat{\boldsymbol{\theta}}_n - \hat{\hat{\boldsymbol{\theta}}}_n)^{k+1}_p \end{matrix}\right]' \sqrt{n}(\hat{\boldsymbol{\theta}}_n - \hat{\hat{\boldsymbol{\theta}}}_n) + o_p(1). \quad (8.88)$$

Now, as $\hat{\boldsymbol{\theta}}_n$ and $\hat{\hat{\boldsymbol{\theta}}}_n$ are solutions of the likelihood equations on Θ and Θ_0, respectively, it follows from the proof of Theorem 7.5 that

$$I(\boldsymbol{\theta}_0)\sqrt{n}(\hat{\boldsymbol{\theta}}_n - \boldsymbol{\theta}_0) = +\frac{1}{\sqrt{n}}\, DL_n(\boldsymbol{\theta}_0) + o_p(1),$$

$$I_{22}\sqrt{n}(\hat{\hat{\boldsymbol{\theta}}}_n - \boldsymbol{\theta}_0)^{k+1}_p = +\frac{1}{\sqrt{n}}\,(DL_n(\boldsymbol{\theta}_0))^{k+1}_p + o_p(1). \quad (8.89)$$

The first relation in (8.89) may be expressed as

$$I_{11}\sqrt{n}(\hat{\boldsymbol{\theta}}_n)^1_k + I_{12}\sqrt{n}(\hat{\boldsymbol{\theta}}_n - \boldsymbol{\theta}_0)^{k+1}_p = -\frac{1}{\sqrt{n}}(DL_n(\boldsymbol{\theta}_0))^1_k + o_p(1),$$

$$I_{21}\sqrt{n}(\hat{\boldsymbol{\theta}}_n)^1_k + I_{22}\sqrt{n}(\hat{\boldsymbol{\theta}}_n - \boldsymbol{\theta}_0)^{k+1}_p = -\frac{1}{\sqrt{n}}(DL_n(\boldsymbol{\theta}_0))^{k+1}_p + o_p(1). \quad (8.90)$$

Comparing the last relations in (8.89) and (8.90), we get

$$I_{22}\sqrt{n}(\hat{\boldsymbol{\theta}}_n - \hat{\hat{\boldsymbol{\theta}}}_n)^{k+1}_p = -I_{21}\sqrt{n}(\hat{\boldsymbol{\theta}}_n)^1_k + o_p(1). \quad (8.91)$$

Using this, the second block of rows of the matrix within square brackets in (8.88) may be taken to be null. The (8.88) becomes

$$\lambda_n = n(\hat{\boldsymbol{\theta}}_n)^{1'}_k \left(I_{11} - I_{12}I_{22}^{-1}I_{21}\right)(\hat{\boldsymbol{\theta}}_n)^1_k + o_p(1). \quad (8.92)$$

To establish (8.84), it is now enough to show that (see (8.87)),

$$I_{11}I^{11} - I_{12}I_{22}^{-1}I_{21}I^{11} = I_k, \quad (8.93)$$

where I_k is the $k \times k$ identity matrix. But $I(\boldsymbol{\theta}_0)I^{-1}(\boldsymbol{\theta}_0) = I_p$, from which we get

$$I_{11}I^{11} + I_{12}I^{12} = I_k,$$
$$I_{21}I^{11} + I_{22}I^{21} = 0. \quad (8.94)$$

Substituting this in (8.93), the left side of (8.93) becomes

$$I_k - I_{12}I^{21} + I_{12}I_{22}^{-1}I_{22}I^{21} = I_k. \quad (8.95)$$

(b) The proof of part (b) follows along the lines of the computations (8.77)–(8.79).

$$\square$$

Remark 8.7 (Amplification of (8.87)–(8.88)). Let the $k \times k$-matrix

$$V(\boldsymbol{\theta}_0) = \overbrace{\mathrm{Grad}\,\pi(\boldsymbol{\theta}_0)}^{(k\times p)\text{-matrix}}\, \overbrace{I^{-1}(\boldsymbol{\theta}_0)}^{p\times p}(\mathrm{Grad}\,\pi(\boldsymbol{\theta}_0))'.$$

If $\pi_i(\boldsymbol{\theta}) = \theta_i$, $1 \leq i \leq k$, then $\operatorname{Grad} \pi(\boldsymbol{\theta}_0) = \begin{bmatrix} I_k & 0 \end{bmatrix}_{k \times (p-k)}$

$$
V(\boldsymbol{\theta}_0) = \begin{bmatrix} I_k & 0 \end{bmatrix} \begin{bmatrix} I^{11} & I^{12} \\ I^{21} & I^{22} \end{bmatrix} \begin{bmatrix} I_k \\ 0 \end{bmatrix}_{(p-k) \times k} = \begin{bmatrix} I^{11} & I^{12} \end{bmatrix} \begin{bmatrix} I_k \\ 0 \end{bmatrix}
$$

$$
= I^{11}.
$$

$$
V^{-1}(\boldsymbol{\theta}_0) = \left(I^{11} \right)^{-1}.
$$

Then

$$
W_n = n \left[\left(\hat{\boldsymbol{\theta}}_n \right)_1^k \right]' (I^{11})^{-1} (\hat{\boldsymbol{\theta}}_n)_1^k.
$$

The log of the likelihood ratio

$$
\lambda_n = 2 \log f_n(\mathbf{X}; \hat{\boldsymbol{\theta}}_n) - 2 \log f_n(\mathbf{X}; \hat{\hat{\boldsymbol{\theta}}}_n))
$$

$$
= 2 \underbrace{(\hat{\boldsymbol{\theta}}_n - \hat{\hat{\boldsymbol{\theta}}}_n)' \operatorname{grad} \ell(\boldsymbol{\theta})\big|_{\theta = \hat{\hat{\theta}}_n}}_{=0} + \sum_{i,j=1}^{k} n(\hat{\boldsymbol{\theta}}_n - \hat{\hat{\boldsymbol{\theta}}}_n)_i (\hat{\boldsymbol{\theta}}_n - \hat{\hat{\boldsymbol{\theta}}}_n)_j
$$

$$
\cdot \underbrace{\frac{1}{n} \frac{\partial^2 \ell(\boldsymbol{\theta})}{\partial \theta_i \partial \theta_j}\bigg|_{\theta_0}}_{=I(\theta_0) + o_p(1)} + o_p(1) \ (\text{under } P_{\theta_0})
$$

$$
= n(\hat{\boldsymbol{\theta}}_n - \hat{\hat{\boldsymbol{\theta}}}_n)' I(\boldsymbol{\theta}_0)(\hat{\boldsymbol{\theta}}_n - \hat{\hat{\boldsymbol{\theta}}}_n) + o_p(1).
$$

Corollary 8.1. *The Pitman ARE of the likelihood ratio test (8.83) relative to Wald's test (8.72) is one.*

Example 8.4 (Multinomial Models and the Chi-square Test). A population is divided into $M + 1$ categories and a random sample of size n is drawn from it (without replacement). Let $\theta^{(j)}$ denote the probability of an observation to belong to the j-th category $(j = 1, 2, \ldots, M + 1)$, and assume $\theta^{(j)} > 0$ for all j. Write $\boldsymbol{\theta} = (\theta^{(1)}, \ldots, \theta^{(M)})'$. From Example 7.1, Chap. 7, we know that (i) the MLE of $\boldsymbol{\theta}$ is $\hat{\boldsymbol{\theta}}_n = (\hat{\theta}_n^{(1)}, \ldots, \hat{\theta}_n^{(M)})$, $\hat{\theta}_n^{(M+1)} := 1 - \sum_1^M \hat{\theta}_n^{(j)}$, where $\hat{\theta}_n^{(j)}$ is the proportion in the sample belonging to the j-th category $(1 \leq j \leq M)$, and (ii) $\sqrt{n}(\hat{\boldsymbol{\theta}}_n - \boldsymbol{\theta}) \xrightarrow{\mathcal{L}} N(0, I^{-1}(\boldsymbol{\theta}))$, where the (i, j) element of $I^{-1}(\boldsymbol{\theta})$ is

$$
\sigma_{ij}(\boldsymbol{\theta}) := \begin{cases} -\theta^{(i)}\theta^{(j)} & \text{if } i \neq j, \\ \theta^{(i)}(1 - \theta^{(i)}) & \text{if } i = j \end{cases} \tag{8.96}
$$

A widely used alternative to (8.72) or (8.83) for testing (8.65) in this case is the so-called *frequency chi-square test*, originally due to Karl Pearson:

$$
\text{Reject } H_0 \text{ iff} \quad \sum_{j=1}^{M+1} \frac{n^2(\hat{\theta}_n^{(j)} - \hat{\hat{\theta}}_n^{(j)})^2}{n\hat{\hat{\theta}}_n^{(j)}} > \chi_{1-\alpha}^2(M). \tag{8.97}
$$

We will show that the *statistic on the left differs from the likelihood ratio statistic λ_n by a quantity which is $o_p(1)$, under H_0.* First note that

$$
f_n(\mathbf{X}; \boldsymbol{\theta}) = \prod_{j=1}^{M} \left(\theta^{(j)} \right)^{\nu_j}, \qquad \nu_j := n\hat{\theta}_n^{(j)}.
$$

Hence

$$\lambda_n = 2L_n(\hat{\boldsymbol{\theta}}_n) - 2L_n(\hat{\hat{\boldsymbol{\theta}}}_n) = 2\sum_{j=1}^{M} \nu_j \log \frac{\hat{\boldsymbol{\theta}}_n^{(j)}}{\hat{\hat{\boldsymbol{\theta}}}_n^{(j)}}$$

$$= 2\sum_{j=1}^{M} \nu_j \log \left(1 + \frac{\hat{\boldsymbol{\theta}}_n^{(j)}}{\hat{\hat{\boldsymbol{\theta}}}_n^{(j)}} - 1\right)$$

$$= 2\sum_{j=1}^{M} \nu_j \left\{ \frac{\hat{\boldsymbol{\theta}}_n^{(j)}}{\hat{\hat{\boldsymbol{\theta}}}_n^{(j)}} - 1 - \frac{1}{2}\left(\frac{\hat{\boldsymbol{\theta}}_n^{(j)}}{\hat{\hat{\boldsymbol{\theta}}}_n^{(j)}} - 1\right)^2 + o_p(n^{-1}) \right\}, \tag{8.98}$$

since $|\log(1+x) - (x - \frac{1}{2}x^2)| \le (\frac{4}{3})|x|^3$ if $|x| \le \frac{1}{2}$, and

$$\left|\frac{\hat{\boldsymbol{\theta}}_n^{(j)}}{\hat{\hat{\boldsymbol{\theta}}}_n^{(j)}} - 1\right|^3 \le 4n^{-\frac{3}{2}} \left\{ \left|\frac{\sqrt{n}(\hat{\boldsymbol{\theta}}_n^{(j)} - \boldsymbol{\theta}_0^{(j)})}{\hat{\hat{\boldsymbol{\theta}}}_n^{(j)}}\right|^3 + \left|\frac{\sqrt{n}(\hat{\hat{\boldsymbol{\theta}}}_n^{(j)} - \boldsymbol{\theta}_0^{(j)})}{\hat{\hat{\boldsymbol{\theta}}}_n^{(j)}}\right|^3 \right\}. \tag{8.99}$$

Since $\sqrt{n}(\hat{\boldsymbol{\theta}}_n^{(j)} - \boldsymbol{\theta}_0^{(j)})/\hat{\hat{\boldsymbol{\theta}}}_n^{(j)}$, $\sqrt{n}(\hat{\hat{\boldsymbol{\theta}}}_n^{(j)} - \boldsymbol{\theta}_0^{(j)}/\hat{\hat{\boldsymbol{\theta}}}_n^{(j)}$ converges in distribution under P_{θ_0}, the expression within curly brackets in (8.99) is bounded in probability, so that (8.99) is $o_p(n^{-1})$.

From (8.98) we get

$$\lambda_n = 2\sum_{j=1}^{M} n\hat{\boldsymbol{\theta}}_n^{(j)} \left\{ \frac{\hat{\boldsymbol{\theta}}_n^{(j)}}{\hat{\hat{\boldsymbol{\theta}}}_n^{(j)}} - 1 - \frac{1}{2}\left(\frac{\hat{\boldsymbol{\theta}}_n^{(j)}}{\hat{\hat{\boldsymbol{\theta}}}_n^{(j)}} - 1\right)^2 \right\} + o_p(1)$$

$$= \sum_{j=1}^{M} \left[2n\left(\hat{\boldsymbol{\theta}}_n^{(j)} - \hat{\hat{\boldsymbol{\theta}}}_n^{(j)}\right)\left(\frac{\hat{\boldsymbol{\theta}}_n^{(j)} - \hat{\hat{\boldsymbol{\theta}}}_n^{(j)}}{\hat{\hat{\boldsymbol{\theta}}}_n^{(j)}}\right) \right.$$

$$\left. + 2n\left(\hat{\boldsymbol{\theta}}_n^{(j)} - \hat{\hat{\boldsymbol{\theta}}}_n^{(j)}\right) - \frac{\hat{\boldsymbol{\theta}}_n^{(j)}}{\hat{\hat{\boldsymbol{\theta}}}_n^{(j)}} \frac{n\left(\hat{\boldsymbol{\theta}}_n^{(j)} - \hat{\hat{\boldsymbol{\theta}}}_n^{(j)}\right)^2}{\hat{\hat{\boldsymbol{\theta}}}_n^{(j)}} \right] + o_p(1)$$

$$= 2\sum_{j=1}^{M} \frac{n\left(\hat{\boldsymbol{\theta}}_n^{(j)} - \hat{\hat{\boldsymbol{\theta}}}_n^{(j)}\right)^2}{\hat{\hat{\boldsymbol{\theta}}}_n^{(n)}}$$

$$+ 2n\sum_{j=1}^{M} \left(\hat{\boldsymbol{\theta}}_n^{(j)} - \hat{\hat{\boldsymbol{\theta}}}_n^{(j)}\right) - \sum_{j=1}^{M} \frac{n\left(\hat{\boldsymbol{\theta}}_n^{(j)} - \hat{\hat{\boldsymbol{\theta}}}_n^{(j)}\right)^2}{\hat{\hat{\boldsymbol{\theta}}}_n^{(j)}}(1 + o_p(1)) + o_p(1)$$

$$= \sum_{j=1}^{M} \frac{n\left(\hat{\boldsymbol{\theta}}_n^{(j)} - \hat{\hat{\boldsymbol{\theta}}}_n^{(j)}\right)^2}{\hat{\hat{\boldsymbol{\theta}}}_n^{(j)}} + 0 + o_p(1), \tag{8.100}$$

using the facts (i) $\sum \hat{\boldsymbol{\theta}}_n^{(j)} = 1 = \sum \hat{\hat{\boldsymbol{\theta}}}_n^{(j)}$, (ii) $\hat{\boldsymbol{\theta}}_n^{(j)} / \hat{\hat{\boldsymbol{\theta}}}_n^{(j)} \to 1$ in probability, and (iii) $n(\hat{\boldsymbol{\theta}}_n^{(j)} - \hat{\hat{\boldsymbol{\theta}}}_n^{(j)})^2 / \hat{\hat{\boldsymbol{\theta}}}_n^{(j)}$ is bounded in probability. This completes the proof of the italicized statement above.

With a little extra effort it may be proved that, for every $\varepsilon > 0$,

$$P_{\boldsymbol{\theta}_n}\left(\left| \lambda_n - \sum_{j=1}^{M} \frac{n\left(\hat{\boldsymbol{\theta}}_n^{(j)} - \hat{\hat{\boldsymbol{\theta}}}_n^{(j)}\right)^2}{\hat{\hat{\boldsymbol{\theta}}}_n^{(j)}} \right| > \varepsilon \right) \longrightarrow 0 \quad \text{as } n \to \infty. \tag{8.101}$$

From this it follows that *the Pitman ARE of the frequency chi-square test relative to the likelihood ratio test is one.* Thus for this example, the Wald test, the likelihood ratio test, and the frequency chi-square test are all asymptotically equivalent from the point of view of the Pitman ARE.

We now discuss *Rao's scores test* which is an important alternative to Wald's and likelihood ratio tests under the hypothesis of Theorem 8.3. First consider a simple null hypothesis: $H_0 : \boldsymbol{\theta} = \boldsymbol{\theta}_0$ ($H_1 : \boldsymbol{\theta} \neq \boldsymbol{\theta}_0$). Under H_0, as in the proof of Theorem 7.5,

$$U_n := \frac{1}{\sqrt{n}} \text{grad} \log f_n(\mathbf{X}; \boldsymbol{\theta})\big|_{\boldsymbol{\theta}=\boldsymbol{\theta}_0} \xrightarrow{\mathscr{L}} N(\mathbf{0}, I(\boldsymbol{\theta}_0)), \tag{8.102}$$

where N is k-dimensional normal. Hence

$$Q_n := U_n' I^{-1}(\boldsymbol{\theta}_0)U_n \xrightarrow{\mathscr{L}} Z_k^0, \tag{8.103}$$

where Z_k^0 has a χ_k^2 distribution, and H_0 is rejected if $Q_n > \chi_{1-\alpha}^2(k)$. Unlike the likelihood ratio and Wald's tests, this test does not require computation of the MLE. For testing a composite $H_0 : \boldsymbol{\theta} \in \Theta_0$, $\boldsymbol{\theta}_0$ in (8.102), (8.103) is replaced by the MLE $\hat{\boldsymbol{\theta}}_n$ under H_0. As in Corollary 8.1, one can prove that the ARE of Rao's scores test relative to Wald's test (and, therefore, to the likelihood ratio test) is one. For details see Rao (1973), Serfling (1980), Sen and Singer (1979) and van der Vaart (1998). For higher order comparisons among the three tests see Mukherjee and Reid (2001). Also see Brown et al. (2001) for a comparison of coverage errors for confidence intervals for the binomial proportion based on Wald's and Rao's (score) tests.

8.4 Tests for Goodness-of-Fit

It is a common practice in statistics to see if a random sample of observations $X_1, \ldots, X_n$ may fit a distribution F—a parametric model specified up to perhaps an unknown finite dimensional parameter θ. That is, test if the observations may be considered to have been drawn from F. For continuous data, one may, e.g., test if F is Normal $N(\mu, \sigma^2)$, with $\boldsymbol{\theta} = (\mu, \sigma^2)$ unknown. Similarly, one may test if the number of accidents per week at a particular traffic intersection follows the Poisson distribution $\mathscr{P}(\lambda)$, $\theta = \lambda > 0$ unknown.

Before considering a number of goodness-of-fit tests, we recall the follow widely used notion or index.

Definition 8.2. The *p-value* of a test is the smallest level of significance (or size) at which the null hypothesis H_0 would be rejected by the observed value of the test statistic.

Informally, the p-value of a test is the probability of having a discrepancy from the null hypothesis as much (or more) as observed, if the null hypothesis were true. Thus a very small p-value may be taken as a strong evidence against H_0, whereas a relatively large p-value provides some evidence in support of H_0.

A classical goodness-of-fit procedure is by the *frequency chi-square*. Here the range of the distribution is divided into a finite number, say k, of disjoint classes, or consecutive intervals, and the number of observations n_j in the jth class is compared with the 'expected' number of observations E_j in the class under H_0. Here E_j equals $n\hat{\theta}_n^{(j)}$, where $\hat{\theta}_n^{(j)}$ is the estimated probability of the jth class computed by using a good estimate, usually the MLE of the unknown parameters of the model. *Pearson's chi-square statistic* $\sum_{j=1}^{k}(n_j - E_j)^2/E_j^2$ has asymptotically a chi-square distribution χ_q^2, as $n \to \infty$, where the degrees of freedom *(d.f.)* q equals $k-1-r$, r being the number of unknown parameters required to specify F. (See the derivation of the frequency chi-square test in the preceding section.) For details see Chernoff and Lehmann (1954), Rao (1973) and Sen and Singer (1979). If the chi-square statistic has a large observed value, i.e., if the p-value is small, one rejects H_0; otherwise the model is not rejected. Generally, this test is *not consistent;* for two different distributions may assign the same probability to each of the k classes, unless the model is discrete with only k different values.

To present one of the earliest *consistent* goodness-of-fit tests due to Kolmogorov and Smirnov, and also derive later tests due to Cramér and von Mises and by Anderson and Darling, we now provide an informal introduction to the *functional central limit theorem and Brownian motion.*[2]

Consider a sequence of i.i.d. random variables X_n $(n = 1, 2, \dots)$ with mean μ and finite variance $\sigma^2 > 0$. Changing to $Z_n = X_n - \mu$, and writing $S_n = Z_1 + \cdots + Z_n$ and $S_0 = 0$, one obtains, by the classical CLT,

$$S_n/\sqrt{n} \xrightarrow{\mathscr{L}} N(0, \sigma^2) \quad \text{as } n \to \infty. \tag{8.104}$$

Consider now the stochastic processes Y_n $(n = 1, 2, \dots)$ on $[0, \infty)$ defined by

$$Y_n(t) = S_{[nt]}/\sqrt{n}, \quad t \in [0, \infty), \ ([nt] := \text{integer part of } nt). \tag{8.105}$$

It is a (random) *step function*, with $Y_n(0) = 0$, $Y_n(1/n) = S_1/\sqrt{n}$, $Y_n(j/n) = S_j/\sqrt{n}$, and $Y_n(t)$ is constant for $j/n \leq t < (j+1)/n$. Then $Y_n(0) = 0$, and for any fixed t, $0 < t < \infty$, $Y_n(t)$ is asymptotically Normal with mean zero and variance $[nt]\sigma^2/n \approx t\sigma^2$. That is,

$$Y_n(t) \xrightarrow{\mathscr{L}} N(0, t\sigma^2), \quad \text{as } n \to \infty,$$

and if $0 < t_1 < t_2 < \cdots < t_k$, then as $n \to \infty$, for $i \leq j$,

$$\text{cov}(Y_n(t_i), Y_n(t_j)) = \left(\frac{1}{n}\right) \text{cov}(Z_1 + \cdots + Z_{[nt_i]}, Z_1 + \cdots + Z_{[nt_j]}) =$$

$$= \left(\frac{1}{n}\right) \text{cov}(Z_1 + \cdots + Z_{[nt_i]}, Z_1 + \cdots + Z_{[nt_j]}) =$$

$$= \left(\frac{1}{n}\right) \text{var}(Z_1 + \cdots + Z_{[nt_i]}) = ([nt_i]/n)\sigma^2 \longrightarrow t_i\sigma^2.$$

---

[2] See, e.g., Bhattacharya and Waymire (2007), Chap. 11, or Billingsley (1968), Sects. 10, 11.

Therefore, by the multivariate CLT (by taking linear combinations of $Y_n(t_i)$, $1 \leq i \leq k$, and applying Lindeberg's CLT),

$$(Y_n(t_1), Y_n(t_2), \ldots, Y_n(t_k)) \xrightarrow{\mathscr{L}} N(0, ((\min\{t_i, t_j\}\sigma^2)), \quad \text{as } n \to \infty. \qquad (8.106)$$

Since k and $0 < t_1 < t_2 < \cdots < t_k$ are arbitrary, one may think of this as convergence of the processes $Y_n = \{Y_n(t) : t \in [0, \infty)\}$ to a *Gaussian process B* on $[0, \infty)$ satisfying

$$E(B(t)) = 0, \quad \text{cov}(B(s), B(t)) = E(B(s)B(t)) = \min\{s, t\}\sigma^2 \quad \text{(for all } s, t \geq 0). \qquad (8.107)$$

We will call such a B the *Brownian motion with zero mean and dispersion σ^2*. It is simple to check either using the corresponding fact for Y_n, or directly using (8.107) that Brownian motion has *independent increments*, i.e., if $0 < t_1 < t_2 < \cdots < t_k$ then

$$B(t_i) - B(t_{i-1}), \ i = 1, \ldots, k,$$

$$\text{are } k \text{ independent } N(0, (t_i - t_{j-1})\sigma^2) \text{ random variables } (t_0 = 0). \quad (8.108)$$

Indeed, (8.108) is equivalent to (8.107) for a Gaussian process, since a Gaussian process is determined by its mean and covariance (functions). If $\sigma^2 = 1$, then the Brownian motion is said to be a *standard Brownian motion*. It was proved by Wiener in 1923 that *Brownian paths are* (or, can be taken to be) *continuous*. The distribution of B on $C[0, \infty)$ is called the *Weiner measure*. Here $C(0, \infty)$ is the space of all real-valued continuous functions on the half line $[0, \infty)$. It is a metric space in which convergence $f_n \to f$ means uniform convergence of f_n to f on bounded intervals. The usual sigma-field on it is the Borel sigma-field generated by its open sets. Although Y_n is a step function since it has the value $S_j/\sqrt{n}$ for $j/n \leq t < (j+1)/n)$, one may still write (in an appropriate sense)

$$Y_n \xrightarrow{\mathscr{L}} B \quad \text{as } n \to \infty. \qquad (8.109)$$

Another possibility is to linearly interpolate Y_n on $j/n < t < (j+1)/n$, for all $j = 0, 1, \ldots$, yielding a continuous process $\widetilde{Y}_n(t)$ (with polygonal paths) which agrees with $Y_n(t)$ at points $t = j/n$,

$$\widetilde{Y}_n(t) = S_j/\sqrt{n} + n((t - j/n)(S_{j+1}/\sqrt{n} - S_j/\sqrt{n})$$
$$= S_j/\sqrt{n} + n((t - j/n)Z_{j+1}/\sqrt{n} \text{ for } j/n \leq t \leq (j+1)/n \ (j = 0, 1, 2, \ldots).$$

Then, in the usual sense of weak convergence of probability measures on metric spaces,

$$\widetilde{Y}_n \xrightarrow{\mathscr{L}} B, \quad \text{as } n \to \infty. \qquad (8.110)$$

That is, if h is a real-valued bounded continuous function on $C[0, \infty)$, then

$$E(h(\widetilde{Y}_n)) \longrightarrow E(h(B)) \quad \text{as } n \to \infty. \qquad (8.111)$$

Also, if h is a continuous real-valued function on $C[0, \infty)$, then

$$h(\widetilde{Y}_n)) \xrightarrow{\mathscr{L}} h(B) \quad \text{as } n \to \infty. \qquad (8.112)$$

As an example,

$$\max\{Y_n(t); 0 \le t \le 1\} = \max\{\widetilde{Y}_n(t); 0 \le t \le 1\}$$
$$= \max\{S_j/\sqrt{n} : j = 0, 1, \ldots, n\} \xrightarrow{\mathscr{L}} \max\{B(t) : 0 \le t \le 1\}. \qquad (8.113)$$

Using a simple symmetric random walk S_n (with $Z_j = +1$ or -1 with probabilities $1/2$ each), the limiting distribution function on the left can be computed directly (from binomial probabilities), thus determining the distribution of $\max\{B(t) : 0 \le t \le 1\}$. On the other hand, the convergence (8.113) holds for partial sums S_j of an arbitrary mean zero sequence Z_j. Hence the limiting distribution for $\max\{S_j/\sqrt{n} : j = 0, 1, \ldots, n\}$ is determined for arbitrary partial sum processes (up to a scalar factor σ^2, which can be taken into account simply by standardizing). The second assertion is referred to as the *invariance principle,* while the convergence (8.110) is called the *functional central limit theorem.*

We will apply the functional central limit theorem (FCLT) to derive the Kolmogorov-Smirnov goodness-of-fit test: $H_0 : Q = Q_0$, based on i.i.d. real-valued observations $X_1, \ldots, X_n$ from an unknown continuous distribution Q (on the real line $\mathbb{R}$). Here Q_0 is a given hypothesized distribution for the observations.

Theorem 8.5 (Kolmogorov-Smirnov One-Sample Statistic). *Let $F_n(t) = n^{-1} \sum_{1 \le j \le n} 1_{\{X_j \le t\}}$ be the empirical distribution function. Also let F be the distribution function of Q. If F is continuous on $\mathbb{R}$, then*

$$\sqrt{n}\sup\{|F_n(t) - F(t)| : t \in \mathbb{R}\} \xrightarrow{\mathscr{L}} \max\{|B(t) - tB(1)| : t \in [0,1]\}, \quad as. \, t \to \infty. \qquad (8.114)$$

Here B is a standard Brownian motion.

Proof. First, by the CLT for proportions (or sums of Bernoulli random variables), $1_{\{X_j \le t\}}$, $\sqrt{n}(F_n(t) - F(t)) \xrightarrow{\mathscr{L}} N(0, F(t)(1 - F(t))$. If $t_1 < t_2, \cdots < t_k$, then

$$\sqrt{n}(F_n(t_1) - F(t_1), F_n(t_2) - F(t_2), \ldots, F_n(t_k) - F(t_k)) \xrightarrow{\mathscr{L}} N(0, \Gamma(t_1, t_2, \ldots, t_k)),$$

where the (i, i)-element of the symmetric matrix Γ is $\mathrm{var}(1_{\{X_j \le t_i\}}) = F(t_i)(1 - F(t_i))$, and the (i, j)-elements are $\mathrm{Cov}(1_{\{X_j \le t_i\}}, 1_{\{X_j \le t_j\}}) = F(t_i)(1 - F(t_j))$ for $i < j$. Note that the above equation may be checked by taking a linear combination of the k components on the left and applying the classical one-dimensional CLT to the summands $c_1 1_{\{X_j \le t_1\}} + c_2 1_{\{X_j \le t_2\}} + \cdots + c_k 1_{\{X_j \le t_k\}}$ $(j = 1, \ldots, n)$. By the arguments leading to (8.109), one now has

$$\sqrt{n}(F_n(\cdot) - F(\cdot)) \xrightarrow{\mathscr{L}} W, \qquad (8.115)$$

where W is a Gaussian process on $\mathbb{R}$, with $E(W(t)) = 0$ for all t, $\mathrm{var}(W(t)) = F(t)(1 - F(t))$, and $\mathrm{Cov}(W(s), W(t)) = F(s)(1 - F(t))$ for $s < t$. In particular,

$$\sup\left\{\sqrt{n}|F_n(t) - F(t)| : t \in \mathbb{R}\right\} \xrightarrow{\mathscr{L}} \sup\{|W(t)| : t \in \mathbb{R}\}. \qquad (8.116)$$

To simplify (8.116) for computational as well as theoretical purposes, we now consider the random variables $U_j = F(X_j)$, $1 \le j \le n$, and show that

$$U_j = F(X_j), \ 1 \le j \le n, \ \text{are i.i.d. uniform on } [0, 1]. \qquad (8.117)$$

To see this first assume that the continuous function F is strictly increasing on the range of X_j (i.e., on the smallest interval (a, b) (finite or infinite) such that $P(a < X_j < b)) = 1$). Then $P(U_j \leq u) = P(F(X_j) \leq u) = P(X_j \leq F^{-1}(u)) = F(F^{-1}(u)) = u$, for all $0 < u < 1$ and, clearly, $P(U < 0) = 0$ and $P(U > 1) = 0$. This argument extends to the case where F may not be strictly increasing on (a, b). For if u is such that $F^{-1}(u)$ is an interval $[c, d]$, so that $F(t) = u$ on $[c, d]$ and $F(t) < u$ for $t < c$ and $F(t) > u$ for $t > b$, then one has the equality of the sets $\{F(X_j) \leq u\}$ and $\{X_j \leq d\}$, so that $P(F(X_j) \leq u) = P(X_j \leq d) = F(d) = u$.

Applying the same arguments to U_j $(1 \leq j \leq n)$ as we used for X_j $(1 \leq j \leq n)$ above, and writing the empirical distribution function of U_j $(1 \leq j \leq n)$ as G_n, it follows that

$$\sqrt{n}\{(G_n(u) - u) : 0 \leq u \leq 1\} \xrightarrow{\mathscr{L}} B^*, \tag{8.118}$$

where B^* is a Gaussian process on $[0, 1]$, with $E(B^*(u)) = 0$ for all $0 \leq u \leq 1$, and $\operatorname{Cov}(B^*(s), B^*(t)) = s(1 - t)$ for $0 \leq s \leq t \leq 1$. Checking means, variances and covariances, we easily see that the process B^* has the same distribution as the Gaussian process

$$\{B(u) - sB(1) : 0 \leq s \leq 1\}, \tag{8.119}$$

where $B(\cdot)$ is the *standard Brownian motion*. Hence B^* is identified as the process (8.119) and is known as the *Brownian Bridge*.

The analog of (8.116) is then

$$D_n \equiv \sup\{\sqrt{n}|G_n(u) - u| : u \in [0, 1]\} \xrightarrow{\mathscr{L}} \sup\{|B^*(u)| : u \in [0, 1]\} = D, \text{ say.} \tag{8.120}$$

Now, in view of (8.117), one has

$$F_n(t) = n^{-1} \sum_{1 \leq j \leq n} \mathbf{1}_{\{X_j \leq t\}} = n^{-1} \sum_{1 \leq j \leq n} \mathbf{1}_{\{F(X_j) \leq F(t)\}} = n^{-1} \sum_{1 \leq j \leq n} \mathbf{1}_{\{U_j \leq F(t)\}} = G_n(F(t)),$$

so that

$$\sup\{\sqrt{n}|F_n(t) - F(t)| : t \in \mathbb{R}\} = \sup \sqrt{n}|G_n(F(t)) - F(t)| : t \in \mathbb{R}\}$$
$$= \sup\{\sqrt{n}|G_n(u) - u| : u \in [0, 1]\}. \tag{8.121}$$

From (8.120) and (8.121) we arrive at (8.114). $\qquad\qquad\square$

The distribution of D on the right side of (8.120) is known, and is given by[3]

$$P(D \leq d) = 1 - 2 \sum_{1 \leq k < \infty} (-1)^{k+1} \exp\{-2k^2 d^2\}, \quad d \in [0, \infty). \tag{8.122}$$

The following Corollary is an immediate consequence of the theorem above.

Corollary 8.2 (Kolmogorov-Smirnov Goodness-of-Fit Test). *Let F_0 be the distribution function of Q_0. Consider the Kolmogorov-Smirnov test for $H_0 : Q = Q_0$ to reject H_0 iff*

$$D_n > d_{1-\alpha} \tag{8.123}$$

---

[3] See Billingsley (1968), p. 85.

where D_n is as given by the left side of (8.114) with $F = F_0$, and $d_{1-\alpha}$ is the $(1 - \alpha)$th quantile of the distribution (8.122). Then the test has asymptotic level of significance α. □

Remark 8.8. Note that the limiting distribution (8.122) under H_0 does not depend on F_0.

By similar arguments one may derive the following results (See Serfling 1980, Chap. 6, and van der Vaart 1998, Chaps. 12, 19) for the *Cramér-von Mises statistic*

$$C_n = n \int (F_n(t) - F(t))^2 dF(t), \tag{8.124}$$

and the *Anderson-Darling statistic*

$$A_n = n \int \frac{(F_n(t) - F(t))^2}{F(t)(1 - F(t))} \, dF(t). \tag{8.125}$$

Theorem 8.6. *Under the hypothesis of Theorem 8.5 the following hold:*

$$C_n \xrightarrow{\mathscr{L}} \int_0^1 B^{*2}(t)dt, \quad A_n \xrightarrow{\mathscr{L}} \int_0^1 \frac{B^{*2}(t)}{t(1 - t)} \, dt, \tag{8.126}$$

where $B^*(t)$, $0 \le t \le 1$, is the Brownian bridge. □

The limiting distributions in (8.126) do not depend on F provided F is continuous. Let $c_{1-\alpha}$ and $a_{1-\alpha}$ be the $(1 - \alpha)$-th quantiles of $\int_0^1 (B^*(t))^2 dt$ and $\int_0^1 \{(B^*(t))^2/t(1 - t)\}dt$, respectively.

Corollary 8.3. *Let F_0 be a continuous distribution function. For the null hypothesis $H_0 : F = F_0$, (a) the Cramér-von Mises test: Reject H_0 iff $C_n > C_{1-\alpha}$ and (b) the Anderson-Darling test: Reject H_0 iff $A_n > a_{1-\alpha}$, are of asymptotic level α, where C_n and A_n are computed with $F = F_0$.*

The following expressions for the statistics D_n, C_n and A_n facilitate their computation:

$$D_n = \sqrt{n} \max_{1 \le i \le n} \max \left\{ \frac{i}{n} - U_{(i)}, U_{(i)} - \frac{i - 1}{n} \right\},$$

$$C_n = \frac{1}{12n} + \sum_{i=1}^{n} \left(U_{(i)} - \frac{2i - 1}{n} \right)^2,$$

$$A_n = -n - \frac{1}{n} \sum_{i=1}^{n} (2i - 1) \left\{ \log U_{(i)} + \log(1 - U_{(n-i+1)}) \right\} \tag{8.127}$$

where $U_{(i)} = F_0(X_{(i)})$, and $X_{(1)} < X_{(2)} < \cdots < X_{(n)}$ is the ordering of the observations $X_1, X_2, \ldots, X_n$ (See Exercise 8.9 for the verification of the expression for D_n in (8.127)). For the expressions for C_n and D_n and much more see D'Agostino and Stephens (1986).

Usually, the models one is required to check for validity are parametric models with unknown parameters, e.g., the Normal model $N(\mu, \sigma^2)$ with μ and σ^2 unknown. For such *composite goodness-of-fit* tests, in the expression $F(t) = F(t; \theta)$ one replaces the unknown parameter(s) θ by a good estimator $\hat{\theta}_n$, which we take to be the MLE (See (8.121), (8.124), (8.125)). This changes the asymptotic distributions, as may be seen from the following expansion around the true parameter value θ_0:

$$\sqrt{n}(F_n(t) - F(t; \hat{\theta}_n)) = \sqrt{n}(F_n(t) - F(t; \theta_0))$$

$$-\sqrt{n}(\hat{\theta}_n - \theta_0)\frac{\partial F(t; \theta)}{\partial \theta}\bigg|_{\theta=\theta_0} + o_p(1). \tag{8.128}$$

Under regularity conditions, the second term on the right is asymptotically Normal and therefore cannot be ignored. For a detailed discussion of and comprehensive references to the goodness-of-fit literature, see Dasgupta (2008), Chaps. 26–28.

The Cramér-von Mises and Anderson-Darling tests and more recent tests such as the one due to Shabiro and Wilk (1965), mostly outperform the Kolmogorov-Smirnov test since the latter is less sensitive to probabilities at the tail than the other tests.

Finally, note that the above procedures based on the empirical process cannot test the goodness-of-fit of discrete distributions such as the Poisson. Hence, although the frequency chi-square test is not consistent, it provides a reasonable and widely used procedure in this case.

8.5 Nonparametric Inference for the Two-Sample Problem

One of the most important problems in statistics is to decide if two populations, or distributions, are different, based on random samples from them. Does one brand of a certain commodity last longer than another? Has the position of the earth's magnetic poles, say the South Pole, shifted from the Quaternary period (2.59 million years ago) to the modern era? To answer the first type of questions on one-dimensional distributions, in Sect. 8.2 procedures such as those based on ranks were compared to the nonparametric procedure based on the two-sample t-statistic. The second type of questions concern multivariate distributions for which no natural ordering of data is available. The particular example mentioned here involves observations lying on the unit sphere S^2; a surface of dimension two. The analysis based on Fréchet means appears in Chap. 12.

In general, a multi-dimensional version of t^2, namely a chi-square statistic, may be used to test effectively if two distributions have different mean vectors.

Let $X_1, \ldots, X_m$ and $Y_1, \ldots, Y_n$ be two independent random samples from two populations with distributions Q_x and Q_y with means μ_x and μ_y, and finite variances σ_x^2 and σ_y^2. For testing a hypothesis concerning $\mu_x - \mu_y$, or estimating it we use the two-sample t-statistic

$$t = \{\overline{X} - \overline{Y} - (\mu_x - \mu_y)\}/\sqrt{(s_x^2/m + s_y^2/n)}, \quad s_x^2 = \sum_{1 \leq j \leq m} (X_j - \overline{X})^2/(m-1),$$

$$s_y^2 = \sum_{1 \leq j \leq n} (Y_j - \overline{Y})^2/(n-1). \tag{8.129}$$

We will denote by $t^{\sim}$ the statistic obtained from (8.129) on replacing s_x^2, s_y^2 by σ_x^2 and σ_y^2, respectively.

Proposition 8.2. *The statistic t in (8.129) converges in distribution to $N(0,1)$, as $m \to \infty$ and $n \to \infty$.*

Proof. Write

$$t^\sim = U - V = \sum_{1 \le j \le m+n} \zeta_{j,m,n}$$

where

$$\zeta_{j,m,n} = (1/m)\{X_j - \mu_x\}/\sqrt{(\sigma_x^2/m + \sigma_y^2/n)} \quad \text{for } 1 \le j \le m,$$

$$\zeta_{m+j,m,n} = -(1/n)\{Y_j - \mu_y\}/\sqrt{(\sigma_x^2/m + \sigma_y^2/n)} \quad \text{for } 1 \le j \le n.$$

Let $\gamma_{j,n}^2 = E\zeta_{j,m,n}^2$, so that $\sum_{1 \le j \le m+n} \gamma_{j,n}^2 = 1$. To apply Lindeberg's CLT, we need to show that, for every given $\varepsilon > 0$,

$$\sum_{1 \le j \le m+n} E\left[\zeta_{j,m,n}^2 : |\zeta_{j,m,n}| > \varepsilon\right] \longrightarrow 0 \quad \text{as } m \to \infty \text{ and } n \to \infty, \qquad (8.130)$$

where $E(Z : A)$ denotes $E(Z\mathbf{1}_A)$. Now the left side of (8.130) equals

$$\sum_{1 \le j \le m} E\left[\left\{\frac{1}{m}(X_j - \mu_x)/\sqrt{\frac{\sigma_x^2}{m} + \frac{\sigma_y^2}{n}}\right\}^2 : \left|\frac{1}{m}(X_j - \mu_x)/\sqrt{\frac{\sigma_x^2}{m} + \frac{\sigma_y^2}{n}}\right| > \varepsilon\right]$$

$$+ \sum_{1 \le j \le n} E\left[\left\{\frac{1}{m}(Y_j - \mu_y)/\sqrt{\frac{\sigma_x^2}{m} + \frac{\sigma_y^2}{n}}\right\}^2 : \left|\frac{1}{n}(Y_j - \mu_y)/\sqrt{\frac{\sigma_x^2}{m} + \frac{\sigma_y^2}{n}}\right| > \varepsilon\right]$$

$$\le \sum_{1 \le j \le m} E\left[\left\{\frac{1}{m}(X_j - \mu_x)/\sqrt{(\sigma_x^2/m)}\right\}^2 : \left|\frac{1}{m}(X_j - \mu_x)/\sqrt{(\sigma_x^2/m)}\right| > \varepsilon\right]$$

$$+ \sum_{m+1 \le j \le m+n} E\left[\left\{\frac{1}{n}(Y_j - \mu_y)/\sqrt{(\sigma_y^2/n)}\right\}^2 : \left|\frac{1}{n}(Y_j - \mu_y)/\sqrt{(\sigma_y^2/n)}\right| > \varepsilon\right]$$

$$\le mE\left[\left\{\frac{1}{m}(X_1 - \mu_x)/\sqrt{(\sigma_x^2/m)}\right\}^2 : \left|\frac{1}{m}(X_1 - \mu_x)/\sqrt{(\sigma_x^2/m)}\right| > \varepsilon\right]$$

$$+ nE\left[\left\{\frac{1}{m}(Y_1 - \mu_y)/\sqrt{(\sigma_x^2/m + \sigma_y^2/n)}\right\}^2 : \left|\frac{1}{n}(Y_j - \mu_y)/\sqrt{(\sigma_y^2/n)}\right| > \varepsilon\right]$$

$$= E\left[\{(X_1 - \mu_x/\sigma_x)^2 : |(X_1 - \mu_x)/\sigma_x| > \sqrt{m}\,\varepsilon\}\right]$$

$$+ E\left[\{(Y_1 - \mu_y/\sigma_y)^2 : |(Y_1 - \mu_y| > \sqrt{n}\,\varepsilon)\}\right]$$

$$= \int_{\{|x| > \sqrt{m}\varepsilon\}} x^2 Q_x^\sim(dx) + \int_{\{|y| > \sqrt{n}\varepsilon\}} y^2 Q_y^\sim(dy) \longrightarrow 0 \text{ as } m \to \infty \text{ and } n \to \infty,$$

where $Q_x^\sim$ is the distribution of $(X_1 - \mu_x)/\sigma_x$ and $Q_y^\sim$ is the distribution of $(Y_1 - \mu_y)/\sigma_y$. This shows that $t^\sim \overset{\mathscr{L}}{\longrightarrow} N(0,1)$, as $m \to \infty$ and $n \to \infty$. Application of Slutsky's Lemma now shows that $t \overset{\mathscr{L}}{\longrightarrow} N(0,1)$, as $m \to \infty$ and $n \to \infty$. $\qquad \square$

One rejects $H_0 : \mu_x = \mu_y$ if $|t| > z_{1-\alpha}$ where $z_{1-\alpha}$ is the $(1-\alpha)$-th quantile of $N(0,1)$. One may also reject $H_0 : Q_x = Q_y$ if $|t| > z_{1-\alpha}$.

We next turn to the case of *the two-sample multi-dimensional problem.* Here $X_1, \ldots, X_m$, and $Y_1, \ldots, Y_n$ are independent random samples from two distributions Q_x and Q_y on $\mathbb{R}^k$ having respective means (vectors) μ_x and μ_y, and finite non-singular covariance matrices $\sum_x = ((\sigma_{r,s,x}))_{1 \leq r,s \leq k}$ and $\sum_y = ((\sigma_{r,s,y}))_{1 \leq r,s \leq k}$. Then one has the following result. We regard all vectors as column vectors, unless specified otherwise. Recall the notation $\chi^2(k)$ for the chi-square distribution with k degrees of freedom, and $\chi^2_{1-\alpha}(k)$ for the $(1-\alpha)$th quantile of a χ^2_k distribution.

Proposition 8.3. *As $m \to \infty$ and $n \to \infty$, one has*

$$[\overline{X} - \overline{Y} - (\mu_x - \mu_y)] \widehat{\sum}_{m,n}^{-1} [\overline{X} - \overline{Y} - (\mu_x - \mu_y)] \xrightarrow{\mathscr{L}} \chi^2_k,$$

where $\widehat{\sum}_{m,n} = [(1/m)\widehat{\sum}_x + (1/n)\widehat{\sum}_y]$, *and* $\widehat{\sum}_x = ((\hat{\sigma}_{r,s,x}))_{1 \leq r,s \leq k}$ *and* $\widehat{\sum}_y = ((\hat{\sigma}_{r,s,y}))_{1 \leq r,s \leq k}$ *are the sample covariance matrices with elements*

$$\hat{\sigma}_{r,s,x} = (m-1)^{-1} \sum_{1 \leq j \leq m} (X_j^{(r)} - \overline{X}^{(r)})(X_j^{(s)} - \overline{X}^{(s)}),$$

$$X_j = (X_j^{(1)}, \ldots, X_j^{(k)})', \quad \overline{X}^{(r)} = m^{-1} \sum_{1 \leq j \leq m} X_j^{(r)};$$

$$\hat{\sigma}_{r,s,y} = (n-1)^{-1} \sum_{1 \leq j \leq n} (Y_j^{(r)} - \overline{Y}^{(r)})(Y_j^{(s)} - \overline{Y}^{(s)}),$$

$$Y_j = (Y_j^{(1)}, \ldots, Y_j^{(k)})', \quad \overline{Y}^{(r)} = n^{-1} \sum_{1 \leq j \leq n} Y_j^{(r)}.$$

Proof. Note that, if $Z = (Z^{(1)}, \ldots, Z^{(k)})'$ is a k-dimensional standard Normal random vector $N(0, I_k)$, where I_k is the $k \times k$ identity matrix, then $|Z|^2 = (Z^{(1)})^2 + \cdots + (Z^{(k)})^2$ has the chi-square distribution χ^2_k. More generally, if $Z = (Z^{(1)}, \ldots, Z^{(k)})$ is a k-dimensional Normal distribution $N(0, \sum)$, where $\sum = ((\sigma_{rs}))$ is a $k \times k$ positive definite (covariance) matrix, and $\sum^{-1} = ((\sigma^{rs}))$, then

$$Z' \sum^{-1} Z = \sum_{1 \leq r,s \leq k} \sigma^{rs} Z^{(r)} Z^{(s)} \quad \text{has a } \chi^2_k \text{ distribution.}$$

Now, as in the case of $k = 1$, one has the multidimensional CLT

$$\left(\frac{\sum_x}{m} + \frac{\sum_y}{n}\right)^{-\frac{1}{2}} [\overline{X} - \overline{Y} - (\mu_x - \mu_y)] \xrightarrow{\mathscr{L}} N(0, I_k), \quad \text{as } m \to \infty \text{ and } n \to \infty.$$

$$(8.131)$$

Here, for a symmetric positive definite matrix A, $A^{-1/2}$ is the symmetric positive definite matrix satisfying $A^{-1/2}A^{-1/2} = A^{-1}$. To prove (8.131), denote the random vector on the left side of (8.131) by $W_n = (W_n^{(1)}, \ldots, W_n^{(k)})'[$. It is enough to prove that every linear combination $\sum_{1 \leq r \leq k} c_r W_n^{(r)}$ of the k coordinates of W_n converges in distribution to the corresponding Normal distribution of $\sum_{1 \leq r \leq k} c_r Z^{(r)}$, where $Z = (Z^{(1)}, \ldots, Z^{(k)})$ is Normal $N(0, I_k)$. The last convergence follows from Lindeberg's CLT the same way as in the case of $k = 1$. Hence

$$W_n' W_n = [\overline{X} - \overline{Y} - (\mu_x - \mu_y)]' \left(\frac{1}{m}\Sigma_x + \frac{1}{n}\Sigma_y\right)^{-1} [\overline{X} - \overline{Y} - (\mu_x - \mu_y)] \xrightarrow{\mathscr{L}} \chi^2_k.$$

$$(8.132)$$

Finally, applying the general version of Slutsky's Lemma, using the fact that $\hat{\sigma}_{r,s,x} \rightarrow \sigma_{r,s,x}$ and $\hat{\sigma}_{r,s,y} \rightarrow \sigma_{r,s,y}$ almost surely for all r, s, as $m \rightarrow \infty$ and $n \rightarrow \infty$, one obtains the desired result. It may be noted that the elements of the inverse $(\sum_x /m + \sum_y /n)^{-1}$ are continuous functions of the elements of $(\sum_x /m + \sum_y /n)$ (on the set where this matrix is nonsingular). Thus the Proposition follows from (8.132) and this consistency. $\qquad\square$

The above proposition may be used to obtain an ellipsoidal confidence region D for $\mu_x - \mu_y$ of asymptotic level $1 - \alpha$ given by

$$P[\mu_x - \mu_y \in D = \{c \in \mathbb{R}^k : [\overline{X} - \overline{Y} - c]' \hat{\sum}_{m,n}^{-1} [\overline{X} - \overline{Y} - c] \leq \chi_{1-\alpha}^2(k)] \longrightarrow 1 - \alpha,$$
$$\text{as } m \rightarrow \infty \text{ and } n \rightarrow \infty.$$

A test, of asymptotic size α, rejects $H_0 : \mu_x = \mu_y$ iff the origin 0 does not belong to D, i.e., iff

$$[\overline{X} - \overline{Y}]' \hat{\sum}_{m,n}^{-1} [\overline{X} - \overline{Y}] > \chi_{1-\alpha}^2(k). \tag{8.133}$$

Note that in the case $k = 1$, $T^2 = t^2$ is an asymptotic chi-square statistic with degrees of freedom 1. We used t there since it can be used to obtain one-sided tests and confidence intervals as well.

Although two different (multivariate) distributions may have the same means, i.e., the test (8.133) is not consistent for testing $Q_x = Q_y$, in many high-dimensional problems the test is usually quite effective in discriminating two distributions with different features (See the section on Fréchet means in Chap. 12, for example).

We next turn to the two-sample Kolmogorov-Smirnov test, which is consistent. Suppose $X_1, \ldots, X_m$ and $Y_1, \ldots, Y_n$ are real-valued independent random samples from distributions Q_x and Q_y, respectively. We wish to test $H_0 : Q_x = Q_y$, (or, equivalently, that the distribution function F_x of X_j equals the distribution function F_y of Y_j).

Theorem 8.7 (Kolmogorov-Smirnov Two-Sample Statistic). *Suppose F_x and F_y are continuous, $m \rightarrow \infty$ and $n \rightarrow \infty$, and $m/(m+n) \rightarrow \theta$, $0 < \theta < 1$. Let $F_{m,x}(t)$, $F_{n,y}(t)$ be the empirical distribution functions of $X_j s$ ($1 \leq j \leq m$) and $Y_j s$ ($1 \leq j \leq n$), respectively. If $F_x = F_y$, then*

$$D_{m,n} \equiv \left(\frac{mn}{(m+n)}\right)^{\frac{1}{2}} \sup\{|F_{m,x}(t) - F_{n,y}(t)| : t \in \mathbb{R}\} \xrightarrow{\mathscr{L}} \sup\{|B^*(u)| : u \in [0,1]\} = D,$$
$$\tag{8.134}$$

say, where B^* is a Brownian bridge.

Proof. Assume $F_x = F_y = F$, say. By the proof of Theorem 8.5,

$$\sqrt{m}(F_{m,x}(t) - F(t)) = \sqrt{m}(G_{m,x}(F(t)) - F(t)) \xrightarrow{\mathscr{L}} B_1^*(F(t)),$$
$$\sqrt{n}(F_{n,y}(t) - F(t)) = \sqrt{n}(G_{n,y}(F(t)) - F(t)) \xrightarrow{\mathscr{L}} B_2^*(F(t)), \ (t \in \mathbb{R}), \tag{8.135}$$

where (i) $F_{m,x}(t) = m^{-1} \sum_{1 \leq j \leq m} \mathbf{1}\{U_{j,x} \leq F(t)\}$, $U_{j,x} = F(X_j)$ ($1 \leq j \leq m$) are independent uniform on $[0,1]$, (ii) $F_{n,y}(t) = n^{-1} \sum_{1 \leq j \leq n} \mathbf{1}\{U_{j,y} \leq F(t)\}$, $U_{j,y} = F(Y_j)$ ($1 \leq j \leq n$) are independent uniform on $[0,1]$, (iii) B_1^* and B_2^* are independent Brownian bridges.

Hence

$$\left(\frac{mn}{(m+n)}\right)^{\frac{1}{2}}\left((F_{m,x}(t))-F_{n,y}(t)\right)$$

$$=\left(\frac{n}{(m+n)}\right)^{\frac{1}{2}}m^{\frac{1}{2}}(F_{m,x}(t))-F(t))-\left(\frac{m}{(m+n)}\right)^{\frac{1}{2}}n^{\frac{1}{2}}(F_{n,y}(t))-F(t))$$

$$\overset{\mathscr{L}}{\longrightarrow}\theta B_1^*(F(t))-(1-\theta)B_2^*(F(t))=B^*(F(t)),\ \text{say},\tag{8.136}$$

where, by checking means and covariances, it is seen that $B^*=\theta B_1^*-(1-\theta)B_2^*$ is a Brownian bridge. Arguing as above, this says

$$\left(\frac{mn}{(m+n)}\right)^{\frac{1}{2}}\left((F_{m,x}(\cdot))-F_{n,y}(\cdot)\right)\overset{\mathscr{L}}{\longrightarrow}B^*(F(\cdot)),\tag{8.137}$$

in the sense of convergence in distribution of the stochastic process on $[0,\infty)$ on the left to the one on the right. Therefore, writing $u=F(t)$, one has

$$D_{m,n}\equiv\sup\left\{\left|\left(\frac{mn}{(m+n)}\right)^{\frac{1}{2}}\left((F_{m,x}(t))-F_{n,y}(t)\right)\right|:t\in\mathbb{R}\right\}$$

$$\overset{\mathscr{L}}{\longrightarrow}\sup\{|B^*(F(t))|:t\in\mathbb{R}\}$$
$$=\sup\{|B^*(u)|:u\in[0,1]\}=D,\quad\text{say}.\tag{8.138}$$

Corollary 8.4. *Suppose F_x and F_y are continuous, and m and $n\to\infty$ as specified in Theorem 8.7. Then a test of asymptotic level of significance α for $H_0:F_x=F_y$ is given by*

$$Reject\ H_0\ iff\ D_{m,n}>d_{1-\alpha},$$

where $d_{1-\alpha}$, is the $(1-\alpha)$th quantile of the distribution of D.

Remark 8.9. Note that (8.138) provides a nonparametric test for the equality of two continuous distributions.

Remark 8.10. Since the distribution function of D in Theorem 8.7 is continuous, one can dispense with the assumption in this theorem and in Corollary 8.4 that $m/(m+n)\to\theta\in(0,1)$, and simply require $m\to\infty$ and $n\to\infty$.

We conclude this section with a description of a simple method for multiple testing known as the *Bonferroni method.*

To illustrate the Bonferroni method, consider the testing of the equality of two multivariate means $\mu_x=(\mu_x^{(1)},\ldots,\mu_x^{(k)})'$ and $\mu_y=(\mu_y^{(1)},\ldots,\mu_y^{(k)})'$ of two populations based on random samples taken from them. Suppose the level of significance for the test of the null hypothesis $H_0:\mu_x=\mu_y$ is set at α. One may try to test the component null hypotheses $\mu_x^{(i)}=\mu_y^{(i)}$ $(i=1,\ldots,k)$ by k separate tests. Then the *Bonferroni principle says that H_0 is to be rejected if at least one of the null hypotheses $H_0^{(i)}:\mu_x^{(i)}=\mu_y^{(i)}$, is rejected at a level of significance α/k $(i=1,\ldots,k)$.* To see that under this procedure the level of significance for H_0 is no more than α, let $R^{(i)}$ be the event that the null hypothesis $H_0^{(i)}$ is rejected, and suppose that its level of significance is set at $\alpha^{(i)}$. Then the probability (under H_0) that H_0 is rejected equals P_{H_0} (at least one $R^{(i)}$ occurs among

$i = 1, \ldots, k) \leq \sum_{1 \leq i \leq k} P_{H_0}(R^{(i)}) = \sum_{1 \leq i \leq k} \alpha^{(i)} = \alpha$, if $\alpha^{(i)} = \alpha/k$ for each i. Thus the Bonferroni principle is *conservative* in that the actual level of significance is less than or equal to α under this procedure. As an illustration, if $k = 3$ and $\alpha = 0.05$, then the Bonferroni procedure is to test each of the three component hypotheses at a level of significance $(0.05)/3 = 0.01666\ldots$, and reject H_0 if at least one of the component hypotheses is rejected at level $0.01666\ldots$.

For estimating the p-value using the Bonferroni principle, let p-min denote the smallest p-value of the k tests (for $H_0^{(i)}$, $i = 1, \ldots, k$). Then the *p-value for H_0 is no more than k (p-min). That is, p-value of $H_0 \leq k(p\text{-}min)$*. For suppose one sets $\alpha = k \, (p - \min)$, and applies the procedure described in the preceding paragraph, namely, to reject H_0 if at least one of the tests is rejected at a level of significance α/k; then the test with the smallest p-value $(= p\text{-min})$ meets that rejection criterion.

There are modifications/corrections of the Bonferroni principle which provide sharper estimates of the actual p-value. But we will consider those in Chap. 13.

8.6 Large Sample Theory for Stochastic Processes

This section provides an introduction to semiparametric inference for an important class of time series models, and to an extension of the asymptotic properties of the maximum likelihood estimators for stationary ergodic processes. Because the linear time series here may be viewed as coordinates of Markov processes, we begin with a brief review of ergodic Markov processes.

Recall that a *Markov process X_n* in discrete time n $(n = 0, 1 \ldots)$ is a sequence of random variables defined on some probability space $(\Omega, \mathscr{F}, P)$ taking values on a (measurable) *state space S* (with a sigma-field $\mathscr{S}$) and governed by a *transition probability function* $p(x, B)$, which is the probability that the process moves to a set B in one step (or one unit of time), starting from a state x in S. The *Markov property* is the following:

$$p(X_n, B) = P(X_{n+1} \in B \mid X_0, \ldots, X_n) = P(X_{n+1} \in B \mid X_n) \quad (\forall B \in \mathscr{S}, \text{ and } \forall n).$$
$$(8.139)$$

Thus the conditional distribution of the "future" X_{n+1} given the "past" $X_0, \ldots, X_{n-1}$ and "present" X_n depends only on the "present" state X_n. This may also be expressed as

$$E[f(X_{n+1}) \mid X_0, \ldots, X_n] = E[f(X_{n+1}) \mid X_n]$$
$$= \int f(y)p(X_n, dy) \, \forall \text{ bounded measurable function } f \text{ on } S.$$
$$(8.140)$$

Here $p(x, dy)$ denotes the *distribution of X_{n+1}, given $X_n = x$*. One may check by iteration that the Markov property (8.139) or (8.140) implies (and is equivalent to) the more general property

$$P(X_{n+k} \in B \mid X_0, \ldots, X_n) = P(X_{n+k} \in B \mid X_n) \quad \forall \, k = 1, 2, \ldots, \text{and } \forall \, n.$$
$$(8.141)$$

The probability on the right provides the *k-step transition probability* $p^{(k)}(x, B)$, which may be iteratively obtained from the (one-step) transition probability $p(x, dy)$. For example,

$$p^{(2)}(x, B) = P(X_{n+2} \in B \mid X_n = x) = E[P(X_{n+2} \in B \mid X_n, X_{n+1}) \mid X_n)]_{X_n=x} =$$
$$= E[P(X_{n+2} \in B \mid X_{n+1}) \mid X_n = x]$$
$$= E[p(X_{n+1}, B) \mid X_n = x] = \int p(y, B)p(x, dy). \qquad (8.142)$$

The second equality is a property of the conditional expectation, while the Markov property is used for the third equality. One may express $p^{(k)}(x, B)$ similarly as

$$p^{(k)}(x, B) = \int p^{(k-1)}(y, B)p(x, dy). \qquad (8.143)$$

Also, the Markov property implies that, for every (measurable) subset C of S^{k+1}.

$$P[(X_n, X_{n+1}, \ldots, X_{n+k}) \in C \mid X_n = x] = P[(X_0, X_1, \ldots, X_k) \in C \mid X_0 = x] \quad \forall \; n. \qquad (8.144)$$

That is, the conditional distribution of $(X_n, X_{n+1}, \ldots, X_{n+k})$, given $X_n = x$, does not depend on n.

Remark 8.11. In the case $p(x, dy)$ has, for every x, a density $p(x, y)$ with respect to some sigma-finite measure μ one may express the Markov property by simply writing down the joint probability density of $(X_{n+1}, X_{n+2}, \ldots, X_{n+k})$ at a point $(y_1, y_2, \ldots, y_k)$, given $(X_0, X_1, \ldots, X_n)$, as $p(x, y_1)p(y_1, y_2) \ldots p(y_{k-1}, y_k)$ on $\{X_n = x\}$.

Definition 8.3. A probability measure π on S is said to be an *invariant probability*, or a *steady state distribution*, for a Markov process with transition probability $p(x, dy)$ if

(i) $\int p(y, B)\pi(dy) = \pi(B) \quad \forall B \in \mathscr{S}$, or (equivalently),

(ii) $\int Tf(y)\pi(dy) \equiv \int f(z)p(y, dz)\pi(dy) = \int f(z)\pi(dz) \, \forall$ bounded measurable f
on S.

Note that in the case $p(x, dy)$ has a density $p(x, y)$, an *invariant probability density* $\pi(y)$ satisfies

$$\int p(x, y)\pi(x)\mu(dx) = \pi(y) \quad \text{for all } y \text{ (outside a set of } \mu\text{-measure zero).}$$

In Definition 8.3 (ii), we have used the notation Tf to denote the function $Tf(y) = \int f(z)p(y, dz)$. Note that the left side of Definition 8.3 (i) says that X_1 has the same distribution as X_0 if X_0 *has distribution* π. Similarly, the left side of Definition 8.3 (ii) equals $\int [Ef(X_1 \mid X_0 = y)]\pi(dy)$ which is $Ef(X_1)$ if X_0 has distribution π, while the right side is $Ef(X_0)$). By the same argument if X_1 has distribution π, then X_2 has distribution π, and so on, implying that X_n has distribution π for all n, if X_0 has distribution π. Indeed more is true. If X_0 has distribution π, then the distribution of $(X_0, X_1, \ldots, X_k)$ is the same as the distribution of $(X_n, X_{n+1}, \ldots, X_{n+k})\ \forall\, n$. This follows from (8.144) and the fact that X_n has the same distribution as X_0, namely, π. Hence, if X_0 has distribution π, then the *after-n process* $X_n^+ = (X_n, X_{n+1}, \ldots, X_{n+k}, \ldots)$ has the same distribution as $X_0^+ = (X_0, X_1, \ldots, X_k, \ldots)$. This last property is referred to as the *stationarity of the process* $\{X_n : n = 0, 1, \ldots\}$.

Example 8.5 ((AR(1) Model). Let a stochastic process $X_n\ (n = 0, 1, \ldots)$ be defined on the state space $S = R$ by the recursion

$$X_{n+1} = g(X_n) + \varepsilon_{n+1} \quad (n = 0, 1, \ldots), \text{ and } X_0 \text{ is given,} \tag{8.145}$$

where g is a given (measurable) function on $\mathbb{R}$, and $\varepsilon_n\ (n = 1, 2, \ldots)$ is an i.i.d. sequence. Assume X_0 is independent of $\{\varepsilon_n : n = 1, 2, \ldots\}$. Then $\{X_n : n = 0, 1, \ldots\}$ is a Markov process with transition probability function given by

$$p(x, B) = P(X_1 \in B \mid X_0 = x) = P(g(x) + \varepsilon_1 \in B) = P(\varepsilon_1 \in B - g(x)),$$

where $B - g(x) = \{y - g(x) : y \in B\}$. A special case of (8.145) is the *autoregressive model of order 1, or AR(1) model,*

$$X_{n+1} = \alpha + \beta X_n + \varepsilon_{n+1} \quad (n = 0, 1, \ldots), \quad X_0 \text{ is independent of } \{\varepsilon_n : n = 1, 2, \ldots\},$$
$$\tag{8.146}$$

where $\{\varepsilon_n : n = 1, 2, \ldots\}$ is an i.i.d. sequence satisfying

$$E\varepsilon_n = 0, \quad 0 < E\varepsilon_n^2 = \sigma^2 < \infty. \tag{8.147}$$

We now show that this Markov process has a unique steady state or invariant probability if the following *stability condition* is satisfied:

$$|\beta| < 1. \tag{8.148}$$

To prove the existence of a unique invariant probability, we now demonstrate that

$$X_n \xrightarrow{\ \mathscr{L}\ } \frac{\alpha}{(1 - \beta)} + Z, \quad \text{where } Z = \sum_{0 \le j < \infty} \beta^j \varepsilon_{j+1}. \tag{8.149}$$

To see this, use (8.146) to get successively,

$$X_1 = \alpha + \beta X_0 + \varepsilon_1, \ \ X_2 = \alpha + \beta X_1 + \varepsilon_2 = \alpha + \beta\alpha + \beta^2 X_0 + \beta\varepsilon_1 + \varepsilon_2,$$
$$X_3 = \alpha + \beta X_2 + \varepsilon_3 = \alpha + \beta\alpha + \beta^2\alpha + \beta^3 X_0 + \beta^2\varepsilon_1 + \beta\varepsilon_2 + \varepsilon_3, \ldots$$
$$X_j = \alpha + \beta\alpha + \beta^2\alpha + \cdots + \beta^{j-1}\alpha + \beta^j X_0 + \beta^{j-1}\varepsilon_1 + \beta^{j-2}\varepsilon_2 + \cdots + \beta\varepsilon_{j-1} + \varepsilon_j,$$
$$\stackrel{\mathscr{L}}{=} \alpha + \beta\alpha + \beta^2\alpha + \cdots + \beta^{j-1}\alpha + \beta^j X_0 + \sum_{0 \le r \le j-1} \beta^r \varepsilon_{r+1}. \tag{8.150}$$

The last equality in distribution follows from the fact $\{\varepsilon_n : n = 1, 2, \ldots\}$ is an i.i.d. sequence (independent of X_0). Now (8.149) follows from (8.150), since $\{\varepsilon_n : n = 1, 2, \ldots\}$ is an i.i.d. sequence. We now appeal to the following general result.

Proposition 8.4. *Let* $\{X_n : n = 0, 1, \ldots\}$ *be a Markov process on a metric space* S *with a transition probability* $p(x, dy)$ *such that* X_n *converges in distribution to the same probability* π, *as* $n \to \infty$, *whatever be the initial sate* X_0. *Assume also* *that $x \to p(x, dy)$ is weakly continuous, i.e., the function $Tf(x) = \int f(y)p(x, dy)$ is continuous for every bounded continuous f. Then π is the unique invariant probability for the Markov process.*

Proof. Whatever the initial state X_0, one has

$$Ef(X_{n+1}) = E[E(f(X_{n+1}) \mid X_n)] = ETf(X_n) \longrightarrow \int Tf(y)\pi(dy) \text{ as } n \to \infty. \tag{8.151}$$

This follows from the definition of convergence in distribution, since X_n converges to π in distribution and Tf is a bounded continuous function. On the other hand, X_{n+1} converges in distribution to π, so that the left side of (8.151) converges to $\int f(y)\pi(dy)$. We, therefore, have

$$\int Tf(y)\pi(dy) = \int f(y)\pi(dy) \quad \forall \text{ bounded continuous } f.$$

From this the criterion (ii) in Definition 8.3 follows (for all bounded measurable f). For (a) the set of all bounded continuous functions is dense in $L^1(\pi)$—the set of all (equivalence classes of) functions on $\mathbb{R}$ integrable with respect to π, which includes all bounded measurable functions,[4] and (b) $f \to Tf$ is a contraction on $L^1(\pi)$, i.e., if f, g are in $L^1(\pi)$ then by the invariance of π,

$$\|Tf - Tg\|_1 \equiv \int |Tf(y) - Tg(y)|\pi(dy) = E|E(f(X_1) - g(X_1) \mid X_0)|$$
$$\leq E\,E(|f(X_1) - g(X_1)| \mid X_0))$$
$$= E|f(X_1) - g(X_1)| = \int |f(y) - g(y)|\pi(dy) \equiv \|f - g\|_1.$$

To prove *uniqueness* of the invariant probability π, let γ be any invariant probability of the Markov process, which implies that if X_0 has distribution γ then X_n has distribution γ for all n. But, no matter what X_0 is, X_n converges in distribution to π as $n \to \infty$. This implies that γ cannot be anything other than π. $\square$

Corollary 8.5. *Under the hypotheses (8.147) and (8.148), the Markov process X_n in (8.146) has the unique invariant probability given by the distribution π of the random variable $\alpha/(1 - \beta) + Z$, where $Z = \sum_{0 \leq j < \infty} \beta^j \varepsilon_{j+1}$. The mean of π is $\mu = \alpha/(1 - \beta)$, and the variance of π is $\delta = \sigma^2/(1 - \beta^2)$.*

For purposes of inference the following simple calculations are needed (see (8.150)).

$$E(X_n) = \alpha + \beta\alpha + \beta^2\alpha + \cdots + \beta^{n-1}\alpha + \beta^n E(X_0)$$
$$= \beta^n E(X_0) + \alpha(1 - \beta^n)/(1 - \beta);$$
$$\text{var}(X_n) = \text{var}(\alpha + \beta\alpha + \beta^2\alpha + \cdots + \beta^{n-1}\alpha + \beta^n X_0 + \beta^{n-1}\varepsilon_1 + \beta^{n-2}\varepsilon_2$$
$$+ \cdots + \beta\varepsilon_{n-1} + \varepsilon_n) =$$
$$= \beta^{2n}\text{var}(X_0) + (\beta^{2(n-1)} + \beta^{2(n-2)} + \cdots + \beta^2 + 1)\sigma^2$$
$$= \beta^{2n}\text{var}(X_0) + \sigma^2(1 - \beta^{2n})/(1 - \beta^2);$$

---

[4] See, e.g., Bhattacharya and Waymire (2007), p. 180.

$$\mathrm{Cov}(X_n, X_{n+j}) = \mathrm{Cov}(X_n, \alpha + \beta\alpha + \beta^2\alpha + \cdots + \beta^{j-1}\alpha + \beta^j X_n + \beta^{j-1}\varepsilon_{n+1}$$
$$+ \beta^{j-2}\varepsilon_{n+2}\cdots + \beta\varepsilon_{n+j-1} + \varepsilon_{n+j})$$
$$= \beta^j \mathrm{var}(X_n) = \beta^j\{\beta^{2n}\mathrm{var}(X_0) + \sigma^2(1 - \beta^{2n})/(1 - \beta^2\}. \quad (8.152)$$

We now turn to the problem of estimating α and β from observations $X_0, X_1, \ldots, X_n$. The least squares estimates are given, as usual, by

$$\underset{\alpha,\beta}{\arg\min} \sum_{1 \leq j \leq n} [X_j - (\alpha + \beta X_{j-1})]^2, \quad (8.153)$$

whose solution is (See (6.14), (6.16))

$$\hat{\beta} = \left[\sum_{1 \leq j \leq n} X_j(X_{j-1} - \overline{X}) \right] \Big/ \left[\sum_{1 \leq j \leq n} (X_{j-1} - \overline{X})^2 \right], \quad \hat{\alpha} = \overline{Y} - \hat{\beta}\overline{X}, \text{ where}$$

$$\overline{X} = \sum_{1 \leq j \leq n} X_{j-1}/n, \text{ and } \overline{Y} = \sum_{1 \leq j \leq n} X_j/n. \quad (8.154)$$

Theorem 8.8. *For the AR(1) model (8.146) assume in addition to (8.147) and (8.148) that either (a) X_0 has the invariant distribution, or (b) $E\varepsilon_j^4 < \infty$ and $E(X_0)^4 < \infty$. Then*

$$\sqrt{n}(\hat{\beta} - \beta) \xrightarrow{\mathscr{L}} N(0, 1 - \beta^2), \quad (8.155)$$

and

$$\sqrt{n}(\hat{\alpha} - \alpha, \hat{\beta} - \beta)' \xrightarrow{\mathscr{L}} N((0,0)', \Gamma), \quad (8.156)$$

where

$$\Gamma = \begin{bmatrix} \sigma^2 + \alpha^2(1 + \beta)/(1 - \beta) & -\alpha\sigma^2(1 + \beta) \\ -\alpha\sigma^2(1 + \beta) & 1 - \beta^2 \end{bmatrix} \quad (8.157)$$

Proof. We will first prove (8.155). By (8.154),

$$\sqrt{n}(\hat{\beta} - \beta) = \left[n^{-\frac{1}{2}} \sum_{1 \leq j \leq n} (X_{j-1} - \overline{X})\varepsilon_j \right] \Big/ \left[n^{-1} \sum_{1 \leq j \leq n} (X_{j-1} - \overline{X})^2 \right]. \quad (8.158)$$

We will first show that the denominator on the right converges in probability to the variance of the invariant distribution π,

$$\left[n^{-1} \sum_{1 \leq j \leq n} (X_{j-1} - \overline{X})^2 \right] \longrightarrow \delta = \sigma^2/(1 - \beta^2) \quad \text{in probability, as } n \to \infty.$$
$$(8.159)$$

First, writing $\mu = \alpha/(1 - \beta)$, one has (See (8.152)),

$$E\overline{X} = n^{-1} \sum_{0 \leq j \leq n-1} \{\beta^j E X_0 + \alpha(1 - \beta^j)/(1 - \beta)\}$$
$$= \mu + \{n^{-1}(1 - \beta^n)/(1 - \beta)\}E X_0 - n^{-1}\alpha(1 - \beta^n)/(1 - \beta)^2$$
$$= \mu + O(1/n), \quad E\overline{X} - \mu = O(1/n);$$

$$E(\overline{X} - \mu)^2 = E(\overline{X} - E\overline{X})^2 + (E\overline{X} - \mu)^2$$
$$= \operatorname{var}(\overline{X}) + O(n^{-2}),$$

$$\operatorname{var}(\overline{X}) = n^{-2}\left\{\sum_{0 \le j \le n-1} \operatorname{var}(X_j) + 2 \sum_{0 \le j \le n-1} \sum_{1 \le r \le n-1-j} \operatorname{Cov}(X_j, X_{j+r})\right\}$$

$$= n^{-2} \sum_{0 \le j \le n-1} \{\beta^{2j}\operatorname{var}(X_0) + \sigma^2/(1-\beta^2) - \sigma^2\beta^{2j}/(1-\beta^2)\}$$

$$+ 2n^{-2}\left\{\sum_{0 \le j \le n-1} \sum_{1 \le r \le n-1-j} \beta^r\left[\beta^{2j}\operatorname{var}(X_0) + (1-\beta^{2r})\sigma^2/(1-\beta^2)\right]\right\}$$

$$= O(1/n).$$

Using $EX_j - \mu = \beta^j(EX_0 - \mu)$ and $\operatorname{var}(X_j) = \beta^{2j}\operatorname{var}(X_0) + \sigma^2(1-\beta^{2j})/(1-\beta^2)$ (See (8.152)), one has $E|(EX_j - \mu)(X_j - EX_j)| \le |\beta^j| |EX_0 - \mu|\{\beta^{2j}\operatorname{var}(X_0) + \sigma^2(1-\beta^{2j})/(1-\beta^2)\}^{1/2}$, so that

$$n^{-1} \sum_{0 \le j \le n-1} (X_j - \overline{X})^2 =$$

$$= n^{-1} \sum_{0 \le j \le n-1} (X_j - \mu)^2 - (\overline{X} - \mu)^2 = n^{-1} \sum_{0 \le j \le n-1} (X_j - \mu)^2 + O_p(n^{-1}) =$$

$$= n^{-1} \sum_{0 \le j \le n-1} \{(X_j - EX_j)^2 + (EX_j - \mu)^2 + 2(EX_j - \mu)(X_j - EX_j)\} + O_p(n^{-1})$$

$$= n^{-1} \sum_{0 \le j \le n-1} \operatorname{var}(X_j) + O_p(n^{-1}) = \sigma^2/(1-\beta^2) + O_p(n^{-1}), \tag{8.160}$$

yielding (8.159). Note that in the stationary case $EX_j = \mu$. Next,

$$E\left|\left[n^{-\frac{1}{2}} \sum_{1 \le j \le n} (X_{j-1} - \overline{X})\varepsilon_j\right] - n^{-\frac{1}{2}} \sum_{1 \le j \le n} (X_{j-1} - \mu)\varepsilon_j\right|$$

$$= n^{-\frac{1}{2}} E\left|(\overline{X} - \mu) \sum_{1 \le j \le n} \varepsilon_j\right|$$

$$\le n^{-\frac{1}{2}}(E(\overline{X} - \mu)^2))^{\frac{1}{2}}(n\sigma^2)^{\frac{1}{2}} = O\left(n^{-\frac{1}{2}}\right),$$

so that $n^{-1/2}\sum_{1 \le j \le n}(X_{j-1} - \overline{X})\varepsilon_j] - n^{-1/2}\sum_{1 \le j \le n}(X_{j-1} - \mu)\varepsilon_j \to 0$ in probability as $n \to \infty$. Hence, by (8.158) and (8.160)

$$\sqrt{n}(\hat{\beta} - \beta) = [n^{-\frac{1}{2}} \sum_{1 \le j \le n} (X_{j-1} - \mu)\varepsilon_j]/\delta + o_p(1), \quad \text{where } o_p(1) \to 0 \text{ in probability.}$$

$$\tag{8.161}$$

Now $[n^{-1/2}\sum_{1 \le j \le n}(X_{j-1} - \mu)\varepsilon_j]/\delta$ is a *martingale*. For, writing $\mathscr{F}_{j-1}$ for the sigma-field of events generated by $\{X_0, X_1, \ldots, X_{j-1}\}$, one has, by independence of $\{X_0, X_1, \ldots, X_{j-1}\}$ and ε_j,

$$E[(X_{j-1} - \mu)\varepsilon_j \mid \mathscr{F}_{j-1}] = (X_{j-1} - \mu)E(\varepsilon_j \mid \mathscr{F}_{j-1}) = 0 \quad \text{for all } j.$$

If X_0 has the invariant distribution, the process $\{(X_n, \varepsilon_n) : n = 0, \dots\}$ is stationary, and $(X_{j-1} - \mu)\varepsilon_j$ are stationary ergodic martingale differences. In this case the martingale CLT applies and (8.155) is proved. Otherwise we may check the Lindeberg-type conditions for the *martingale CLT* to hold.[5] First,

$$\sum_{1 \le j \le n} E\left[(n^{-\frac{1}{2}}(X_{j-1} - \mu)\varepsilon_j)^2 \mid \mathscr{F}_{j-1}\right] = n^{-1} \sum_{1 \le j \le n} (X_{j-1} - \mu)^2 E(\varepsilon_j^2 \mid \mathscr{F}_{j-1}) =$$

$$= \sigma^2 n^{-1} \sum_{1 \le j \le n} (X_{j-1} - \mu)^2 \longrightarrow \sigma^2 \delta \text{ in probability as } n \to \infty, \tag{8.162}$$

using (8.159) in the last step. It remains to show that, for every $\theta > 0$,

$$\sum_{1 \le j \le n} E\left[(n^{-\frac{1}{2}}(X_{j-1} - \mu)\varepsilon_j)^2 \mathbf{1}\{|n^{-\frac{1}{2}}(X_{j-1} - \mu)\varepsilon_j| > \theta\} \mid \mathscr{F}_{j-1}\right] \to 0 \text{ in probability.}$$

$$\tag{8.163}$$

By Chebyshev's inequality, the left side of (8.163) is bounded above by

$$n^{-1} \sum_{1 \le j \le n} (E[(((X_{j-1} - \mu)\varepsilon_j)^2)^2 \mid \mathscr{F}_{j-1}])^{\frac{1}{2}}$$

$$\cdot E[(\mathbf{1}\{|n^{-\frac{1}{2}}(X_{j-1} - \mu)\varepsilon_j\}| > \theta\} \mid \mathscr{F}_{j-1})]^{\frac{1}{2}}$$

$$\le n^{-1} \sum_{1 \le j \le n} \{(X_{j-1} - \mu)^4\}^{\frac{1}{2}} \gamma \cdot \{E[(X_{j-1} - \mu)^2 \varepsilon_j^2 \mid \mathscr{F}_{j-1}]\}^{\frac{1}{2}}/(\theta\sqrt{n})$$

$$\le n^{-1} \sum_{1 \le j \le n} (X_{j-1} - \mu)^2 \gamma\sigma|X_{j-1} - \mu|/(\theta\sqrt{n})$$

$$= \left(n^{-1} \sum_{1 \le j \le n} |X_{j-1} - \mu|^3\right) \gamma\sigma/(\theta\sqrt{n})$$

$$\le \left\{n^{-1} \sum_{1 \le j \le n} (X_{j-1} - \mu)^4\right\}^{\frac{3}{4}} \gamma\sigma/(\theta\sqrt{n}) \quad [\gamma := (E\varepsilon_j^4)^{\frac{1}{2}}]. \tag{8.164}$$

By calculations using the expression for X_j in (8.150) (Exercise 8.14), one has

$$\varlimsup_n E\left(n^{-1} \sum_{1 \le j \le n} (X_{j-1} - \mu)^4\right) < \infty. \tag{8.165}$$

Now (8.163) follows from (8.164), (8.165). This proves (8.155). The proof of (8.156) may be given by expressing an arbitrary linear combination of $\sqrt{n}(\hat{\alpha} - \alpha)$ and $\sqrt{n}(\hat{\beta} - \beta)$ as a martingale (plus a negligible term), and appealing to the martingale central limit theorem, noting that $\sqrt{n}(\hat{\alpha} - \alpha) = n^{-\frac{1}{2}} \sum_{1 \le j \le n} \varepsilon_j + \sqrt{n}(\hat{\beta} - \beta)\mu + o_p(1)$ (Exercise 8.14). $\qquad \square$

Under the assumptions of Theorem 8.8, one can obtain CLT-based classical confidence intervals and tests for the parameters, using sample estimates of the

---

[5] See, e.g., Bhattacharya and Waymire (2009), pp. 507–511.

dispersions (see (8.154), (8.155), (8.159)) (Exercise 8.15). Bootstrap confidence intervals and tests for the parameters may be derived by resampling from the estimated residuals

$$\{\hat{\varepsilon}_j := X_j - \hat{\alpha} - \hat{\beta} X_{j-1} : j = 1, \ldots, n\}.$$

See Bose (1988) and Lahiri (2003).

Example 8.6 (AR(p) Model). We next consider the more general *autoregressive model of order* $p \geq 1$, or the *AR(p) model.* Here the sequence of real-valued random variables $\{X_n : n = 0, 1, 2, \ldots\}$ are recursively defined by

$$X_n = \alpha + \sum_{1 \leq r \leq p} \beta_r X_{n-r} + \varepsilon_n \quad (n = p, p+1, \ldots), \quad (\beta_p \neq 0), \tag{8.166}$$

given the *initial set of p values* $(X_0, \ldots, X_{p-1})$. The sequence $\{\varepsilon_n : n = p, p+1, \ldots\}$ is i.i.d. and independent of $(X_0, \ldots, X_{p-1})$, and satisfy (8.147). Since X_n depends on the past p values in the sequence, in addition to the random error ε_n, it is not Markov. However, the vector sequence $\{Y_n = (X_{n-p+1}, X_{n-p+2}, \ldots, X_n)' : n = p - 1, p, \ldots\}$ is Markov and satisfies the recursion

$$Y_n = \alpha^{\sim} + B Y_{n-1} + \varepsilon_n^{\sim} \quad (n = p, p+1, \ldots), \tag{8.167}$$

where the p-dimensional vectors $\alpha^{\sim}$ and $\varepsilon_n^{\sim}$ and the $p \times p$ matrix B are defined as $\alpha^{\sim} = (0, 0, \ldots, 0, \alpha)'$, $\varepsilon_n^{\sim} = (0, 0, \ldots, 0, \varepsilon_n)' \ (n = p, p+1, \ldots)$,

$$B = \begin{bmatrix} 0 & 1 & 0 & 0 & 0 & & 0 & 0 \\ 0 & 0 & 1 & 0 & 0 & & 0 & 0 \\ \cdots & \cdots & \multicolumn{5}{c}{\cdots\cdots\cdots\cdots\cdots} \\ 0 & 0 & 0 & & & & 0 & 1 \\ \beta_p & \beta_{p-1} & & & & \cdots & \beta_2 & \beta_1 \end{bmatrix} \tag{8.168}$$

Note that the initial value $Y_{p-1} = (X_0, X_1, \ldots, X_{p-1})'$ is independent of the i.i.d. sequence $\{\varepsilon_n^{\sim} : n = p, p+1, \ldots\}$. The state space of the Markov process $\{Y_n : n = p, p+1, \ldots\}$ is $S = \mathbb{R}^p$. A convenient way of expressing (8.167), for purposes of the analysis that follows, is

$$Y_n = B Y_{n-1} + \zeta_n, \quad \zeta_n := \alpha^{\sim} + \varepsilon_n^{\sim} \quad (n = p, p+1, \ldots), \tag{8.169}$$

The random vectors ζ_n are still i.i.d. and independent of Y_{p-1}, although their mean vector is $\alpha^{\sim}$ which is not zero. For stability, i.e., for convergence of the Markov process to a unique invariant probability irrespective of the initial state, a necessary and sufficient condition turns out to be

$$\text{Maximum modulus of the eigenvalues of } B \text{ is less than 1.} \tag{8.170}$$

Expanding the determinant $\det(B - \lambda I_p)$ by its last row, one gets the polynomial (in λ) $\det(B - \lambda I_p) = (-1)^p \{\lambda^p - \beta_1 \lambda^{p-1} - \beta_2 \lambda^{p-2} - \cdots - \beta_p\}$. Thus (8.170) says that all its roots lie in the interior of the unit circle of the complex plane. One can show that (8.170) is equivalent to (see, e.g., Bhattacharya and Waymire 2009, pp. 168–172):

$$\text{There exists a positive integer } m \text{ such that } \|B^m\| < 1. \tag{8.171}$$

The relations (8.170) and (8.171) are called *stability conditions*. Successive iterations of (8.169) lead to

$$Y_n = B^{n-p+1}Y_{p-1} + \sum_{p \leq r \leq n} B^{n-r}\zeta_r. \tag{8.172}$$

Using and proceeding as the case AR(1), one arrives at (Exercise 8.16).

Proposition 8.5. *Assume, as before, that the i.i.d. ε_n satisfy (8.147) and are independent of Y_{p-1}, and that (8.170) (or, equivalently, (8.171) holds. Then the Markov process converges in distribution to a unique invariant distribution π, irrespective of its initial state Y_{p-1}, given by the distribution of*

$$\mathbf{Z} = \sum_{0 \leq n < \infty} B^n \zeta_n = b^\sim + \sum_{0 \leq n < \infty} B^n \varepsilon_n^\sim, \tag{8.173}$$

where all the elements of $b^\sim$ are $\alpha(1 - \sum_r \beta_r)^{-1}$. Each coordinate of $E\mathbf{Z}$ equals $\alpha(1 - \sum \beta_r)^{-1}$, while the covariance matrix of $\mathbf{Z}$ is $\sigma^2 V$, where $V = ((\sum_{0 \leq n < \infty} b_{ip}^{(n)} b_{i'p}^{(n)}))_{i,i'=1,\ldots,p}$, and $b_{i.i'}^{(n)}$ is the (i, i') element of the matrix B^n.

Given observations $X_0, \ldots, X_n$ from the AR(p) model (8.166), the least square estimators of the parameters are obtained algebraically the same way as in linear regression models (see (6.152), (6.153)) and one has

$$\hat{\beta} = C_n^{-1} c_{0,n}, \quad \hat{\alpha} = \overline{X}_{p,n} - \sum_{1 \leq r \leq p} \hat{\beta}_r \overline{X}_{p-r,n}, \tag{8.174}$$

where C_n is a $p \times p$ matrix, $c_{0,n}$ is a $p \times 1$ (column) vector, and $\overline{X}_{p,n}$, $\overline{X}_{p-r,n}$ are lag averages, define by

$$\overline{X}_{p-r,n} = (n-p+1)^{-1} \sum_{p \leq j \leq n} X_{j-r}(r = 1, \ldots, p); \quad \overline{X}_{p,n} = (n-p+1)^{-1} \sum_{p \leq j \leq n} X_j.$$

$$C_n = ((c_{r,s,n}))_{1 \leq r,s \leq p},$$

$$c_{r,s,n} = (n-p+1)^{-1} \sum_{p \leq j \leq n} (X_{j-r} - \overline{X}_{p-r,n})(X_{j-s} - \overline{X}_{p-s,n}),$$

$$c_{0,n} = (c_{0,1,n}, c_{0,2,n}, \ldots, c_{0,p,n})'. \tag{8.175}$$

Proceeding as in the case $p = 1$, one shows that C_n converges to $\mathrm{Cov}(Z) = \sigma^2 V$ (Exercise 8.17), and

$$\sqrt{n}(\hat{\beta} - \beta) =$$

$$= C_n^{-1}\left(\frac{1}{\sqrt{n}}\right) \sum_{p \leq j \leq n} ((X_{j-1} - \mu)\varepsilon_j, (X_{j-2} - \mu)\varepsilon_j, \ldots, (X_{j-p} - \mu)\varepsilon_j)' + o_p(1)$$

$$= (\sigma^2 V)^{-1}\left(\frac{1}{\sqrt{n}}\right) \sum_{p \leq j \leq n} ((X_{j-1} - \mu)\varepsilon_j, (X_{j-2} - \mu)\varepsilon_j, \ldots, (X_{j-p} - \mu)\varepsilon_j)' +$$

$$+ o_p(1) \xrightarrow{\mathscr{L}} N(0, V^{-1}), \tag{8.176}$$

by the martingale CLT, noting that every linear combination of $(X_{j-1} - \mu)\varepsilon_j$, $(X_{j-2} - \mu)\varepsilon_j, \ldots, (X_{j-p} - \mu)\varepsilon_j$ is a martingale difference sequence, and the covariance matrix of its components is $\sigma^2 \mathrm{Cov}(Z) = \sigma^4 V$. One may also show that the least squares estimate of σ^2, namely,

$$\hat{\sigma}^2 = (n-p+1)^{-1} \sum_{p \leq j \leq n} \hat{\varepsilon}_j^2, \quad \hat{\varepsilon}_j := X_j - \hat{\alpha} - \sum_{1 \leq r \leq p} \hat{\beta}_r X_{j-r}, \qquad (8.177)$$

is consistent (Exercise 8.17). We have arrived at the following useful result.

Theorem 8.9. *Assume (8.170) holds for the $AR(p)$ model (8.166) with i.i.d. errors satisfying (8.147), and that $X_0, \ldots, X_{p-1}$ and ε_n have finite fourth moments, or that their joint distribution is the invariant distribution (of Z). Then $\sqrt{n} C_n^{1/2}((\hat{\beta} - \beta)/\hat{\sigma}$ converges in distribution to $N(0, I_p)$ as $n \to \infty$.*

Note that $(\hat{\alpha}, \hat{\beta})$ as given by (8.174) is now easily shown to be jointly asymptotically Normal, and by computing variances and covariances of $\hat{\alpha}$ and the components of $\hat{\beta}$, one can find the (joint) Normal distribution of $\sqrt{n}(\hat{\alpha} - \alpha, \hat{\beta} - \beta)$. Classical CLT-based confidence regions for β and for (α, β) may now be constructed (Exercise 8.18). One may also obtain bootstrap confidence regions by resampling from the estimated residuals in (8.177) (see Bose 1988; Lahiri 2003).

Remark 8.12. If one assumes that the errors ε_n are Normal $N(0, \sigma^2)$, a common assumption in time series analysis (see Brockwell and Davis 2002), then the least squares estimates $\hat{\alpha}$, $\hat{\beta}$ are the MLEs, conditionally given the initial values (Exercise 8.20).

Remark 8.13. It is customary in texts on time series to express the polynomial equation: determinant $(B - \lambda I_p) = 0$, for the $AR(p)$ model, namely, $\lambda^p = \beta_1 \lambda^{p-1} + \beta_2 \lambda^{p-2} + \cdots + \beta_p$, in terms of the variable $\zeta = \lambda^{-1}$. The stability condition (8.170) then becomes: *the polynomial $\beta(\zeta) = 1 - \sum_{1 \leq r \leq p} \beta_r \zeta^r$ (in ζ) has all its roots in $|\zeta| > 1$, i.e., outside the unit circle in the complex plane.*

Our final example is the *autoregressive-moving average model ARMA(p,q)*, where to the new innovation (error) in each period is added a linear combination of the past q errors.

Example 8.7 (ARMA(p,q) Model). For $p \geq 1$, $q \geq 1$, and a sequence of mean zero (and finite variance) i.i.d. sequence $\{\varepsilon_n : n \geq 1\}$ define

$$X_n = \alpha + \sum_{1 \leq r \leq p} \beta_r X_{n-r} + \sum_{1 \leq r \leq q} \theta_r \varepsilon_{n-r} + \varepsilon_n, \quad (n \geq p), \quad (\beta_p \neq 0 + \theta_q \neq 0). \quad (8.178)$$

If the initial states $\{X_0, X_1, \ldots, X_{p-1}\}$ are independent of $\{\varepsilon_n : n \geq p\}$, the sequence $\{Y_n = (X_{n-p+1}, X_{n-p+2}, \ldots, X_n, \varepsilon_{n-q+1}, \varepsilon_{n-q}, \ldots, \varepsilon_n)' : n = p - 1, p, p + 1, \ldots\}$ is Markov, as follows from the representation

$$Y_n = \alpha^{\sim} + H Y_{n-1} + \zeta_n, \qquad (8.179)$$

where $\alpha^{\sim} = (0, \ldots, 0, \alpha, 0, \ldots, 0)'$ with $p - 1$ zero's before α and q zero's after it; $\zeta_n = (0, \ldots, 0, \varepsilon_n, 0, \ldots, 0, \varepsilon_n)'$, and the $(p+q) \times (p+q)$ matrix H is of the form

$$H = \begin{bmatrix} B & C \\ O_{q \times p} & D \end{bmatrix}.$$

Here B is the $p \times p$ matrix (8.168); the $p \times q$ matrix C has all elements zero, except for its last (p-th) row which is $(\theta_q, \ldots, \theta_1)$; the $q \times p$ matrix $O_{q \times p}$ has all zero elements; the i-th row of the $q \times q$ matrix D has 1 as its $(i+1)$-th element and zeros for the rest $(i = 1, \ldots, q-1)$, its last (q-th) row has all zeros. For example, if $p = 3$, $q = 3$, then H is given by

$$
\begin{bmatrix}
0 & 1 & 0 & 0 & 0 & 0 \\
0 & 0 & 1 & 0 & 0 & 0 \\
\beta_3 & \beta_2 & \beta_1 & \theta_3 & \theta_2 & \theta_1 \\
0 & 0 & 0 & 0 & 1 & 0 \\
0 & 0 & 0 & 0 & 0 & 1 \\
0 & 0 & 0 & 0 & 0 & 0
\end{bmatrix}
$$

Expanding by the last row, determinant $(H - \lambda I_{p+q}) = (-\lambda)^q \times$ determinant of $(B - \lambda I_p)$. Therefore the eigenvalues of H all lie inside the unit circle in the complex plane, iff the same is true of B. Iterations of (8.179) yield

$$
Y_n = H^{n-p+1} Y_{p-1} + \sum_{p \leq r \leq n} H^{n-r} \xi_r, \qquad (\xi_r = \zeta_r + \alpha^{\sim}). \tag{8.180}
$$

Proposition 8.6. *If the eigenvalues of the matrix B in (8.168) all lie in the interior of the unit circle in the complex plane, then the Markov process (8.179) has a unique invariant distribution π given by the distribution of $Z = \sum_{0 \leq n < \infty} H^n \xi_n$, whose mean vector is the constant vector equal to $\alpha/(1 - \sum_r \beta_r)$ in each coordinate, and the covariance matrix is $\sigma^2 V$, $V = \sum_{0 \leq n < \infty} H^n A (H')^n$ with $\sigma^2 A$ as the covariance matrix of ζ_r and A has (p, p), $(p+q, p+q)$, $(p, p+q)$ and $(p+q, p)$ elements equal to 1 and all other elements zero. Also, no matter what the initial state Y_{p-1} is, the process converges in distribution to π.*

Note that the variance $\gamma(0)$, say, of X_n under the stationary distribution is $\sigma^2 V_{pp}$ (or $\sigma^2 V_{ii}$, for any i, $1 \leq i \leq p$), and one obtains $\gamma(r) := \operatorname{cov}(X_{n-r}, X_n)$ as the element $\sigma^2 V_{1,1+r}$ for $1 \leq r \leq p-1$. Also, $\rho(q) := \operatorname{cov}(X_n, \varepsilon_n) = \sigma^2$ and $\rho(r) := \operatorname{cov}(X_n, \varepsilon_{n-q+r}) = \sigma^2 V_{p,p+r}$ for $1 \leq r < q$. Clearly, $\operatorname{cov}(X_n, \varepsilon_{n+r}) = 0$ for $r > 0$. One may derive $\gamma(r) := \operatorname{cov}(X_{n-r}, X_n)$ for $r > p - 1$ recursively, using (8.178), namely,

$$
\gamma(s) = \sum_{1 \leq r \leq p} \beta_r \gamma(s - r) + \sum_{1 \leq r \leq \min\{s,q\}} \theta_r \rho(q + s - r) \quad (s > p). \tag{8.181}
$$

However, as illustrated by Example 8.8 below (also see Exercise 8.21), it is generally much simpler to compute the covariances $\gamma(s)$ using (8.178) (and (8.181)) for all $s \geq 0$, than by using the expression for V in Proposition 8.6 for the computation of $\gamma(r)$ for $r = 0, \ldots, p - 1$ and $\rho(r)$ for $r = 0, \ldots, q$ first and then (8.181). The function $r \to \gamma(r)$ is referred to as the *autocovariance function* or *ACVF*.

In the stable AR(p) and, more generally, ARMA(p, q) models *one may subtract the mean from the original variables X_n or from the augmented Markovian Y_n to* simplify the equations. For example, (8.178), (8.179) may then be expressed as

$$
X_n = \sum_{1 \leq r \leq p} \beta_r X_{n-r} + \sum_{1 \leq r \leq q} \theta_r \varepsilon_{n-r} + \varepsilon_n \qquad (n \geq p - q), \ (\beta_p \neq 0 + \theta_q \neq 0),
$$

$$
Y_n = H Y_{n-1} + \zeta_n. \tag{8.182}
$$

If one also assumes that the initial data $Y_0 = (X_0, \ldots, X_{p-1})$ has the distribution π, then iterations such as (8.172) lead to the so-called (infinite order) *moving average representations* of stationary time series:

$$X_n = \sum_{0 \le r < \infty} \psi_r \varepsilon_{n-r}. \tag{8.183}$$

The coefficients ψ_r may be computed from (8.181) in different ways. One convenient formalism is to use the *one-step backward operator* $\boldsymbol{B} : \boldsymbol{B}U_n = U_{n-1}$. (We use slanted and bold $\boldsymbol{B}$ to distinguish it from the matrix B in the models above). Then (8.181) may be expressed as

$$X_n = \sum_{1 \le r \le p} \beta_r \boldsymbol{B}^r X_n + \sum_{1 \le r \le q} \theta_r \boldsymbol{B}^r \varepsilon_n + \varepsilon_n. \tag{8.184}$$

Write $\beta(z)$, $\theta(z)$ for the polynomials $\beta(z) = 1 - \sum_{1 \le r \le p} \beta_r z^r$, $\theta(z) = 1 + \sum_{1 \le r \le q} \theta_r z^r$. By Remark 8.13, the stability condition of the process means that the zeros of $\beta(z)$ all lie outside the unit circle in the complex plane, and $\psi(z) := \theta(z)/\beta(z)$ may be expanded in a convergent series expansion around $z = 0$, cancelling out common zeros of $\beta(z)$ and $\theta(z)$, namely, $\psi(z) = \sum_{0 \le r < \infty} \psi_r z^r$. One then has $\beta(\boldsymbol{B})X_n = \theta(\boldsymbol{B})\varepsilon_n$, or

$$X_n = \psi(\boldsymbol{B})\varepsilon_n = \sum_{0 \le r < \infty} \psi_r \varepsilon_{n-r}. \tag{8.185}$$

If one assumes the *invertibility* condition that *all the zeros of $\theta(z)$ also lie outside the unit circle,* one may similarly invert (8.185) to write

$$\sum_{0 \le r < \infty} \chi_r z^r = \beta(z)/\theta(z),$$

$$\varepsilon_n = [\beta(\boldsymbol{B})/\theta(\boldsymbol{B})]X_n = \sum_{0 \le r < \infty} \chi_r X_{n-r},$$

$$\frac{1}{\theta(z)} = \sum_{0 \le j < \infty} a_j z^j,$$

$$\varepsilon_n = \sum_{0 \le j < \infty} a_j \boldsymbol{B}^j \beta(\boldsymbol{B})X_n = \sum_{0 \le j < \infty} a_j (X_{n-j} - \beta_1 X_{n-j-1} - \cdots - \beta_p X_{n-j-p}). \tag{8.186}$$

For the estimation of the stationary ergodic ARMA(p, q) model, one may first estimate the mean μ by $\overline{X}$ and the autocovariances $\gamma(r)$ by

$$\hat{\gamma}(r) = n^{-1} \sum_{0 \le j \le n-r} (X_j - \overline{X})(X_{j+r} - \overline{X}), \quad (r = 0, 1, 2, \ldots), \quad \overline{X} = n^{-1} \sum_{0 \le j \le n} X_j. \tag{8.187}$$

In general, one requires $\hat{\gamma}(r)$ $(r = 0, 1, \ldots, p+q)$ values to estimate the $p+q+1$ parameters β_r $(r = 1, \ldots, p)$, θ_r $(r = 1, \ldots, q)$ and σ^2 by this rather elementary method. The estimates (8.187) are easily shown to be consistent, and, therefore, the estimates of the $p+q+1$ parameters, using (8.181) are consistent. This "method of moments" is also known as the *Yule-Walker method.* Assume now that $E\varepsilon_n^4 < \infty$. By using general central limit theorems under dependence, such as that for martingales with stationary ergodic differences (See Bhattacharya and Waymire 2009, pp. 507–513), one may show that the estimates $\hat{\gamma}(r)$, for any finite set of values of r, are (jointly) Normal. By the delta method one may then derive the asymptotic (joint) Normality of $\hat{\beta}_r$ $(r = 1, \ldots, p)$, $\hat{\theta}_r$ $(r = 1, \ldots, q)$. This method is rather messy, and generally inefficient, and there exist more efficient, but computationally elaborate procedures in the literature (see Brockwell and Davis 2002). For a comprehensive account of bootstrap methods we refer to Lahiri (2003), especially Sect. 8.5. Here is a brief outline of the latter section. By (8.186), writing $\beta_0 = -1$, $\theta_0 = 1$, and $-\sum_{0 \leq k \leq p} \beta_k X_{r-k} = \sum_{0 \leq k \leq q} \theta_k \varepsilon_{r-k}$, one may represent the innovations ε_n as

$$
\begin{aligned}
\varepsilon_i &= \sum_{0 \leq j < \infty} a_j (X_{i-j} - \beta_1 X_{i-j-1} - \cdots - \beta_p X_{i-j-p}) \\
&= \sum_{0 \leq j < i_0 + 1} a_j (X_{i-j} - \beta_1 X_{i-j-1} - \cdots - \beta_p X_{i-j-p}) + R_{i_0, i} \\
R_{i_0, i} &:= \sum_{j \geq i_0 + 1} a_j (X_{i-j} - \beta_1 X_{i-j-1} - \cdots - \beta_p X_{i-j-p}) \\
&= \sum_{j \geq i_0 + 1} a_j \sum_{0 \leq r \leq q} \theta_r \varepsilon_{i-j-r}.
\end{aligned}
$$

$$(8.188)$$

Note that $a_i \to 0$ exponentially fast with i, in view of $\|H^m\| < 1$ for some m (see (8.171)), so that one may ignore the term $R_{i_0, i}$ in (8.188) for large enough i. Since the coefficients a_i are polynomial in θ_j (use the identity $(1 + \sum_{1 \leq r \leq q} \theta_r z^r)(\sum_{0 \leq j < \infty} a_j z^j) = 1$), consistent and asymptotically Normal estimates (for example, the Yule-Walker estimates $\hat{\theta}_r$ of θ_r) lead to consistent and asymptotically Normal estimates $\hat{a}_j$ of a_j $(j = 1, \ldots, i)$. Together with similar estimates $\hat{\beta}_k$ of β_k one uses (8.188), after deleting the term $R_{i_0, i}$, to obtain estimates $\hat{\varepsilon}_i$ of the residuals ε_i, $i = 1, 2, \ldots, i_0$ for a large enough i_0. To make its mean zero (in bootstrap sampling), let $\hat{\varepsilon}_{i,n} = \hat{\varepsilon}_i - \bar{\varepsilon}$, where $\bar{\varepsilon} = (1/i_0) \sum_{1 \leq i \leq i_0} \hat{\varepsilon}_i$. The bootstrap observations X_i^* are now recursively obtained using the relation $X_i^* = \sum_{1 \leq k \leq p} \hat{\beta}_k X_{i-k}^* + \sum_{0 \leq k \leq q} \hat{\beta}_k \hat{\varepsilon}_{i-k,n} + \varepsilon_i^*$, for $i \geq 1 - \max\{p, q\}$ and setting $X_i^* = 0$, $\varepsilon_i^* = 0$ for $i \leq -\max\{p, q\}$.

Example 8.8 (ARMA(1,1)). Assume the stability condition $|\beta_1| < 1$ and the invertibility condition $|\theta_1| < 1$. Here, after subtracting the mean $\mu = \alpha/(1 - \beta_1)$ from the original sequence, the mean zero stationary AR(1, 1) sequence satisfies

$$
X_n = \beta_1 X_{n-1} + \theta_1 \varepsilon_{n-1} + \varepsilon_n,
$$

where, the i.i.d. ε_n have mean zero, variance $\sigma^2 > 0$ and a finite fourth moment. The covariance $\gamma(r)$ of X_{n-r} and X_n is obtained by taking covariances of both sides of (8.185) with X_{n-r}, yielding $\gamma(0) = \beta_1 \gamma(1) + \sigma^2(1 + \theta_1(\beta_1 + \theta_1))$, and $\gamma(1) = \beta_1 \gamma(0) + \theta_1 \sigma^2$, to solve for both $\gamma(0)$ and $\gamma(1)$. One thus has

$$
\begin{aligned}
\gamma(0) &= \sigma^2(1 + 2\beta_1 \theta_1 + \theta_1^2)/(1 - \beta_1^2), \quad \gamma(1) = \beta_1 \gamma(0) + \theta_1 \sigma^2; \\
\gamma(r) &= \beta_1 \gamma(r-1) = \beta_1^{r-1} \gamma(1) \text{ for } r > 1.
\end{aligned}
$$

$$(8.189)$$

One may use the sample estimates $\hat{\gamma}(r)$ of $\gamma(r)$ in (8.187) to estimate the three parameters $\beta_1, \theta_1, \sigma^2$ by the "method of moments":

$$\hat{\sigma}^2(1 + 2\hat{\beta}_1\hat{\theta}_1 + \hat{\theta}_1^2)/(1 - \hat{\beta}_1^2) = \hat{\gamma}(0), \quad \hat{\beta}_1\hat{\gamma}(0) + \hat{\theta}_1\hat{\sigma}^2 = \hat{\gamma}(1), \quad \hat{\beta}_1\hat{\gamma}(1) = \hat{\gamma}(2).$$
$$(8.190)$$

As mentioned above, $(\hat{\gamma}(0), \hat{\gamma}(1)\,\hat{\gamma}(2))$ has a three-dimensional asymptotic Normal distribution. It then follows, by the delta method, that $(\hat{\beta}_1, \hat{\theta}_1)$ is asymptotically bivariate Normal, and its covariance matrix may be computed using that of $(\hat{\gamma}(0), \hat{\gamma}(1), \hat{\gamma}(2))$. Also, for purposes of bootstrapping as described above, one has

$$\frac{1}{\theta(z)} = (1 + \theta_1 z)^{-1} = 1 - \theta_1 z + \theta_1^2 z^2 + \cdots + (-1)^j \theta_1^j z^j + \ldots \,;$$

$$a_0 = 1, \qquad a_j = (-1)^j \theta_1^j \ (j \geq 1). \tag{8.191}$$

We next turn to an extension of Theorem 7.5 to maximum likelihood estimators in parametric models for stationary ergodic stochastic processes.

Let $\{X_n : n \geq 1\}$ be a stochastic process with values in a measurable space $\mathscr{X}$ (with σ-field $\mathscr{S}$), defined on a probability space $(\Omega, \mathscr{F}, P_{\boldsymbol{\theta}})$. Here $\boldsymbol{\theta} \in \Theta$—an open subset of $\mathbb{R}^p$. Assume that μ_n $(n \geq 1)$ are σ-finite measures on $(\mathscr{X}, \mathscr{S})$ such that under P_θ the distribution of $(X_1, \ldots, X_n)$ is absolutely continuous with respect to the product measure $\mu_1 \times \cdots \times \mu_n$ with density $f_n(\mathbf{x}_1^n; \theta)$ $(\mathbf{x}_1^n := (x_1, \ldots, x_n)')$. Write h_n for the conditional p.d.f. of X_n, given $\mathbf{X}_1^{n-1} \equiv (X_1, \ldots, X_{n-1})'$. That is,

$$h_n(x_n; \mathbf{x}_1^{n-1}, \theta) = f_n(\mathbf{x}_1^n; \theta)/f_{n-1}(\mathbf{x}_1^{n-1}; \theta), \qquad (n \geq 2)$$
$$h_1(x_1; \theta) = f_1(x_1; \theta). \tag{8.192}$$

For $n \geq 2$, h_n is defined whenever the denominator in (8.192) is positive. If $f_{n-1}(\mathbf{x}_1^{n-1}; \theta) = 0$, then $f_n(\mathbf{x}_1^n; \theta) = 0$ (for all x_n outside a set of μ_n-measure zero). Therefore, one may define $h_n(x_n; \mathbf{x}_1^{n-1}, \theta)$ arbitrarily in this case. We make the following assumptions:

$(\mathbf{D}_1)$: $\theta \to f_n(\mathbf{x}_1^n; \theta)$ is thrice continuously differentiable on Θ, for all $\mathbf{x}_1^n$ outside a set of zero $\mu_1 \times \cdots \times \mu_n$-measure $(n \geq 1)$.

$(\mathbf{D}_2)$:

$$\text{(i)} \quad 0 \equiv \frac{\partial}{\partial \theta_r} \int h_n(x_n; \mathbf{x}_1^{n-1}, \theta)\mu_n(dx_n)$$
$$= \int \frac{\partial}{\partial \theta_r} h_n(x_n : \mathbf{x}_1^{n-1}, \theta)\mu_n(dx_n) \qquad (1 \leq r \leq p, \, n \geq 1).$$

$$\text{(ii)} \quad 0 \equiv \frac{\partial^2}{\partial \theta_r \partial \theta_{r'}} \int h_n(x_n; \mathbf{x}_1^{n-1}, \theta)\mu_n(dx_n)$$
$$= \int \frac{\partial^2}{\partial \theta_r \partial \theta_{r'}} h_n(x_n; \mathbf{x}_1^{n-1}; \theta)\mu_n(dx_n) \qquad (1 \leq r, r' \leq p; \, n \geq 1).$$

$(\mathbf{D}_3)$: $\qquad E_\theta \left(\frac{\partial \log f_n(\mathbf{X}_1^n; \theta)}{\partial \theta_r} \right)^2 < \infty \qquad (1 \leq r \leq p, \, n \geq 1).$

$(\mathbf{D}_4)$: $\qquad$ For each $\theta_0 \in \Theta$ there exists $\delta = \delta(\theta_0) > 0$ such that

$$\sup \left\{ \left| \frac{1}{n} \sum_{j=1}^n \frac{\partial^3 \log h_j(X_j; \mathbf{X}_1^{j-1}, \theta)}{\partial \theta_r \partial \theta_{r'} \partial \theta_{r''}} \right| : |\theta - \theta_0| \leq \delta \right\}$$

is bounded by a function $g_n(\mathbf{X}_1^n)$ such that $\overline{\lim}_{n \to \infty} E_{\theta_0} g_n(\mathbf{X}_1^n) < \infty$.

$(\mathbf{D}_5)$:

$$\text{(i)} \quad \frac{1}{n}\sum_{j=1}^{n}\frac{\partial}{\partial\theta_r}\log h_j(X_j;\mathbf{X}_1^{j-1},\theta) \xrightarrow{P_\theta} 0, \qquad (1\le r\le p);$$

$$\text{(ii)} \quad \frac{1}{n}\sum_{j=1}^{n}\frac{\partial^2}{\partial\theta_r\partial\theta_{r'}}\log h_j(X_j;\mathbf{X}_1^{j-1},\theta) \xrightarrow{P_\theta} -I_{r,r'}(\theta), \quad (1\le r,r'\le p),$$

where $I(\theta):=((I_{r,r'}(\theta)))$ is, for each $\theta\in\Theta$, a positive definite matrix.

Theorem 8.10. *Under the assumptions (D_1)–(D_5), for each $\theta_0\in\Theta$ there exists a measurable sequence $\hat\theta_n:\Omega\to\Theta$ such that $\hat\theta_n$ is $\mathscr{F}_n$-measurable and*

(i) $P_{\theta_0}(\hat\theta_n$ is a solution of the likelihood equation $\partial f_n(\mathbf{x}_1^n;\theta)/\partial\theta_r = 0,\ 1\le r\le p) \longrightarrow 1$ as $n\to\infty$,

(ii) $\hat\theta_n \xrightarrow{P_{\theta_0}} \theta_0$.

Proof. The proof follows exactly as the first part of the proof of Theorem 7.5. (See (7.49)–(7.66).) $\qquad\qquad\qquad\qquad\qquad\qquad\qquad\qquad\qquad\qquad\square$

In addition to the above assumptions if one assumes also that

$$(\mathbf{D}_6): \quad \frac{1}{\sqrt{n}}\,\mathrm{Grad}\log f_n(\mathbf{X}_1^n;\theta_0) \;\equiv\; \frac{1}{\sqrt{n}}\sum_{j=1}^{n}\mathrm{Grad}\log h_j(X_j;\mathbf{X}_1^{j-1},\theta_0) \xrightarrow{\mathscr{L}}$$

$N(0,I(\theta_0))$

under P_θ, then the proof of the second part of Theorem 7.5 also goes over.

Theorem 8.11. *Let θ_0 be the true parameter value. Under the assumptions (D_1)–(D_6) every $\hat\theta_n$ satisfying (i), (ii) of Theorem 8.10 converges in law to $N(0,I^{-1}\theta_0))$.*

Definition 8.4. Assume (D_1)–(D_6). An estimator $\hat\theta_n$ of θ is said to be *asymptotically efficient* if $\sqrt{n}(\hat\theta_n-\theta_0) \xrightarrow{\mathscr{L}} N(0,I^{-1}(\theta_0))$, under P_{θ_0}, for every $\theta_0\in\Theta$.

Remark 8.14. Suppose $\widetilde\theta_n$ is an unbiased estimator of θ, and $E_\theta|\widetilde\theta_n|^2 < \infty$. Then Theorem 7.4 yields

$$\sum\nolimits_n(\theta) \ge I_n^{-1}(\theta), \tag{8.193}$$

where $\sum_n(\theta)$ is the dispersion matrix of $\widetilde\theta_n$ (under P_θ), and

$$\frac{1}{n}I_n(\theta)$$

$$= \frac{1}{n}\left(\left(E_\theta\left[\frac{\partial\log f_n(\mathbf{X}_1^n;\theta)}{\partial\theta_r}\frac{\partial\log f_n(\mathbf{X}_1^n;\theta)}{\partial\theta_{r'}}\right]\right)\right)$$

$$= \frac{1}{n}\left(\left(E_\theta\left[\left\{\sum_{j=1}^{n}\frac{\partial\log h_j(X_j;\mathbf{X}_1^{j-1},\theta)}{\partial\theta_r}\right\}\left\{\sum_{j=1}^{n}\frac{\partial\log h_j(X_j;\mathbf{X}_1^{j-1},\theta)}{\partial\theta_{r'}}\right\}\right]\right)\right)$$

$$= \frac{1}{n}\left(\left(\sum_{j=1}^{n}E_\theta\left[\frac{\partial\log h_j(X_j;\mathbf{X}_1^{j-1},\theta)}{\partial\theta_r}\cdot\frac{\partial\log h_j(X_j;\mathbf{X}_1^{j-1},\theta)}{\partial\theta_{r'}}\right]\right)\right), \tag{8.194}$$

since $E_\theta\left(\frac{\partial \log h_j(X_j;\mathbf{X}_1^{j-1},\theta)}{\partial\theta_r}\Big|\mathscr{F}_{j-1}\right) = 0 \;\forall\; j$. Suppose now that $\left(\frac{\partial \log h_j(X_j\mathbf{X}_1^{j-1},\theta)}{\partial\theta_r}\right)^2$ $(j \geq 1)$ is a uniformly integrable sequence for every r. Then (D_6) implies that the sum in (8.194) converges to $I(\theta)$. Together with (8.193), this leads to

$$\lim_{n\to\infty} n\sum_n (\theta) \geq I^{-1}(\theta). \tag{8.195}$$

This justifies the definition of asymptotic efficiency. It may be noted that, under uniform integrability of $\sqrt{n}(\tilde\theta_n - \theta)$, $E_\theta\sqrt{n}(\tilde\theta_n - \theta) \to 0$ (by (D_5)). Therefore, $E_\theta\tilde\theta_n = \theta + o(n^{-\frac{1}{2}})$, and (8.195) holds without the requirement of unbiasedness of $\tilde\theta_n$.

Remark 8.15. Observe that $\partial \log h_j(X_j;\mathbf{X}_1^{j-1},\theta)/\partial\theta_r$ $(j \geq 1)$ is a *martingale difference sequence* (under P_θ), by (D_2). Therefore one may often apply martingale limit theorems to verify (D_4)–(D_6). For example, if $\{X_n : n \geq 0\}$ is a *stationary ergodic Markov process* then $h_j(X_j;\mathbf{X}_1^{j-1},\theta) \equiv g(X_j, X_{j-1};\theta)$ for an appropriate g (the so-called transition probability density of the Markov process), so that $\partial \log g(X_j, X_{j-1};\theta)/\partial\theta_r$ $(j \geq 1)$ is a stationary ergodic martingale difference sequence, and (D_5) holds by *Birkhoff's ergodic theorem* and (D_6) holds by the *Billingsley–Ibragimov martingale CLT.*[6]

Example 8.9 (AR(p) Model). Let $p > 1$, $\{\varepsilon_j : j \geq p\}$ an *i.i.d.* sequence of mean zero squared integrable random variables, $X_0, X_1, \ldots, X_{p-1}$ square integrable random variables independent of the sequence $\{\varepsilon_j : j \geq p\}$. Define, recursively, the p-th order autoregressive process (or, the $AR(p)$ process)

$$X_n = \alpha + \beta_1 X_{n-1} + \beta_2 X_{n-2} + \cdots + \beta_p X_{n-p} + \varepsilon_n \qquad (n \geq p). \tag{8.196}$$

Assume first that ε_j are i.i.d. $N(0,\sigma^2)$. Then the log of the conditional p.d.f. of X_j given $X_{j-p}, \ldots, X_{j-i}$ is

$$\log h_j = \frac{1}{2}\log 2\pi - \frac{1}{2}\log\sigma^2 - \frac{1}{2\sigma^2}\left(X_j - \alpha - \beta_1 X_{j-1} - \cdots - \beta_p X_{j-p}\right)^2,$$

and

$$\frac{\partial \log h_j}{\partial\alpha} = \frac{1}{\sigma^2}\left(X_j - \alpha - \beta_1 X_{j-1} - \cdots - \beta_p X_{j-p}\right),$$

$$\frac{\partial \log h_j}{\partial\beta_r} = \frac{1}{\sigma^2}X_{j-r}\left(X_j - \alpha - \beta_1 X_{j-1} - \cdots - \beta_p X_{j-p}\right), \quad 1 \leq r \leq p,$$

$$\frac{\partial \log h_j}{\partial\sigma^2} = -\frac{1}{2\sigma^2} + \frac{1}{2\sigma^4}\left(X_j - \alpha - \beta_1 X_{j-1} - \cdots - \beta_p X_{j-p}\right)^2. \tag{8.197}$$

The likelihood equations are (for observations $X_p, \ldots, X_n$, given $X_0, \ldots, X_{p-1}$)

$$0 = \frac{1}{\sigma^2}\sum_{j=p}^n \left(X_j - \alpha - \beta_1 X_{j-1} - \cdots - \beta_p X_{j-p}\right) \quad \left(= \tfrac{1}{\sigma^2}\textstyle\sum_{j=p}^n \varepsilon_j\right);$$

$$0 = \frac{1}{\sigma^2}\sum_{j=p}^n X_{j-r}\left(X_j - \alpha - \beta_1 X_{j-1} - \cdots - \beta_p X_{j-p}\right) \quad \left(= \tfrac{1}{\sigma^2}\textstyle\sum_p^n X_{j-r}\varepsilon_j\right),$$

---

[6] See Billingsley (1968), p. 206, or Bhattacharya and Waymire (2009), p. 511.

$$(1 \leq r \leq p);$$

$$0 = -\frac{n-p+1}{2\sigma^2} + \frac{1}{2\sigma^4} \sum_{p}^{n} (X_j - \alpha - \beta_1 X_{j-1} - \cdots - \beta_p X_{j-p})^2$$

$$\left(= -\frac{1}{2\sigma^2} + \frac{1}{2\sigma^4} \sum_{p}^{n} \varepsilon_j^2 \right).$$

$$(8.198)$$

The solutions are given by

$$
\begin{pmatrix} \hat{a} \\ \hat{\beta}_1 \\ \vdots \\ \hat{\beta}_p \end{pmatrix}
=
\begin{pmatrix}
1 & s_1 s_2 \ldots s_p \\
s_1 & s_{11} s_{12} \ldots s_{1p} \\
s_2 & s_{21} s_{22} \ldots _{2p} \\
\cdot & \cdots \\
s_p & s_{p1} s_{p2} \ldots s_{pp}
\end{pmatrix}^{-1}
\begin{pmatrix} s_0 \\ s_{01} \\ s_{02} \\ \vdots \\ s_{0p} \end{pmatrix},
\qquad (8.199)
$$

$$\hat{\sigma}^2 = \sum_{j=p}^{n} \left(X_j - \hat{a} - \hat{b}_1 X_{j-1} - \cdots - \hat{b}_j X_{j-p} \right)^2 / (n-p+1),$$

where

$$s_r = \frac{1}{n-p+1} \sum_{j=p}^{n} X_{j-r} \qquad (0 \leq r \leq p),$$

$$s_{rr'} = \frac{1}{n-p+1} \sum_{j=p}^{n} X_{j-r} X_{j-r'} \qquad (0 \leq r, r' \leq p). \qquad (8.200)$$

Write

$$\varepsilon_j' = a + \varepsilon_j, \quad \gamma_j (0, 0, \ldots, 0, \varepsilon_j)' \in \mathbb{R}^p \qquad (j \geq p), \qquad (8.201)$$

and

$$X_j = \sum_{r=1}^{p} \beta_r X_{j-r} + \varepsilon_j',$$

$$Y_j = (X_{j-p+1}, X_{j-p+2}, \ldots, X_j)' \qquad (j \geq p). \qquad (8.202)$$

Then Y_j, $j \geq p$, is a Markov process,

$$Y_j = B Y_{j-1} + \gamma_j, \qquad (8.203)$$

where B is the $p \times p$ matrix (8.168). *All eigenvalues of B have magnitude less than one.* Now the eigenvalues of B are the roots of the polynomial equation $\det(B - \lambda I) = 0$, i.e.,

$$-\lambda^p + \beta_1 \lambda^{p-1} + \beta_2 \lambda^{p-2} + \cdots + \beta_p = 0. \qquad (8.204)$$

In the stable case the Markov process converges in distribution to a unique invariant distribution π, whatever the initial state $Y_j = (X_0, X_1, \ldots, X_{p-1})'$. Clearly π is Normal whose mean vector and dispersion matrix are given in Proposition 8.5. Then all the conditions (D_1)–(D_6) hold. For example (see (8.198)),

$$\left(\frac{1}{\sqrt{n-p-1}} \sum_{j=p}^{n} \varepsilon_j, \quad \frac{1}{\sqrt{n-p+1}} \sum_{j=p}^{n} X_{j-1}\varepsilon_j, \ldots, \right.$$

$$\left. \frac{1}{\sqrt{n-p+1}} \sum_{j=p}^{n} X_{j-p}\varepsilon_j, \quad \frac{1}{\sqrt{n-p+1}} \sum_{j=p}^{n} (\varepsilon_j^2 - \sigma^2) \right)$$

is asymptotically Normal, by the martingale central limit theorem. Thus $\sqrt{n}(\hat{\alpha}-\alpha, \hat{\beta} - \beta, \hat{\sigma}^2 - \sigma^2) \to N(0, I^{-1}(\theta))$, where $I(\theta)$ is the information matrix.

Remark 8.16. Finally, consider a *nonlinear autoregressive model of order p,* or NLAR(p),

$$X_{n+1} = f(X_{n-p+1}, \ldots, X_n) + g(X_{n-p+1}, \ldots, X_n)\varepsilon_{n+1}, \quad (n = p-1, p, \ldots), \tag{8.205}$$

where ε_n, $n \geq p$, are i.i.d. with mean zero and variance $\sigma^2 > 0$, and independent of $X_0, X_1, \ldots, X_{p-1}$. The unknown real-valued functions f and g on $\mathbb{R}^p$ are such that the Markov process $Y_n = (X_{n-p+1}, \ldots, X_n)'$, $n \geq p$, (on the state space $\mathbb{R}^p$) is ergodic. For the case $p = 1$ an asymptotically optimal estimate of f by the so-called kernel method (See Chap. 10), and a consistent bootstrap estimate of its distribution, were obtained by Frankel et al. (2002). Independently of this, and more generally, Hwang (2002) derived an asymptotically optimal estimate of f and a consistent bootstrap estimate of the distribution of this estimate. She also found interesting conditions on a misspecified order $p' > p$ such that the estimate of f is still consistent. The problem of validity of approximating long memory processes by short memory ARMA processes are considered in Hosking (1984), Chan and Palma (1998), and Basak et al. (2001).

8.7 Notes and References

Ferguson (1996) and Serfling (1980) may be used as general references for this chapter. Pitman asymptotic relative efficiency (ARE) is due to Pitman (1948) and our treatment follows Serfling (1980), Chap. 10. A broader, more sophisticated and elegant approach to Pitman ARE is due to LeCam based on his notion of *contiguity.* See LeCam and Yang 1990 or van der Vaart 1998). An entirely different notion of asymptotic relative efficiency is due to Bahadur (1960). Here one looks at the asymptotic exponential rate at which the p-value under $H_1 : \theta > \theta_0$ goes to zero in probability (as $n \to \infty$) for a test statistic T_n which rejects $H_0 : \theta = \theta_0$ for large values of T_n (e.g., $T_n = \sqrt{(\overline{X} - \theta_0)}$). While the Pitman efficiency looks at alternatives at a distance $O(n^{-1/2})$ of θ_0 and compares the rejection probabilities of two tests under such alternatives within the range of Normal approximation, the *Bahadur ARE* looks at rejection probabilities under fixed alternatives θ in the large deviation domain. A fine exposition of this is given in Serfling (1980), Chap. 10.

A standard reference for inference for time series is Brockwell and Davis (2002). For results and procedures for bootstrapping under dependence and for time series, see Lahiri (2003). For many econometric time series modeled as AR(1), AR(p) or ARMA(p, q), an appropriate assumption would be to have eigenvalues of the matrix B in (8.168) to be inside the circle, but with some very close to the boundary of the circle. For example, in AR(1) one may think of scaling β_1 as $1 - b/n$ for some $b > 0$. The asymptotic distribution theory of the ordinary least squares estimate (OLS) then presents challenges. Basic work on this phenomenon under broad assumptions is due to Chan and Wei (1987, 1988). Also, see Phillips (1987). This so-called *unit root problem* is now an important area in econometrics.

Exercises for Chap. 8

Exercises for Sects. 8.1, 8.2

Ex. 8.1. Consider the family of p.d.f.'s $\{f(x - \theta) : \theta \in \mathbb{R}^1\}$ in Example 8.1.

(a) For the one-sided alternative $H_1 : \theta > 0$, show that $e_P(T_n, t) = 4\sigma_f^2 f^2(0)$.
(b) Consider the tests of $H_0 : \theta = 0$, $H_1 : \theta \neq 0$.

$$\delta_{1,n} : \text{Reject } H_0 \text{ iff } |t| \equiv \left| \frac{\overline{X}}{s} \right| > a_n,$$

$$\delta_{2,n} : \text{Reject } H_0 \text{ iff } \left| \frac{1}{n} \sum_{j=1}^{n} \mathbf{1}_{\{X_j > 0\}} - \frac{1}{2} \right| > b_n,$$

having asymptotic size α. Prove that

$$e_P(\delta_{2,n}, \delta_{1,n}) = 4\sigma_f^2 f^2(0).$$

Ex. 8.2. (a) Show that for every pair of random variables X, Y, every distribution function $G(x)$, and every constant $\varepsilon > 0$, one has

$$\sup_{x \in \mathbb{R}^1} |\text{Prob}(Y \leq x) - \text{Prob}(X \leq x)|$$
$$\leq P(|X - Y| > \varepsilon) + 2 \sup_x |P(X \leq x) - G(x)| + \sup_x |G(x + \varepsilon) - G(x)|. \tag{8.206}$$

(b) Use (8.206), (8.43) and (8.44) to prove (8.45), under the assumptions (B) and (8.29).

Ex. 8.3. Let $\varphi(\mathbf{X})$ be the *symmetrization* of $T = T(\mathbf{X})$ as in Remark 8.2, with $\mathbf{X} = (X_1, \ldots, X_n)'$ and X_i's i.i.d. If $E|T|^p < \infty$ for some $p \geq 1$, then show that $E|\varphi(\mathbf{X}) - \theta|^p \leq E|T - \theta|^p$, where $\theta = ET$. [Hint: $\varphi(\mathbf{X}) = E(T \mid \mathscr{F})$, where $\mathscr{F}$ is the sigma-field generated by permutations of indices $(1, 2, \ldots, n)$ of $X_1, X_2, \ldots, X_n$.]

Ex. 8.4. Prove Proposition 8.1.

Exercises for Sect. 8.3

Ex. 8.5 (Mendel's Experiment). In Mendel's experiment in pea breeding, possible types of progeny were (1) round-yellow, (2) round green, (3) wrinkled yellow and (4) wrinkled green. According to Mendel's theory these were to occur in respective proportions (H_0) $p_1 = 9/16$, $p_2 = p_3 = 3/16$, $p_4 = 1/16$. In his experiment Mendel observed the respective numbers of progeny (out of a total $n = 556$), $n_1 = 315$, $n_2 = 108$, $n_3 = 101$, $n_4 = 32$. Find the p-values using

(a) Pearson's frequency chi-square test,
(b) Wald's test,
(c) the likelihood ratio test,
(d) Rao's scores test.

Ex. 8.6 (Hardy-Weinberg Model). Test the Hardy-Weinberg model for probabilities of the three genotypes of a single gene with two alleles: $p_1 = \theta^2$, $p_2 = 2\theta(1 - \theta)$, $p_3 = (1 - \theta)^2$, $0 < \theta < 1$, and observed frequencies n_1, n_2, n_3, using

(a) Pearson's frequency chi-square test,
(b) Wald's test,
(c) the likelihood ratio test,
(d) Rao's scores test.

Ex. 8.7. Test the *Fisher linkage model* of Example 4.15, using

(a) Pearson's frequency chi-square test,
(b) Wald's test,
(c) the likelihood ratio test,
(d) Rao's scores test.

Ex. 8.8 (Chi-square Test for Independence in Two-Way Contingency Tables). A population is classified according to two categorical variables A and B. The variable A has k classes $A_1, \ldots, A_k$ while B has m classes $B_1, \ldots, B_m$. One wishes to test if the two classifications are independent of each other, namely, $H_0 : p_{ij} = p_{i.}p_{.j}$ $(1 \leq i \leq k, 1 \leq j \leq m)$, where p_{ij} is the proportion in the population belonging to class A_i and class B_j, $p_{i.} = \sum_j p_{ij}$ is the proportion belonging to class A_i of the categorical variable A, $p_{.j} = \sum_i p_{ij}$ is the proportion belonging to class B_j of the categorical variable B.

(a) Write down the frequency chi-square test.
(b) Apply the test (a) to the following data concerning heart disease in male federal employees. Researchers classified 356 volunteer subjects according to their socio-economic status (SES)—A and their smoking habits—B.

Smoking habit	SES			Total
	High	Middle	Low	
Current	51	22	43	116
Former	92	21	28	141
Never	68	9	22	99
Total	211	52	93	356

Exercises for Sects. 8.4, 8.5

Ex. 8.9. Verify the expression (8.127) for D_n.

Ex. 8.10. The numbers k of micro-organisms of a certain type found within each of $n = 60$ squares of a hemocytometer are as follows.

k	0	1	2	3	4
Observed frequency	28	20	8	3	1

Test if the data follow a Poisson distribution (Example taken from the website vassarstats. net).

Ex. 8.11. It is common in insurance to assume that the times of arrival of claims follow a homogeneous Poisson process. Given n successive arrivals of claims $0 < t_1 < t_2 < \cdots < t_n$, derive the following tests for this model.

(a) Kolmogorov-Smirnov
(b) Carmér-von Mises
(c) Anderson-Darling

Ex. 8.12. Suppose the intervals between successive arrivals in the preceding exercise are i.i.d. *Pareto* with p.d.f. $f(u; \alpha) = \alpha(1 + u)^{-\alpha-1}$ $(\alpha > 0)$, α unknown.

(a) Find the MLE of α.
(b) Draw a random sample of size $n = 50$ from this *Pareto distribution* with $\alpha = 2$, and carry out the goodness-of-fit tests (a)–(c) above for the data to have come from a Pareto distribution (with α unknown).
(c) Use tests (a)–(c) for the data in (b) to have come from an exponential distribution..

Ex. 8.13. The following data on 32 skulls are taken from the book *A Handbook of Small Data Sets* (Hand et al. 1994), which reproduced them from Moran (1923). The 17 type A skulls came from Sikkim and neighboring areas of Tibet. The 15 type B skulls were picked up on a battlefield in the Lhasa district and were believed to be those of native soldiers from the eastern province of Khams. It was thought at the time that the Tibetans from Khams might be survivors of a particular fundamental human type, unrelated to the Mongolian and Indian types which surrounded them. The five measurements on each skull are as follows:
 $X_1 =$ greatest length of skull, $X_2 =$ greatest horizontal breadth of skull, $X_3 =$ height of skull, $X_4 =$ upper face height, $X_5 =$ face breadth, between outermost points of cheekbones.
 Test if the skulls A and Skulls B belonged to the same type of humans. [Hint: Use Proposition 8.3.]

Exercises for Sect. 8.6

Ex. 8.14. To complete the proof of Theorem 8.8,

(a) prove (8.165) under the hypothesis of Theorem 8.8(b),
(b) prove (8.156) using the martingale CLT as suggested at the end of the proof of Theorem 8.8.

Ex. 8.15. Under the hypothesis of Theorem 8.8 construct a confidence interval for β with asymptotic level $1 - \theta(0 < \theta < 1)$.

A skulls					B skulls				
X_1	X_2	X_3	X_4	X_5	X_1	X_2	X_3	X_4	X_5
190.5	152.5	145.0	73.5	136.5	182.5	136.0	138.5	76.0	134.0
172.5	132.0	125.5	63.0	121.0	179.5	135.0	128.5	74.0	132.0
167.0	130.0	125.5	69.5	119.5	191.0	140.5	140.5	72.5	131.5
169.5	150.5	133.5	64.5	128.0	184.5	141.5	134.5	76.5	141.5
175.0	138.5	126.0	77.5	135.5					
177.5	142.5	142.5	71.5	131.0	181.0	142.0	132.5	79.0	136.5
179.5	142.5	127.5	70.5	134.5	173.5	136.5	126.0	71.5	136.5
179.5	138.0	133.5	73.5	132.5	188.5	130.0	143.0	79.5	136.0
173.5	135.5	130.5	70.0	133.5	175.0	153.0	130.0	76.5	142.0
162.5	139.0	131.0	62.0	126.0	196.0	142.5	123.5	76.0	134.0
178.5	135.0	136.0	71.0	124.0	200.0	139.5	143.5	82.5	146.0
171.5	148.5	132.5	65.0	146.5	185.0	134.5	140.0	81.5	137.0
180.5	139.0	132.0	74.5	134.5	174.5	143.5	132.5	74.0	136.5
183.0	149.0	121.5	76.5	142.0	195.5	144.0	138.5	78.5	144.0
169.5	130.0	131.0	68.0	119.0	197.0	131.5	135.0	80.5	139.0
172.0	140.0	136.0	70.5	133.5	182.5	131.0	135.0	68.5	136.0
170.0	126.5	134.5	66.0	118.5					

Ex. 8.16. Verify Proposition 8.5 using (8.172).

Ex. 8.17. Assume the hypothesis of Theorem 8.9 for the AR(p) model.

(a) Prove that $\hat{\gamma}(r)$ is a consistent estimate of $\gamma(r)$ for $r = 0, 1, \ldots, p$.
(b) Prove that $\hat{\sigma}^2$ in (8.177) is a consistent estimator of σ^2.
(c) Prove that C_n converges in probability to $\sigma^2 V$.

Ex. 8.18. In addition to the hypothesis of Theorem 8.9, assume ε_n are Normal $N(0, \sigma^2)$, and then prove that Z in (8.173) is Normal.

Ex. 8.19. Construct a confidence region for $\boldsymbol{\beta}$ of asymptotic level $1 - \theta$ in the stable AR(p) model.

Ex. 8.20. Assume that the error ε_n in the AR(p) model (8.166) are i.i.d. $N(0, \sigma^2)$. Show that the least squares estimates $\hat{\alpha}$, $\hat{\boldsymbol{\beta}}$ are then the MLEs, conditionally given $X_0, \ldots, X_{p-1}$.

Ex. 8.21. Consider a stable and invertible stationary ARMA$(1, 2)$ model.

(a) Show that $\hat{\gamma}(r)$ is a consistent estimate of $\gamma(r)$ for $r = 0, 1, \ldots$, and that $(\hat{\gamma}(0), \hat{\gamma}(1), \hat{\gamma}(2), \hat{\gamma}(3))$ is asymptotically Normal.
(b) Derive the analogs of (8.189) for $\gamma(0)$, $\gamma(1)$, $\gamma(2)$, $\gamma(3)$, and show that the Yule-Walker ("method of moments") estimator of $(\beta_1, \theta_1, \theta_2)'$ derived using these equations is asymptotically Normal.

Ex. 8.22 (Trend Removal). Let $X_n = f(n) + Z_n$, where $\{Z_n : n \geq 0\}$ is a stationary time series, and f is a deterministic function on the set of integers $\mathbb{Z}_t = \{0, 1, \ldots\}$. Show that if f is linear then $\Delta X_n \equiv X_n - X_{n-1}$, $n \geq 1$, is a stationary time series and, more generally, if f is a polynomial of order k then $\Delta^k X_n$, $n \geq k$, is stationary.

Ex. 8.23 (Trend Estimation). Let $X_n = f(n) + Z_n$, as in Exercise 8.22. Use the Dow Jones Utilities Index for 100 days from the internet, or from August 28 to December 18 of 1972 as given in Brockwell and Davis (1987), p. 499, for the problems below.

(a) Assuming f is linear, use the method of least squares to estimate f, and plot $\widehat{Z}_n := X_n - \hat{f}(n)$ to see if it looks stationary.

(b) Assume f is quadratic, estimate it by the method of least squares, and plot $\widehat{Z}_n = X_n - \hat{f}(n)$ to check visually if this "trend-removed" process looks stationary.

(c) Assuming $\{Z_n, n \geq 0\}$ is a stable mean zero stationary AR(1) process, show that the estimated coefficients of f in (a) and (b) are consistent and asymptotically Normal.

References

Bahadur, R. R. (1960). On the asymptotic efficiency of tests and estimates. *Sankhya, 22*, 229–252.

Basak, G. K., Chan, N. H., & Palma, W. (2001). The approximation of long memory processes by an ARMA model. *Journal of Forecasting, 20*, 367–389.

Bhattacharya, R., & Ranga Rao, R. (2010). *Stochastic processes with applications.* SIAM Classics in Applied Mathematics (Vol. 61). Philadelphia: SIAM.

Bhattacharya, R., & Waymire, E. C. (2009). *Stochastic processes with applications.* SIAM classics in applied mathematics series.

Billingsley, P. (1968). *Convergence of probability measures.* New York: Wiley.

Bose, A. (1988). Edgeworth correction by bootstrap in autoregressions. *Annals of Statistics, 16*, 1709–1722.

Brockwell, P.J., & Davis, R. A. (2002). *Introduction to time series and forecasting.* New York: Springer.

Brown, L., Cai, T., & DevsGupta, A. (2001). Interval estimation for a binomial proportion. *Statistical Science, 16*(2), 101–133.

Chan, N. H., & Palma, W. (1998). State space modeling of long-memory processes. *The Annals of Statistics, 26*, 719–740.

Chan, N. H., & Wei, C. Z. (1987). Asymptotic inference for nearly nonstationary *AR*(1) processes. *The Annals of Statistics, 15*(3), 1050–1063.

Chan, N. H., & Wei, C. Z. (1988). Limiting distributions of least squares estimates of unstable autoregressive processes. *The Annals of Statistics, 16*(1), 367–401.

Chernoff, H., & Lehmann, E. L. (1954). The use of maximum likelihood estimates in chisquare tests for goodness of fit. *The Annals of Mathematical Statistics, 25*, 579–586.

D'Agostino, R., & Stephens, M. (1986). *Goodness of fit techniques.* New York: Marcel Dekker

Ferguson, T. S. (1967). *Mathematical statistics: A decision theoretic approach.* New York: Academic Press.

Ferguson, T. S. (1996). *A course in large sample theory.* London: Taylor & Francis.

Frankel, J., Kreiss, J. P., & Mammen, E. (2002). Bootstrap kernel smoothing in nonlinear time series. *Bernoulli, 8*, 1–37.

Hand, D. J., Daly, F., Lunn, A. D., McConway, K. J., & Ostrowski, E. (1994). *A handbook of small data sets*. London: Chapman & Hall.

Hodges, J. L., & Lehmann, E. L. (1956). The efficiency of some nonparametric competitors of the t-test. *Annals of Mathematical Statistics, 27*(2), 324–335.

Hodges, J. L., & Lehmann, E. L. (1960). Comparison of the normal scores and Wilcoxon tests. *Proceedings of Fourth Berkeley Symposium on Mathematical Statistics and Probability* (Vol. 1, pp. 307–317).

Hosking, J. R. M. (1984). Modelling persistence in hydrological time series using fractional differencing. *Water Resources Research, 20*, 1898–1908.

Hwang, E. (2002). *Nonparametric estimation for nonlinear autoregressive processes*. Ph.D. Thesis, Indiana University, Bloomington.

Lahiri, S. N. (2003). *Resampling methods for dependent data*. New York: Springer.

LeCam, L., & Yang, G. (1990). *Asymptotics in statistics. Some basic concepts*. New York: Springer.

Moran, G. M. (1923). A first study of the Tibetan skull. *Biometrika, 14*, 193–260.

Mukherjee, R., & Reid, N. (2001). Comparison of test statistics via expected lengths of associated confidence intervals. *Journal of Statistical Planning and Inference, 97*(1), 141–151.

Neyman, J., & Pearson, E. S. (1928). On the use and interpretation of certain test criteria for purposes of statistical inference: Part I. *Biometrika, 20A*, 175–240.

Phillips, P. C. B. (1987). Towards a unified asymptotic theory for autoregression. *Biometrika, 74*(3), 535–547.

Pitman, E. J. G. (1948). *Lecture notes on nonparametric statistical inference*. New York: Columbia University.

Sen, P. R., & Singer, J. (1979). *Large simple methods in statistics: An introduction with applications*. New York: Chapman and Hall.

Serfling, R. (1980). *Approximation theorems of mathematical statistics*. New York: Wiley.

Shapiro, S. S., & Wilk, M. B. (1965). An analysis of variance test for normality. *Biometrika, 52*, 591–611.

van der Vaart, A. (1998). *Asymptotic statistics*. Cambridge: Cambridge University Press.

Wald, A. (1943). Tests of statistical hypotheses concerning several parameters when the number of observations is large. *Transactions of the American Mathematical Society, 54*(3), 426–482.

Wilks, S. S. (1938). The large-sample distribution of the likelihood ratios for testing composite hypotheases. *The Annals of Mathematical Statistics, 9*, 60–62.

Chapter 9
The Nonparametric Bootstrap

Abstract This chapter introduces Efron's nonparametric bootstrap, with applications to linear statistics, and semi-linear regression due to Bickel and Freedman.

9.1 What is "Bootstrap"? Why Use it?

We describe in this section an important methodology due to Efron (1979) to estimate distributions of statistics from data by resampling.

Suppose one needs to construct a confidence interval for a parameter θ based on an estimator $\hat{\theta}_n$ constructed from i.i.d. observations $X_1, X_2, \ldots, X_n$ from a distribution P. (That is, X_i's have a common distribution P.) If $\sqrt{n}(\hat{\theta}_n - \theta) \xrightarrow{\mathscr{L}} N(0, \sigma^2)$ as $n \to \infty$ and $\hat{\sigma}_n^2$ is a consistent estimator of σ^2 based on $X_1, \ldots, X_n$, then a confidence interval of *approximate size* $1 - \alpha$ is given by

$$\left[\hat{\theta}_n - z_{1-\frac{\alpha}{2}} \frac{\hat{\sigma}_n}{\sqrt{n}} , \ \hat{\theta}_n + z_{1-\frac{\alpha}{2}} \frac{\hat{\sigma}_n}{\sqrt{n}} \right], \tag{9.1}$$

where z_δ is such that $\mathrm{Prob}(Z \leq z_\delta) = \delta$, Z having the standard normal distribution $N(0, 1)$. The targeted size, or coverage probability, $1 - \alpha$ is called the *nominal coverage*. Since (9.1) is based on an asymptotic result, the *coverage error* ($=$ nominal coverage $-$ true coverage probability) may be significant, especially when the sample size n in an actual experiment is not very large.

Let $T_n(P)$ be a function of i.i.d. observations $X_1, \ldots, X_n$ (from a distribution P) and P. For example, in the case of real-valued X_i, $T_n(P)$ may be $\sqrt{n}(\overline{X}_n - \theta)/\hat{\sigma}_n$, which not only involves X_i's ($1 \leq i \leq n$), but also P via $\theta := \int xP(dx) \equiv$ mean of P. First estimate P by its consistent and unbiased estimator $\widehat{P}_n = \frac{1}{n} \sum_{i=1}^n \delta_{X_i} \equiv$ the *empirical distribution* (based on $X_1, \ldots, X_n$). Note that δ_{X_i} is the point mass at X_i. Hence $\widehat{P}_n(B) = \frac{1}{n} \sum_{i=1}^n 1_B(X_i) \equiv$ proportion of those observations which lie in B (for all measurable B): (1) $E\widehat{P}_n(B) = E1_B(X_i) = O \cdot (1 - P(B)) + 1 \cdot P(B) = P(B)$ *(unbiasedness)*, (2) $\widehat{P}_n(B) \to E1_B(X_i) = P(B)$ a.s. as $n \to \infty$, by the strong law of large numbers *(consistency)*. We first consider the *percentile bootstrap*. To arrive at the bootstrap estimate follow the steps below.

© Springer-Verlag New York 2016
R. Bhattacharya et al., *A Course in Mathematical Statistics and Large Sample Theory*, Springer Texts in Statistics, DOI 10.1007/978-1-4939-4032-5_9

Step 1: Take a *random sample with replacement of size n from the empirical* $\widehat{P}_n$. Denote this by $X_1^*, X_2^*, \ldots, X_n^*$.

Step 2: Calculate $T_n^*(\widehat{P}_n)$, substituting $X_1^*, \ldots, X_n^*$ for $X_1, \ldots, X_n$, and P by $\widehat{P}_n$ in the functional form $T_n(P)$. For example, if $T_n(P) = (\overline{X}_n - \theta)$, $T_n^*(\widehat{P}_n) = (\overline{X}_n^* - \overline{X}_n)$ where $\overline{X}_n^* = \frac{1}{n}\sum_{i=1}^n X_i^*$. Note that here $\theta = \int xP(dx)$, $\overline{X}_n = \int x\widehat{P}_n(dx)$.

Step 3. Repeat independently Steps 1 and 2 a very large number of times, say M, obtaining M independent values of the *bootstrapped statistic* $T_n^*(\hat{P}_n)$l. (Typically M is between 500 and 2000).

Step 4. Find the lower and upper $\frac{\alpha}{2}$-quantiles of the M values in Step 3. Call these $q_{\frac{\alpha}{2}}^*$, $q_{1-\frac{\alpha}{2}}^*$, respectively.

Step 5. The bootstrap confidence interval for θ is

$$\left[q_{\frac{\alpha}{2}}^*, q_{1-\frac{\alpha}{2}}^*\right]. \tag{9.2}$$

To see that (9.2) is a confidence interval for θ with an asymptotic level $1 - \alpha$ (almost surely, as $n \to \infty$), note that a classical confidence interval is given by the interval $[\hat{\theta}_n - z_{1-\alpha/2}\hat{\sigma}_n, \hat{\theta}_n - z_{\alpha/2}\hat{\sigma}_n] = [\hat{\theta}_n + z_{\alpha/2}\hat{\sigma}_n, \hat{\theta}_n + z_{1-\alpha/2}\hat{\sigma}_n] = [l, u]$, say. Now the bootstrap version θ_n^* of $\hat{\theta}_n$ is, under the empirical $P^* = \widehat{P}_n$, asymptotically Normal $N(\hat{\theta}_n, \hat{\sigma}_n^2)$, so that the $\alpha/2$-th and $(1 - \alpha/2)$-th quantiles of θ_n^*, $q_{\alpha/2}^*$ and $q_{1-\alpha/2}^*$ say, are asymptotically equal to $\hat{\theta}_n + z_{\alpha/2}\hat{\sigma}_n = l$ and $\hat{\theta}_n + z_{1-\alpha/2}\hat{\sigma}_n = u$, respectively.

When the standard error $\hat{\sigma}_n$ of $\hat{\theta}_n$ is known in closed form, one may use the studentized or pivoted statistic $T_n = (\hat{\theta}_n - \theta)/\hat{\sigma}_n$, which is asymptotically standard Normal $N(0, 1)$. The usual CLT-based symmetric confidence interval for θ is given by

$$[\hat{\theta}_n + z_{\frac{\alpha}{2}}\hat{\sigma}_n, \hat{\theta}_n + z_{1-\frac{\alpha}{2}}\hat{\sigma}_n] = [\hat{\theta}_n - z_{1-\frac{\alpha}{2}}\hat{\sigma}_n, \hat{\theta}_n - z_{\frac{\alpha}{2}}\hat{\sigma}_n], \tag{9.3}$$

using $P(|T_n| \leq z_{1-\alpha/2}) = 1 - \alpha$. The corresponding *pivotal bootstrap confidence interval* is based on the resampled values of $T_n^* = (\hat{\theta}_n^* - \hat{\theta}_n)/\hat{\sigma}_n^*$, where $\hat{\sigma}_n^*$ is the bootstrap estimate of the standard error obtained by steps analogous to those described in the preceding paragraph. Let $c_{\alpha/2}^*$ be such that $P^*(|T_n^*| \leq c_{\alpha/2}^*) = 1 - \alpha$. The bootstrap pivotal confidence interval for θ is then

$$\left[\hat{\theta}_n - c_{\frac{\alpha}{2}}^* \hat{\sigma}_n^*, \; \hat{\theta}_n + c_{\frac{\alpha}{2}}^* \hat{\sigma}_n^*\right]. \tag{9.4}$$

Suppose $\hat{\theta}_n$ is based on i.i.d. observations $X_1, \ldots, X_n$, whose common distribution has a density (or a nonzero density component), and that it is a smooth function of sample means of a finite number of characteristics of X, or has a stochastic expansion (Taylor expansion) in terms of these sample means (such as the MLE in regular cases). It may then be shown that the *coverage error* of the CLT-based interval (9.3) is $O(n^{-1})$, while that based on (9.4) is $O(n^{-3/2})$, a major advantage of the bootstrap procedure. The coverage error of the percentile interval (9.2) is $O(n^{-1/2})$, irrespective of whether the distribution of X is continuous or discrete. Chapter 11, Part III provides a rigorous treatment of coverage errors.

One of the compelling arguments in favor of using the percentile bootstrap is that it does not require an analytical computation of the standard error $\hat{\sigma}_n$ of the estimator $\hat{\theta}_n$. In many problems such analytic computations are complex and difficult. Moreover, one may obtain a bootstrap estimate of the standard error simply as the standard deviation of the bootstrap estimates $\hat{\theta}^*$.

9.2 When Does Bootstrap Work?

9.2.1 Linear Statistics, or Sample Means

Let $X_1, X_2, \ldots, X_n$ be i.i.d. observations from a distribution P. Consider a *linear statistic* $T_n = \frac{1}{n}\sum_{i=1}^{n} g(X_i)$ for some real-valued measurable function g, which is used to estimate $Eg(X_1) = T(P)$. Let $\widehat{P}_n$ denote the empirical, $\widehat{P}_n = \frac{1}{n}\sum_{i=1}^{n}\delta_{X_i}$. Let $X_{n1}^*, \ldots, X_{nn}^*$ be a bootstrap sample, i.e., conditionally given $\widehat{P}_n$, $X_{n1}^*, X_{n2}^*, \ldots, X_{nn}^*$ are i.i.d. with common distribution $\widehat{P}_n$. For simplicity of notation, we will drop n from the subscript of the bootstrap observations and write X_i^* in place of X_{ni}^*. Let $\mathscr{L}(V)$ denote the *law, or distribution,* of a random variable V. Let $d_\infty(Q_1, Q_2)$ denote the *Kolmogorov distance* between two probability measures Q_1, Q_2 on $\mathbb{R}$ with distribution functions F_1, F_2,

$$d_\infty(Q_1, Q_2) = \sup_x |F_1(x) - F_2(x)|. \tag{9.5}$$

Theorem 9.1. *Assume* $0 < \sigma^2 \equiv \operatorname{var} g(X_1) < \infty$. *Then, with probability one,*

$$d_\infty\left(\mathscr{L}^*\left(\sqrt{n}(T_n - T(P))\right), \mathscr{L}\left(\sqrt{n}(T_n^* - T(\widehat{P}_n))\right)\right) \longrightarrow 0 \quad as\ n \to \infty,$$

where $\mathscr{L}^*$ *denotes law under* $\widehat{P}_n$.

Proof. We will prove the result under the additional assumption $E|g(X_1)|^3 < \infty$. (The general case is proved under Remark 9.1.) It is enough to consider the case $g(x) = x$. Write $\mu = EX_1$, $\sigma^2 = \operatorname{var} X_1$, $\rho_3 = E|X_1 - \mu|^3$. We will show that the $\widehat{P}_n$-distribution of $\sqrt{n}(\overline{X}^* - \overline{X})$ approximates the P-distribution of $\sqrt{n}(\overline{X} - \mu)$ (almost surely). Since the latter distribution converges to $\Phi_{\sigma^2}(x)$, it is enough to show that the $\widehat{P}_n$-distribution of $\sqrt{n}(\overline{X}^* - \overline{X})$ converges to Φ_{σ^2} (almost surely, or in probability). Now, by the Berry–Esséen Theorem,[1]

$$\left|\widehat{P}_n\left(\sqrt{n}(\overline{X}^* - \overline{X}) \le x\right) - \Phi_{\hat{\sigma}^2}(x)\right| \le \ell_n^*, \tag{9.6}$$

where $\hat{\sigma}^2 = \frac{1}{n}\sum_{i=1}^{n}(X_i - \overline{X})^2$, and

$$\ell_n^* = \frac{n^{-\frac{1}{2}}\frac{1}{n}\sum_{i=1}^{n}|X_i - \overline{X}|^3}{\hat{\sigma}^3} \longrightarrow 0 \quad \text{a.s. as } n \to \infty.$$

To see this note that $\hat{\sigma}^2 = \frac{1}{n}\sum_{i=1}^{n}[(X_i - \mu)^2 + (\overline{X} - \mu)^2 - 2(X_i - \mu)(\overline{X} - \mu)] = \frac{1}{n}\sum_{i=1}^{n}(X_i - \mu)^2 - (\overline{X} - \mu)^2 \to E(X_1 - \mu)^2 = \sigma^2$ a.s. Also, $\frac{1}{n}\sum_{i=1}^{n}|X_i - \overline{X}|^3 \le \frac{1}{n}\sum_{i=1}^{n}2^{3-1}(|X_i - \mu|^3 + |\overline{X} - \mu|^3) \to 4\rho_3$ a.s., using the elementary inequality

$$|a + b|^p \le 2^{p-1}\left(|a|^p + |b|^p\right) \quad \forall\, p \ge 1.$$

Hence $\ell_n^* \to 0$ a.s. Next, $\frac{1}{\sqrt{2\pi\hat{\sigma}^2}}e^{-z^2/2\hat{\sigma}^2} \longrightarrow \frac{1}{\sqrt{2\pi\sigma^2}}e^{-z^2/2\sigma^2}$ as $n \to \infty$, (for all ω, outside a set N of probability zero). Hence, by Scheffé's Theorem, $\Phi_{\hat{\sigma}^2}(x) \to \Phi_{\sigma^2}(x)$ uniformly for all x, outside N). $\qquad\square$

---

[1] See, e.g., Bhattacharya and Rao (1976, pp. 110).

Remark 9.1. Write $\overline{g} = \frac{1}{n}\sum_{i=1}^{n} g(X_i)$, $s_n^2 = \frac{1}{n}\sum_{i=1}^{n}(g(X_i) - \overline{g})^2$. Now, conditionally given $\widehat{P}_n$, $Y_{n,i}^* := (g(X_{ni}^*) - \overline{g})$ is a triangular array of i.i.d. random variables with mean zero and variance s_n^2 $(1 \le i \le n; \, n \ge 1)$. Also, for each $\varepsilon > 0$,

$$E^* \sum_{i=1}^{n} \left(\left(\frac{Y_{n,i}^*}{\sqrt{n}}\right)^2 \cdot \mathbf{1}_{\left\{\left|\frac{Y_{n,i}^*}{\sqrt{n}}\right| > \varepsilon\right\}} \right) = E^* Y_{n,1}^{*2} \mathbf{1}_{\{|Y_{n,1}^*| > \varepsilon \sqrt{n}\}}$$

$$= \frac{1}{n} \sum_{i=1}^{n} (g(X_i) - \overline{g})^2 \mathbf{1}_{\{|g(X_i) - \overline{g}| > \varepsilon \sqrt{n}\}} \quad (9.7)$$

whose expectation

$$E\left[(g(X_1) - \overline{g})^2 \cdot \mathbf{1}_{\{|g(X_1) - \overline{g}| > \varepsilon \sqrt{n}\}} \right]$$

goes to zero. This implies that there exists a sequence $\varepsilon_n \downarrow 0$ such that the last expectation goes to zero with ε_n in place of ε. One may now apply the Lindeberg–Feller central limit theorem. Since $s_n^2 \to \sigma^2$ by SLLN, it follows that the Kolmogorov distance between the (bootstrap) distribution of $T_n^* = \Sigma_{i=1}^n Y_{n,i}^*$ (under $\widehat{P}_n$) and the distribution (under P) of $T_n = \overline{g} - Eg(X_1)$ goes to zero, a.s.

Remark 9.2. It may be shown that for linear statistics T_n, the *bootstrap approximation* of the distribution is valid (i.e., *consistent* in the Kolmogorov distance) if and only if T_n is asymptotically normal. A similar phenomenon holds for more general statistics (See Giné and Zinn 1989, 1990). Thus, roughly speaking, bootstrap works only when classical Gaussian approximation works (with relatively minor exceptions). Various counterexamples highlight the inconsistency of the bootstrap estimate when the conditions for the validity of classical asymptotics break down (See, e.g., Athreya 1987).

9.2.2 Smooth Functions of Sample Averages

Let $Z_i = (f_1(X_i), f_2(X_i), \ldots, f_k(X_i))$ be a vector of sample characteristics of the ith observation $(1 \le i \le n)$. Here f_j are real-valued measurable functions such that $Ef_j^2(X_1) < \infty$ for $1 \le j \le k$. Let

$$\overline{Z} = \frac{1}{n} \sum_{i=1}^{n} Z_i = \left(\frac{1}{n} \sum_{i=1}^{n} f_1(X_i), \ldots, \frac{1}{n} \sum_{i=1}^{n} f_k(X_i) \right),$$

$$\mu := EZ_i = (Ef_1(X_1), \ldots, Ef_k(X_1)). \quad (9.8)$$

Consider a statistic $T_n = H(\overline{Z})$, where H is a continuously differentiable function in a neighborhood of μ in $\mathbb{R}^k$. We have seen earlier (using the delta method) that $\widehat{T}_n - H(\mu)$ equals $(\overline{Z} - \mu) \cdot (\text{grad } H)(\mu) + o_p(n^{-\frac{1}{2}})$. Therefore, one needs to look only at the linear statistic $(\overline{Z} - \mu) \cdot (\text{grad } H)(\mu)$. The same is true for $T_n^* - T_n = H(\overline{Z}^*) - H(\overline{Z})$, conditionally given the empirical $\widehat{P}_n = \frac{1}{n}\sum_{i=1}^n \delta_{Z_i}$. Hence, by Remark 9.1 above,

$$d_\infty(\mathscr{L}^*(T_n^* - T_n), \mathscr{L}(T_n - H(\mu))) \longrightarrow 0 \text{ in probability.} \quad (9.9)$$

9.2.3 Linear Regression

Consider the regression model

$$Y \equiv \begin{pmatrix} Y_1 \\ Y_2 \\ \vdots \\ Y_n \end{pmatrix} = X\beta + \varepsilon, \tag{9.10}$$

where X is the known (nonrandom) $n \times p$ *design matrix*,

$$X = \begin{bmatrix} X_{11} & X_{12} & \cdots & X_{1p} \\ X_{21} & X_{22} & \cdots & X_{2p} \\ & & \cdots & \\ & & \cdots & \\ X_{n1} & X_{n2} & \cdots & X_{np} \end{bmatrix} \tag{9.11}$$

$\beta = (\beta_1\beta_2\ldots\beta_p)'$ is the $p \times 1$ column vector of unknown parameters to be estimated, and $\varepsilon = (\varepsilon_1, \varepsilon_2, \ldots, \varepsilon_n)'$ is the $n \times 1$ column vector of i.i.d. random variables satisfying

$$E\varepsilon_i = 0, \quad 0 < \sigma^2 \equiv E\varepsilon_i^2 < \infty. \tag{9.12}$$

We will make the standard assumption that X is of full rank p $(n \geq p)$. Then the least square estimator $\hat{\beta}$ of β is given by

$$\begin{aligned} \hat{\beta} &= (X'X)^{-1}X'Y \equiv (X'X)^{-1}X'(X\beta + \varepsilon) \\ &= \beta + (X'X)^{-1}X'\varepsilon, \end{aligned} \tag{9.13}$$

so that $\hat{\beta} - \beta = (X'X)^{-1}X'\varepsilon$, and the covariance matrix of $\hat{\beta}$ is given by

$$\operatorname{cov}\hat{\beta} = (X'X)^{-1}X'\sigma^2 IX(X'X)^{-1} = \sigma^2(X'X)^{-1}. \tag{9.14}$$

The estimated residuals are given by

$$\hat{\varepsilon}_{is} := \hat{\varepsilon}_i - \bar{\hat{\varepsilon}}, \quad \text{with } \hat{\varepsilon}_i = Y_i - X_i\hat{\beta} \quad (1 \leq i \leq n) \text{ and}$$

$$\bar{\hat{\varepsilon}} = \frac{1}{n}\sum_{i=1}^{n}\hat{\varepsilon}_i, \tag{9.15}$$

where X_i is the ith row vector of X

$$X_i = (X_{i1}, X_{i2}, \ldots, X_{ip}) \qquad (1 \leq i \leq n). \tag{9.16}$$

The *shifted empirical distribution of the residuals* is $\widetilde{P}_n := \frac{1}{n}\sum_{i=1}^{n}\delta_{\hat{\varepsilon}_{is}}$. Let

$$T_n = (X'X)^{\frac{1}{2}}(\hat{\beta} - \beta) \equiv (X'X)^{-\frac{1}{2}}X'\varepsilon$$

$$T_n^* = (X'X)^{\frac{1}{2}}(\beta^* - \hat{\beta}) = (X'X)^{-\frac{1}{2}}X'\varepsilon^*, \tag{9.17}$$

Where $\varepsilon^* = (\varepsilon_1^*, \ldots, \varepsilon_n^*)'$ with ε_i^* $(1 \leq i \leq n)$ i.i.d. with common distribution $\widetilde{P}_n$ (conditionally given the estimated residuals $\hat{\varepsilon}_i$ $(1 \leq i \leq n)$), and

$$\beta^* = \hat{\beta} + (X'X)^{-1}X'\varepsilon^*. \tag{9.18}$$

We will show that the bootstrapped distribution of T_n^* is consistent for the distribution of T_n, under an appropriate condition on X.

For this let us introduce the *Mallows distance* d_2 on the space $\mathscr{P}\,(\mathbb{R}^k)$ of probability measures on (the Borel sigma-field of) $\mathbb{R}^k$:

$$d_2(P_1, P_2) = (\inf E\|U - V\|^2)^{\frac{1}{2}} \qquad (P_1, P_2 \in \mathscr{P}(\mathbb{R}^k)), \tag{9.19}$$

where the infimum is over all random vectors U, V (defined on some probability space, arbitrary) such that U has distribution P_1 and V has distribution P_2. Equivalently, one could define $d_2^2(P_1, P_2) = \inf \int_{\mathbb{R}^{2k}} \|u - v\|^2 Q(du\,dv)$, where the infimum is over all distributions Q on $\mathbb{R}^{2k} = \mathbb{R}^k \times \mathbb{R}^k$, such that the distribution of the vectors of the first k coordinates u is P_1 and that of the last k coordinates v is P_2. It is known (See Bickel and Freedman 1981) that

$$d_2(P_n, P) \longrightarrow 0 \iff \text{(i) } P_n \to P \text{ weakly,}$$
$$\text{and} \quad \text{(ii) all first and second order moments of } P_n \text{ converge}$$
$$\text{to those of } P. \tag{9.20}$$

For the following results due to Bickel and Freedman (1983), let $\pi_n(P)$ denote the distribution of T_n under a common distribution P of the i.i.d. residuals ε_i, and $\pi_{n,c}(P)$ the corresponding distribution of $c'T_n$, for $c \in \mathbb{R}^p$.

Proposition 9.1. *Let $X'X$ be nonsingular, and P and Q two probability measures for the residuals ε_i such that $E\varepsilon_i = 0$, $E\varepsilon_i^2 < \infty$ under both P and Q. Then*

$$d_2(\pi_n(P), \pi_n(Q)) \le \sqrt{p}\, d_2(P, Q), \tag{9.21}$$

and

$$d_2(\pi_{n,c}(P), \pi_{n,c}(Q)) \le \|c\| d_2(P, Q).$$

Proof. Let (U_i, V_i), $1 \le i \le n$, be i.i.d. with $\mathscr{L}(U_i) = P$, $\mathscr{L}(V_i) = Q$. Write $U = (U_1, \ldots, U_n)'$, $V = (V_1, \ldots, V_n)'$, $Z = X(X'X)^{-\frac{1}{2}}$. Let 'inf' below be the infimum over all such U, V. Then

$$d_2^2(\pi_n(P), \pi_n(Q)) \le \inf E\|Z'(U - V)\|^2$$
$$= \inf \sum_{r=1}^{p} E \left(\sum_{i=1}^{n} Z_{ir}(U_i - V_i) \right)^2 = \left(\sum_{r=1}^{p} \sum_{i=1}^{n} Z_{ir}^2 \right) \inf E(U_i - V_i)^2$$
$$= (\text{Trace of } Z'Z) d_2^2(P, Q) = p d_2^2(P, Q),$$

since $Z'Z$ is the $p \times p$ identity matrix. This proves (9.21). Similarly,

$$d_2^2(\pi_{n,c}(P), \pi_{n,c}(Q)) \le \inf E(c'Z'(U - V))^2$$
$$= \inf E(c'Z'(U - V) \cdot (U - V)'Zc)$$
$$= \inf(c'Z'\{E(U - V)(U - V)'\}Zc)$$
$$= \inf c'Z'E(U_1 - V_1)^2 I_{n \times n} Zc = \inf c'Z'E(U_1 - V_1)^2 Zc$$
$$= (c'Z'Zc) d_2^2(P, Q) = c'c\, d_2^2(P, Q).$$

$\square$

Theorem 9.2. *Suppose that $X'X$ is nonsingular for all sufficiently large n, $E\varepsilon_i = 0$, $0 < \sigma^2 = E\varepsilon_i^2 < \infty$. Then*

$$d_2(\pi_n(\widetilde{P}_n), \pi_n(P)) \longrightarrow 0 \text{ in probability, if } \frac{p+1}{n} \to 0. \tag{9.22}$$

Proof. First note that, writing $\bar{\varepsilon} = \frac{1}{n}\sum_{i=1}^n \varepsilon_i$, $\bar{\hat{\varepsilon}} = \frac{1}{n}\sum_{i=1}^n \hat{\varepsilon}_i$,

$$d_2^2(\widetilde{P}_n, \widehat{P}_n) \leq n^{-1}\sum_{i=1}^n \left[(\hat{\varepsilon}_i - \bar{\hat{\varepsilon}}) - \varepsilon_i\right]^2$$

$$= n^{-1}\sum_{i=1}^n \left[\hat{\varepsilon}_i - \varepsilon_i - (\bar{\hat{\varepsilon}} - \bar{\varepsilon})\right]^2 = n^{-1}\sum_{i=1}^n \left[\hat{\varepsilon}_i - \varepsilon_i - (\bar{\hat{\varepsilon}} - \bar{\varepsilon})\right]^2 + \bar{\varepsilon}^2$$

$$\leq n^{-1}\sum_{i=1}^n (\hat{\varepsilon}_i - \varepsilon_i)^2 + \bar{\varepsilon}^2$$

$$= n^{-1}\sum_{i=1}^n (X_i(\hat{\beta} - \beta))^2 + \bar{\varepsilon}^2$$

$$= n^{-1}\sum_{i=1}^n \sum_{r,r'=1}^p X_{ir}X_{ir'}(\hat{\beta}_r - \beta_r)(\hat{\beta}_{r'} - \beta_{r'}) + \bar{\varepsilon}^2$$

$$= \sum_{r,r'=1}^p \left(n^{-1}\sum_{i=1}^n X_{ir}X_{ir'}\right)(\hat{\beta}_r - \beta_r)(\hat{\beta}_{r'} - \beta_{r'}) + \bar{\varepsilon}^2$$

$$= \frac{1}{n}(\hat{\beta} - \beta)'X'X(\hat{\beta} - \beta) + \bar{\varepsilon}^2 = \frac{1}{n}\varepsilon'X(X'X)^{-1}X'X(X'X)^{-1}X'\varepsilon + \bar{\varepsilon}^2$$

$$= \frac{1}{n}\varepsilon'X(X'X)^{-1}X'\varepsilon + \bar{\varepsilon}^2. \tag{9.23}$$

Therefore,

$$Ed_2^2(\widetilde{P}_n, \widehat{P}_n) \leq \frac{\sigma^2}{n}(\text{Trace of } X(X'X)^{-1}X') + \frac{\sigma^2}{n} = \frac{\sigma^2}{n}(p+1). \tag{9.24}$$

To see this, let $A = X(X'X)^{-1}X'$. Then $A' = A$ and $A^2 = A$. Hence the eigenvalues of A are 0 or 1, and the number of 1's equals the rank of A. Clearly Rank $A \leq p$. But $AX = X$, the p column vectors of X are eigenvectors of A with eigenvalues and X has rank p. Hence Rank $A = p$.

Using (9.24) and the fact (See (9.20))

$$d_2(\widehat{P}_n, P) \longrightarrow 0 \text{ in probability}, \tag{9.25}$$

one arrives at

$$d_2(\widetilde{P}_n, P) \longrightarrow 0 \text{ in probability if } \frac{p+1}{n} \longrightarrow 0. \tag{9.26}$$

Remark 9.3. If one lets p be fixed (as is usually the case), then the bootstrap estimate of the distribution of $T_n = (X'X)^{\frac{1}{2}}(\hat{\theta} - \theta)$ is consistent if $X'X$ is nonsingular. On the other hand, T_n is asymptotically Normal if and only if (i) $X'X$ is nonsingular *and* (ii) the maximum diagonal element of $A = X(X'X)^{-1}X'$ goes to zero as

$n \to \infty$ (See Theorem 6.4, Chap. 6). Thus the bootstrap estimate may be consistent even in the case the CLT does not hold. This is a rather exceptional example of the consistency of the bootstrap in the absence of a valid Normal approximation (See Remark 9.2).

9.3 Notes and References

The main sources for this chapter are Efron (1979), who introduced the bootstrap, and Efron and Tibshirani (1994) where one can find detailed procedures for various bootstrapping techniques and many applications with real data. Section 9.2.3 is due to Bickel and Freedman (1981, 1983). More extensive references to the bootstrap appear in Chap. 11.

Exercises for Chap. 9

Exercises for Sects. 9.1, 9.2

Ex. 9.1. We reproduce from Diaconis and Efron (1983) the following data of $Y = LSAT$ (the average score of the entering class of the law school on a national test), and $X = GPA$ (the average grade-point average for the class), of $n = 15$ randomly chosen law schools from a population of $N = 82$ participating law schools in the U.S.

School															
$Y = LSAT$	576	635	558	578	666	580	555	661	651	605	653	575	545	572	594
$X = GPA$	3.39	3.30	2.81	3.03	3.44	3.07	3.00	3.43	3.36	3.13	3.12	2.74	2.76	2.88	2.96

(a) Compute the sample correlation coefficient $r = r_{X,Y}$.
(b) Take $M = 200$ (bootstrap) samples of size $n = 15$ each from the above data and compute the corresponding 200 bootstrap values r_j^* $(j = 1, \ldots, 200)$ of the correlation coefficient.
(c) Construct a 90 % percentile bootstrap confidence interval for the population coefficient of correlation $\rho = \rho_{X,Y}$. [True value of ρ is 0.761.]
(d) Find the bootstrap estimate of the standard error of r.
(e) Construct a classical CLT based confidence interval for ρ based on the above sample of size $n = 15$ [Hint: Use (6.67) to estimate the standard error of r].

Ex. 9.2. Consider the data in Exercise 9.1, and assume that the regression of Y on X is linear: $Y = \alpha + \beta x + \varepsilon$, with ε having mean zero and finite variance.

(a) Find the least squares estimates $\hat{\alpha}$, $\hat{\beta}$.
(b) Find a 90 % percentile bootstrap confidence interval for β, and one for α.

Ex. 9.3. Consider the AR(2) model $X_n = 1 + \frac{3}{2}X_{n-1} - \frac{9}{16}X_{n-2} + \varepsilon_n$ $(n \geq 2)$, with ε_n $(n \geq 2)$ i.i.d. $N(0,1)$ and independent of (X_0, X_1).

(a) Show that the model satisfies the stability condition (8.170) and find the stationary distribution π of the Markov process $Y_n = (X_{n-1}, X_n)'$, $n \geq 1$. In particular, show that π is Gaussian.
(b) Show that under π, $\{X_n : n \geq 0\}$ is a stationary Gaussian process.
(c) Draw a random sample $(X_0, X_1, \ldots, X_{100})$ of the time series from the above AR(2) model.
(d) Suppose one is told that the data in (c) came from a stable stationary A(2) model $X_n = \alpha + \beta_1 X_{n-1} + \beta_2 X_{n-2} + \varepsilon_n$ with an i.i.d. Gaussian $N(0, \sigma^2)$. Compute the maximum likelihood estimators $\hat{\alpha}$, $\hat{\beta}_1$, $\hat{\beta}_2$, $\hat{\sigma}^2$ of α, β_1, β_2, σ^2.
(e) Construct a classical (i.e., CLT based) 95 % confidence interval for β_2 in (d) above.
(f) Construct a 95 % percentile bootstrap confidence interval for β_2, by (i) resampling from the *estimated residuals* $\hat{\varepsilon}_j$ $(2 \leq j \leq 100)$ given by (8.177), with sample size 90, (ii) plugging the bootstrap sample ε_j^* $(2 \leq n \leq 100)$ in the formula $X_n^* = \hat{\alpha} + \hat{\beta}_1 X_{n-1}^* + \varepsilon_n^*$ $(2 \leq n \leq 100)$, beginning with $X_0^* = X_0$, $X_1^* = X_1$, to obtain a bootstrap sample $(X_0^*, X_1^*, \ldots, X_{100}^*)$ of the sequence $(X_0, X_1, \ldots, X_{100})$, (iii) computing the bootstrap value of β_2^* of the statistic $\hat{\beta}_2$ using (d), and (iv) repeating (i)–(iii) many times (say $M = 500$ times) to obtain the quantiles $q_{\alpha/2}^*$, $q_{1-\alpha/2}^*$ $(\alpha = 0.025)$ of the M values of β_2^* so obtained.

References

Athreya, K. B. (1987). Bootstrap of the mean in the infinite variance case. *Annals of Statistics, 15*(2), 724–731.

Bhattacharya, R., & Waymire, E. C. (1990). *Stochastic processes with applications.* Philadelphia: SIAM.

Bickel, P. J., & Freedman, D. A. (1981). Some asymptotic theory for the Bootstrap. *Annals of Statistics, 9*(6), 1196–1217.

Bickel, P. J., & Freedman, D. A. (1983). Bootstrapping regression models with many parameters. In *A festschrift for Erich L. Lehmann* (pp. 28–48). Belmont: Wadsworth.

Efron, B. (1979). Bootstrap methods: Another look at the jackknife. *Annals of Statistics, 7*, 1–26.

Efron, B., & Tibshirani, T. (1994). *An introduction to the bootstrap.* New York: Chapman and Hall.

Giné, E., & Zinn, J. (1989). Necessary conditions for the bootstrap of the mean. *Annals of Statistics, 17*(2), 684–691.

Giné, E., & Zinn, J. (1990). Bootstrapping general empirical measures. *Annals of Statistics, 18*(2), 851–869.

Chapter 10
Nonparametric Curve Estimation

Abstract This chapter provides an introduction to nonparametric estimations of densities and regression functions by the kernel method.

10.1 Nonparametric Density Estimation

Let $X_1, X_2, \ldots, X_n$ be a random sample from a distribution P on the real line. A consistent estimator of P is the *empirical* $\widehat{P}_n = \frac{1}{n} \sum_{i=1}^{n} \delta_{X_i}$, assigning mass $\frac{1}{n}$ to each of the points $X_1, X_2, \ldots, X_n$. By the Strong Law of Large Numbers, $\widehat{P}_n(B) \equiv \frac{1}{n} \sum_{i=1}^{n} \mathbf{1}_B(X_i) \to E\mathbf{1}_B(X_1) = P(B)$ almost surely, as $n \to \infty$, for each Borel set B. With a little effort one can prove (the *Glivenko–Cantelli Theorem*):

$$\sup_{x \in \mathbb{R}} \left| \widehat{F}_n(x) - F(x) \right| \longrightarrow 0 \quad \text{as } n \to \infty, \text{ almost surely.} \tag{10.1}$$

Here $\widehat{F}_n(x) \equiv \widehat{P}_n((-\infty, x])$ is the cumulative distribution function of $\widehat{P}_n$ and $F(x)$ that of P. If F is continuous on $\mathbb{R}$ then one can prove, in addition to (10.1), that

$$\sup_{x \in \mathbb{R}} \sqrt{n} \left| \widehat{F}_n(x) - F(x) \right| \xrightarrow{\mathscr{L}} W, \tag{10.2}$$

where $W = \max\{|B_t^*| : 0 \le t \le 1\}$, $\{B_t^* : 0 \le t \le 1\}$ being the so-called *Brownian Bridge* (See, e.g., Bhattacharya and Waymire 1990, pp. 37–39). One may use (10.2) to obtain a confidence band for $F(\cdot)$.

Suppose now that F is absolutely continuous with a density f. Since $\widehat{P}_n$ is discrete, to estimate f one may use the density of the random variable $\widehat{X} + hZ$ where $\widehat{X}$ has the distribution $\widehat{P}_n$ (conditionally, given $X_1, \ldots, X_n$) and Z is independent of $\widehat{X}$ and has a nice density, say, K, and h is a small positive number, called the *bandwidth* satisfying

$$h \equiv h_n \longrightarrow 0 \quad \text{as } n \to \infty. \tag{10.3}$$

Note that $\widehat{X} + hZ$ has the density

$$\hat{f}_n(x) = \frac{1}{n} \sum_{i=1}^{n} K_h(x - X_i), \tag{10.4}$$

R. Bhattacharya et al., *A Course in Mathematical Statistics and Large Sample Theory*, Springer Texts in Statistics, DOI 10.1007/978-1-4939-4032-5_10

where K_h is the density of hZ, namely,

$$K_h(y) = \frac{1}{h} K\left(\frac{y}{h}\right). \tag{10.5}$$

To see that $\hat{f}_n$ is the density of $\widehat{X} + hZ$, note that the latter has the distribution function

$$\widetilde{\mathrm{Prob}}(\widehat{X} + hZ \le x) = \widetilde{E}[\widetilde{\mathrm{Prob}}(hZ \le x - \widehat{X}|\widehat{X})]$$
$$= \frac{1}{n} \sum_{i=1}^{n} \int_{-\infty}^{x - X_i} K_h(u)du, \tag{10.6}$$

where the superscript $\tilde{\ }$ indicates that the probabilities are computed given $X_1, \ldots, X_n$. Differentiating (10.6) w.r.t. x one arrives at (10.4).

Alternatively, one may think of (10.4) as obtained by spreading out the point mass δ_{X_i} with a density centered at X_i and concentrating most of the density near X_i $(1 \le i \le n)$. Now we show that the *bias* $E\hat{f}_n(x) - f(x) \to 0$. For this write

$$E\hat{f}_n(x) = E\frac{1}{n} \sum_{i=1}^{n} \frac{1}{h} K\left(\frac{x - X_i}{h}\right) = E\frac{1}{h} K\left(\frac{x - X_1}{h}\right)$$
$$= \frac{1}{h} \int_{-\infty}^{\infty} K\left(\frac{x - y}{h}\right) f(y)dy$$
$$= \int_{-\infty}^{\infty} K(v)f(x - vh)dv \longrightarrow f(x) \text{ as } h \downarrow 0, \tag{10.7}$$

if f is bounded and is continuous at x (by the Lebesgue Dominated Convergence Theorem). Also,

$$\mathrm{var}(\hat{f}_n(x)) = \frac{1}{n} \mathrm{var}\, K_h(x - X_1)$$
$$= \frac{1}{n} \left\{ \frac{1}{h^2} EK^2\left(\frac{x - X_1}{h}\right) - \left(E\frac{1}{h} K\left(\frac{x - X_1}{h}\right)\right)^2 \right\}$$
$$= \frac{1}{nh} \int_{-\infty}^{\infty} K^2(v)f(x - vh)dv - \frac{1}{n} \left(\int_{-\infty}^{\infty} K(v)f(x - vh)dv\right)^2 \longrightarrow 0, \tag{10.8}$$

if (1) f is bounded, (2) f is continuous at x, (3) $K^2(v)$ is integrable,

$$nh \longrightarrow \infty \quad \text{and} \quad h \longrightarrow 0 \quad (\text{as } n \to \infty). \tag{10.9}$$

Thus, under the hypothesis (1)–(3) and (10.9), one has (by (10.7) and (10.8))

$$E\left(\hat{f}_n(x) - f(x)\right)^2 = \mathrm{var}\,\hat{f}_n(x) + (\mathrm{Bias}\,\hat{f}_n(x))^2 \longrightarrow 0 \quad \text{as } n \to \infty. \tag{10.10}$$

In other words, under the above assumptions, $\hat{f}_n(x) \to f(x)$ *in probability* as $n \to \infty$.

Note that the convergences (10.7) and (10.8) do not really require boundedness of f on all of $\mathbb{R}$. For example, if one takes K to have a compact support then it is enough to require that f is continuous at x. We have proved that under mild assumptions the so-called *kernel estimator* (with *kernel* K) is a consistent estimator of f at every point of continuity of f.

By choosing an appropriately smooth and *symmetric* kernel K one may make the error of approximation $\hat{f}_n(x) - f(x)$ reasonably small.

A measure of the (squared) error of approximation is provided by the so-called *mean integrated squared error,* or MISE, given by

$$\text{MISE}(\hat{f}_n) = \int_{\mathbb{R}} E[\hat{f}_n(x) - f(x)]^2 dx = \int_{\mathbb{R}} [\text{var}\,\hat{f}_n(x) + (\text{Bias}\,\hat{f}_n(x)]^2 dx. \tag{10.11}$$

Write

$$c_1 = \int v^2 K(v) dv, \quad c_2 = \int K^2(v) dv, \quad c_3 = \int (f''(x))^2 dx, \tag{10.12}$$

and assume c_1, c_2, c_3 are finite and that

$$\int K(v) dv = 1, \quad \int v K(v) dv = 0. \tag{10.13}$$

Now it follows from (10.7) that

$$E\hat{f}_n(x) = \int_{\mathbb{R}} K(v) \left[f(x) - vhf'(x) + \frac{v^2 h^2}{2} f''(x) + o(h^2) \right] dv$$

$$= f(x) + \frac{c_1 h^2}{2} f''(x) + o(h^2),$$

$$(\text{Bias}\,\hat{f}_n)(x) = \frac{c_1 h^2}{2} f''(x) + o(h^2), \tag{10.14}$$

if f'' is continuous and bounded. Then

$$\int_{\mathbb{R}} (\text{Bias}\,\hat{f}_n)^2(x) dx = \frac{c_3 c_1^2 h^4}{4} + o(h^4). \tag{10.15}$$

Next, by (10.8) and (10.9),

$$\text{var}\,\hat{f}_n(x) = \frac{c_2 f(x)}{nh} + O\left(\frac{1}{n}\right),$$

$$\int_{\mathbb{R}} \text{var}\,\hat{f}_n(x) dx = \frac{c_2}{nh} + O\left(\frac{1}{n}\right), \quad \text{as } n \to \infty. \tag{10.16}$$

Hence

$$\text{MISE}(\hat{f}_n) = \frac{c_1^2 c_3}{4} h^4 + \frac{c_2}{nh} + o(h^4) + O\left(\frac{1}{n}\right). \tag{10.17}$$

Neglecting the two smaller order terms, the asymptotically optimal choice of the bandwidth h, for a given kernel K as above, is obtained by

$$h_n = \arg\min_h \left\{ \frac{c_1^2 c_3}{4} h^4 + \frac{c_2}{nh} \right\} = \left(\frac{c_2}{c_1^2 c_3} \right)^{\frac{1}{5}} n^{-\frac{1}{5}}. \tag{10.18}$$

The corresponding asymptotically minimal MISE is

$$\text{MISE } \hat{f}_n = \frac{c_4}{n^{4/5}} + o\left(n^{-4/5}\right) \quad c_4 := \frac{5}{4}\left(c_1^{2/5} c_2^{4/5} c_3^{1/5}\right). \tag{10.19}$$

We have arrived at the following result.

Theorem 10.1. *Assume f'' is continuous and bounded. Then for any choice of a symmetric kernel K satisfying (10.13), and $0 < c_i < \infty$ $(i = 1, 2, 3)$, the asymptotically optimal bandwidth h is given by the extreme right side of (10.18), and the asymptotically minimal MISE is given by (10.19).*

From the expression (10.4) it follows that $\hat{f}_n(x)$ is, for each n, a sum of i.i.d. random variables. By the Lindeberg CLT it now follows that, under the hypothesis of Theorem 10.1 one has (Exercise 10.1)

$$\frac{\hat{f}_n(x) - E\hat{f}_n(x)}{\sqrt{\text{var}\hat{f}_n(x)}} \xrightarrow{\mathscr{L}} N(0, 1), \quad \text{if } f(x) > 0. \tag{10.20}$$

Also check, using (10.14), (10.15) and (10.18), that

$$\frac{E\hat{f}_n(x) - f(x)}{\sqrt{\text{var}\hat{f}_n(x)}} \longrightarrow c_3^{-\frac{1}{2}} \frac{f''(x)}{\sqrt{f(x)}} = \gamma, \quad \text{say, if } f(x) > 0. \tag{10.21}$$

Hence

$$\frac{\hat{f}_n(x) - f(x)}{\sqrt{\text{var}\hat{f}_n(x)}} \xrightarrow{\mathscr{L}} N(\gamma, 1) \quad \text{if } f(x) > 0. \tag{10.22}$$

To remove the asymptotic bias γ, one may choose a slightly sub-optimal bandwidth $h_n = o(n^{-\frac{1}{5}})$ (Exercise 10.2). Since $\text{var}\hat{f}_n(x)$ involves $f''(x)$, for setting confidence regions for $f(x)$, one may resort to bootstrapping (See Hall 1992).

Remark 10.1. It has been shown by Epanechnikov (1969) that the constant c_4 in MISE is minimized (under the hypothesis of Theorem 10.1) by the kernel

$$K(v) = \frac{3}{4\sqrt{5}}\left(1 - \frac{1}{5}v^2\right) \mathbf{1}_{\{|v| \leq \sqrt{5}\}}. \tag{10.23}$$

However, the loss of efficiency is rather small if, instead of (10.23), one chooses any symmetric kernel with high concentration, such as the (standard) Normal density or the triangular density (See Lehmann 1999, pp. 415, 416).

Remark 10.2 (Optimal Choice of Bandwidth and Cross Validation). The "optimal" h_n given by (10.18) is not usable in practice because it involves the second derivative of the unknown density f (see (10.12)). Following (Tsybakov 2009, pp. 28, 29), we now describe a practical choice of an optimal bandwidth $h(CV)$ given by

$$h(CV) = \underset{h > 0}{\arg\min}\, CV(h),$$

$$CV(h) := \int \hat{f}_n^2(x)dx - \frac{2}{n}\sum_{i=1}^{n} \hat{f}_{n,-i}(X_i), \tag{10.24}$$

where $\hat{f}_{n,-i}(x) \equiv ((n-1)h)^{-1} \sum_{j \neq i} K(\frac{x-X_j}{h})$ is the kernel estimate of $f(x)$ using the $n-1$ observations $\{X_j : j \neq i\}$ omitting X_i. The quantity $CV(h)$ in (10.24) is called the *cross-validation*. To prove this optimality observe that $MISE(\hat{f}_n) = E \int (\hat{f}_n(x) - f(x))^2 dx = E \int \hat{f}_n^2(x)dx - 2E \int f(x)\hat{f}_n(x)dx + \int f^2(x)dx$. Of the three terms, $\int f^2(x)dx$ does not involve h, and $\int \hat{f}_n^2(x)dx$ is an unbiased estimate of its expectation. We now seek an unbiased estimate of the middle term $E \int f(x)\hat{f}_n(x)dx$, and this is provided by $\hat{A} = n^{-1} \sum_{i=1}^n \hat{f}_{n,-i}(X_i)$ (See Exercise 10.4). Thus $E(CV(h)) + \int f^2(x)dx = MISE(\hat{f}_n)$, so that the functions $E(CV(h))$ and $MISE(\hat{f}_n)$ have the same minimizing h.

The asymptotic theory presented above has extensions to the multidimensional case. We provide a brief sketch of the arguments here, leaving the details to Exercise 10.3. Let f be a probability density function on $\mathbb{R}^d$, and let K be a symmetric kernel density with finite second moments $c_{1,i,j} = \int v_i v_j K(\mathbf{v})d\mathbf{v}$, and with $c_2 = \int K^2(\mathbf{v})d\mathbf{v}$. Then if f'' is continuous and bounded, one uses the kernel estimate

$$\hat{f}_n(\mathbf{x}) = \frac{1}{nh^d} \sum_{i=1}^n K\left(\frac{\mathbf{x} - X_i}{h}\right), \tag{10.25}$$

based on i.i.d. observations $X_1, \ldots, X_n$ with p.d.f. f. Then

$$E\hat{f}_n(x) = E\frac{1}{h^d} K\left(\frac{\mathbf{x} - X_1}{h}\right) = \frac{1}{h^d} \int_{\mathbb{R}^d} K\left(\frac{\mathbf{x} - \mathbf{v}}{h}\right) f(\mathbf{v})dv_1 \ldots dv_d$$

$$= \int_{\mathbb{R}^d} K(\mathbf{u})f(\mathbf{x} - h\mathbf{u})d\mathbf{u}$$

$$= \int_{\mathbb{R}^d} K(\mathbf{u}) \left[f(\mathbf{x}) - h\mathbf{u} \cdot \text{grad } f(\mathbf{x}) + \frac{h^2}{2} \sum u_i u_j \frac{\partial^2 f(\mathbf{x})}{\partial x_i \partial x_j} + o(h^2)\right] d\mathbf{u}$$

$$= f(\mathbf{x}) + \frac{h^2}{2} \sum c_{1,i,j} \frac{\partial^2 f(\mathbf{x})}{\partial x_i \partial x_j} + o(h^2),$$

so that

$$\text{Bias } \hat{f}_n(\mathbf{x}) = E\hat{f}_n(\mathbf{x}) - f(\mathbf{x}) = \frac{h^2}{2} \sum c_{1,i,j} \frac{\partial^2 f(\mathbf{x})}{\partial x_i \partial x_j} + o(h^2). \tag{10.26}$$

Also,

$$E\left(\frac{1}{h^d} K\left(\frac{\mathbf{x} - X_1}{h}\right)\right)^2 = \frac{h^d}{h^{2d}} \int_{\mathbb{R}^d} K^2(\mathbf{u}) \left[f(\mathbf{x}) - h\mathbf{u} \cdot \text{grad } f(\mathbf{x})\right.$$

$$\left. + \frac{h^2}{2} \sum u_i u_j \frac{\partial^2 f(\mathbf{x})}{\partial x_i \partial x_j} + o(h^2)\right] d\mathbf{u}$$

$$= \frac{1}{h^d} c_2 f(\mathbf{x}) + o\left(\frac{h^2}{h^d}\right),$$

$$\text{var}\left(\frac{1}{h^d} K\left(\frac{\mathbf{x} - X_1}{h}\right)\right) = \frac{1}{h^d} c_2 f(\mathbf{x}) + O(1)$$

$$\text{var}(\hat{f}_n(\mathbf{x})) = \frac{1}{nh^d} c_2 f(\mathbf{x}) + o\left(\frac{1}{nh^d}\right). \tag{10.27}$$

Hence

$$E(\hat{f}_n(\mathbf{x}) - f(\mathbf{x}))^2 = (\text{Bias } \hat{f}_n(\mathbf{x}))^2 + \text{var}(\hat{f}_n(\mathbf{x}))$$

$$= \frac{1}{nh^d} c_2 f(\mathbf{x}) + \frac{h^4}{4} \left(\sum_{i,j} c_{1,i,j} \frac{\partial^2 f(\mathbf{x})}{\partial x_i \partial x_j} \right)^2$$

$$+ o(h^4) + o\left(\frac{1}{nh^d} \right). \tag{10.28}$$

Therefore,

$$\text{MISE} \, (\hat{f}_n) = \int_{\mathbb{R}^d} E(\hat{f}_n(\mathbf{x}) - f(\mathbf{x}))^2 d\mathbf{x} = \frac{c_2}{nh^d} + \frac{h^4}{4} \tilde{c}_3 + o(h^4) + o\left(\frac{1}{nh^d} \right), \tag{10.29}$$

where

$$\tilde{c}_3 = \int_{\mathbb{R}^d} \left(\sum_{i,j} c_{1,i,j} \frac{\partial^2 f(\mathbf{x})}{\partial x_i \partial x_j} \right)^2 d\mathbf{x}.$$

As before, the asymptotically optimal bandwidth is given by

$$h_n = \arg\min_h \left\{ \frac{h^4}{4} \tilde{c}_3 + \frac{c_2}{nh^d} \right\} = \left(\frac{c_2 d}{\tilde{c}_3} \right)^{\frac{1}{d+4}} n^{-\frac{1}{d+4}}, \tag{10.30}$$

and the asymptotically minimal MISE is

$$\text{MISE} \, (\hat{f}_n) = C_2^{\frac{4}{d+4}} \tilde{C}_3^{\frac{d}{d+4}} \left(\frac{1}{4} d^{\frac{4}{d+4}} + d^{-\frac{d}{d+4}} \right) n^{-\frac{4}{d+4}} + o\left(n^{-\frac{4}{d+4}} \right). \tag{10.31}$$

Multi-dimensional versions of (10.20)–(10.22) may now be derived (Exercise 10.3).

10.2 Nonparametric Regression-Kernel Estimation

We now turn to the problem of nonparametric estimation of a regression function $f(x) = E[Y \mid X = x]$, based on observations (X_j, Y_j), $j = 1, \ldots, n$. Assume first that the regressor X is also stochastic and that (X_j, Y_j), $j = 1, \ldots, n$, are i.i.d. with a density $g_1(x)$ of X and a conditional distribution $G_2(dy \mid x)$ of Y, given $X = x$. Then, with a suitable symmetric kernel K, one has, by the SLLN, the following convergence almost surely as $n \to \infty$, for every x and every $h > 0$:

$$n^{-1} h^{-1} \sum_{1 \le j \le n} Y_j K \left(\frac{x - X_j}{h} \right) \longrightarrow h^{-1} E Y_j K \left(\frac{x - X_j}{h} \right)$$

$$= \int y h^{-1} K \left(\frac{x - u}{h} \right) g_1(u) G_2(dy \mid u) du. \tag{10.32}$$

As $h \downarrow 0$, under appropriate continuity conditions the last integral converges to $\int y g_1(x) G_2(dy \mid x)$, since $h^{-1} K((x - u)/h) du$ converges in distribution to the point mass at x, namely, δ_x. Thus the left side of (10.32) approximates $g_1(x) E[Y \mid X = x] = g_1(x) f(x)$ for small h and large n. Since, by the preceding section,

$n^{-1}h^{-1}\sum_{1\leq j\leq n}K((x-X_j)/h)\to g_1(x)$ as $n\to\infty$ and $h\downarrow 0$, one obtains the *Nadaraya-Watson estimator* of $f(x)$, namely,

$$n^{-1}h^{-1}\sum_{1\leq j\leq n}Y_j K\left(\frac{x-X_j}{h}\right)\Big/ n^{-1}h^{-1}\sum_{1\leq j\leq n}K\left(\frac{x-X_j}{h}\right)$$

$$=\sum_{1\leq j\leq n}Y_j K\left(\frac{x-X_j}{h}\right)\Big/\sum_{1\leq j\leq n}K\left(\frac{x-X_j}{h}\right). \tag{10.33}$$

Note that the argument goes through if X_js satisfy the convergence of the empirical measure $n^{-1}\sum\delta x_j$ to a probability distribution $g_1(x)dx$ as $n\to\infty$. This is also satisfied by a *non-stochastic design of X*, with the same convergence. For example, if one takes X_js to be equidistributed in $[0,1]$, $X_j = j/n$ $(j=1,\ldots,n)$, then the denominator on the left in (10.32) converges to $g_1(x)=1$, while the numerator converges to $f(x)g_1(x)=f(x)$.

We will now derive in detail the properties of a kernel estimator due to Gasser and Müller (1984) instead of the Nadaraya-Watson estimator. Both kernel estimators are asymptotically optimal among nonparametric estimators in a sense to be defined later. However, the former seems to perform a little better in simulation studies, and is more suitable with non-equidistant design points of (a non-stochastic) X. The proof of the asymptotic optimality of the Nadaraya-Watson estimator is similar, and is left as an exercise (Exercise 10.5).

First consider a non-stochastic X with equidistant design points labeled as $x_j = j/n$ $(j=1,\ldots,n)$. Also denote $x_0 = 0$. The domain over which the function f is estimated is taken to be $[0,1]$ without any essential loss of generality. Let K be a smooth symmetric probability density with support $[-1,1]$ (e.g., the rescaled Epanechnikov kernel (see Remark 10.1)). Assume the nonparametric regression model

$$Y_j = f(x_j)+\varepsilon_j \quad (j=1,\ldots,n),\ (x_j = j/n,\ (j=1,\ldots,n)\in[0,1]), \tag{10.34}$$

where ε_j are i.i.d., satisfying

$$E\varepsilon_j = 0, \quad 0 < E\varepsilon_j^2 = \sigma^2 < \infty. \tag{10.35}$$

Theorem 10.2. *Let f be twice continuously differentiable on $[0,1]$. Consider the estimator of f given by*

$$f_h(x) = h^{-1}\sum_{1\leq j\leq n}Y_j\int_{(x_{j-1},x_j]}K\left(\frac{x-u}{h}\right)du, \tag{10.36}$$

where the kernel K is a probability density which is symmetric and twice continuously differentiable with support $[-1,1]$. (a) Then, as $h\downarrow 0$ and $n\to\infty$, one has

(i) $\quad Ef_h(x) = f(x)+\left(\frac{c_1}{2}\right)h^2 f''(x)+o(h^2)+O(n^{-1}),\quad (h < x < 1-h),$

(ii) $\quad \mathrm{Var}(f_h(x)) \leq \sigma^2 c_4(nh)^{-1},\quad (0 < x < 1). \tag{10.37}$

Here c_1 is as in (10.12), and $c_4 = 2(\max\{K^2(u): u\in[-1,1]\})$.

(b) *The expected squared error of $f_h(x)$, for $h < x < 1-h$, attains its minimal rate $O(n^{-4/5})$ with $h = b_3 n^{-1/5}$ for any $b_3 > 0$.*

Proof. (a) By the labeling used, $EY_j = f(x_j)$, and

$$Ef_h(x) = h^{-1} \sum_{1 \le j \le n} f(x_j) \int_{(x_{j-1},x]} K((x-u)/h)du,$$

and

$$\left| Ef_h(x) - h^{-1} \int_{[0,1]} K((x-u)/h)f(u)du \right|$$

$$= h^{-1} \left| \sum_{1 \le j \le n} \int_{(x_{j-1},x_j]} K((x-u)/h)(f(x_j) - f(u))du \right|$$

$$\le ch^{-1}n^{-1} \int_{[0,1]} K((x-u)/h)du$$

$$= ch^{-1}n^{-1}h \int_{[-1,1]} K(v)dv = cn^{-1} \quad [c = \max\{|f'(u)| : u \in [0,1]\}]. \quad (10.38)$$

Now, as in the calculations (10.14), and (10.15), with c_1, c_2, c_3 as in (10.12), one has

$$f(x) - h^{-1} \int_{[0,1]} K((x-u)/h)f(u)du = (c_1h^2/2)f''(x) + o(h^2), \quad (10.39)$$

which, together with (10.38), yields the first relation in (10.37).

Next, by the mean value theorem, there exist $v_j \in [x_{j-1}, x_j]$ such that

$$\mathrm{Var}(f_h(x)) = h^{-2}\sigma^2 \sum_{1 \le j \le n} \left[\int_{(x_{j-1},x_j]} K((x-u)/h)du \right]^2$$

$$= h^{-2}\sigma^2 \sum_{1 \le j \le n} K^2((x-v_j)/h)(x_j - x_{j-1})^2$$

$$= n^{-2}h^{-2}\sigma^2 \sum_{1 \le j \le n} K^2((x-v_j)/h).$$

Since $(x - v_j)/h \in [-1,1]$ only if $v_j \in [x-h, x+h]$, and there are at most $2nh$ nonzero summands in the last sum, one derives the desired inequality

$$\mathrm{Var}(f_h(x)) \le 2n^{-1}h^{-1}\sigma^2 \left(\max\{K^2(u) : u \in [-1,1]\} \right). \quad (10.40)$$

(b) It follows from (10.37) that for $h < x < 1 - h$, if one takes $h = b_3 n^{-1/5}$ for some $b_3 > 0$, then

$$\text{Expected squared error of } f_h(x) = (\text{Bias of } f_h(x))^2 + \mathrm{Var}(f_h(x)) = O\left(n^{-\frac{4}{5}}\right). \quad (10.41)$$

With a more precise estimation of $\mathrm{Var}(f_h(x))$ one can show that (See Eubank 1999, pp. 165–166) that

$$\mathrm{Var}(f_h(x)) = c_2 h^{-1}h^{-1}\sigma^2 + O(n^{-2}h^{-2}). \quad (10.42)$$

Hence the right side of (10.42) is minimized by $h = b(x)n^{-1/5}$ for some appropriate constant $b(x)$. This completes the proof of (b). $\square$

Remark 10.3. The proof of Theorem 10.2 is easily extended to the case of *arbitrary design* points x_j such that $a/n < x_{i+1} - x_i < b/n$ for all $i = 0, 1, \ldots, n$ $(x_0 = 0)$, for any pair of positive constants $a < b$. If the *regressor X is stochastic*, then the proof (with X and ε independent) extends to the case where X has a density which is continuous and strictly positive on $(0, 1)$.

Remark 10.4. The arguments above show that, under the hypothesis of Theorem 10.2, the asymptotic optimal rate is given by $O(n^{-4/5})$, over the class of all kernels that may be used. It has actually been proved by Stone (1980), that nonparametric estimators of the regression function f cannot have an integrated squared error of smaller rate than $O(n^{-4/5})$, no matter whether kernel or other methods are used, provided f is required to be only twice continuously differentiable.

As in the case of density estimation, one can establish the CLT for $f_h(x)$, namely,

$$[\hat{f}_n(x) - E\hat{f}_n(x)]/[\text{Var}(\hat{f}_n(x)]^{\frac{1}{2}} \longrightarrow N(0, 1) \text{ in distribution as } n \to \infty, \quad (10.43)$$

where $\hat{f}_n(x) := f_h(x)$ with $h = b_3 n^{-1/5}$ for some $b_3 > 0$, and

$$[\hat{f}_n(x) - f(x)]/[\text{Var}(\hat{f}_n(x)]^{\frac{1}{2}} \longrightarrow N(\gamma(x), 1) \text{ in distribution as } n \to \infty, \quad (10.44)$$

where $\gamma(x) = c'f''(x)$ arises from the bias term (10.37) and the variance term (10.42). Also, just as in the case of density estimation (See Exercise 10.2), by choosing the bandwidth to go to zero at a slightly faster rate than the optimal rate, namely, $h = o(n^{-1/5})$, one may prove the useful result

$$[f_h(x) - f(x)]/[\text{Var}(f_h(x)]^{\frac{1}{2}} \longrightarrow N(0, 1) \text{ in distribution if } h = o(n^{-1/5}), n \to \infty.$$
$$(10.45)$$

Remark 10.5. In the case of multiple regression, i.e., with a *d-dimensional covariate* X, one can show in the same manner as in Sect. 10.1, that the optimal rate of the integrated squared error is $O(n^{-4/(d+4)})$, attained with bandwidth $h = c''n^{-1/(d+4)}$. Simulation studies exhibit dramatic deteriorations in the performance of nonparametric estimators of density and regression as d increases, a phenomenon often referred to as the *curse of dimensionality* (See, e.g., Wasserman 2003, p. 319).

Remark 10.6. It may be noted that the bias term in (10.37) is computed only for $h < x < 1 - h$. Unfortunately, for x beyond this range one does not get the expected squared error to be as small as $O(n^{-4/5})$, but rather $O(n^{-3/4})$. For a discussion of this so-called *boundary effect,* see Eubank (1999, p. 170).

Finally, we turn to the important problem of the data-driven choice of the optimal bandwidth h, which is not immediately available from the above discussion. The most popular procedure for this is known as *cross-validation,* analogous to that for density estimation (see Remark 10.2). In this method one seeks h which minimizes the sum of squares $\sum_{1 \le i \le n}[Y_i - f_h^{(-i)}(x_i)]^2 = CV(h)$, where $f_h^{(-i)}$ is the kernel estimate of f using $n - 1$ observation pairs (x_j, Y_j), omitting the i-th one. Note that $Y_i - f_h^{(-i)}(x_i)$ is the error of predicting Y by the kernel method at the design point (or covariate value) x_i, using the remaining $n - 1$ observations. Also, $n^{-1}CV(h)$ is the estimate of the expected (integrated) prediction error using bandwidth h. Thus $h = h_n$ aims at minimizing this error. In Eubank (1999, pp. 37–46), one can find a modified version called *generalized cross-validation, or GCV*, as well. That this choice is asymptotically optimal is also shown by the following result of C. Stone for the case of density estimation: If $\hat{f}_{n,h}$ denotes the estimate (10.4) for a given h, and $\hat{f}_{n,h_n}$ that with $h = h_n$, then

$$MISE(\hat{f}_{n,h_n})/\inf_h MISE(\hat{f}_{n,h}) \longrightarrow 1 \quad \text{in probability as } n \to \infty. \qquad (10.46)$$

10.3 Notes and References

A very readable text on the subject of this chapter is Eubank (1999). A more comprehensive account of nonparametric density estimation is given in Silverman (1986). A precise study of error rates may be found in Tsybakov (2009).

For *monotone increasing* regression in bioassay, Bhattacharya and Kong (2007) study the asymptotics of nonparametric quantile estimates using a continuous extension of the so-called *pool adjusted violators algorithm (PAV)*. Bhattacharya and Lin (2010, 2011, 2013) improve upon this by providing asymptotically optimal procedures for bioassay and environmental risk assessment, and also provide extensive data analysis and simulations for a comparative study of a number of different estimation methods. Lin (2012) develops a general nonparametric theory of regression in this context under order restrictions. Dette et al. (2005) and Dette and Scheder (2010) use an *inverse kernel method* to design an asymptotically optimal procedure for nonparametric quantile estimation in the order restricted context.

We have not discussed in this chapter the important method of nonparametric regression using *splines*, because of some complications involving the statistical inference involved. See Eubank (1999) for a fine treatment of optimal smoothing splines, proving in particular the following result. Let the true regression curve belong to the Sobolev class $W_2^2[0,1]$, which is the completion of $C^2[0,1]$ in $L^2[0,1]$. Then the optimal estimate in $W_2^2[0,1]$ is the cubic spline with knots at the data points (See Wahba 1990). Kong and Eubank (2006) derived the optimal monotone cubic spline estimate and applied it to the quantal bioassay problem.

Exercises for Chap. 10

Exercises for Sects. 10.1, 10.2

Ex. 10.1. Let the hypothesis of Theorem 10.1 hold.

(a) Derive (10.19) and (10.22).

(b) Show that $\{(\hat{f}_n(x) - f(x))/\sqrt{\text{var}\,\hat{f}_n(x)} : x \text{ such that } f(x) > 0\}$ converges in law to a Gaussian process, and compute the mean and covariance function of this Gaussian process.

Ex. 10.2. Assume that $h \equiv h_n = o(n^{-1/5})$ and (10.9) holds, and prove that with this bandwidth (10.22) holds with $\gamma = 0$, under the hypothesis of Theorem 10.1.

Ex. 10.3. Consider a pdf f on $\mathbb{R}^d$, having continuous and bounded second derivatives.

(a) State and prove the analog of Theorem 10.1.
(b) Derive the multidimensional versions of (10.20)–(10.22).

Ex. 10.4. Prove that, in Remark 10.2, $E\widehat{A} = E \int f(x)\hat{f}_n(x)dx$, where $\widehat{A} = n^{-1}\sum_{i=1}^{n} \hat{f}_{n,-i}(X_i)$. [Hint: $E\widehat{A} = E\hat{f}_{n,-1}(X_1) = \frac{1}{n}EK((X_1 - X_2)/h) = \frac{1}{h}\int K((x-y)/h)f(x)f(y)dxdy$. Also, $E\int f(x)\hat{f}_n(x)dx = \frac{1}{h}\int f(x)EK\left(\frac{x-X_1}{h}\right)dx = \frac{1}{h}\int f(x)K\left(\frac{x-y}{h}\right)f(x)f(y)dxdy$.]

Ex. 10.5. Extend Theorem 10.2 to the Nadaraya-Watson estimator (10.33).

References

Bhattacharya, R., & Kong, M. (2007). Consistency and asymptotic normality of the estimated effective dose in bioassay. *Journal of Statistical Planning and Inference, 137*, 643–658.

Bhattacharya, R., & Lin, L. (2010). An adaptive nonparametric method in benchmark analysis for bioassay and environmental studies. *Statistics and Probability Letters, 80*, 1947–1953.

Bhattacharya, R., & Lin, L. (2011). Nonparametric benchmark analysis in risk assessment: A comparative study by simulation and data analysis. *Sankhya Series B, 73*(1), 144–163.

Bhattacharya, R., & Lin, L. (2013). Recent progress in the nonparametric estimation of monotone curves—with applications to bioassay and environmental risk assessment. *Computational Statistics and Data Analysis, 63*, 63–80.

Dette, H., Neumeyer, N., & Pliz, K. F. (2005). A note on nonparametric estimation of the effective dose in quantal bioassay. *Journal of the American Statistical Association, 100*, 503–510.

Dette, H., & Scheder, R. (2010). A finite sample comparison of nonparametric estimates of the effective dose in quantal bioassay. *Journal of Statistical Computation and Simulation, 80*(5), 527–544.

Epanechnikov, V. A. (1969). Non-parametric estimation of a multivariate probability density. *Theory of Probability and its Applications, 14*, 153–158.

Eubank, R. (1999). *Nonparametric regression and spline smoothing.* New York: Chapman and Hall.

Gasser, T., & Müller, R. G. (1984). Estimating regression functions and their derivatives by the kernel method. *Scandinavian Journal of Statistics, 11*, 171–185.

Hall, P. (1992). On bootstrap confidence intervals in nonparametric regression. *Annals of Statistics, 20*(2), 695–711.

Kong, M., & Eubank, R. (2006). Monotone smoothing with application to dose-response curve. *Communications in Statistics Simulation and Computation, 35*(4), 991–1004.

Lehmann, E. L. (1999). *Element of large-sample theory.* Springer, New York.

Lin, L. (2012). Nonparametric inference for bioassay (Ph.D Thesis). University of Arizona.

Silverman, B. (1986). *Density estimation for statistics and data analysis.* New York: Chapman and Hall.

Stone, C. (1980). Optimal rates of convergence for nonparametric estimators. *Annals of Statistics, 8*(6), 1348–1360.

Tsybakov, A. B. (2009). *Introduction to nonparametric estimation.* New York: Springer.

Wahba, G. (1990). *Spline models for observational data.* CBMS-NSF Series (Vol. 59). Philadelphia: SIAM.

Wasserman, L. (2003). *All of statistics: A concise course in statistical inference.* New York: Springer.

Part III
Special Topics

Chapter 11
Edgeworth Expansions and the Bootstrap

Abstract This chapter outlines the proof of the validity of a properly formulated version of the formal Edgeworth expansion, and derives from it the precise asymptotic rate of the coverage error of Efron's bootstrap. A number of other applications of Edgeworth expansions are outlined.

11.1 Cramér Type Expansion for the Multivariate CLT

Let X_j $(j \geq 1)$ be a sequence of k-dimensional random vectors with (common) distribution Q, mean (vector) 0 and a nonsingular covariance matrix $\mathbf{V}$. Assume that $\rho_s := E\|X_1\|^s < \infty$ for some integer $s \geq 3$. The X_j's have the common characteristic function (c.f.) $\widehat{Q}(\xi) := E(\exp\{i\xi \cdot X_j\})$, $\xi \in \mathbb{R}^k$. Let $S_n = X_1 + \cdots + X_n$. Then the c.f. of the distribution Q_n of $n^{-\frac{1}{2}} S_n$ is

$$\widehat{Q}_n(\xi) = \widehat{Q}^n\left(\xi/\sqrt{n}\right), \tag{11.1}$$

The *cumulant generating function (c.g.f.)* of Q in a neighborhood of $\xi = 0$ may be expressed by a Taylor expansion as

$$\log \widehat{Q}(\xi) = -\xi \cdot V\xi/2 + \sum_{r=3}^{s} i^r \lambda_r(\xi)/r! + o(|\xi|^s), \tag{11.2}$$

where $\lambda_r(\xi)$ is the r-th cumulant of the random variable $\xi \cdot X_1$. Note that

$$\lambda_r(\xi) = \sum_{|\beta|=r} (\lambda_\beta/\beta!)\xi^\beta. \tag{11.3}$$

Here $\beta = (\beta_1, \ldots, \beta_k) \subset Z_+^k$, $|\beta| = \beta_1 + \cdots + \beta_k$. Also, λ_β is the β-th cumulant of X_1 (i.e., $\lambda_\beta = i^{-|\beta|}(\mathbf{D}^\beta \log \widehat{Q})(0))$, $\beta! = \beta_1!\beta_2!\ldots\beta_k!$ and $(i\xi)^\beta = (i\xi_1)^{\beta_1}\ldots(i\xi_k)^{\beta_k}$. Note that for, $|\beta| = 3$, $\lambda_\beta = \mu_\beta = EX_1^\beta = EX_1^{\beta_1}\ldots X_k^{\beta_k}$, where $\mathbf{D} = (D_1, D_2, \ldots, D_k)$, $D_j = \partial/\partial x_j$, $(-\mathbf{D})^\beta = (-1)^{|\beta|} D_1^{\beta_1}\ldots D_k^{\beta_k} \ \forall \beta \in \mathbf{Z}_+^k$, and $\varphi_{\mathbf{V}}$ is the density of the k-dimensional Normal distribution with mean 0 and covariance matrix $\mathbf{V}$.

© Springer-Verlag New York 2016

R. Bhattacharya et al., *A Course in Mathematical Statistics and Large Sample Theory*, Springer Texts in Statistics, DOI 10.1007/978-1-4939-4032-5_11

The c.g.f. of Q_n may then be expressed as

$$\log \widehat{Q}_n(\xi) = n \log \widehat{Q}(\xi/\sqrt{n})$$

$$= -\xi \cdot V\xi/2 + \sum_{r=3}^{s} n^{-(r-2)/2} i^r \lambda_r(\xi)/r! + o\left(n^{-(s-2)/2}\right),$$

so that

$$\widehat{Q}_n(\xi) = \exp\{-\xi \cdot V_\xi\} \exp\left\{ \sum_{r=3}^{s} n^{-(\gamma-2)/2} i^r \frac{\lambda_r(\xi)}{r!} \right\} + o\left(n^{-(s-2)/2}\right). \qquad (11.4)$$

Expanding the second factor on the right in powers of $n^{-\frac{1}{2}}$, one has

$$\widehat{Q}_n(\xi) = \exp\{-\xi \cdot V_\xi/2\} \left[1 + \sum_{r=1}^{s-2} n^{-r/2} \widetilde{\mathbf{P}}_r(i\xi) \right] + o\left(n^{-(s-2)/2}\right), \qquad (11.5)$$

where $P_r(i\xi)$ is a polynomial in $i\xi$. For example,

$$\widetilde{\mathbf{P}}_1(i\xi) = \frac{i^3 \lambda_3(\xi)}{3!} = \frac{i^3}{3!} E(\xi \cdot X_1)^3 = \sum_{|\beta|=3} \frac{\lambda_\beta}{\beta!} (i\xi)^\beta, \qquad (11.6)$$

The 'formal' density of Q_n is then, by Fourier inversion,

$$\psi_{s-2,n}(x) = \left[1 + \sum_{r=1}^{s-2} n^{-r/2} \widetilde{\mathbf{P}}_r(-\mathbf{D}) \right] \varphi_{\mathbf{V}}(x). \qquad (11.7)$$

In particular,

$$\widetilde{\mathbf{P}}_1(-D)\varphi_{\mathbf{V}}(x) = - \sum_{|\beta|=3} \frac{\mu_\beta}{\beta!} \mathbf{D}^\beta \varphi_{\mathbf{V}}(x). \qquad (11.8)$$

In general, one has the following theorem. Here $\mathscr{B}(\mathbb{R}^k)$ is the Borel sigma-field of $\mathbb{R}^k$ (See Theorem 19.4 and Corollary 19.6 in Bhattacharya and Rao 1976).

Theorem 11.1. *Suppose the p-fold convolution Q^{*p} (i.e., the distribution of $X_1 + \cdots + X_p$) has a nonzero absolutely continuous component for some positive integer p. Then, if $\rho_s < \infty$ for some integer $s \geq 3$, one has*

$$\sup_{B \in \mathscr{B}(\mathbb{R}^k)} |Q_n(B) - \Psi_{s-2,n}(B)| = o(n^{-(s-2)/2}), \qquad (11.9)$$

where $\Psi_{s-2,n}$ is the finite signed measure whose density is $\psi_{s-2,n}$ given by (11.7).

11.2 The Formal Edgeworth Expansion and Its Validity

A crucial factor underlying the expansion of the distribution Q_n of the normalized mean of i.i.d. random variables (or, vectors) is that the r-th cumulant (or v-th cumulant with $|v| = r$) of Q_n is $o(n^{-(r-2)/2})$. Nearly a hundred years ago, Edgeworth (1905) proposed an analogous expansion of distributions of more general statistics. In this section we sketch the main ideas needed to validate a properly formulated version of Edgeworth's formal expansion. The details may be found in Bhattacharya (1977), Bhattacharya and Ghosh (1978), or Bhattacharya and Denker (1990).

Let X_j, $1 \leq j \leq n$, be i.i.d. observations with values in $\boldsymbol{R}^m$. Many important statistics may be expressed as, or approximated adequately by, a smooth function of a finite number, say, k of averaged sample characteristics $(1/n)\sum_{j=1}^{n} f_i(X_j)$, $1 \leq i \leq k$. For simplicity we will consider only real-valued statistics, in which case f_i's are real-valued Borel measurable functions on $\mathbb{R}^m$.

As simple examples, one may consider (1) the t-statistic, which is a function of $\overline{X}$ and $(1/n)\sum_{1 \leq j \leq n} X_j^2$ (Here $m = 1, k = 2, f_1(x) = x,\ f_2(x) = x^2$), and (2) the sample correlation coefficient, which is a function of $\overline{X}^{(1)}$, $\overline{X}^{(2)}$, $(1/n)\sum_{1 \leq j \leq n} X_j^{(1)} X_j^{(2)}, (1/n)\sum_{1 \leq j \leq n}(X_j^{(1)})^2, (1/n)\sum_{1 \leq j \leq n}(X_j^{(2)})^2$. (Here $X_j = (X_j^{(1)}, X_j^{(2)})$, $m = 2$, $k = 5$, $f_1(x) = x^{(1)}$, $f_2(x) = x^{(2)}$, $f_3(x) = x^{(1)} x^{(2)}$, $f_4(x) = (x^{(1)})^2$, $f_5(x) = (x^{(2)})^2$.) Other important statistics to which the present theory applies include M-estimators (e.g., maximum likelihood estimators) under appropriate smoothness conditions. (See Bhattacharya and Ghosh 1978).

In general, define

$$\overline{Z} = (1/n) \sum_{1 \leq j \leq n} Z_j,\ Z_j := (f_1(X_j),\ f_2(X_2),\ldots, f_k(X_j),$$

$$\mu = EZ_n = E\overline{Z} = (Ef_1(X_j),\ Ef_2(X_j),\ldots, Ef_k(X_j)). \tag{11.10}$$

We will consider statistics of the form $H(\overline{Z})$, where H is a real-valued, $(s-1)$-times continuously differentiable function on a neighborhood of $\mu \in \mathbb{R}^k$ containing a ball $\{z \in \mathbb{R}^k : ||z - \mu|| \leq \delta\}$ for some $\delta > 0$. As in the preceding section, $s \geq 3$. By a Taylor expansion around μ,

$$H(\overline{Z}) - H(\mu) = (\operatorname{grad} H)(\mu) \cdot (\overline{Z} - \mu)$$
$$+ \sum_{|\beta|=2}^{s-1} (\mathbf{D}^\beta H)(\mu)(\overline{Z} - \mu)^\beta / \beta! + R(\overline{Z}, \mu)$$
$$= H_{s-1}(\overline{Z} - \mu) + R(\overline{Z}, \mu),\ \text{say}, \tag{11.11}$$

where the remainder satisfies

$$|R(\overline{Z}, \mu)| \leq c_{13}||\overline{Z} - \mu||^{s-1}\ \text{ on } \{||\overline{Z} - \mu|| \leq \delta\},$$
$$|R(\overline{Z}, \mu)| = o(||\overline{Z} - \mu||^{s-1})\ \text{ as } ||\overline{Z} - \mu|| \to 0. \tag{11.12}$$

Consider the normalized statistics

$$W_n := \sqrt{n}\,(H(\overline{Z}) - H(\mu)) = W_n' + \sqrt{n}\,R(\overline{Z}, \mu),$$
$$W_n' := \sqrt{n}\,H_{s-1}(\overline{Z} - \mu)$$
$$= \operatorname{grad} H(\mu) \cdot \sqrt{n}\,(\overline{Z} - \mu) + \sum_{|\beta|=2}^{s-1} n^{-(|\beta|-2)/2}(\mathbf{D}^\beta H)(\mu)\frac{(\sqrt{n}\,(\overline{Z} - \mu))^\beta}{\beta!}.$$

$$\tag{11.13}$$

An important fact concerning polynomials in $\overline{Z} - \mu$ is that their cumulants decay at the same asymptotic rate as those of $\overline{X} - \mu$ (James 1958 and Leonov and Shiryaev 1959). In particular, the r-th cumulant of $H_{s-1}(\overline{Z} - \mu)$ is $O(n^{-r})$, and that of W_n', is therefore $O(n^{-r/2})$, if sufficiently many moments of $\|\overline{Z} - \mu\|$ are finite. Omitting terms of smaller order than $n^{-(s-2)/2}$, the "approximate" cumulants $K_{r,n}$, say, of W_n' may then be expressed as (See Bhattacharya and Ghosh 1978, or Bhattacharya and Denker 1990)

$$K_{1,n} = \sum_{r'=0}^{[(s-3)/2]} n^{-(2r'+1)/2} b_{1,r'}$$

$$K_{2,n} = b_{2,0} + \sum_{r'=1}^{[(s-2)/2]} n^{-r'} b_{2,r'}$$

$$K_{r,n} = \begin{cases} \sum_{r'=(r-3)/2}^{[(s-3)/2]} n^{(-2r'+1)/2} b_{r,r'} & \text{if } r \text{ is odd, } 3 \le r \le s, \\ \sum_{r'=(r-2)/2}^{[(s-2)/2]} n^{-r'} b_{r,r'} & \text{if } r \text{ is even, } 4 \le r \le s. \end{cases} \tag{11.14}$$

Here $b_{r,r'}$ are constants involving derivatives of H at μ and cumulants (or moments) of $Z_1 - \mu$. Under the assumptions $(\operatorname{grad} H)(\mu) \neq 0$ and *covariance matrix* $\mathbf{V}$ of Z_1 is nonsingular, one has

$$\sigma^2 := b_{2,0} = (\operatorname{grad} H)(\mu) \cdot \mathbf{V}(\operatorname{grad} H)(\mu) > 0. \tag{11.15}$$

The 'formal' approximation to the characteristic function of W_n (or of W_n') is then obtained in a manner quite analogous to that for $\sqrt{n}(\overline{X} - \mu)$ (See, e.g., Bhattacharya and Ranga Rao 1976), keeping only terms up to order $n^{-(s-2)/2}$. We write this expansion as

$$\exp\{-\sigma^2 \xi^2/2\} \exp\left\{ K_{1,n}(i\xi) - (K_{2,n} - \sigma^2)\xi^2/2 + \sum_{3 \le r \le s} \frac{K_{r,n}}{r!}(i\xi)^r \right\}$$

$$= \exp\{-\sigma^2 \xi^2/2\} \left[1 + \sum_{r=1}^{s-2} n^{-r/2} \pi_r(i\xi) \right] + o(n^{-(s-2)/2}). \tag{11.16}$$

The Fourier inversion of the last expansion is the 'formal' *Edgeworth expansion of the density* of W_n (or W_n'), which we write as

$$\tilde{\psi}_{s-2,n}(x) = \left[1 + \sum_{1 \le r \le s-2} n^{-r/2} \pi_r(-\mathbf{D}) \right] \varphi_{\sigma^2}(x). \tag{11.17}$$

We now make the following assumption.

(A) The distribution of the underlying observations X_j *has a nonzero abso-lutely continuous component whose density is positive in a ball B in $\mathbb{R}^m$ such that the functions $f_1, f_2, \ldots, f_k$ are continuously differentiable on B and $1, f_1, f_2, \ldots, f_k$ are linearly independent as elements of the vector space of real-valued continuous functions on B.*

Remark 11.1. Under (A), Q^{*k} (the distribution of $Z_1 + Z_2 + \cdots + Z_k$) has a nonzero absolutely continuous component (w.r.t. Lebesgue measure on $\mathbb{R}^k$), so that Theorem 11.1 applies (See Bhattacharya and Ghosh 1978; Bhattacharya and Denker 1990).

We can now state the main result of this section.

Theorem 11.2. *Suppose (1) $E|f_i(X_1)|^s < \infty$, $1 \le i \le k$, for some integer $s \ge 3$, (2) H is $(s-1)$-times continuously differentiable in a neighborhood of $\mu = EZ_1$, (3) $\mathrm{grad}\, H(\mu) \ne 0$ and $V = \mathrm{Cov}\, Z_1$ is nonsingular, (4) (A) holds. Then*

$$\sup_{B \in \mathscr{B}(\mathbb{R})} \left| P(W_n \in B) - \int_B \tilde{\psi}_{s-2,n}(x)\, dx \right| = o(n^{-(s-2)/2}). \tag{11.18}$$

Proof (Sketch). We sketch the proof of Theorem 11.2 in a number of steps.

Step 1. First, $P(\|\overline{Z} - \mu\| > \delta) \equiv P(\|\sqrt{n}\,(\overline{Z} - \mu)\| > \delta\sqrt{n}))$, is easily shown to be $o(n^{-(s-2)/2})$. One may then restrict attention to the set $B_n := \{z \in \mathbb{R}^k : \|z\| \le \delta\sqrt{n}\}$ to integrate $\psi_{s-2,n}(z)$ in (11.7) over the two sets $\{g_n(z) \in B\} \cap B_n$ and $\{h_n(z) \in B\} \cap B_n$, where $g_n(z) = \sqrt{n}\,(H(\mu + n^{-1/2}z) - H(\mu))$, $h_n(z) = \sqrt{n}\, H_{s-1}(n^{-1/2}z)$ so that $\{W_n \in B\} = \{g_n(\sqrt{n}(\overline{Z} - \mu)) \in B\}$ and $\{W_n' \in B\} = \{h_n(\sqrt{n}(\overline{Z} - \mu)) \in B\}$. From multivariate calculus, using (11.11)–(11.13), it then follows that

$$\sup_{B \in \mathscr{B}(\mathbb{R})} |P(W_n \in B) - P(W_n' \in B)| = o(n^{-(s-2)/2}) \tag{11.19}$$

Step 2. In view of (11.19), it is enough to consider W_n' instead of W_n in (11.13). To apply Theorem 11.1, consider the class $\mathscr{A}$ (which depends on n) $\subset \mathscr{B}(\mathbb{R}^k)$ comprising sets of the form $A := \{z \in \mathbb{R}^k : h_n(z) \in B\}$, $B \in \mathscr{B}(\mathbf{R})$. By a change of variables $z \to y = (y_1, \ldots, y_k)$ where $y_1 = h_n(z)$ and integrating out $y_2, \ldots, y_k$ one obtains (See Remark 11.1).

$$\sup_{B \in \mathscr{B}(\mathbb{R})} \left| P(W_n' \in B) - \int_{\{z \in \mathbb{R}^k : h_n(z) \in B\}} \psi_{s-2,n}(z)\, dz \right|$$

$$\le \sup_{A \in \mathscr{B}(\mathbb{R}^k)} \left| Q_n(A) - \int_A \psi_{s-2,n}(z)\, dz \right| = o(n^{-(s-2)/2}). \tag{11.20}$$

Step 3. In view of (11.20), in order to prove (11.18) it remains to show that (uniformly $\forall\, B \in \mathscr{B}(\mathbb{R})$)

$$\int_A \psi_{s-2,n}(z)\, dz = \int_B \tilde{\psi}_{s-2,n}(x)\, dx + o(n^{-(s-2)/2}), \tag{11.21}$$

where $A = \{z \in \mathbb{R}^k : h_n(z) \in B\}$. By a change of variables $z \to T(z) := y$ as before, with $y_1 \equiv T_1(z) = h_n(z)$, the density of y_1, induced from $\psi_{s-2,n}(z)\, dz$, is given in powers of $n^{-1/2}$ by $\overline{\psi}_{s-2,n}(x) + o(n^{-(s-2)/2})$,

$$\overline{\psi}_{s-2,n}(x) := \left[1 + \sum_{r=1}^{s-2} n^{-r/2} q_r(x) \right] \varphi_{\sigma^2}(x), \tag{11.22}$$

where $q_r(x)$, $1 \leq r \leq s - 2$, are certain polynomials in x. Note that since the dominant term in $\psi_{s-2,n}(z)$ is $\varphi_{\mathbf{V}}(z)$ and the dominant term in $T_1(z) \equiv h_n(z)$ is $z \cdot (\text{grad } H)(\mu)$, the density φ_{σ^2} of $Z \cdot (\text{grad } H)(\mu)$ (under $\varphi_{\mathbf{V}}(z)\,dz$) is the dominant term in (11.22). More generally, since $\psi_{s-2,n}$ and $\overline{\psi}_{s-2,n}$ both have Gaussian decays at the tails, one has (by the same transformation argument),

$$\int_{\mathbb{R}} x^r \overline{\psi}_{s-2,n}(x)\,dx = \int_{\mathbb{R}^k} h_n^r(z)\psi_{s-2,n}(z)\,dz + o(n^{-(s-2)/2}) \ \forall \ r = 0,1,2,\ldots.$$

$$(11.23)$$

Assume for the moment that Z_j has finite moments of all orders. Then, by construction, the β-th cumulant of $\psi_{t-2,n}(z)\,dz$ (namely, $i^{-|\beta|}(\mathbf{D}^\beta \log \tilde{\psi}_{t-2,n})(0)$) matches the β-th cumulant of Q_n (i.e., of $\sqrt{n}(\overline{Z} - \mu)$) up to order $n^{-(t-2)/2}$, $|\beta| \leq t$. This implies

$$\int_{\mathbb{R}^k} z^\beta \psi_{t-2,n}(z)\,dz = \int_{\mathbb{R}^k} z^\beta Q_n(dz) + o(n^{-(t-2)/2}), \ \forall \ |\beta| \leq t. \qquad (11.24)$$

Letting $t = s$, and $t > s$, in turn, (11.24) leads to

$$\int z^\beta \psi_{s-2,n}(z)\,dz = \int z^\beta Q_n(dz) + o(n^{-(s-2)/2}) \ \ \forall \ \beta, \qquad (11.25)$$

since $\psi_{t-2,n}$ equals $\psi_{s-2,n}$ up to a polynomial multiple of $\varphi_{\mathbf{V}}$ of order $0(n^{-(s-1)/2})$, if $t > s$. Using (11.25) in (11.23), and noting that $h_n^r(z)$ is a polynomial in z, one then has

$$\int_{\mathbb{R}} x^r \overline{\psi}_{s-2,n}(x)\,dx = \int_{\mathbb{R}^k} h_n^r(z)Q_n(dz) + o(n^{-(s-2)/2})$$

$$\equiv EW_n'^r + o(n^{-(s-2)/2}) \ \ \forall \ r = 0,1,2,\ldots. \qquad (11.26)$$

By construction of $\tilde{\psi}_{s-2,n}$ on the other hand,

$$\int_{\mathbb{R}} x^r \tilde{\psi}_{s-2,n}(x)\,dx = EW_n''^r + o(n^{-(s-2)/2}) \ \ \forall \ r = 0,1,2,\ldots, \qquad (11.27)$$

using the same argument as above, namely, the r-th cumulant of $\tilde{\psi}_{t-2,n}(z)\,dz$ matches that of W_n' up to order $n^{-(t-2)/2}$ for all r, $0 \leq r \leq t$. Comparing (11.26) and (11.27), and noting that the left sides have no terms which are $o(n^{-(s-2)/2})$, we have

$$\int_{\mathbb{R}} x^r \tilde{\psi}_{s-2,n}(x)\,dx = \int_{\mathbb{R}} x^r \overline{\psi}_{s-2,n}(x)\,dx \ \ \forall \ r = 0,1,2,\ldots, \qquad (11.28)$$

which yields $\tilde{\psi}_{s-2,n} \equiv \overline{\psi}_{s-2,n}$ and implies, in particular, (11.21). $\qquad \square$

In order to apply Theorem 11.2, one needs first to compute the "approximate" cumulants of $W_n' = \sqrt{n}\, H_{s-1}(\overline{Z} - \mu)$. In the Appendix we carry this out for approximations with errors of the order $O(n^{-3/2})$. Here we consider two simple examples.

Example 11.1 (The Sample Mean). Here $m = 1, k = 1, f_1(x) = x, H(z) = z, W_n = \sqrt{n}\,(\overline{X} - \mu) = W_n'$. In this case the r-th cumulant of $W_n = W_n'$ is

$K_{r,n} = n^{-(r-2)/2}K_r$ $(r \geq 2)$, where K_r is the r-th cumulant of X_j; $K_{1,n} = 0$, $K_{2,n} = \sigma^2 = EX_j^2$, $K_{3,n} = n^{-1/2}\mu_3$, $K_{4,n} = n^{-1}(\mu_4 - 3\sigma^2)$, etc., with $\mu_r := E(X_j - \mu)^r$. The Edgeworth expansion in this case is the same as the Cramér expansion (11.7) (with $k = 1$) and Theorem 11.2 is valid.

Example 11.2 (Student's t). Here $m = 1$, $k = 2$, $f_1(x) = x$, $f_2(x) = x^2$, $EX_j = 0$, $EX_j^2 = \sigma^2$, and $H(z^{(1)}, z^{(2)}) = z^{(1)}/\sqrt{z^{(2)} - (z^{(1)})^2}$. Then (See, e.g., Hall 1983, or Qumsiyeh 1989), $K_{1,n} = (-1/2\mu_3)n^{-1/2}+O(n^{-3/2})$, $K_{2,n} = 1+((7/4)\mu_3^2+3)n^{-1}+O(n^{-2})$, $K_{3,n} = (-2\mu_3)n^{-1/2}+O(n^{-3/2})$, $K_{4,n} = -2(\mu_4 - 6\mu_3^2 - 6)n^{-1}+O(n^{-2})$, with $\mu_r := E(X_j/\sigma)^r$. The expansion (11.17) up to order n^{-1} (and error $O(n^{-3/2})$) is

$$\tilde{\psi}_{2,n}(x) = \varphi(x) + n^{-1/2}\left(\frac{\mu_3}{6}\right)\frac{d}{dx}[(2x^2 + 1)\varphi(x)]$$

$$+ n^{-1}\frac{d}{dx}\left\{\left[(-3 - 2\mu_3^2)x + (6 - \mu_4 + 8\mu_3^2)(3x - x^3)\right.\right.$$

$$\left.\left. + \frac{\mu_3^2}{18}(-15x + 10x^3 - x^5)\right]\varphi(x)\right\}. \tag{11.29}$$

The expansion of the *distribution function of* $t = \sqrt{n}\,\overline{X}/s$, with $s^2 = \frac{1}{n}\sum X_j^2 - \overline{X}^2$, is obtained by removing the d/dx symbol from (11.29) and replacing the standard normal density $\varphi(x) \equiv \varphi_1(x)$ by its distribution function $\Phi(x)$. Note that the usual t-statistic is $\sqrt{\frac{n-1}{n}}\,t$, for which the expansion (with error $O(n^{-3/2})$) is obtained by a simple change of variables.

The final result of this section applies to statistics such as $(H(\overline{Z}) - H(\mu))^2 = G(\overline{Z})$, say, for which $(\text{grad}\,H)(\mu) \neq 0$, but $(\text{grad}\,G)(\mu) = 0$. More generally, consider the statistic

$$\mathbf{U}_n := 2nG(\overline{Z}) \equiv 2n(H(\overline{Z}) - H(\mu))^2$$

$$= \sum_{1 \leq r,r' \leq k} (D_r D_{r'}G)(\mu)(\sqrt{n}\,(\overline{Z} - \mu))^{(r)}(\sqrt{n}\,(\overline{Z} - \mu))^{(r')}$$

$$+ \sum_{|\beta|=3}^{s} n^{-(|\beta|-2)/2}(\mathbf{D}^{\beta}G)(\mu)\frac{(\sqrt{n}\,(\overline{Z} - \mu))^{\beta}}{\beta!} + 2nR'(\overline{Z}, \mu)$$

$$= \mathbf{U}_n' + 2nR'(\overline{Z}, \mu), \text{ say,} \tag{11.30}$$

where $|R'(z, \mu)| = o(||z - \mu||^s)$ as $||z - \mu||\to 0$, so that $2nR'(\overline{Z}, \mu) = o_p(n^{-(s-2)/2})$, and $Z^{(r)}$ is the r-th coordinate of Z. Here we assume $G(\mu) = 0$ and $(\text{grad}\,G)(\mu) = 0$, G is s-times continuously differentiable and $E||Z_j||^s < \infty$ for some integer $s \geq 3$. It is important to note that the moments and, therefore, cumulants of $\mathbf{U}_n'$ involve *only* powers of n^{-1}. If the Hessian matrix of H at μ, namely, $(((D_r D_{r'}H)(\mu)))_{1\leq r,r'\leq k}$ is nonzero, then the *limiting distribution of* $\mathbf{U}_n$ (or $\mathbf{U}_n'$) is that of

$$\mathbf{U} := \sum_{1 \leq r,r' \leq k} (D_r D_{r'}G)(\mu)Y^{(r)}Y^{(r)'}$$

where Y is k-dimensional normal with mean zero and covariance $\mathbf{V} \equiv \text{Cov}Z_j$. For likelihood ratio tests, Wald's tests and Rao's score tests, etc., the asymptotic distribution of $\mathbf{U}_n$ is the chi-square distribution with d degrees of freedom (d.f.),

$1 \leq d \leq k$. In such cases, one may proceed as in the case of W_n, W_n' to derive a formal Edgeworth expansion

$$\eta_{r,n}(u) := f_{\chi_d^2}(u)[1 + \sum_{r'=1}^{r} n^{-r'} g_{r'}(u)],$$

$$r := [(s-2)/2] \text{ (integer part of } (s-2)/2). \tag{11.31}$$

where $f_{\chi_d^2}$ is the density of the chi-square distribution with d degrees of freedom, and $g_{r'}, 1 \leq r' \leq r$, are polynomials. The following result is due to Chandra and Ghosh (1979). Let $\mathscr{B}(\mathbb{R}^+)$ denote the Borel sigma-field on $\mathbb{R}^+ := \{u \in \mathbb{R} : u \geq 0\}$.

Theorem 11.3. *Let $E\|Z_j\|^s < \infty$ for some integer $s \geq 3$, and let $V := CovZ_j$ be nonsingular. Suppose G is s-times continuously differentiable in a neighborhood of $\mu = EZ_j$, $G(\mu) = 0$, $\operatorname{grad} G(\mu) = 0$, and the limiting distribution of $\mathbf{U}_n$ is $f_{\chi_d^2}(u)\, du$. (a) If, in addition, the hypothesis (A) stated before Remark 11.1 holds, then*

$$\sup_{B \in \mathscr{B}(\mathbb{R}^+)} \left| P(U_n \in B) - \int_B \eta_{[(s-2)/2],n}(u)\, du \right| = o(n^{-(s-2)/2}). \tag{11.32}$$

The proof of Theorem 11.3 follows the line of proof of Theorem 11.2, noting that the "approximate" cumulants of $\mathbf{U}_n$ and (U_n') only involve powers of n^{-1} (See Chandra and Ghosh 1979 and Bhattacharya and Denker 1990).

Remark 11.2. For many purposes, if not most, the asymptotic expansions in Theorems 11.2 and 11.3, are needed only for distribution functions of W_n and U_n, which hold under *Cramér's condition:*

$$\limsup_{|\xi| \to \infty} \left| E e^{i\xi \cdot Z_1} \right| < 1. \tag{11.33}$$

More generally, we have the following version of Theorem 11.2 (See Bhattacharya and Rao 1976; Bhattacharya 1987).

Theorem 11.4. *Suppose (1) $E|f_i(X_1)|^s < \infty$, $1 \leq i \leq k$, for some integer $s \geq 3$, (2) H is $(s-1)$-times continuously differentiable in a neighborhood of $\mu = EZ_1$, (3) $\operatorname{grad} H(\mu) \neq 0$ and $V = \operatorname{Cov} Z_1$ is nonsingular, (4) the distribution Q of Z_1 satisfies Cramér's condition (11.33). Then*

$$\sup_{B \in \mathscr{C}} |P(W_n \in B) - \int_B \tilde{\psi}_{s-2,n}(x)\, dx| = o(n^{-(s-2)/2}). \tag{11.34}$$

for every class $\mathscr{C} \subset \mathscr{B}(\mathbb{R})$ such that, for some $a > 0$,

$$\sup_{B \in \mathscr{C}} \int_{(\partial B)^\varepsilon} \varphi_{\sigma^2}(x)dx = 0(\varepsilon^\alpha) \quad \text{as } \varepsilon \downarrow 0. \tag{11.35}$$

Remark 11.3. Consider the problem of constructing a symmetric confidence interval for a real valued parameter θ, of the form $|\hat{\theta}_n - \theta| \leq n^{-1/2}c$ (for some $c > 0$), where $\hat{\theta}_n$ is of the form $H(\overline{Z})$, or adequately approximated by such, one may use Theorem 11.4 to show that

$$P(|\hat{\theta}_n - \theta| \leq n^{-1/2}c) \equiv P((\sqrt{n}\,(\hat{\theta}_n - \theta))^2 \leq c^2) = P(Y^2 \leq c^2) + O(n^{-1})$$

$$\equiv P(|Y| \leq c) + O(n^{-1}), \tag{11.36}$$

where Y is a standard normal random variable. Thus with $c = z_{\gamma/2}$ (the upper $\gamma/2$-point of the standard normal distribution), the coverage error of a nominal $(1 - \gamma)$-confidence interval is $O(n^{-1})$, provided $s \geq 4$, and the other hypotheses of Theorem 11.3 are satisfied.

The theory presented in this section applies to the estimation and testing in parametric families under appropriate regularity conditions (Bhattacharya and Ghosh 1978; Bhattacharya and Denker 1990; Chandra and Ghosh 1979), in providing more accurate coverage probabilities of confidence regions and levels of significance of tests. If one desires to adjust the confidence bounds or the critical points of tests so as to have specified levels, (up to an error $O(n^{-1})$, $O(n^{-3/2})$, etc.) one may resort to the inversion of the expansions leading to the so-called *Cornish–Fisher expansions* (See Hall 1983, or Bhattacharya and Denker 1990). See Sect. 8.5 for nonparametric two-sample tests based on sample mean vectors and an asymptotic chi-square statistic.

Expansions for regression statistics based on independent observations are given in Qumsiyeh (1989) and Ivanov and Zwanzig (1983).

In conclusion, we remark that Chibishov (1973) also obtained asymptotic expansions such as $\overline{\psi}$ (See (11.22)), but not the validity of the formal Edgeworth expansion.

11.3 Bootstrap and Edgeworth Expansion

The nonparametric bootstrap due to Efron (1979) is now an essential tool in statistical applications. Its two most important impacts have been (1) in the setting of confidence bounds and critical levels for estimation and testing in nonparametric models, even in the absence of an estimate of the standard deviation of the statistic, and (2) in the estimation of the above standard deviation. In both cases the computer based procedures are simple and automatic. It was first pointed out by Singh (1981) that, in addition to the simplicity of the procedure, bootstrap provides an asymptotically better estimate of the distribution of the standardized mean than its classical normal approximation, provided the observations are from a nonlattice distribution. Later authors have shown that bootstrap performs better in estimating the distribution of more general studentized, or *pivotal*, statistics, if the observations come from a distribution with some smoothness satisfying Cramér's condition, or having a density (Beran 1987; Bhattacharya 1987; Hall 1988). In this section we present a precise analysis of the efficacy of the bootstrap estimation of distributions of pivotal statistics.

Consider statistics $H(\overline{Z})$ as in the preceding section, with $\overline{Z}$ as given in (11.10). Let $F_n = (1/n)\sum_{1 \leq j \leq n} \delta_{Z_j}$ denote the *empirical distribution* of the random vector Z, given $\{Z_j : 1 \leq j \leq n\}$, while $F = Q$ denotes its true distribution. The *bootstrap distribution* of the statistic $W_n = \sqrt{n}\,(H(\overline{Z}) - H(\mu))$ is the distribution (conditionally given $Z_j, 1 \leq j \leq n$) of $W_n^* := \sqrt{n}\,(H(\overline{Z}^*) - H(\overline{Z}))$ where $\overline{Z}^* = (1/n)\sum_{1 \leq j \leq n} Z_j^*$, $\{Z_j^*; 1 \leq j \leq n\}$ being a random sample from the empirical F_n. We will write P and P^* for probabilities under F and F_n, respectively. We will consider *studentized* or *pivotal statistics*, i.e., assume

$$\sigma^2 := (\operatorname{grad} H)(\mu) \cdot V(\operatorname{grad} H)(\mu) = 1. \tag{11.37}$$

The densities of the (formal) *empirical Edgeworth expansions* of the distributions of $\sqrt{n}\,(\overline{Z} - \mu)$ and $W_n = \sqrt{n}\,(H(\overline{Z}) - H(\mu))$ are denoted by $\psi_{s-2,n}$ and $\tilde{\psi}_{s-2,n}$, respectively (See (11.7), (11.17)), while we define

$$
\psi^*_{s-2,n}(z) = \left[1 + \sum_{r=1}^{s-2} n^{-r/2}\mathbf{P}^*_r(-\mathbf{D})\right] \varphi_{\mathbf{V}^*}(z)
$$

$$
\tilde{\psi}_{s-2,n}(x) = \left[1 + \sum_{r=1}^{s-2} n^{-r/2}\pi^*_r(-\mathbf{D})\right] \varphi(x), \tag{11.38}
$$

where the superscript $*$ indicates that all population moments are replaced by corresponding sample moments.

Theorem 11.5. *Under the hypothesis of Theorem 11.2, almost surely,*

$$
\sup_{B\in\mathscr{C}} \left| P^*(W^*_n \in B) - \int_B \tilde{\psi}^*_{s-2,n}(x)\,dx \right| = o(n^{-(s-2)/2}), \tag{11.39}
$$

where $\mathscr{C}$ satisfies (11.19) (with $\sigma^2 = 1$).

Proof (Sketch). The main difference between the proof of this theorem and that of Theorem 11.2 lies in deriving an expansion of the distribution Q^*_n of $\sqrt{n}\,(\overline{Z}^* - \overline{Z})$ under F_n. The rest of the proof showing the validity of the formal Edgeworth expansion for the distribution of W^*_n is entirely analogous to the corresponding proof for Theorem 11.2. Now because F_n is discrete, $\lim_{\|\xi\|\to\infty} |\hat{F}_n(\xi)| = 1$, and Cramer's condition does not hold for F_n. But as we show now (See (11.43)), $|\hat{F}_n(\xi)|$ is bounded away from one on $[c, e^{n\delta}]$ for every $c > 0$ and $\delta > 0$ depending on c. For this, first one may use Bernstein' s inequality (See Serfling 1980, p. 95) to get

$$
P(|\hat{F}_n(\xi) - \hat{F}(\xi)| \geq a) \leq 4\exp\{-na^2/10\}. \tag{11.40}
$$

By approximating points in the set $A_n = \{\|\xi\| \leq e^{n\delta}\}$ by the set B_n of lattice points in A_n of the form $\xi = \gamma m$ for some $m \in \mathbf{Z}^k$ and some $\gamma > 0$, one has for every $\delta > 0$,

$$
P\left(\sup_{\xi\in B_n} |\hat{F}_n(\xi) - \hat{F}(\xi)| \geq a\right) \leq \left\{\frac{e^{n\delta}}{\gamma} + 2\right\}^k 4\exp\{-na^2/10\}
$$

$$
\leq c_{18}(k,\gamma)\exp\left\{-n\left(\frac{a^2}{10} - k\delta\right)\right\},
$$

$$
\sup_{\|\xi\|\leq\exp\{n\delta\}} \left|\hat{F}_n(\xi) - \hat{F}(\xi)\right| \leq \sup_{\xi\in B_n} |\hat{F}_n(\xi) - \hat{F}(\xi)|
$$

$$
+ \sqrt{k}\,\gamma\left(E\|Z_1\| + \frac{1}{n}\sum_{1}^{n} \|Z_j\|\right). \tag{11.41}
$$

For $0 < \delta < a^2/10k$, using the Borel-Cantelli Lemma it follows from the first inequality that with probability one, $\sup\{|\hat{F}_n(\xi) - \hat{F}(\xi)| := \xi \in B_n\} < a$ for all sufficiently large n. From the second inequality in (11.41) it now follows that

$$
\limsup_{n\to 0}\ \sup_{\|\xi\|\leq e^{n\delta}} |\hat{F}_n(\xi) - \hat{F}(\xi)| \leq a + 2\sqrt{k}\,\gamma E\|Z_1\| \quad \text{a.s.} \tag{11.42}
$$

Let now $a := (1/4)(1 - \sup\{|\hat{F}(\xi)| : ||\xi|| \geq c\})$, $\gamma = a/(2\sqrt{k}\,E||Z_1||)$, to get

$$\limsup_{n\to\infty} \sup_{c\leq||\xi||\leq\exp\{n\delta\}} |\hat{F}_n(\xi)| \leq a + 2\sqrt{k}\gamma E||Z_1|| + \sup_{||\xi||\geq c} |\hat{F}(\xi)|$$

$$= \frac{1}{2}(1 - \sup\{|\hat{F}(\xi)| : ||\xi|| \geq c\}) + \sup_{||\xi||\geq c} |\hat{F}(\xi)| = \delta', \text{ say, a.s.} \qquad (11.43)$$

Since $\delta' < 1$, this estimate of Babu and Signh (1984) enables one to proceed as in the proof of Cramér's Theorem 11.4 (See Remark 11.2), replacing $\hat{Q}$ by $\hat{F}_n$, $\hat{Q}_n$ by the c.f. of $\sqrt{n}\,(\overline{Z}^* - \overline{Z})$ under F_n, and population moments by sample moments throughout. Since the estimate (11.43) holds only up to the upper bound $e^{n\delta}$ for $||\xi||$, one needs to show that

$$\int_{||\xi||>\exp\{n\delta\}} |\hat{K}(\varepsilon\xi)|\,d\xi = \varepsilon^{-k}\int_{||\xi||>\varepsilon e^{n\delta}} |\hat{K}(\xi)|\,d\xi = o(n^{-(s-2)/2}) \qquad (11.44)$$

by choosing a kernel K (with support in the unit ball) such that $\hat{K}(\xi)$ decays fast as $||\xi||\to\infty$. For example, one may have $|\hat{K}(\xi)| \leq c\exp\{-c'||\xi||^{1/2}\}$ (See Bhattacharya and Rao 1976, Corollary 10.4). For a detailed proof see Bhattacharya (1987) or Bhattacharya and Denker (1990).

Theorem 11.5 allows us to determine the precise asymptotic accuracy of the bootstrap estimate of the distribution of W_n. To see this, write the formal Edgeworth expansion $\tilde{\psi}_{s-2,n}$ of the 'density' of W_n and the corresponding expansion $\tilde{\psi}^*$ for W_n^* as

$$\tilde{\psi}_{s-2,n(x)} = \left[1 + \sum_{r=1}^{s-2} n^{-r/2}q_r(x)\right]\varphi(x)$$

$$\tilde{\psi}^*_{s-2,n(x)} = \left[1 + \sum_{r=1}^{s-2} n^{-r/2}q_r^*(x)\right]\varphi(x). \qquad (11.45)$$

Then, by Theorems 11.2, 11.5, and noting that sample moments converge to population moments a.s.,

$$\mathbf{P}^*(W_n^* \leq x) - \mathbf{P}(W_n \leq x) = \frac{1}{\sqrt{n}}\int_{-\infty}^{x} (q_1^*(y) - q_1(y))\varphi(y)\,dy$$

$$+ \frac{1}{n}\int_{-\infty}^{\infty} (q_2^*(y) - q_2(y))\varphi(y)\,dy + o(n^{-1}) \text{ a.s.}$$

$$= \frac{1}{\sqrt{n}}(p_1(x) - p_1^*(x))\varphi(x) + o(n^{-1}) \text{ a.s.} \qquad (11.46)$$

where $p_1(x)$ is a polynomial (in x) whose coefficients are polynomials in the moments of Z_1, while $p_1^*(x)$ is obtained on replacing population moments by corresponding sample moments. Hence $p_1^*(x) - p_1(x)$ is of the form $H(\overline{Y}; x) - H(v; x)$ where v is the vector of population moments of Z_1 and $\overline{Y}$ is the corresponding vector of sample moments. By Theorem 11.2 (or, just the *delta method*), if sufficiently many moments of Z_1 are finite,

$$\sqrt{n}[p_1^*(x) - p_1(x)]\varphi(x) \xrightarrow{\mathscr{L}} N(0, \sigma_b^2(x)) \quad \text{as } n\to\infty \qquad (11.47)$$

where $\sigma_b^2(x) \geq 0$ decays to zero fast as $x \to \infty$, in view of the presence of the exponential factor $\varphi^2(x)$. Hence we have

$$n(\mathbf{P}^*(W_n^* \leq x) - \mathbf{P}(W_n \leq x)) \overset{\mathscr{L}}{\to} N(0, \sigma_b^2(x)) \tag{11.48}$$

uniformly for all x. We have then proved the following result.

Theorem 11.6. *Assume the hypothesis of Theorem 11.2 with $s = 4$. Also assume that W_n is pivotal, i.e., (11.37) holds, and that $E\|Z_1\|^8 < \infty$. Then (11.48) holds.*

Remark 11.4. The relation (11.48) implies that the error of bootstrap approximation is $O(n^{-1})$ a.s. In the same manner, one can show that

$$n^{3/2}(P^*(\mathbf{U}_n^* \leq u) - P(\mathbf{U}_n \leq u)) \overset{\mathscr{L}}{\to} N(0, \theta_b^2(u)) \tag{11.49}$$

for some $\theta_b^2(u)$ which decay fast to zero as $u \to \infty$. For this one uses Theorem 11.3 and the corresponding version for the expansion $\eta_{r,n}^*$ for $\mathbf{U}_n^*$, with $s = 5$, to get (See (11.31), (11.32))

$$P^*(U_n^* \leq u) - P(U_n \leq u) = n^{-1}(g_1(u) - g_1^*(u))f_{\chi_d^2}(u) + O_p(n^{-3/2}). \tag{11.50}$$

Hence (11.49) holds, arguing as in the case of (11.48). In particular, we have the following result.

Corollary 11.1. *(a) Under the hypothesis of Theorem 11.6, the coverage error of the bootstrap for symmetric confidence intervals is $O_p(n^{-3/2})$. (b) Assume the hypothesis of Theorem 11.3 with $s = 5$, and also assume $E\|Z_1\|^{10} < \infty$. Then the coverage error of the bootstrap with test statistics U_n is $O_p(n^{-3/2})$.*

A similar analysis shows that for nonpivotal statistics $W_n = \sqrt{n}\,(H(\overline{Z}) - H(\mu))$ having a normal asymptotic distribution with mean 0 and variance $\sigma^2 > 0$,

$$n^{1/2}(P^*(W_n^* \leq x) - P(W_n \leq x)) \overset{\mathscr{L}}{\to} N(0, \delta^2(x)) \tag{11.51}$$

for an appropriate $\delta^2(x)$. For this, one assumes the hypothesis of Theorem 11.2 with $s = 3$, so that

$$P(W_n \leq x) = \Phi_{\sigma^2}(x) + n^{-1/2}p_1(x)\varphi_{\sigma^2}(x) + o(n^{-1/2})$$

$$P^*(W_n^* \leq x) = \Phi_{\hat{\sigma}^2}(x) + n^{-1/2}p_1^*(x)\varphi_{\hat{\sigma}^2}(x) + o_p(n^{-1/2}). \tag{11.52}$$

If z_λ^* is the λ-th quantile of the bootstrap distribution of W_n^* then, by (11.51),

$$1 - \lambda = P^*(z_{\lambda/2}^* \leq W_n^* \leq z_{1-\lambda/2}^*)$$

$$= P(z_{\lambda/2}^* \leq W_n \leq z_{1-\lambda/2}^*) + o_p(n^{-1/2}), \tag{11.53}$$

i.e.,

$$P(H(\bar{z}) - n^{-1/2}z_{1-\lambda/2}^*, \leq H(\mu) \leq H(\overline{Z}) - n^{-1/2}z_{\lambda/2}^*)$$

$$= 1 - \lambda + o_p(n^{-1/2}). \tag{11.54}$$

In other words, $[H(\overline{Z}) - n^{-1/2}z_{1-\lambda/2}^*,\ H(\overline{Z}) - n^{-1/2}z_{\lambda/2}^*]$ is a confidence interval for $H(\mu)$ with a coverage error $o_p(n^{-1/2})$. Note that, without the knowledge of the asymptotic variance σ^2 and studentization, one can not use $H(\overline{Z})$ to obtain a valid classical confidence interval for $H(\mu)$.

11.4 Miscellaneous Applications

In this section we briefly touch upon a number of applications of the theory, in addition to applications already discussed.

11.4.1 Cornish-Fisher Expansions

By inverting the expansions of the distribution function

$$\tilde{\Psi}_{s-2,n}(x) := \int_{-\infty}^{x} \tilde{\psi}_{s-2,n}(y)\, dy$$

of W_n, and of $\int_{-\infty}^{u} \eta_{r,n}(u)\, du$ of $\mathbf{U}_n$, one may refine critical points for tests and confidence regions to provide a desired level (of significance or confidence) up to an error $O(n^{-r})$ for $r = 1, 3/2$, etc. This idea goes back to Cornish and Fisher (1937). See Hall (1983) and Bhattacharya and Denker (1990) for general derivations. As an example, consider the t-statistic of Example 11.1 in Sect. 3. Write G_n for its distribution function and Φ for that of the standard normal, and let $z_{p,n}$ and z_p be their respective p-th quantities, $G_n(z_{p,n}) = p$, $\Phi(z_p) = p$. Then $z_{p,n} = z_{p,n}^{(1)} + O(n^{-1})$, and $z_{p,n} = z_{p,n}^{(2)} + O(n^{-3/2})$, where (See Bhattacharya and Denker 1990, p. 46)

$$z_{p,n}^{(1)} = z_p - n^{-1/2}\frac{\mu_3}{6}(2z_p^2 + 1),$$

$$z_{p,n}^{(2)} = z_{p,n}^{(1)} + n^{-1}\frac{\mu_3^2}{18}z_p^5 - \left(\frac{\mu_4}{12} + \frac{\mu_3^2}{9} + \frac{1}{2}\right)z_p^3$$

$$- \left(\frac{\mu_3^3}{6} + \frac{\mu_4}{4}\right)z_p + \frac{7\mu_3^2}{72}(2z_p^3 + z_p),$$

$$\mu_r := E\left(\frac{X_1 - \mu}{\sigma}\right)^r.$$

Now write, $\hat{\mu}_r$ for the sample moment $1/n\sum_{j=1}(X_j - \overline{X})^r/s^r$, and

$$y_{p,n}^{(1)} := z_p - n^{-1/2}\frac{\hat{\mu}_3}{6}(2z_p^2 + 1) = \hat{z}_{p,n}^{(1)}, \text{ say,}$$

$$P(W_n \le y_{p,n}^{(1)}) = p + O(n^{-1}).$$

Next let $\hat{z}_{p,n}^{(2)}$ denote $z_{p,n}^{(2)}$ with μ_r replaced by $\hat{\mu}_r$ ($2 \le r \le 4$). Then define

$$y_{p,n}^{(2)} := y_{p,n}^{(1)} - n^{-1}(\hat{z}_{p,n}^{(2)} - \hat{z}_{p,n}^{(1)}),$$

to get

$$\mathbf{P}(W_n \le y_{p,n}^{(2)}) = p + O(n^{-3/2}).$$

In this manner one can approximate the p-th quantile by a sample estimate to provide any level of accuracy desired, if sufficiently many moments of X_j are finite.

11.4.2 Higher Order Efficiency

In regular parametric models, the MLE (maximum likelihood estimator) of a parameter θ attains the Cramér-Rao lower bound for variance of unbiased estimators in an asymptotic sense, and is therefore called *efficient,* a terminology due to R.A. Fisher. There are, however, other estimators of θ which are also efficient in this sense. To discriminate among all such estimators, a notion of *second order efficiency* was introduced by Fisher (1925) and further developed by Rao (1961, 1962, 1963) and other later authors. See Ghosh (1994), and Bhattacharya and Denker (1990) for a detailed account. Using Edgeworth expansions it may be shown that a bias correction of order $n^{-1/2}$ makes a first order efficient estimator second order efficient, and a bias correction of order n^{-1} of the latter yields a third order efficient estimator, and so on (Pfanzagl 1980, 1985). Similar results for tests are provided in Bickel et al. (1981).

11.4.3 Computation of Power in Parametric Models

Asymptotic expansions of power under contiguous alternatives, similar to the expansion in Theorem 11.3 but with the chi-square density term replaced by the density of a noncentral chi-square distribution, was derived in Chandra and Ghosh (1980). Also see Groeneboom and Oosterhoff (1981).

11.4.4 Convergence of Markov Processes to Diffusions

Konakov and Mammen (2005) have obtained an asymptotic expansion of the density of a Markov chain converging to a diffusion on $\mathbb{R}^k$, with an error $O(n^{-1-\delta})$ for some $\delta > 0$. Here a version of Theorem 11.1 which holds for independent but non i.i.d. summands is needed, as given in Bhattacharya and Rao (1976, Theorem 19.3).

11.4.5 Asymptotic Expansions in Analytic Number Theory

A famous classical problem in number theory considered by Gauss and Landau is to provide precise asymptotic rates of the errors in estimating by the volume of a ball (or, ellipsoid) in $\mathbb{R}^k$ the number of lattice points (in $\mathbf{Z}^k$) lying in the ball as the radius increases to infinity ($k > 1$). Esseen (1945) showed that Landau' s estimates may be derived by using Cramér type expansions for lattice random vectors. See, e.g., Bhattacharya and Rao (1976, Chap. 5), for expansions in the lattice case, and Bhattacharya (1977) for a simple exposition of Landau's estimates. Important progress has been recently made by Bentkus and Götze (1999) in this problem using a combination of probabilistic and analytic techniques.

In a different direction, Kubilius (1964) derived an asymptotic expansion, similar to Cramér's of the number of prime factors (counting multiplicity) of an integer m, as $m \to \infty$.

11.4.6 Asymptotic Expansions for Time Series

Cramér-type expansions have been extended to the case of dependent sequences, with an exponentially decaying dependence over time, by Götze and Hipp (1983). The Edgeworth expansions for smooth statistics based on such time series follow more or less the same way as in the independent case considered in Sect. 11.2 (See, e.g., Götze and Hipp 1994; Bose 1988). An excellent exposition of the Götze–Hipp theory, with complete but simplified proofs, is given in Jensen (1986). It should be mentioned that for Markovian sequences a fairly complete extension of Cramér-type expansions was given earlier by Nagaev (1957) (Also see Jensen 1989). An extension of the Götze–Hipp results under less restrictive polynomial decay rates of dependence was obtained by Lahiri (1996). Second order accuracy of general pivotal bootstrap estimates have been derived by Götze and Kunsch (1996) and others (See Lahiri 1996 for an up to date account).

11.5 Notes and References

Asymptotic expansions for distributions of normalized sums $\sqrt{n}(\overline{X} - \mu)/\sigma$ of i.i.d. random variables were stated formally without proof by Chebyshev (1890). Cramér (1928), (1937) provided the first rigorous derivation of Chebyshev type expansions under the so-called *Cramér's condition*. One dimensional expansions for the lattice case were obtained by Esseen (1945). A fine account of these may be found in the monograph by Guedenko and Kolmogorov (1954). Multidimensional expansions of this theory and later developments are detailed in Bhattacharya and Ranga Rao (1976).

Independently of Chebyshev, Edgeworth (1905) proposed asymptotic expansions for distribution functions of general statistics. Edgeworth's innovative idea and corresponding expansions were validated in Bhattacharya and Ghosh (1978).

Efron (1979) provided his novel simulation methodology termed the (nonparametric) *bootstrap*, originally conceived as a convenient way in the computer age for the construction of critical points of tests and confidence intervals. His percentile bootstrap does not require the computation of the standard error of the statistic, often an onerous task for applied statisticians. But the bootstrap may be used separately to compute the standard error, and when used to pivot the statistic it yielded a more accurate estimate of the distribution. It was shown by Singh (1981) that in the non-lattice case the bootstrap approximation of the distribution function of $\sqrt{n}(\overline{X} - \mu)/\sigma$ had an error $o_p(n^{-1/2})$ as opposed to the $O(n^{-1/2})$ error of the classical CLT-based approximation. Since σ is generally unknown, for statistical purposes one needs to estimate the distribution of the studentized statistic $\sqrt{n}(\overline{X} - \mu)/s$. Babu and Signh (1984) extended Singh's result to this case. That bootstrap approximation of distributions of general smooth pivotal statistics for continuous data improve upon the CLT-based approximation was proved independently, using the expansions in Bhattacharya and Ghosh (1978), by Beran (1987), Bhattacharya (1987), Hall (1988) and Bhattacharya and Qumsiyeh (1989). There does not seem to be much evidence, theoretical or otherwise, that the bootstrap does better than the classical procedure if the underlying distribution is lattice. For precise comparisons for the standardized sample mean in the lattice case, one may compare the relation (23.12) in Bhattacharya and Ranga Rao (1976) and the asymptotic limit in the lattice case in Singh (1981). See Bhattacharya and Chan (1996) for some numerical results in the lattice case; the Appendix in the present chapter is also due to them. Hall (1992) is a good source for the mathematical theory of Edgeworth expansions in relation to the bootstrap. Athreya (1987) showed that the usual bootstrap estimate of the distribution function of a properly normalized $\overline{X} - \mu$ is not consistent unless the i.i.d. observations belonged to the domain of Normal attraction, a result extended further by Gine and Zinn (1989). This may be contrasted with the result of Bickel and Freedman (1981) in Chap. 9, Remark 9.3, showing that in the linear multiple regression problem there are cases where the bootstrap estimate of the distribution of the vector of regression coefficients is consistent although the latter is not asymptotically Normal.

Efron and Tibshirani (1994) present a readable and very useful account of the bootstrap with diverse applications illustrated by data analyses. Theoretical extensions of the methodology to dependent data and time series are given in Lahiri (1996). Finally, we mention Yoshida (1997) for a novel asymptotic expansion for martingales using Malliavin calculus.

Appendix: Approximate Moments and Cumulants of W_n

Write $H(\overline{Z}) - H(\mu) = G(\overline{Z}) + O_p(n^{-3/2})$, where

$$G(\overline{Z}) = (\overline{Z} - \mu) \cdot \operatorname{grad} H(\mu) + \frac{1}{2!} \sum_{1 \leq i_1, i_2 \leq k} D_{i_1 i_2}(\overline{Z}^{(i_1)} - \mu^{(i_1)})(\overline{Z}^{(i_2)} - \mu^{(i_2)})$$

$$+ \frac{1}{3!} \sum_{1 \leq i_1, i_2, i_3 \leq k} D_{i_1 i_2 i_3}(\overline{Z}^{(i_1)} - \mu^{(i_1)})(\overline{Z}^{(i_2)} - \mu^{(i_2)})(\overline{Z}^{(i_3)} - \mu^{(i_3)}).$$

$$[(\overline{Z} - \mu) \cdot \operatorname{grad} H(\mu) = \sum_i (D_i H)(\mu)(\overline{Z}^{(i)} - \mu^{(i)})].$$

Notation $(D_i H)(z) = (\partial H/\partial z^{(i)})(z)$, $(D_{i_1 i_2} H)(z) = (D_{i_1} D_{i_2} H)(z)$, $D_i = (D_i H)(\mu)$, $D_{i_1 i_2} = (D_{i_1 i_2} H)(\mu)$, $D_{i_1 i_2 i_3} = (D_{i_1 i_2 i_3} H)(\mu)$, etc., $\sigma_{i_1 i_2} = E(Z_j^{(i_1)} - \mu^{(i_1)})(Z_j^{(i_2)} - \mu^{(i_2)}) = \mu_{i_1 i_2}$, $\mu_{i_1 i_2 i_3} = E(Z_j^{(i_1)} - \mu^{(i_1)}) \cdot (Z_j^{(i_2)} - \mu^{(i_2)})(Z_n^{(i_3)} - \mu^{(i_3)})$, etc., $m_{r,n} := EG(\overline{Z})^r$, $\mu_{r,n} = E(\sqrt{n}G(\overline{Z}))^r = n^{r/2}m_{r,n}$. We will compute $\mu_{r,n}$ up to $O(n^{-3/2})$ $(1 \leq r \leq 6)$.

$$m_{1,n} = \frac{a_1}{n} + O(n^{-2}), \quad a_1 = \frac{1}{2!} \sum_{i_1, i_2} D_{i_1 i_2} \sigma_{i_1 i_2}, \tag{11.55}$$

$$m_{2,n} = \frac{b_1}{n} + \frac{b_2}{n^2} + O(n^{-3}),$$

$$b_1 = \sum_{i_1, i_2} D_{i_1} \cdot D_{i_2} \sigma_{i_1 i_2} = E((Z_j - \mu)\operatorname{grad} H(\mu))^2,$$

$$b_2 = \sum_{i_1,i_2,i_3} D_{i_1} D_{i_2 i_3} \mu_{i_1 i_2 i_3}$$

$$+ \frac{1}{3} \sum_{i_1,i_2,i_3,i_4} D_{i_1} D_{i_2 i_3 i_4} (\sigma_{i_1 i_2} \sigma_{i_3 i_4} + \sigma_{i_1 i_3} \sigma_{i_2 i_4} + \sigma_{i_1 i_4} \sigma_{i_2 i_3})$$

$$+ \frac{1}{4} \sum_{i_1,i_2,i_3,i_4} D_{i_1 i_2} D_{i_3 i_4} (\sigma_{i_1 i_2} \sigma_{i_3 i_4} + \sigma_{i_1 i_3} \sigma_{i_2 i_4} + \sigma_{i_1 i_4} \sigma_{i_2 i_3}), \qquad (11.56)$$

$$m_{3,n} = \frac{c_1}{n^2} + O(n^{-3}),$$

$$c_1 = \sum_{i_1,i_2,i_3} D_{i_1} D_{i_2} D_{i_3} \mu_{i_1 i_2 i_3}$$

$$+ \frac{3}{2} \sum_{i_1,i_2,i_3,i_4} D_{i_1} D_{i_2} D_{i_3 i_4} (\sigma_{i_1 i_2} \sigma_{i_3 i_4} + \sigma_{i_1 i_3} \sigma_{i_2 i_4} + \sigma_{i_1 i_4} \sigma_{i_2 i_3}), \qquad (11.57)$$

$$m_{4,n} = \frac{d_1}{n^2} + \frac{d_2}{n^3} + O(n^{-4}), \ \text{where}$$

$$d_1 = 3b_1^2, \ \text{and}$$

$$d_2 = E[(Z_1 - \mu) \cdot \operatorname{grad} H(\mu)]^4 - 3b_1^2$$

$$+ 2 \sum_{i_1,i_2} D_{i_1 i_2} \bigg[E((Z_1 - \mu) \cdot \operatorname{grad} H(\mu))^3 \sigma_{i_1 i_2}$$

$$+ 3E((Z_1 - \mu) \cdot \operatorname{grad} H(\mu))^2$$

$$\cdot E\{(Z_1^{(i_1)} - \mu^{(i_1)})(Z_1^{(i_2)} - \mu^{(i_2)})((Z_1 - \mu) \cdot \operatorname{grad} H(\mu))\}$$

$$+ 3E\{(\operatorname{grad} H(\mu) \cdot (Z_1 - \mu))(Z_1^{(i)} - \mu^{(i_1)})\}$$

$$\cdot E\{(\operatorname{grad} H(\mu) \cdot (Z_1 - \mu))^2 \cdot (Z_1^{(i_2)} - \mu^{(i_2)})\}$$

$$+ 3E\{(\operatorname{grad} H(\mu) \cdot (Z_1 - \mu))(Z_1^{(i_2)} - \mu^{(i_2)})\}$$

$$\cdot E\{(\operatorname{grad} H(\mu) \cdot (Z_1 - \mu))^2 (Z_1^{(i_1)} - \mu^{(i_1)})\} \bigg]$$

$$+ \frac{2}{3} \sum_{i_1,i_2,i_3} D_{i_1 i_2 i_3} \bigg[3E((Z_1 - \mu) \cdot \operatorname{grad} H(\mu))^2$$

$$\cdot \bigg\{ E((Z_1^{(i_1)} - \mu^{(i_1)})(Z_1 - \mu) \cdot \operatorname{grad} H(\mu)))\sigma_{i_2 i_3}$$

$$+ E((Z_1^{(i_2)} - \mu^{(i_2)})((Z_1 - \mu) \cdot \operatorname{grad} H(\mu)))\sigma_{i_1 i_3}$$

$$+ E((Z_1^{(i_3)} - \mu^{(i_3)})((Z_1 - \mu) \cdot \operatorname{grad} H(\mu)))\sigma_{i_1 i_2} \bigg\}$$

$$+ 6E((Z_1^{(i_1)} - \mu^{(i_1)})((Z_1 - \mu) \cdot \operatorname{grad} H(\mu)))$$

$$\cdot E((Z_1 - \mu) \cdot \operatorname{grad} H(\mu))(Z_1^{(i_2)} - \mu^{(i_2)})$$

$$\cdot E((Z_1 - \mu) \cdot \operatorname{grad} H(\mu))(Z_1^{(i_3)} - \mu^{(i_3)}) \bigg]$$

$$+ \frac{3}{2} \sum_{i_1,i_2,i_3,i_4} D_{i_1 i_2} D_{i_3 i_4} \bigg[E(\operatorname{grad} H(\mu) \cdot (Z_1 - \mu))^2$$

$$\cdot \{\sigma_{i_1 i_2} \sigma_{i_3 i_4} + \sigma_{i_1 i_3} \sigma_{i_2 i_4} + \sigma_{i_1 i_4} \sigma_{i_2 i_3}\}$$

$$+ E((\operatorname{grad} H(\mu) \cdot (Z_1 - \mu))(Z_1^{(i_1)} - \mu^{(i_1)}))$$

$$\cdot \Big\{ E((\operatorname{grad} H(\mu) \cdot (Z_1 - \mu))(Z_1^{(i_2)} - \mu^{(i_2)}))\sigma_{i_3 i_4}$$

$$+ E((\operatorname{grad} H(\mu) \cdot (Z_1 - \mu))(Z_1^{(i_3)} - \mu^{(i_3)}))\sigma_{i_2 i_4}$$

$$+ E((\operatorname{grad} H(\mu) \cdot (Z_1 - \mu))(Z_1^{(i_4)} - \mu^{(i_4)})\sigma_{i_2 i_3} \Big\}$$

$$+ E((\operatorname{grad} H(\mu) \cdot (Z_1 - \mu))(Z_1^{(i_2)} - \mu^{(i_2)}))$$

$$\cdot \Big\{ E((\operatorname{grad} H(\mu) \cdot (Z_1 - \mu))(Z_1^{(i_1)} - \mu^{(i_1)}))\sigma_{i_3 i_4}$$

$$+ E((\operatorname{grad} H(\mu) \cdot (Z_1 - \mu))(Z_1^{(i_3)} - \mu^{(i_3)}))\sigma_{i_1 i_4}$$

$$+ E((\operatorname{grad} H(\mu) \cdot (Z_1^{(i_4)} - \mu^{(i_4)}))\sigma_{i_1 i_3} \Big\}$$

$$+ E((\operatorname{grad} H(\mu) \cdot (Z_1 - \mu))(Z_1^{(i_3)} - \mu^{(i_3)}))$$

$$\cdot \Big\{ E((\operatorname{grad} H(\mu) \cdot (Z_1 - \mu))(Z_1^{(i_1)} - \mu^{(i_1)}))\sigma_{i_2 i_4}$$

$$+ E((\operatorname{grad} H(\mu) \cdot (Z_1 - \mu))(Z_1^{(i_2)} - \mu^{(i_2)}))\sigma_{i_1 i_4}$$

$$+ E((\operatorname{grad} H(\mu) \cdot (Z_1 - \mu))(Z_1^{(i_4)} - \mu^{(i_4)}))\sigma_{i_1 i_2} \Big\}$$

$$+ E((\operatorname{grad} H(\mu) \cdot Z_1 - \mu))(Z_1^{(i_4)} - \mu^{(i_4)}))$$

$$\cdot \Big\{ E((\operatorname{grad} H(\mu) \cdot Z_1 - \mu))(Z_1^{(i_1)} - \mu^{(i_1)}))\sigma_{i_2 i_3}$$

$$+ E((\operatorname{grad} H(\mu) \cdot (Z_1 - \mu))(Z_1^{(i_2)} - \mu^{(i_2)}))\sigma_{i_1 i_3}$$

$$+ E((\operatorname{grad} H(\mu) \cdot (Z_1^{(i_3)} - \mu^{(i_3)}))\sigma_{i_1 i_2} \Big\} \Big], \tag{11.58}$$

$$m_{5,n} = \frac{e_1}{n^3} + O(n^{-4}),$$

$$e_1 = 10 E((Z_1 - \mu) \cdot \operatorname{grad} H(\mu))^3 \cdot E((Z_1 - \mu) \cdot \operatorname{grad} H(\mu))^2$$

$$+ \frac{5}{2} \sum_{i_1, i_2} D_{i_1 i_2} E(Z_1 - \mu) \cdot \operatorname{grad} H(\mu))^2$$

$$\cdot \Big\{ 3\sigma_{i_1 i_2} (E(Z_1 - \mu) \cdot \operatorname{grad} H(\mu))^2$$

$$+ 12 E((Z_1 - \mu) \cdot \operatorname{grad} H(\mu))(Z_1^{(i_1)} - \mu^{(i_1)}))$$

$$\cdot E((Z_1 - \mu) \cdot \operatorname{grad} H(\mu))(Z_1^{(i_2)} - \mu^{(i_2)})) \Big\}, \tag{11.59}$$

$$m_{6,n} = \frac{f_1}{n^3} + O(n^{-4}),$$

$$f_1 = 15[E(Z_1 - \mu) \cdot \operatorname{grad} H(\mu))^2]^3. \tag{11.60}$$

"Approximate" cumulants of W_n are then given by

$$k_{1,n} = \mu_{1,n} = \sqrt{n}\, m_{1,n} = n^{-1/2} a_1 + O(n^{-3/2})$$

$$= \frac{n^{-1/2}}{2} \sum_{i_1,i_2} D_{i_1 i_2} \sigma_{i_1 i_2} + O(n^{-3/2})$$

$$k_{2,n} = \mu_{2,n} - \mu_{1,n}^2 = n(m_{2,n} - m_{1,n}^2)$$

$$= b_1 + \frac{b_2 - a_1^2}{n} + O(n^{-2}),$$

$$k_{3,n} = \mu_{3,n} - 3\mu_{2,n}\mu_{1,n} + 2\mu_{1,n}^3 = n^{-1/2}(c_1 - 3b_1 a_1) + O(n^{-3/2}),$$

$$k_{4,n} = \mu_{4,n} - 4\mu_{3,n}\mu_{1,n} - 3\mu_{2,n}^2 + 12\mu_{2,n}\mu_{1,n}^2 - 6\mu_{1,n}^4$$

$$= d_1 + \frac{d_2}{n} - \frac{4a_1 c_1}{n} - 3\left(b_1^2 + \frac{2b_1 b_2}{n}\right) + 12\frac{a_1^2 b_1}{n} + O(n^{-2})$$

$$= n^{-1}(d_2 - 4a_1 c_1 - 6b_1 b_2 + 12a_1^2 b_1) + O(n^{-2}). \tag{11.61}$$

As an example for Student's t,

$$k_{1,n} = -\frac{1}{2}\mu_3 n^{-1/2} + O(n^{-3/2}),$$

$$k_{2,n} = 1 + n^{-1}(2\mu_3^2 + 3) - n^{-1}\left(\frac{1}{4}\mu_3^2\right) + O(n^{-2})$$

$$= 1 + n^{-1}\left(\frac{7}{4}\mu_3^2 + 3\right) + O(n^{-2})$$

$$k_{3,n} = n^{-1/2}\left(-\frac{7}{2}\mu_3 + \frac{3}{2}\mu_3\right) + O(n^{-3/2}) = -2n^{-1/2}\mu_3 + O(n^{-3/2})$$

$$k_{4,n} = n^{-1}[-2\mu_4 + 28\mu_3^2 + 30 - 7\mu_3^2 - 6(2\mu_3^2 + 3) + 3\mu_3^2] + O(n^{-2})$$

$$= n^{-1}(-2\mu_4 + 12\mu_3^2 + 12) + O(n^{-2})$$

$$= -2n^{-1}(\mu_4 - 6\mu_3^2 - 6) + O(n^{-2}). \tag{11.62}$$

Exercises for Chap. 11

Ex. 11.1. Derive (a) the Edgeworth expansion for the distribution function of the (nonparametric) Student's t, using (11.62), and under appropriate conditions, and (b) prove the analog of (11.49) for the coverage error of the bootstrap approximation for a symmetric confidence interval for the mean based on t.

References

Babu, G. J., & Signh, K. (1984). On one term Edgeworth correction by Efron's bootstrap. *Sankhyá Series A, 46*, 219–232.

Bentkus, V., & Götze, F. (1999). Lattice point problems and distribution of values of quadratic forms. *Annals of Mathematics, 150*, 977–1027.

Beran, R. (1987). Prepivoting to reduce level error of confidence sets. *Biometrika, 74*, 457–468.

Bhattacharya, R. N. (1977) Refinements of the multidimensional central limit theorem and applications. *Annals of Probability, 5*, 1–28.

Bhattacharya, R. N. (1987) Some aspects of Edgeworth expansions in statistics and probability. In M. L. Puri, J. P. Vilaplana, & W. Wertz (Eds.), *New perspectives in theoretical and applied statistics*. New York: Wiley.

Bhattacharya, R. N., & Chan, N. H. (1996). Edgeworth expansions and bootstrap approximations to the distributions of the frequency chisquare. *Sankhya Series A, 58*, 57–68.

Bhattacharya, R. N., & Denker, M. (1990). *Asymptotic statistics*. Boston: Birkhäuser.

Bhattacharya, R. N., & Ghosh, J. K. (1978). On the validity of the formal Edgeworth expansion. *Annals of statistics, 6*, 434–451.

Bhattacharya, R. N., & Qumsiyeh, M. (1989). Second order and L^p-comparisons between the bootstrap and empirical Edgeworth expansion methodologies. *Annals of Statistics, 17*, 160–169.

Bhattacharya, R. N., & Ranga Rao, R. (1976). *Normal approximation and asymptotic expansions*. New York: Wiley; Revised Reprint, SIAM Reprint with a new chapter in SIAM No.64. Malabar: Krieger (2010).

Bickel, P. J., Chibishov, D. N., & van Zwet, W. R. (1981). On efficiency of first and second order. *International Statistics Revised, 49*, 169–175.

Chandra, T. K., & Ghosh, J. K. (1979) Valid asymptotic expansions for the likelihood ratio statistic and other perturbed chi-square alternatives. *Sankhyá Series A, 41*, 22–47.

Chandra, T. K., & Ghosh, J. K. (1980). Valid asymptotic expansions for the likelihood ratio and other statistics under contiguous alternatives. *Sankhyá Series A, 42*, 170–184.

Chebyshev, P. L. (1890). Sur deux théorèmes relatifs aux probabilitiés. *Acta Mathematica, 14*, 305–315.

Chibishov, D. M. (1973). An asymptotic expansion for a class of estimators containing maximum likelihood estimators. *Theory of Probability and Its Applications, 18*, 295–303.

Cornish, E. A., & Fisher, R. A. (1937). Moments and cumulants in the specification of distributions. *Review of the Institute of Mathematical Statistics, 5*, 307–322.

Cramér, H. (1928). On the composition of elementary errors. *Skandinavisk Aktuarietidskr, 11*, 13–74, 141–180.

Cramér, H. (1937). *Random variables and probability distribution*. Cambridge: Cambridge University Press

Edgeworth, F. Y. (1905). The law of error. *Proceedings of the Cambridge Philological Society, 20*, 36–65.

Efron, B., & Tibshirani, R. (1994). *An introduction to the bootstrap*. New York: Chapman and Hall.

Esseen, C. G. (1945). Fourier analysis of distribution functions: A mathematical study of the Laplace-Gaussian law. *Acta Mathematica, 77*, 1–125.

Fisher, R. A. (1925). Theory of statistical estimation. *Proceedings of the Cambridge Philological Society, 22*, 700–725.

Ghosh, J. K. (1994). *Higher order asymptotics*. NSF-CBMS Regional Conference Series in Probability and Statistics (Vol. 4). Hayward: Institute of Mathematical Statistics.

Gine, E., & Zinn, J. (1989). Necessary conditions for the bootstrap of the mean. *Annals of Statistics, 17*(2), 684–691.

Götze, F., & Hipp, C. (1983). Asymptotic expansions for sums of weakly dependent random vectors. *Zeitschrift für Wahrscheinlichkeitstheorie und verwandte Gebiete, 64*, 211–239.

Götze, F., & Hipp, C. (1994). Asymptotic distribution of statistics in time series. *Annals of Statistics, 22*, 2062–2088.

Götze, F., & Kunsch, H. R. (1996). Second order correctness of the blockwise bootstrap for stationary observations. *Annals of Statistics, 24*, 1914–1933.

Groeneboom, P., & Oosterhoff, J. (1981). Bahadur efficiency and small sample efficiency. *International Statistical Review, 49*, 127–142.

Guedenko, B. V., & Kolmogorov, A. N. (1954). *Limit distributions of sums of independent random variables* (K.L. Chung, Trans.). London: Addison-Wesley.

Hall, P. (1983). Inverting an Edgeworth expansion. *Annals of Statistics, 11*, 569–576.

Hall, P. (1988). Theoretical comparison of bootstrap confidence intervals. *Annals of Statistics, 16*, 927–985.

Hall, P. (1992). *The bootstrap and Edgeworth expansions*. New York: Springer.

Ivanov, A. V., & Zwanzig, S. (1983). An asymptotic expansion of the distribution of least squares estimators in the nonlinear regression model. *Mathematische Operationsforschung und Statistik, Series Statistics, 14*, 7–27.

James, G. S. (1958). On moments and cumulants of system of statistics. *Sankhya, 20*, 1–30.

Jensen, J. L. (1986). A note on the work of Götze and Hibb concerning asymptotic expansions for sums of weakly dependent random vectors (Memoirs no. 10). Department of Mathematics, Aarhus University.

Jensen, J. L. (1989). Asymptotic expansions for strongly mixing Harris recurrent Markov chains. *Scandinavian Journal of Statistics, 16*, 47–64.

Konakov, V., & Mammen, E. (2005). Edgeworth-type expansions for transition densities of Markov chains converging to diffutions. *Bernoulli, 11*, 591–641.

Kubilius, J. (1964). *Probabilistic methods in the theory of numbers*. Translations for Mathematical Monographs. Providence: American Mathematical Society.

Lahiri, S. N. (1996). Asymptotic expansions for sums of random vectors under polynomial mixing rates. *Sankhyá Series A, 58*, 206–224.

Leonov, V. P., & Shiryaev, A. N. (1959). On a method of calculation of semi-invariants. *Theory of Probability and Its Applications, 4*, 319–329.

Nagaev, S. V. (1957). Some limit theorems for stationary Markov chains. *Theory of Probability and Its Applications, 71*, 378–405.

Pfanzagl, J. (1980). Asymptotic expansions in parametric statistical theory. *Developments in Statistics (Krishnaiah, P.R., Ed.), 3*, 1–97.

Pfanzagl, J. (1985). *Asymptotic expansions for general statistical models*. Lecture Notes in Statistics (Vol. 31). Berlin: Springer.

Qumsiyeh, M. (1989). Edgeworth expansion in regression models. *Journal of Multivariate Analysis*

Rao, C. R. (1961). Asymptotic efficiency and limiting information. *Proceedings of the Fourth Berkeley Symposium on Mathematical Statistics and Probability, 1*, 531–546.

Rao, C. R. (1962). Efficient estimates and optimum inference procedures in large samples. *Journal of the Royal Statistical Society, Series B, 24*, 45–63.

Rao, C. R. (1963). Criteria of estimation in large samples. *Sankhyá Series A, 25*, 189–206.

Singh, K. (1981). On the asymptotic accuracy of Efron's bootstrap. *Annals of Statistics, 9*, 1187–1195.

Yoshida, N. (1997). Malliavin calculus and asymptotic expansion for martingales. *Probability Theory and Related Fields, 190*, 301–312.

Chapter 12
Fréchet Means and Nonparametric Inference on Non-Euclidean Geometric Spaces

Abstract Fréchet means as minimizers of expected squared distances are used for nonparametric inference on geometric spaces M. Applications are provided to paleomagnetism ($M = S^2$ or spheres) and landmarks based image analysis ($M =$ Kendall's planar shape space).

12.1 Introduction

Among statistical analysis on non-Euclidean spaces, statistics on spheres $S^d = \{x \in \mathbb{R}^{d+1} : |x|^2 = 1\}$, $d \geq 1$, has a long history (See, e.g., Watson 1983 and Mardia and Jupp 2000). Its growth was especially spurred on by the seminal 1953 paper of R.A. Fisher providing conclusive statistical evidence that the earth's magnetic poles had dramatically shifted positions over geological time scales. This was a fundamental contribution to paleomagnetism—a field of earth science devoted to the study of fossil magnetism (See Irving 1964). This theory also has important consequences in the field of tectonics, especially to the older theory that the continents had changed their relative positions over a period of several hundred millions of years. If rock samples in different continents dating back to the same period exhibit different magnetic polarities, that would be a confirmation of the theory of continental drift. See Fisher et al. (1987) for examples of data on so-called remanent magnetism in fossilized rock samples. In Sect. 12.3 we discuss the parametric theory of R.A. Fisher (1953) and compare that with the nonparametric theory based on Fréchet means developed in Sect. 12.2.

Due to advances in modern technology in recent decades, digital images are now available and extensively used in biology, medicine and many other areas of science and technology. An important class of examples are landmarks based images whose analysis was pioneered by D.G. Kendall (1977, 1984) and F. Bookstein (1978). As described in Sect. 12.2, such an image on the plane may be viewed as an orbit under rotation of a point on a sphere of high dimension known as the preshape sphere. The present chapter is devoted to the nonparametric statistical analysis of data on such non-Euclidean spaces M by means of Fréchet means of the samples, whose asymptotic distribution theory is described in the next section.

© Springer-Verlag New York 2016
R. Bhattacharya et al., *A Course in Mathematical Statistics and Large Sample Theory*, Springer Texts in Statistics, DOI 10.1007/978-1-4939-4032-5_12

12.2 Fréchet Means on Metric Spaces

Let (M, ρ) be a metric space and Q a probability measure on the Borel sigma-field of M. Consider a *Fréchet function* of Q defined by

$$F(x) = \int \rho^\alpha(x, y) Q(dy), \quad x \in M. \tag{12.1}$$

for some $\alpha \geq 1$. We will be mostly concerned with the case $\alpha = 2$. Assume that F is finite. A minimizer of F, if unique, serves as a measure of location of Q. In general, the set C_Q of minimizers of F is called the *Fréchet mean set* of Q. In the case the minimizer is unique, one says that the *Fréchet mean exists* and refers to it as the *Fréchet mean* of Q. If $X_1, \ldots, X_n$ are i.i.d. observations with common distribution Q, the Fréchet mean set and the Fréchet mean of the empirical $Q_n = 1/n \sum_{1 \leq j \leq n} \delta_{X_j}$ are named the *sample Fréchet mean set* and the *sample Fréchet mean*, respectively. For a reason which will be clear from the result below, in the case the Fréchet mean of Q exists, a (every) measurable selection from C_{Q_n} is taken to be a sample Fréchet mean.

Remark 12.1. For $M = \mathbb{R}^m$ with the Euclidean norm $|\cdot|$ and distance $\rho(x, y) = |x - y|$, the Fréchet mean for the case $\alpha = 2$ is the same as the usual mean $\int y Q(dy)$, provided $\int |y|^2 Q(dy) < \infty$. The generalization considered here is due to Fréchet (1948).

The following is a general result on Fréchet mean sets C_Q and C_{Q_n} of Q and Q_n and *consistency* of the sample Fréchet mean. It is due to Ziezold (1977) and Bhattacharya and Patrangenaru (2003).

Theorem 12.1. *Let M be a metric space such that every closed and bounded subset of M is compact. Suppose $\alpha \geq 1$ in (12.1) and $F(x)$ is finite. Then (a) the Fréchet mean set C_Q is nonempty and compact, and (b) given any $\varepsilon > 0$, there exists a positive integer valued random variable $N = N(\omega, \varepsilon)$ and a P-null set $\Omega(\varepsilon)$ such that*

$$C_{Q_n} \subseteq C_Q^c = \{x \in M : \rho(x, C_Q) < \varepsilon\} \forall n \geq N, \ \forall \omega \in (\Omega(\varepsilon))^c. \tag{12.2}$$

(c) In particular, if the Fréchet mean of Q exists then the sample Fréchet mean, taken as a measurable selection from C_{Q_n}, converges almost surely to it.

Proof. We give a proof of Theorem 12.1 for a compact metric M, which is the case in many of the applications of interest here. Part (a) is then trivially true. For part (b), for each $\varepsilon > 0$, write

$$\eta = \inf\{F(x) : x \in M\} \equiv F(q) \, \forall q \in C_Q,$$
$$\eta + \delta(\varepsilon) = \inf\{F(x) : x \in M \backslash C_{Q^\varepsilon}\}. \tag{12.3}$$

If $C_Q^\varepsilon = M$, then (12.2) trivially holds. Consider the case $C_Q^\varepsilon \neq M$, so that $\delta(\varepsilon) > 0$. Let $F_n(x)$ be the Fréchet function of Q_n, namely,

$$F_n(x) = \frac{1}{n} \sum_{1 \leq j \leq n} \rho^\alpha(x, X_j).$$

Now use the elementary inequality,

$$|\rho^{\alpha}(x,y) - \rho^{\alpha}(x',y)| \leq \alpha\rho(x,x') \left[\rho^{\alpha-1}(x,y) + \rho^{\alpha-1}(x',y)\right] \leq c\alpha\rho(x,x'),$$

with $c = 2\max\{\rho^{\alpha-1}(x,y),\ x,y \in M\}$, to obtain

$$|F(x) - F(x')| \leq c\alpha\rho(x,x'),\ |F_n(x) - F_n(x')| \leq c\alpha\rho(x,x'),\ \forall\, x, x'. \tag{12.4}$$

For each $x \in M\backslash C_{Q^{\varepsilon}}$ find $r = r(x,\varepsilon) > 0$ such that $c\alpha\rho(x,x') < \delta(\varepsilon)/4\ \forall\, x'$ within a distance r from x. Let $m = m(\varepsilon)$ of balls with centers $x_1, \cdots, x_m$ and radii $r(x_1), \ldots, r(x_m)$ (in $M\backslash C_{Q^{\varepsilon}}$) cover $M\backslash C_{Q^{\varepsilon}}$. By the SLLN, there exist integers $N_i = N_i(\omega)$ such that, outside a P-null set $\Omega_i(\varepsilon)$, $|F_n(x_i) - F(x_i)| < \delta(\varepsilon)/4\ \forall\, n \geq N_i\ (i = 1, \ldots, m)$. Let $N' = \max\{N_i : i = 1, \ldots, m\}$. If $n > N'$, then for every i and all x in the ball with center x_i and radius $r(x_i, \varepsilon)$,

$$F_n(x) > F_n(x_i) - \delta(\varepsilon)/4 > F(x_i) - \delta(\varepsilon)/4 - \delta(\varepsilon)/4$$
$$\geq \eta + \delta(\varepsilon) - \delta(\varepsilon)/2 = \eta + \delta(\varepsilon)/2.$$

Next choose a point $q \in C_Q$ and find $N'' = N''(\omega)$, again by the SLLN, such that, if $n \geq N''$ then $|F_n(q) - F(q)| < \delta(\varepsilon)/4$ and, consequently, $F_n(q) < \eta + \delta(\varepsilon)/4$, outside of a P-null set $\Omega''(\varepsilon)$. Hence (12.2) follows with $N = \max\{N', N''\}$ and $\Omega(\varepsilon) = \{\cup\, \Omega_i(\varepsilon) : i = 1, \ldots, m\} \cup \Omega''(\varepsilon)$. Part (c) is an immediate consequence of part (b).

For noncompact M, the proof of Theorem 12.1 is a little more elaborate and may be found in Bhattacharya and Bhattacharya (2012) or, for the case $\alpha = 2$, in Bhattacharya and Patrangenaru (2003).

In the applications considered here, the space M of observations is a manifold, i.e., a space which is locally like an Euclidean space of dimension d with a differentiable structure. Here is the formal definition.

Definition 12.1. A d-*dimensional differentiable manifold* M is a separable metric space with the following properties:

(i) Every point $p \in M$ has an open neighborhood U_p with a homeomorphism $\psi_p : U_p \longrightarrow B_p$, where B_p is an open subset of $\mathbb{R}^d$.
(ii) The maps ψ_p are smoothly compatible; that is, if $U_p \cap U_q \neq \emptyset$, then $\psi_p \circ \psi_q^{-1}$ is an infinitely differentiable diffeomorphism on $\psi_q(U_p \cap U_q) \subset B_q$ onto $\psi_p(U_p \cap U_q) \subset B_p$.

The pair (U_p, ψ_p) is called a *coordinate neighborhood* of p and $\psi_p(p') = (p'_1, p'_2, \ldots, p'_d)$ are the *local coordinates* of $p' \in U_p$, and the collection of all coordinate neighborhoods is an *atlas* for M.

Example 12.1 (The Sphere S^d). Consider the sphere $M = S^d = \{p \in \mathbb{R}^{d+1}| : |p|^2 = 1\}$. For $p \in S^d$, the tangent space at p is $T_p = T_p(S^d) = \{x \in \mathbb{R}^{d+1} : p.x = 0\}$, where $p.x = \langle p, x \rangle$ is the Euclidean inner product. Note that T_p is isomorphic to $\mathbb{R}^d$, with $y = (y_1, y_2, \ldots, y_d) \in \mathbb{R}^d$ corresponding to $\sum y_i e_i$ where $e_1, \ldots, e_d$ comprise an orthonormal basis of T_p. Let $U_p = \{q \in S^d : |q - p| < 1/2\}$ and ψ_p the projection of U_p into $T_p : q \to q - (q.p)p = \sum y_i e_i$, so that $B_p = \psi_p(U_p)$ considered as a subset of $\mathbb{R}^d$.

An important task in the use of Fréchet means for inference is to choose a proper distance ρ for M. If M is a submanifold of an Euclidean space E^N (of dimension $N > d$) such as a (hyper-)surface, then one may just use the distance inherited from E^N. For example, the sphere $M = S^d = \{p \in \mathbb{R}^{d+1} | : |p|^2 = 1\}$ is a submanifold of $E^{d+1} = \mathbb{R}^{d+1}$, and a natural distance on it may be the *chord distance* inherited from $\mathbb{R}^{d+1}$ as a subset. One may, alternatively, use the geodesic distance on S^d. The *geodesic distance* $\rho = d_g$ between two points p and q is the arc length (of the smaller of the two arcs) along the big circle connecting p and q. For general manifolds the notion of the geodesic distance depends on endowing M with a Riemannian structure, i.e., a metric tensor, which is not considered here. Instead, we will consider the following analog of the chord distance.

Definition 12.2. An *embedding* J of a d-dimensional manifold M is a map of M into an Euclidean space E^N which is (i) a homeomorphism of M onto its image $J(M)$ which is given the relative topology of E^N, and (ii) in local coordinates, infinitely differentiable with the Jacobian (of J) of rank d at every point of M. The *extrinsic distance* $\rho = \rho^J$ on M under J is the (Euclidean) distance on E^N restricted to $J(M)$; that is, $\rho(p,q) = |J(p) - J(q)| \; \forall p, q \in M$, where $|x|^2 = \langle x, x \rangle$ is the squared norm of x in E^N with inner product $\langle \, , \, \rangle$. The Fréchet mean μ_E, say, if it exists as the unique minimizer under the extrinsic distance is called the *extrinsic mean* of Q. In order that the extrinsic mean be well defined we assume that $J(M)$ is *closed*, which is automatic if M is compact.

Consider the case $\alpha = 2$. Letting $Q^J = Q \circ J^{-1}$ denote the distribution induced on E^N from Q by J, and μ^J the usual (Euclidean) mean $\int y Q^J(dy)$ of Q^J, the Fréchet function F^J *under the extrinsic distance under the embedding J may be expressed at a point* $p = J^{-1}(x)$, $x \in J(M)$, *as*

$$F^J(p) = \int_{EN} |x - y|^2 Q^J(dy) = \int_{EN} |x - \mu^J + \mu^J - y|^2 Q^J(dy)$$

$$= |x - \mu^J|^2 + \int_{EN} |y - \mu^J|^2 Q^J(dy) \quad (x = J(p)). \tag{12.5}$$

The following is a simple but very useful result concerning the Fréchet mean under the extrinsic distance. Let $P = P^J$ denote the orthogonal projection on E^N onto $J(M)$. That is, $P(x) = P^J(x)$ is the point of $J(M)$ closest to x in the Euclidean distance, provided there is a unique such point of $J(M)$. From (12.5) it immediately follows that the minimum of $F^J(p)$ is attained at p whose image x, if unique, minimizes $|x - \mu^J|^2$ over $J(M)$. We have then proved the following result (Bhattacharya and Patrangenaru 2003). In particular, the uniqueness of the projection is the necessary and sufficient condition for the existence of the extrinsic mean.

Proposition 12.1. *Assume that $\int_{EN} |y|^2 Q^J(dy) < \infty$ and there is a unique point in $J(M)$ closest to μ^J, namely, $P(\mu^J)$. Then $P(\mu^J)$ is the image of the Fréchet mean μ_E in $J(M)$; that is, $\mu_E = J^{-1}(P(\mu^J))$.*

The next task is to find the asymptotic distribution of the sample extrinsic Fréchet mean μ_{nE}, whose image $J(\mu_{nE})$ under J is $P(\mu_n^J)$ with $\mu_n^J = \overline{Y}$ as the mean of the empirical Q_n^J of $Y_i = J(X_i)$ based on i.i.d. observations X_i on M with common distribution Q $(1 \leq i \leq n)$. Denote by $d_{\mu^J} P$ the $N \times N$ Jacobian matrix at μ^J of the projection P on a neighborhood of $\mu^J \in E^N \approx R^N$ into $J(M) \subset E^N$ considered as a map on E^N into E^N. Since $J(M)$ is of dimension $d < N$, $d_{\mu^J} P$ is singular and has rank d. Indeed, $d_{\mu^J} P$ maps the tangent space $T_{\mu^J}(E^N) \approx R^N$ onto the tangent space of $J(M)$ at the point $P(\mu^J)$, namely, $T_{P(\mu^J)}(J(M))$. Let $F_1, \ldots, F_d$ be an orthonormal basis of the latter tangent space. Then one has

$$P(\mu_n^J) - P(\mu^J) = (d_{\mu^J} P)(\overline{Y} - \mu^J) + o\left(|\overline{Y} - \mu^J|\right)$$
$$= \sum_{1 \leq i \leq d} \langle (d_{\mu^J} P)(\overline{Y} - \mu^J), F_i \rangle F_i + o(|\overline{Y} - \mu^J|), \qquad (12.6)$$

where $\langle \, , \, \rangle$ is the inner product in E^N. The asymptotic distribution of the image (under J) of the sample extrinsic mean $P(\overline{Y})$ on the tangent space of $J(M)$ at $P(\mu^J)$, given in terns of its coordinates (with respect to the basis $\{F_1, F_2, \ldots, F_d\}$), now follows from (12.6) using the classical CLT for $\sqrt{n}(\overline{Y} - \mu^J)$ (See Bhattacharya and Patrangenaru 2003 and Bhattacharya and Bhattacharya 2012, pp. 38, 39).

Theorem 12.2. *Suppose that the projection P is uniquely defined and continuously differentiable in a neighborhood of μ^J, and that the $N \times N$ covariance matrix V of Y_i is nonsingular. Then*

$$\left(\langle \sqrt{n}(d_{\mu^J} P)(\overline{Y} - \mu^J), F_i \rangle : i = 1, \ldots, d \right) \xrightarrow{\mathcal{L}} N(0, \Gamma), \qquad (12.7)$$

where Γ is the nonsingular $d \times d$ covariance matrix given by $\Gamma = F(d_{\mu^J} P) V (d_{\mu^J} P)' F'$, with the rows of the $d \times N$ matrix F being $F_1, \ldots, F_d$.

This theorem then has the following corollary, using a Slutsky type argument in replacing $d_{\mu^J} P$ by $d_{\overline{Y}} P$ and Γ by $\widehat{\Gamma} = [\widehat{F}(d_{\overline{Y}} P)]' \widehat{V} [\widehat{F}(d_{\overline{Y}} P)]$, where the rows of $\widehat{F}$ are the orthonormal basis $\{\widehat{F}_1, \widehat{F}_2, \ldots, \widehat{F}_d\}$ of the tangent space of $J(M)$ at $P(\overline{Y})$ corresponding smoothly to $\{F_1, F_2, \cdots, F_d\}$, and $\widehat{V}$ is the sample covariance matrix of Y_i $(i = 1, \ldots, n)$.

Corollary 12.1. *Under the hypothesis of Theorem 12.2, one has*

$$n \left[(d_{\overline{Y}} P)(\overline{Y} - \mu^J) \right]' \widehat{\Gamma}^{-1} \left[(d_{\overline{Y}} P)(\overline{Y} - \mu^J) \right] \xrightarrow{\mathcal{L}} \chi^2(d) \text{ as } n \to \infty, \qquad (12.8)$$

and a confidence region for the extrinsic mean μ_E of asymptotic confidence level $1 - \alpha$ is given by

$$\left\{ \mu_E = J^{-1} P(\mu^J) : n \left[(d_{\mu^J} P)(\overline{Y} - \mu^J) \right]' \widehat{\Gamma}^{-1} \left[(d_{\mu^J} P)(\overline{Y} - \mu^J) \right] \leq \chi^2_{1-\alpha}(d) \right\},$$
$$(12.9)$$

where one may replace d_{μ^J} by $d_{\overline{Y}}$.

We next consider the two-sample problem of distinguishing two distributions Q_1 and Q_2 on M, based on two independent samples of sizes n_1 and n_2, respectively: $\{Y_{j_1} = J(X_{j_1}) : j = 1, \ldots, n_1\}$, $\{Y_{j_2} = J(X_{j_2}) : j = 1, \ldots, n_2\}$. Hence the proper null hypothesis is $H_0 : Q_1 = Q_2$. For high dimensional M it is often sufficient to test if the two Fréchet means are equal. For the extrinsic procedure, again consider an embedding J into E^N. Write μ_i for μ_i^J for the population means and $\overline{Y}_i$ for the corresponding sample means on E^N $(i = 1, 2)$. Let $n = n_1 + n_2$, and assume $n_1/n \to p_1$, $n_2/n \to p_2 = 1 - p_1$, $0 < p_i < 1$ $(i = 1, 2)$, as $n \to \infty$. If $\mu_1 \neq \mu_2$ then $Q_1 \neq Q_2$. One may then test $H_0 : \mu_1 = \mu_2$ $(= \mu, \text{ say})$. Since N is generally quite large compared to d, the direct test for $H_0 : \mu_1 = \mu_2$ based on $\overline{Y}_1 - \overline{Y}_2$ is generally not a good test. Instead, we compare the two extrinsic means μ_{E_1} and μ_{E_2} of Q_1 and Q_2 and test for their equality. This is equivalent to testing if $P(\mu_1) = P(\mu_2)$. Then, assuming H_0,

$$n^{\frac{1}{2}} d_{\overline{Y}} P(\overline{Y}_1 - \overline{Y}_2) \longrightarrow N(0, B(p_1 V_1 + p_2 V_2)B') \tag{12.10}$$

in distribution, as $n \to \infty$. Here $\overline{Y} = \frac{n_1}{n}\overline{Y}_1 + \frac{n_2}{n}\overline{Y}_2$ is the *pooled estimate* of the common mean $\mu_1 = \mu_2 = \mu$, say, $B = B(\mu) = F(d_\mu P)$ and V_1, V_2 are the covariance matrices of Y_{j_1} and Y_{j_2}. This leads to the asymptotic chi-square statistic below:

$$n \left[d_{\overline{Y}} P(\overline{Y}_1 - \overline{Y}_2) \right]' \left[\widehat{B}(\hat{p}_1 \widehat{V}_1 + \hat{p}_2 \widehat{V}_2) \widehat{B}^t \right]^{-1} \left[d_{\overline{Y}} P(\overline{Y}_1 - \overline{Y}_2) \right] \longrightarrow \chi^2(d) \tag{12.11}$$

in distribution, as $n \to \infty$. Here $\widehat{B} = B(\overline{Y})$, $\widehat{V}_i$ is the sample covariance matrix of Y_{ji}, and $\hat{p}_i = n_i/n$ $(i = 1, 2)$. One rejects the null hypothesis H_0 at a level of significance $1 - \alpha$ if and only if the observed value of the left side of (12.11) exceeds $\chi^2_{1-\alpha}(d)$.

Example 12.2. Consider again the sphere S^d of Example 12.1, and let J be the inclusion map. Then $P(x) = x/|x|$ $(x \neq 0)$. It is not difficult to check that the Jacobian matrix $d_x P$ is given by

$$d_x P = |x|^{-1} \left[I_{d+1} - |x|^{-2}(xx^t) \right], \quad x \neq 0. \tag{12.12}$$

Let $F(x)$ be a $d \times (d+1)$ matrix whose rows form an orthonormal basis of $T_x(S^d) = \{q \in R^{d+1} : x \cdot q = 0\}$. One may apply Theorem 12.2 and Corollary 12.1. For $d = 2$, and $x = (x_1, x_2, x_3)' \neq (0, 0, \pm 1)'$ and $x_3 \neq 0$, one may choose the two rows of $F(x)$ as $(-x_2, x_1, 0)/\sqrt{x_1^2 + x_2^2}$ and $(x_1, x_2, -(x_1^2 + x_2^2)/x_3)c$ with $c = [x_1^2 + x_2^2 + (x_1^2 + x_2^2)^2/x_3^2]^{1/2}$. For $x = (0, 0, 1)$ one may simply take the two rows of $F(x)$ as $(1, 0, 0)$ and $(0, 1, 0)$. If $x_3 = 0$ and $x_1 \neq 0$, $x_2 \neq 0$, then take the two rows as $(-x_2, x_1, 0)$ and $(0, 0, 1)$. Permuting the indices all cases are now covered.

Example 12.3 (Kendall's Planar Shape Space Σ_2^k). In landmarks based image analysis on the plane one chooses, with expert help, $k > 2$ points or landscapes, not all the same, on an image in the plane $\mathbb{R}^2$. The ordered set of k points—a k-ad—$((x_1, y_1), \ldots, (x_k, y_k))$ is more conveniently viewed as a k-tuple $\mathbf{z} = (z_1, \ldots, z_k)$ in the complex plane $\mathbb{C}$, $z_j = x_j + iy_j$ $(j = 1, \ldots, k)$. The *shape* $\sigma(\mathbf{z})$ of a k-ad is defined to be $\mathbf{z}$ modulo translation, scaling and rotation, in order that images of the same object taken from different locations, different distances and different angles are not distinguished. To rid the k-ad of the effects of location and distance, or scale, one considers $\mathbf{u} = (\mathbf{z} - \langle \mathbf{z} \rangle)/|(\mathbf{z} - \langle \mathbf{z} \rangle)|$, where $\langle \mathbf{z} \rangle = (\frac{1}{k} \sum_{j=1}^{k} z_j)(1, 1, \ldots, 1)$ and the norm $|(c_1, c_2, \ldots, c_k)|$ of a k-tuple of complex numbers $c_j = a_j + ib_j$ $(j = 1, \ldots, k)$ is given by $(\sum_{j=1}^{1} |c_j|^2)^{1/2}$ with $|c_j|^2 = a_j^2 + b_j^2$. Note that this would be the same as the Euclidean norm of the k-tuple of points $((a_1, b_1), \ldots, (a_k, b_k))'$ in $\mathbb{R}^2$. The normed k-tuple $\mathbf{u}$ lies in the complex $(k - 1)$-dimensional hyperplane L in $\mathbb{C}^k$ defined by $\sum_{j=1}^{k} u_j = 0$, which is also the real $2(k - 1)$-dimensional subspace of $(\mathbb{R}^2)^k$ defined by $\sum_1^k v_j = 0$, $\sum_1^k w_j = 0$ where $u_j = v_j + iw_j$. Due to norming, $|\mathbf{u}| = 1$ so that $\mathbf{u}$ lies on the unit sphere in $L \approx \mathbb{C}^{k-1}$ (or, $\approx \mathbb{R}^{2(k-1)}$), called the *preshape sphere* denoted $\mathbb{C}S^{k-1} \approx S^{2k-3}$. In order to rid the k-ad of the angle of observation as well, one considers the *shape* $\sigma(\mathbf{z})$ of a k-ad $\mathbf{z}$ to be its preshape $\mathbf{u}$ modulo rotations in the plane. This is conveniently expressed as the *orbit of* $\mathbf{u}$ *under rotations,* i.e.,

$$\sigma(\mathbf{z}) = \left\{ e^{i\theta} u : -\pi < \theta \le \pi \right\}. \tag{12.13}$$

Since the preshape sphere has (real) dimension $2k - 3$, and $\{e^{i\theta} : -\pi < \theta \le \pi\}$ is one-dimensional, *Kendall's planar shape space* denoted Σ_2^k is a manifold which has dimension $2k - 4$ $(k > 2)$ (See, e.g., Bhattacharya and Bhattacharya 2012, Lemma A3, p. 214). It is preferable to represent the preshape $\mathbf{u}$ as a point $(u_1, u_2, \ldots, u_{k-1}) \in \mathbb{C}^{k-1}$ (or $\mathbb{R}^{2k-2}$). This may be achieved by premultiplying the k-ad by a $k \times (k-1)$ *Helmert matrix* H comprising $k - 1$ column vectors forming an orthonormal basis of the subspace $1^\perp$ of $\mathbb{R}^k$ orthogonal to $(1, 1, \ldots, 1)$. For example, one may take the j-th column of H to be $(a(j), \ldots, a(j)), -ja(j), 0, \ldots, 0)$ where $a(j) = [j(j + 1)]^{-1/2}$ $(j = 1, \ldots, k - 1)$. That is, $\mathbf{x} = (x_1, \ldots, x_k)$ is changed to $\mathbf{x}H = (x_1^0, \ldots, x_{k-1}^0)$, say, and, similarly, $\mathbf{y} = (y_1, \ldots, y_k)$ is changed to $\mathbf{y}H = (y_1^0, \ldots, y_{k-1}^0)$. This translated k-ad $((x_1^0, y_1^0), \ldots, (x_{k-1}^0, y_{k-1}^0))$ then has the complex representation $\mathbf{z}^0 = (z_1^0, \ldots, z_{k-1}^0)$ with $z_j^0 = x_j^0 + iy_j^0$, and the preshape is then $\mathbf{u}^0 = \mathbf{z}^0/|\mathbf{z}^0|$. Note that we only consider k-ads whose k points are not all the same, so that $\mathbf{z}^0 \ne \mathbf{0}$. To avoid excessive notation we will drop the superscript 0 from $\mathbf{z}^0$ and $\mathbf{u}^0$ and write $(k - 1)$-tuples as $\mathbf{z}$ and $\mathbf{u}$, respectively.

A good embedding of the shape space Σ_2^k is the so-called *Veronese–Whitney embedding* J into the (real) vector space $S(k-1, \mathbb{C})$ of all $(k-1) \times (k-1)$ Hermitian matrices $B = B^*$ (B^* is the transpose of the matrix $\overline{B}$ of complex conjugates of elements of B), given by

$$J\sigma(\mathbf{z}) = \mathbf{u}^*\mathbf{u}. \tag{12.14}$$

Note that $(e^{i\theta}\mathbf{u})^*(e^{i\theta}\mathbf{u}) = \mathbf{u}^*\mathbf{u} \; \forall \theta \in (-\pi, \pi]$, so that J is a one-to-one map of Σ_2^k into $S(k - 1, \mathbb{C})$. The space $S(k - 1, \mathbb{C})$ is a vector space with respect to real scalars and with the inner product given by $\langle B, C \rangle = Re(\text{Trace } BC^*)$. If $B = ((\alpha_{jj'} + i\beta_{jj'}))$ and $C = ((\delta_{jj'} + i\gamma_{jj'}))$, then $\langle B, C \rangle = \sum_{j,j'} (\alpha_{jj'}\delta_{jj'} + \beta_{jj'}\gamma_{jj'})$. One may think of an element B of $S(k-1, \mathbb{C})$ to be represented by a real $2(k-1) \times (k-1)$ matrix with the first $k - 1$ rows $\{\alpha_{jj'} : 1 \le j' \le k - 1, 1 \le j \le k - 1\}$ comprising a symmetric $(k - 1) \times (k - 1)$ matrix, and the last $k - 1$ rows $\{\beta_{jj'} : 1 \le j' \le k - 1, 1 \le j \le k - 1\}$ comprising a $(k - 1) \times (k - 1)$ skew symmetric matrix. The inner product $\langle B, C \rangle$ is then simply the Euclidean inner product on this space of $2(k - 1) \times (k - 1)$ real matrices considered as a $2(k - 1) \times (k - 1)$-dimensional vector. Note that the dimension of $S(k - 1, \mathbb{C})$ (or of its representation as real $2(k-1) \times (k-1)$ matrices as described) is $k(k-1)$. Hence $S(k-1, \mathbb{C}) \approx E^N$ with $N = k(k - 1)$.

We next turn to extrinsic analysis on Σ_2^k, using the embedding (12.14). Let μ^J be the mean of $Q \circ J^{-1}$ on $S(k - 1, \mathbb{C})$. Denote by $SU(k - 1)$ the special unitary group of $(k-1) \times (k-1)$ complex matrices T such that $TT^* = I_{k-1}$ and $det(T) = 1$. To compute the projection $P(\mu^J)$, let T be a unitary matrix, $T \in SU(k - 1)$ such that $T\mu^J T^* = D = diag(\lambda_1, \ldots, \lambda_{k-1})$, $\lambda_1 \le \cdots \le \lambda_{k-2} \le \lambda_{k-1}$. For $u \in \mathbb{C}S^{k-1}$, $u^*u \in J(\Sigma_2^k)$, write $v = Tu^*$. Then $Tu^*uT^* = vv^*$, and

$$\|u^* u - \mu_J\|^2 = \|vv^* - D\|^2 = \sum_{i,j} |v_i v_j - \lambda_j \delta_{ij}|^2 \tag{12.15}$$

$$= \sum_j (|v_j|^2 + \lambda_j^2 - 2\lambda_1 v_j|^2)$$

$$= \sum_j \lambda_j^2 + 1 - 2\sum_j \lambda_j |v_j|^2,$$

which is minimized on $J(\Sigma_2^k)$ by $v = (v_1, \ldots, v_{k-1})$ for which $v_j = 0$ for $j = 1, \ldots, k-2$, and $|v_{k-1}| = 1$. That is, the minimizing u^* in (12.15) is a unit eigenvector of μ^J with the largest eigenvalue λ_{k-1}, and $P(\mu^J) = u^* u$. *This projection is unique if and only if the largest eigenvalue of μ^J is simple*, i.e., $\lambda_{k-2} < \lambda_{k-1}$.

Assuming that the largest eigenvalue of μ^J is simple, one may now obtain the asymptotic distribution of the sample extrinsic mean $\mu_{n,E}$, namely, that of $J(\mu_{n,E}) = v_n^* v_n$, where v_n is a unit eigenvector of $\overline{\widetilde{X}} = \sum \widetilde{X}_j / n$ corresponding to its largest eigenvalue. Here $\widetilde{X}_j = J(X_j)$, for i.i.d. observations $X_1, \ldots, X_n$ on Σ_2^k. For this purpose, a convenient orthonormal basis (frame) of $T_p S(k-1, \mathbb{C}) \approx S(k-1, \mathbb{C})$ is the following

$$v_{a,b} = 2^{-\frac{1}{2}} (e_a e_b' + e_b e_a') \text{ for } a < b,\ v_{a,a} = e_a e_a'; \tag{12.16}$$

$$w_{a,b} = i2^{-\frac{1}{2}} (e_a e_b' - e_b e_a') \text{ for } b < a\ (a, b = 1, \ldots, k-1),$$

where e_a is the column vector with all entries zero other than the a-th, and the a-th entry is 1. Let $U_1, \ldots, U_{k-1}$ be orthonormal unit eigenvectors corresponding to the eigenvalues $\lambda_1 \leq \cdots \leq \lambda_{k-2} < \lambda_{k-1}$. Then choosing $T = (U_1, \ldots, U_{k-1}) \in SU(k-1)$, $T\mu^J T^* = D = diag(\lambda_1, \ldots, \lambda_{k-1})$, the columns of $Tv_{a,b}T^*$ and $Tw_{a,b}T^*$ together constitute an orthonormal basis of $S(k-1, \mathbb{C})$. It is not difficult to check that the differential of the projection operator P satisfies

$$(d_{\mu^J} P) T v_{a,b} T^* = \begin{cases} 0 & \text{if } 1 \leq a \leq b < k-1, \\ & \text{or } a = b = k-1, \\ (\lambda_{k-1} - \lambda_a)^{-1} T v_{a,k-1} T^* & \text{if } 1 \leq a < k-1,\ b = k-1; \end{cases} \tag{12.17}$$

$$(d_{\mu^J} P) T w_{a,b} T^* = \begin{cases} 0 & \text{if } 1 \leq a \leq b < k-1, \\ (\lambda_{k-1} - \lambda_a)^{-1} T w_{a,k-1} T^* & \text{if } 1 \leq a < k-1. \end{cases}$$

To check these, take the projection of a linear curve $c(s)$ in $S(k-1, \mathbb{C})$ such that $\dot{c}(0)$ is one of the basis elements $v_{a,b}$, or $w_{a,b}$, and differentiate the projected curve with respect to s. It follows that $\{Tv_{a,k-1}T^*, Tw_{a,k-1}T^* : a = 1, \ldots, k-2\}$ form an orthonormal basis of $T_{P(\mu^J)} J(\Sigma_2^k)$. Expressing $\overline{\widetilde{X}} - \mu^J$ in the orthonormal basis of $S(k-1, \mathbb{C})$, and $d_{\mu^J} P(\overline{\widetilde{X}} - \mu^J)$ with respect to the above basis of $T_{P(\mu^J)} J(\Sigma_2^k)$, one may now apply Theorem 12.2 and Corollary 12.1.

For a two-sample test for $H_0 : Q_1 = Q_2$, one may use (12.11).

12.3 Data Examples

In this section we apply the theory to a number of data sets available in the literature.

Example 12.4 (Paleomagnetism). The first statistical confirmation of the shifting of the earth's magnetic poles over geological times, theorized by paleontologists based on observed fossilised magnetic rock samples, came in a seminal paper by R.A. Fisher (1953). Fisher analyzed two sets of data—one recent (1947–1948) and another old (Quaternary period), using the so-called *von Mises–Fisher model*

$$f(x; \mu, \tau) = c(\tau) \exp\{\tau x.\mu\} \ (x \in S^2), \tag{12.18}$$

Here $\mu(\in S^2)$, is the extrinsic *mean direction,* under the inclusion map J (Exercise 12.1) $(\mu = \mu_E)$, and $\tau > 0$ is the concentration parameter. The maximum likelihood estimate of μ is $\hat{\mu} = \overline{X}/|\overline{X}|$, which is also our sample extrinsic mean (Exercise 12.1). The value of the MLE for the first data set of $n = 9$ observations turned out to be $\hat{\mu} - \hat{\mu}_E = (0.2984, 0.1346, 0.9449)$, where $(0, 0, 1)$ is the geographic north pole. Fisher's 95 % confidence region for μ is $\{\mu \in S^2 : d_g(\hat{\mu}, \mu) \leq 0.1536)\}$, where d_g is the geodesic distance on S^2. The nonparametric confidence region based on $\hat{\mu}_E$, is given by (12.9) and is about 10 % smaller in area than Fisher's region (See Bhattacharya and Bhattacharya 2012, Chap. 2).

The second data set based on $n = 29$ observations from the Quaternary period that Fisher analyzed, using the same parametric model as above, had the MLE $\hat{\mu}_E = \overline{X}/|\overline{X}| = (0.0172, -0.2978, -0.9545)$, almost antipodal of that for the first data set, and with a confidence region of geodesic radius 0.1475 around the MLE. Note that the two confidence regions are not only disjoint, they also lie far away from each other. This provided the first statistical confirmation of the hypothesis of shifts in the earth's magnetic poles, a result hailed by paleontologists (See Irving 1964). Because of the difficulty in accessing the second data set, the nonparametric procedures could not be applied to it. But the analysis of another data set dating from the Jurassic period, with $n = 33$, once again yielded a nonparametric extrinsic confidence region, and about 10 % smaller than the region obtained by Fisher's parametric method (See Bhattacharya and Bhattacharya, Chap. 5, for details).

Example 12.5 (Brain Scan of Schizophrenic and Normal Patients). We consider an example from Bookstein (1991) in which 13 landmarks were recorded on a midsagittal two-dimensional slice from magnetic brain scans of each of 14 schizophrenic patients and 14 normal patients. The object is to detect the deformation, if any, in the shape of the k-ad due to the disease, and to use it for diagnostic purposes. The shape space is Σ_2^{13}. The extrinsic test based on (12.11) has an observed value 95.5476 of the chi-square statistic and a p-value 3.8×10^{-11}. The calculations made use of the analytical computations carried out in Example 12.3. For details of these calculations and others we refer to Bhattacharya and Bhattacharya (2012). This may also be contrasted with the results of parametric inference in the literature for the same data, as may be found in (Dryden and Mardia, 1998, pp. 146, 162–165). Using a isotropic Normal model for the original landmarks data, and after removal of "nuisance" parameters for translation, size and rotation, an F-test known as Goodall's F-test (See Goodall 1991) gives a p-value 0.01. A Monte Carlo test based permutation test obtained by 999 random assignments of the data into two groups and computing Goodall's F-statistic, gave a p-value 0.04. A Hotellings's T^2 test

in the tangent space of the pooled sample mean had a p-value 0.834. A likelihood
ratio test based on the isotropic offset Normal distribution on the shape space
has the value 43.124 of the chi-square statistic with 22 degrees of freedom, and a
p-value 0.005.

Example 12.6 (Corpus Callosum Shapes of Normal and ADHD Children). We consider a planar shape data set, which gives measurements on a group of typically developing children and a group of children suffering the ADHD (Attention deficit hyperactivity disorder). ADHD is one of the most common psychi-
atric disorders for children that can continue through adolescence and adult-
hood. Symptoms include difficulty staying focused and paying attention, difficulty controlling behavior, and hyperactivity (over-activity). ADHD in general has
three subtypes: (1) ADHD hyperactive-impulsive, (2) ADHD-inattentive; (3) Combined hyperactive-impulsive and inattentive (ADHD-combined) (Ramsay 2007).
ADHD-200 Dataset (`http://fcon_1000.projects.nitrc.org/indi/adhd200/`)
is a data set that records both anatomical and resting-state functional MRI data
of 776 labeled subjects across 8 independent imaging sites, 491 of which were obtained from typically developing individuals and 285 in children and adolescents
with ADHD (ages: 7–21 years old). The data was further processed by UNC BIAS
lab (see Huang et al. 2015) to extract the planar Corpus Callosum shape data,
which contains 50 landmarks on the contour of the Corpus Callosum of each subject (see Fig. 12.1 for a plot of the raw landmarks of a normal developing child and
a ADHD child). After quality control, 647 CC shape data out of 776 subjects were
obtained, which included 404 (n_1) typically developing children, 150 (n_2) diagnosed
with ADHD-Combined, 8 (n_3) diagnosed with ADHD-Hyperactive-Impulsive, and
85 (n_4) diagnosed with ADHD-Inattentive. Therefore, the data lie in the space

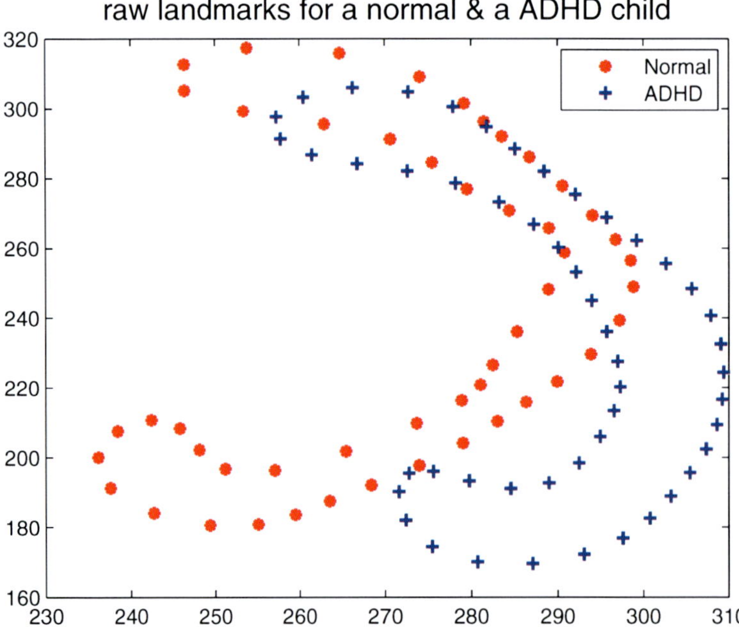

Fig. 12.1 Raw landmarks from the contour of the Corpus Callosum of a typically developing
child and an ADHD child

Σ_2^{50}, which has a high dimension of $2 \times 50 - 4 = 96$. We carry out *extrinsic two sample testings* based on (12.11) between the group of typically developing children and the group of children diagnosed with ADHD-Combined, and also between the group of typically developing children and ADHD-Inattentive children. We construct testing statistics that base on the asymptotic distribution of the extrinsic mean for the planar shapes.

The p-value of the two-sample test between the group of typically developing children and the group of children diagnosed with ADHD-Combined is 5.1988×10^{-11}, which is based on the asymptotic chi-squared distribution given in (12.11). The p-value of the test between the group of typically developing children and the group ADHD-Inattentive children is smaller than 10^{-50}.

More details of the above two examples can be found in `https://stat.duke.edu/~ll162/research/`.

12.4 Notes and References

In addition to the early seminal work of R.A. Fisher (1953) and books by Watson (1983) and N. Fisher et al. (1987) on directional statistics on spheres S^d and axial spaces $\mathbb{R}P^{p^d}$ mentioned in this chapter, Dryden and Mardia (1998) give a comprehensive account of parametric inference on shape manifolds of D.G. Kendall (See Kendall et al. 1999 for a detailed account of these manifolds). Nonparametric inference based on Fréchet means on general manifolds was introduced in the Indiana University Ph.D. dissertation of Vic Patrangenaru (1998), and further developed in Bhattacharya and Patrangenaru (2003, 2005). This theory is much more general than, and was done independently of, the work of Hendriks and Landsman (1996, 1998) on Euclidean submanifolds such as spheres and hypersurfaces of an Euclidean space with the inclusion map as the embedding. The Fréchet mean on a non-Euclidean space depends on the distance chosen. The role of a proper choice of the distance in analyzing complex data was recently emphasized by Holmes (2015).

The data analysis on S^2 in Sect. 12.3 follows Bhattacharya and Bhattacharya (2012), and that on the planar shape space may be found in Bhattacharya and Lin (2016). Earlier, nonparametric tests for uniformity on compact Riemannian manifolds were developed by Beran (1968) and Giné (1975). For certain functional data analysis on manifolds, especially projective shape spaces, one may refer to Musse et al. (2008).

A very recent source on the subject matter of this chapter is the book by Ellingson and Patrangenaru (2015).

Exercises for Chap. 12

Ex. 12.1. (a) Prove that the extrinsic mean of the von Mises–Fisher distribution (12.18) is μ.

(b) Prove that the MLE of μ in (12.18) is the sample extrinsic mean $\overline{X}/|\overline{X}|$.

(c) Using the first data set of 9 observations in Fisher (1953) construct the asymptotic 95 % confidence region for $\mu = \mu_E$ using (12.9).

References

Beran, R. J. (1968). Testing for uniformity on a compact homogeneous space. *Journal of Applied Probability, 5*, 177–195.

Bhattacharya, A., & Bhattacharya, R. (2012). *Nonparametric inference on manifolds: With applications to shape spaces.* IMS Monograph (Vol. #2). Cambridge: Cambridge University Press.

Bhattacharya, R., & Lin, L. (2016). Omnibus CLTs for Fréchet means and nonparametric inference on non-Euclidean spaces. The Proceedings of the American Mathematical Society (in Press).

Bhattacharya, R., & Patrangenaru, V. (2003). Large sample theory of intrinsic and extrinsic sample means on manifolds. *Annals of Statistics, 31*, 1–29.

Bhattacharya, R., & Patrangenaru, V. (2005). Large sample theory of intrinsic and extrinsic sample means on manifolds-II. *Annals of Statistics, 33*, 1225–1259.

Bookstein, F. (1978). *The measurement of biological shape and shape change.* Lecture Notes in Biomathematics. New York: Springer.

Bookstein, F. (1991). *Morphometric tools for landmark data: Geometry and biology.* Cambridge: Cambridge University Press.

Dryden, I., & Mardia, K. V. (1998). *Statistical analysis of shape.* New York: Wiley.

Ellingson, L., & Patrangenaru, V. (2015). *Nonparametric statistics on manifolds and their applications to object data analysis.* New York: Chapman and Hall.

Fisher, R. A. (1953). Dispersion on a sphere. *Proceedings of the Royal Society of London, Series A, 217*, 295–305.

Fisher, N. I., Lewis, T., & Embleton, B. J. J. (1987). *Statistical analysis of spherical data.* Cambridge: Cambridge University Press.

Fréchet, M. (1948). Lés elements aléatoires de nature quelconque dans un espace distancié. *Annales de l'Institute Henri Poincare, 10*, 215–310.

Giné, E. (1975). Invariant tests for uniformity on compact Riemannian manifolds based on Sobolev norms. *Annals of Statistics, 3*, 1243–1266.

Goodall, C. (1991). Procrustes methods in the statistical analysis of shape. *Journal of the Royal Statistical Society, Series B, 53*, 285–339.

Hendriks, H., & Landsman, Z. (1996). Asymptotic tests for mean location on manifolds. *Comptes Rendus de l'Académie des Sciences. Série I, Mathématique, 322*, 773–778.

Hendriks, H., & Landsman, Z. (1998). Mean location and sample mean location on manifolds: Asymptotics, tests, confidence regions. *Journal of Multivariate Analysis, 67*, 227–243.

Holmes, S. (2015). Statistically relevant metrics for complex data. *Joint Mathematics Meetings, 2015*, San Antonio.

Huang, C., Styner, M., & Zhu, H. T. (2015). Penalized mixtures of offset-normal shape factor analyzers with application in clustering high-dimensional shape data. *Journal of the American Statistical Association* (to appear).

Irving, E. (1964). *Paleomagnetism and its application to geological and geographical problems.* New York: Wiley.

Kendall, D. G. (1977). The diffusion of shape. *Advances in Applied Probability, 9*, 428–430.

Kendall, D. G. (1984). Shape Manifolds, Procrustean metrics, and complex projective spaces. *Bulletin of the London Mathematical Society, 16*, 81–121.

Kendall, D. G., Barden, D., Carne, T. K., & Le, H. (1999). *Shape and shape theory.* New York: Wiley.

Mardia, K. V., & Jupp, P. E. (2000). *Directional statistics.* New York: Wiley.

Munk, A., Paige, R., Pang, J., Patrangenaru, V., & Ruymgaart, F. H. (2008). The one and multisample problem for functional data with applications to projective shape analysis. *Journal of Multivariate Analysis, 99,* 815–833.

Patrangenaru, V. (1998). Asymptotic statistics on manifolds and their applications (Ph.D. Thesis). Indiana University, Bloomington.

Ramsay, J. R. (2007). Current status of cognitive-behavioral therapy as a psychosocial treatment for adult attention-deficit/hyperactivity disorder. *Current Psychiatry Reports, 9*(5), 427–433.

Watson, G. S. (1983). *Statistics on spheres.* University Arkansas Lecture Notes in the Mathematical Sciences. New York: Wiley.

Ziezold, H. (1977). On expected figures and a strong law of large numbers for random elements in quasi-metric spaces. In *Transactions of the Seventh Pragure Conference on Information Theory, Statistical Functions, Random Processes and of the Eighth European Meeting of Statisticians* (Vol. A, pp. 591–602).

Chapter 13
Multiple Testing and the False Discovery Rate

Abstract Here is an introduction to the theory of the false discovery rates (FDR) developed by Benjamini and Hochberg (Journal of the Royal Statistical Society, Series B, 57, 289–300, 1995), Benjamini and Yekatieli (Annals of Statistics, 29(4), 1165–1188, 2001) and others, dealing with the problem of testing a large number of hypotheses often based on relatively small or moderate sample sizes.

13.1 Introduction

Statisticians are often confronted with the problem of testing simultaneously m null hypotheses $H_{01}, H_{02}, \ldots, H_{0m}$, $m > 1$, based on some data. If the goal is to determine if all these hypotheses are right, one may take the *global null hypothesis* to be $H_0 = \cap_{1 \leq i \leq m} H_{0i}$. This may be the case in a two-sample problem in which the only objective is to see if the two underlying distributions are the same. For a test with a level of significance α, the classical Bonferroni procedure is to reject H_0 if and only if at least one of the m p-values $p_1, \ldots, p_m$ is smaller than α/m. This, of course, is a very conservative test in protecting H_0, i.e., the actual level of significance of the test is probably far smaller than α. For a test of size α, the Bonferroni test has then a small power. A much improved procedure was suggested by Simes (1986), in which the p-values are ordered as $p_{(1)} \leq p_{(2)} \leq \cdots \leq p_{(m)}$ and H_0 is rejected if and only if $p_{(i)} \leq (i/m)\alpha$ for at least one i. He proved that this test is conservative (i.e., it has size $\leq \alpha$) if the test statistics are i.i.d. and their common distribution (function) is continuous, and he conjectured that this is true more generally. Sarkar (1998) proved the conjecture for a class of positively dependent joint, or multivariate, distributions of the test statistics, with common marginals. Also see Sen (1999a,b) for additional facts and some history on multiple testing. From results due to Benjamini and Yekutieli (2001) derived in Sect. 13.2, Simes' conjectured inequality follows for test statistics $\mathbf{T} = (T_1, T_2, \ldots, T_m)$, whose joint distribution has the property of *positive regression dependency on each element of a subset* of these tests, or *PRDS*, defined as follows. A set D of m-tuples $\mathbf{t} = (t_1, t_2, \ldots, t_m)$ of values of the test statistics is said to be *increasing*, provided $\mathbf{t} \in D$ and $\mathbf{t} \leq \mathbf{s}$ (i.e., $t_i \leq s_i$ for all i) implies $\mathbf{s} \in D$.

© Springer-Verlag New York 2016

R. Bhattacharya et al., *A Course in Mathematical Statistics and Large Sample Theory*, Springer Texts in Statistics, DOI 10.1007/978-1-4939-4032-5_13

> *PRDS holds for a given subset I_0 of $\{1,\ldots,m\}$*
>
> *if for every measurable increasing set D,*
>
> $Prob(\mathbf{T} \in D \mid T_i = t_i)$ *is increasing in* t_i *for every* $i \in I_0$.   (13.1)

This includes the case of independent test statistics T_i, $i = 1,\ldots,m$, as well as many other positively dependent ones.

Of greater interest in this chapter is the determination, with a limited statistical error rate, of those among the m null hypotheses which are false. For example, in microarray experiments a single observation records gene expression levels of thousands of genes for the purpose of locating those genes which may contribute to a certain disease. The experiments are generally expensive and the number of observations made is not large. Here m is of the order of several thousands, and the Bonferroni procedure is obviously inadequate for this purpose only pointing to those genes for which the p-values are less than α/m. In contrast, the procedure for independent test statistics due to Benjamini and Hochberg (1995) and, more general procedures due to Benjamini and Yekutieli (2001) discussed in the next section are more effective. For an example, one may look at the article of Reiner et al. (2003), Identifying differentially expressed genes using false discovery rate controlling procedures, Bioinformatics 19:368–375. The *Benjamini–Hochberg procedure* looks somewhat analogous to that of Simes, but it rejects all k null hypotheses with the smallest p-values, where $k = \max\{i : p_{(i)} \le (i/m)\alpha\}$ (See Theorem 13.1 in the next section), whereas the Benjamini–Yekutieli procedure, valid without any restriction on the nature of the joint distribution of the test statistics, rejects the k null hypotheses with the lowest p-values with $k = \max\{i : p_{(i)} \le (i/c_m m)\alpha\}$, where $c_m = \sum_{1 \le j \le m}(1/j)$ is, for large m, approximately $\log m$ (Theorem 13.2).

In Sect. 13.3, the theory is applied to a set of real data for $m = 75$ two-sample tests with 28 HIV+ patients and 18 controls.

13.2 False Discovery Rate

As mentioned above, in many problems involving multiple tests it is important to identify, within a small statistical error, those hypotheses H_{0i} which are false among the m null hypotheses that are presented ($i = 1, 2, \ldots, m$). As a measure of the effectiveness of such multiple testing the following notion was introduced by Benjamini–Hochberg. Denote by m_0 the number of true hypotheses among the m null hypotheses H_{0i} ($1 \le i \le m$). Let the true null hypotheses be labeled as H_{0i} ($1 \le i \le m_0$), unless this set is empty.

Definition 13.1. Suppose that of the m null hypotheses tested, V true null hypotheses and S false null hypotheses are rejected, with a proportion $Q = V/(V+S)$ of true hypotheses among all hypotheses rejected. The quantity $E(Q)$ is called *the false discovery rate*.

Remark 13.1. If $m_0 = m$, then $Q = 1$, provided $V > 0$. If $V = 0$, then $Q = 0$. Hence in the case all null hypotheses are true, the false discovery rate is the same as the type 1 error, i.e., it is the probability of rejection under the global null hypothesis $H_0 = \cap_{1 \le i \le m} H_{0i}$.

Recall the *Benjamini–Hochberg procedure:*

$$\textit{Reject only the } k \textit{ null hypotheses with the smallest p-values,}$$
$$\textit{where } k = \max\{i : p_{(i)} \leq \left(\tfrac{i}{m}\right)\alpha\}. \tag{13.2}$$

The proofs below are along the lines of those in Benjamini and Yekutieli (2001).

Theorem 13.1. *(a) (Benjamini–Hochberg) Assume that the m test statistics T_i, $i = 1, \ldots, m$, are independent. Then the false discovery rate $E(Q)$ for the procedure (13.2) is no more than $\alpha m_0/m$. (b) (Benjamini–Yekutieli) if the PDRS (13.1) holds with I_0 indexing the set of all true null hypotheses, then also the procedure (13.2) has a false discovery rate no more than $\alpha m_0/m$.*

Proof. We only give the proof of (a). For part (b), See Benjamini and Yekutieli (2001).

Step 1.   Consider the distribution of the p-values, $P_1, P_2, \ldots, P_m$, on the probability space $[0,1]^m$. Let $A(v,s)$ denote the event that v of the true null hypotheses and s false null hypotheses are rejected by the procedure (13.2), and $A(v,s;J)$ the subset of $A(v,s)$ with a specific set of v true null hypotheses indexed by $J \subset \{1, \ldots, m_0\}$. Writing $q_i = (i/m)\alpha$, note that on $A(v,s)$ the p-values of only the $v + s$ hypotheses rejected are less than, or equal to, q_{v+s}. Then on the set $A(v,s;J)$, and for $1 \leq i \leq m_0$, $\{P_i \leq q_{v+s}\}$ holds if and only if $i \in J$. That is, $\{P_i \leq q_{v+s}\} \cap A(v,s;J) = A(v,s;J)$ or $\emptyset$ according as $i \in J$ or $i \notin J$ $(1 \leq i \leq m_0)$. Hence

$$\sum_{1 \leq i \leq m_0} \mathrm{Prob}(\{P_i \leq q_{v+s}\} \cap A(v,s))$$

$$= \sum_{1 \leq i \leq m_0} \sum_{J} \mathrm{Prob}(\{P_i \leq q_{v+s}\} \cap A(v,s;J))$$

$$= \sum_{J} \sum_{i \in J} \mathrm{Prob}(A(v,s;J)) = \sum_{J} v\,\mathrm{Prob}(A(v,s;J)) = v\,\mathrm{Prob}(A(v,s)). \tag{13.3}$$

Writing $\tilde{m} = m - m_0$, it follows from (13.3) that

$$E(Q) = \sum_{0 \leq s \leq \tilde{m}} \sum_{1 \leq v \leq m_0} \frac{v}{v+s}\,\mathrm{Prob}(A(v,s))$$

$$= \sum_{0 \leq s \leq \tilde{m}} \sum_{1 \leq v \leq m_0} \sum_{1 \leq i \leq m_0} \mathrm{Prob}(\{P_i \leq q_{v+s}\} \cap A(v,s))/(v+s)$$

$$= \sum_{1 \leq i \leq m_0} \left[\sum_{0 \leq s \leq \tilde{m}} \sum_{1 \leq v \leq m_0} \sum (v+s)^{-1}\mathrm{Prob}(\{P_i \leq q_{v+s}\} \cap A(v,s)) \right]. \tag{13.4}$$

Step 2.   The event $\{P_i \leq q_{v+s}\} \cap A(v,s)$ may be expressed as $\{P_i \leq q_{v+s}\} \cap C^i(v,s)$, where $C^i(v,s)$ is the event that s false null hypotheses and $v - 1$ true null hypotheses H_{0j} with $j \in \{1, \ldots, m_0\}\backslash\{i\}$ are rejected. For each k, the m_0 sets $C(i|k) = \cup\{C^i(v,s)$: all v,s such that $v + s = k\}$ are disjoint. Now (13.4) may be expressed as

$$E(Q) = \sum_{1 \leq i \leq m_0} \left[\sum_{1 \leq k \leq m} k^{-1}\mathrm{Prob}(\{P_i \leq q_k\} \cap C(i|k)) \right]. \tag{13.5}$$

Step 3. By the hypothesis of independence of T_i $(i = 1, \ldots, m)$, the probability within square brackets on the right side of (13.5) equals the product
$\text{Prob}(P_i \le q_k)\text{Prob}(C(i|k))$. For $i = 1, \ldots, m_0$, $\text{Prob}(P_i \le q_k) \le q_k = (k/m)\alpha$
(Exercise 13.1)). Therefore, the right side of (13.5) is no more than (α/m) times
$\sum_{1 \le i \le m_0} [\sum_{1 \le k \le m} \text{Prob}(C(i|k))]$. By disjointness of the m events $C(i|k)$, for
each i, the inner sum (of the double sum) equals the probability of the union
of the sets over k and is, therefore, no more than one. Hence the double sum is
no more than m_0, and we arrive at the desired result. $\qquad \square$

Observe that the proof above does not make use of the hypothesis of independence of the T_i's for Steps 1 and 2. In particular, (13.5) holds generally,
without any dependency restriction on the distribution of $\mathbf{T} = (T_1, T_2, \ldots, T_m)$.
To bound the right side properly, split the event $\{P_i \le q_k\}$ as the union
$\cup\{P_i \in (\alpha(j - 1)/m, \alpha j/m] : j = 1, \ldots, k\}$, and note that, if the T_i have continuous distribution functions, then

$$\sum_{1 \le k \le m} \text{Prob}(\{P_i \in (\alpha(j - 1)/m, \alpha j/m]\} \cap C(i, k))$$

$$= \text{Prob}(\{P_i \in (\alpha(j - 1)/m, \alpha j/m]\} \cap \{\cup C(i, k) : k = 1, \ldots, m\})$$

$$\le \text{Prob}(\{P_i \in (\alpha(j - 1)/m, \alpha j/m]\}) = \alpha/m. \tag{13.6}$$

The equality in (13.6) is a consequence of the disjointness of the sets $C(i, k)$,
$k = 1, \ldots, m$. The inequality follows from the fact that for each $i = 1, \ldots, m_0$,
$\text{Prob}(\{P_i \in (\alpha(j - 1)/m, \alpha j/m]\}) = \alpha/m$. The relations (13.5), (13.6) then lead
to

$$E(Q) = \sum_{1 \le i \le m_0} \left[\sum_{1 \le k \le m} k^{-1}\text{Prob}(\{P_i \le q_k\} \cap C(i, k)) \right]$$

$$= \sum_{1 \le i \le m_0} \sum_{1 \le k \le m} \sum_{1 \le j \le k} k^{-1}\text{Prob}(\{P_i \in (\alpha(j - 1)/m, \alpha j/m\} \cap C(i, k))$$

$$\le \sum_{1 \le i \le m_0} \sum_{1 \le k \le m} \sum_{1 \le j \le k} j^{-1}\text{Prob}(\{P_i \in (\alpha(j - 1)/m, \alpha j/m\} \cap C(i, k))$$

$$= \sum_{1 \le i \le m_0} \sum_{1 \le j \le m} j^{-1} \sum_{j \le k \le m} \text{Prob}(\{P_i \in (\alpha(j - 1)/m, \alpha j/m\} \cap C(i, k))$$

$$\le \sum_{1 \le i \le m_0} \sum_{1 \le j \le m} j^{-1}\alpha/m = m_0 c_m \alpha/m. \tag{13.7}$$

Since this holds for arbitrary α, replacing α by α/c_m, one arrives at the following
result, under an additional assumption of continuity of the test statistics.

Theorem 13.2 (Benjamini–Yekutieli). *Consider the procedure: Reject all null
hypotheses H_{0i} with the smallest k p-values where $k = \max\{i : p_{(i)} \le i\alpha/(c_m m)\}$,
and $c_m = \sum_{1 \le j \le m}(1/j)$. The false discovery rate of this procedure is no more than
$(m_0/m)\alpha \le \alpha$.*

13.3 An Application to a Diffusion Tensor Imaging Data Set

In this section, we consider an application of multiple testing to a diffusion tensor imaging (DTI) data set from a HIV study. The DTI data set consist of 46 subjects with 28 HIV+ subjects and 18 healthy controls. Diffusion tensors were extracted along the fiber tract of the splenium of the corpus callosum of each subject. The DTI for all the subjects are registered in the same atlas space based on arc lengths, with tensors extracted from 75 locations along the fiber tract of each subject. This data set have been studied in a regression setting in Yuan et al. (2012). We instead consider the problem of multiple testing with $m = 75$ null hypotheses in testing whether there is a difference between control group and the HIV+ group at each of the 75 locations where the DTI are extracted. We carry out the tests using the two procedures introduced in the previous section.

The DTI data are represented by the diffusion matrices which are 3 by 3 positive definite matrices. At each location, a nonparametric testing statistics is constructed. Let $X_{i1}, \ldots, X_{in_1}$ be the sample of DTI data from control group at location i ($i = 1, \ldots, 75$) and $Y_{i1}, \ldots, Y_{in_2}$ be an i.i.d. sample from the HIV positive group at location i, with $\overline{X}_i$ and $\overline{Y}_i$ their corresponding sample means. That is, $\overline{X}_i$ and $\overline{Y}_i$ are the sample mean vectors of dimension 6 for the 6 distinct values of the vectorized data. Let Σ_{X_i} and Σ_{Y_i} be the corresponding sample covariance matrices at location i. Then, for testing the two-sample hypothesis H_{0i} we use the test statistic $(\overline{X}_i - \overline{Y}_i)\Sigma^{-1}(\overline{X}_i - \overline{Y}_i)^T$ with $\Sigma = (1/n_1\Sigma_{X_i} + 1/n_2\Sigma_{Y_i})$, which has the asymptotic chi-square distribution $\chi^2(6)$.

Set the significant level $\alpha = 0.05$. We first carry out the test using the *Benjamini-Hochberg procedure*. That is, reject only the k null hypothesis with the smallest p-values, where $k = \max\{i : p_{(i)} \leq \frac{1}{m}\alpha\}$. We first order the 75 p-values corresponding to the tests carried out at all the locations (see Fig. 13.1 for a plot of the p-values as a function of the arc length (location) before the ordering). The ordered p-values are compared with the vector $\{0.05/75, 0.1/75, \ldots, 0.05\}$, which gives the result $k = 58$. Therefore we reject the 58 null hypotheses corresponding to the first 58 ordered p-values.

We now carry out the tests using the *Benjamini–Yekutieli procedure*. The ordered vector of p-values are compared with the vector $\{i\alpha/(c_m m), i = 1, \ldots, m\}$, where $\alpha = 0.05$, $m = 75$, and $c_m = \sum_{1 \leq j \leq m} 1/j = 4.0021$. With $k = \max\{i : p_{(i)} \leq i\alpha/(c_m m)\}$. The procedure yields $k = 50$. Thus we reject the 50 null hypotheses corresponding to the first 50 ordered p-values.

13.4 Notes and References

The presentation in this chapter depends much on the article by Benjamini and Yekutieli (2001). For a history of multiple testing we refer to this article, and to Sarkar (1998), and Sen (1999a,b). In Sen (1999a,b) one may also find the history of earlier procedures suited especially in clinical trials and their connections with the classic work of Scheffe (1959) on Normal models and the early work of S.N. Roy detailed in Roy et al. (1971). The data example in Sect. 13.3 is taken from Bhattacharya and Lin (2016).

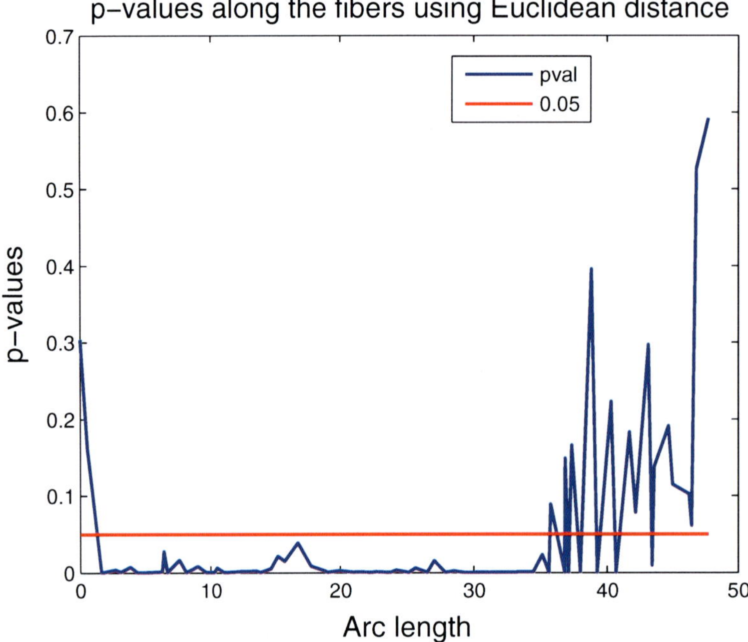

Fig. 13.1 *p*-values along the fiber track

Exercises for Chap. 13

Ex. 13.1. Let P denote the p-value of a test of a null hypothesis H^0 of size no more than α.

(a) Show that $\mathrm{Prob}(P \leq \alpha | H^0) \leq \alpha$, with equality if the test is of exact size α, i.e., $\mathrm{Prob}(\mathrm{Reject}\ H^0 \mid H^0) = \alpha$.
(b) Prove that, if the test statistic T which rejects H^0 if $T > c$ (for c depending on α) has a continuous distribution under H^0, then the p-value has the uniform distribution on $[0, 1]$.

References

Benjamini, Y., & Hochberg, Y. (1995). Controlling the false discovery rate: A practical and powerful approach to multiple testing. *Journal of the Royal Statistical Society, Series B, 57*, 289–300.

Benjamini, Y., & Yekutieli, D. (2001). The control of the false discovery rate in multiple testing under dependency. *Annals of Statistics, 29*(4), 1165–1188.

Bhattacharya, R., & Lin, L. (2016). Omnibus CLTs for Fréchet means and nonparametric inference on non-Euclidean spaces. *Proceedings of the American Mathematicasl Society* (in Press).

Reiner, A., Yekutieli, D., & Benjamini, Y. (2003). Identifying differentially expressed genes using false discovery rate controlling procedures. *Bioinformatics, 19*(3), 368–375.

Roy, S. N., Gnanadesikan, R., & Srivastava, J. N. (1971). *Analysis and design of certain quantitative multiresponse experiments*. New York: Pergamon Press.

Sarkar, S. K. (1998). Some probability inequalities for ordered MTP2 random variables: A proof of Simes conjecture. *Annals of Statistics, 26*, 494–504.

Scheffe, H. (1959). *The analysis of variance*. New York: Wiley.

Sen, P. K. (1999a). Some remarks on Simes-type multiple tests of significance. *Journal of Statistical Planning and Inference, 82*, 139–145.

Sen, P. K. (1999b). Multiple comparisons in interim analysis. *Journal of Statistical Planning and Inference, 82*, 5–23.

Simes, R. J. (1986). An improved Bonferroni procedure for multiple tests of significance. *Biometrika, 73*, 751–754.

Yuan, Y., Zhu, H., Lin, W., & Marron, J. S. (2012). Local polynomial regression for symmetric positive definite matrices. *Journal of the Royal Statistical Society: Series B, 74*, 697–719.

Chapter 14
Markov Chain Monte Carlo (MCMC) Simulation and Bayes Theory

Abstract Markov Chain Monte Carlo is an innovative and widely used computational methodology for an accurate estimation of a distribution, whose direct numerical evaluation is intractable. The main idea is to construct an ergodic Markov chain which is simple to simulate and has the target distribution as its invariant probability. This technique has been indispensable in the estimation of posterior distribution in Bayesian inference.

14.1 Metropolis–Hastings Algorithm

The topic of this chapter is the use of Markov chain theory in the computation of posterior distributions in Bayes theory. First consider the problem of computing the expectation $Eg(Z)$ when Z has a pdf f with respect to a sigma-finite measure μ on a state space $(S, \mathscr{S})$, and g is a measurable function with a finite expectation,

$$Eg(Z) = \int g(y)f(y)\mu(dy). \tag{14.1}$$

Suppose f is not explicitly known, but the *ratio* $f(x)/f(y)$ is tractable. This occurs in mathematical physics where Z has the Gibbs distributions and f is given as $f(y) = r\exp\{h(x)\}$, with h known explicitly (for example, as the Hamiltonian of a system of a large number of particles), but the normalizing constant r, the so-called *partition function*, is not known analytically in computable form, being simply given as the reciprocal of the multiple integral of f over a very large number of coordinates. Here $f(x)/f(y) = \exp\{h(x)\}/\exp\{h(y)\}$ is known explicitly and it is computable. Another example occurs in Bayes theory, where $f(\theta) = \pi(\theta \mid \mathbf{X})$ is the *posterior density* of θ, i.e., the conditional density of θ, given the observation $\mathbf{X}$. Given a parametric family $g(\mathbf{x} \mid \theta)$, $\theta \in \boldsymbol{\Theta}\ (= S)$, of densities of $\mathbf{X}$ with respect to some sigma-finite measure v (e.g., $\mathbf{X} = (X_1, \ldots, X_n)$—a random sample from some distribution parametrized by θ), the Bayesian views $g(\mathbf{x} \mid \theta)$ as the conditional density of $\mathbf{X}$ (at $\mathbf{x}$), given the value of the random variable θ which has a density $\pi(\theta)$, say, with respect to Lebesgue measure μ on the p-dimensional parameter space $\boldsymbol{\Theta} \subset \mathbb{R}^p$. One calls $\pi d\mu$ the *prior distribution* of the random variable θ. In this formulation, the joint density of $(\theta, \mathbf{X})$ is given by $\pi(\theta)g(\mathbf{x} \mid \theta)$, and the posterior density of θ is $\pi(\theta \mid \mathbf{x}) := \pi(\theta)g(\mathbf{x} \mid \theta)/c(\mathbf{x})$, $c(\mathbf{x})$ being the marginal

© Springer-Verlag New York 2016

R. Bhattacharya et al., *A Course in Mathematical Statistics and Large Sample Theory*, Springer Texts in Statistics, DOI 10.1007/978-1-4939-4032-5_14

density of $\mathbf{X}$. That is, $c(\mathbf{x}) = \int \pi(\theta)g(\mathbf{x} \mid \theta)\mu(d\theta)$ which is usually very difficult to compute. On the other hand, $\pi(\theta_1 \mid \mathbf{x})/\pi(\theta_2 \mid \mathbf{x}) = \pi(\theta_1)g(\mathbf{x} \mid \theta_1)/\pi(\theta_2)g(\mathbf{x} \mid \theta_2)$ is explicitly given in an analytical and easily computable form.

For an introduction to Markov chains, also referred to as Markov processes in discrete time, see Chap. 8, Sect. 8.6. While this is enough for an understanding of the results in this chapter, for proofs and a more detailed account, especially suited for the present topic, see Robert and Casella (2004, Chap. 6), which is also a standard reference for MCMC as a whole. Our presentation is also influenced by an exposition given in Wasserman (2003, Chap. 24).

In general for the computation of (14.1), the *Metropolis–Hastings algorithm* constructs a Markov chain $\{X_j : j = 0, 1, 2, \ldots\}$ with a transition probability density function $p(x, y)$ which satisfies the following *detailed balance condition:*

$$f(x)p(x, y) = f(y)p(y, x) \quad \text{for all } x, y. \tag{14.2}$$

On integrating both sides with respect to x, one gets

$$\int f(x)p(x, y)\mu(dx) = f(y) \quad \text{for all } y, \tag{14.3}$$

which implies that f is an invariant density for the Markov process. By ensuring that the choice of $p(x, y)$ is such that this invariant probability is unique and the strong law of large numbers (ergodic theorem) for Markov processes holds, one gets

$$N^{-1} \sum_{1 \leq j \leq N} g(X_j) \longrightarrow \int g(y)f(y)\mu(dy) \quad \text{with probability one, as } N \to \infty. \tag{14.4}$$

Since the number N of realizations of successive states of the Markov chain can be taken to be as large as needed by the simulation procedure, the objective of computing the desired expectation is achieved. We now describe the method for constructing $p(x, y)$. For simplicity, take S to be an open subset of $\mathbb{R}^d$, and μ Lebesgue measure.

Metropolis–Hastings Algorithm

Step 1: *Choose a transition probability density $q(x, y)$ (with respect to μ) of a Markov process on S, i.e., (i) for each $x \in S$, $y \to q(x, y)$ is a probability density and (ii) for each $y \in S$, $x \to q(x, y)$ is measurable.*

Step 2: *Starting with some initial state $X_0 = x_0$, pick Y_0 with density $q(x_0, \cdot)$ and choose X_1 according to the rule*

$$X_1 = \begin{cases} Y_0 \text{ with probability } a(x_0, Y_0), \\ X_0 \text{ with probability } 1 - a(x_0, Y_0), \end{cases} \tag{14.5}$$

where the acceptance ratio $a(x, y)$ *is defined by*

$$a(x, y) = \min\left\{\frac{f(y)}{f(x)} \cdot \frac{q(y, x)}{q(x, y)}, 1\right\}. \tag{14.6}$$

Step 3: *In general, after X_n is chosen, pick $Y_n \overset{\mathscr{L}}{\sim} q(X_n, y)\mu(dy)$ and pick X_{n+1} according to (14.5), (14.6), with X_n in places of X_0 and Y_n in place of Y_0 ($n = 1, 2, \ldots$).*

Theorem 14.1. *Assume $q(x, y) > 0$ for all $x, y \in S$. Then (a) the transition probability density $p(x, y)$ of the Markov chain $\{X_j : j = 0, 1, \dots\}$ has f as its invariant probability density, and (b) the convergence (14.4) holds for every g such that $\int |g(y)| f(y) \mu(dy) < \infty$.*

Proof. (a). We only need to establish the detailed balance relation (14.2). For this fix $x, y \in S$ ($x \neq y$), and suppose that $f(x)q(x, y) < f(y)q(y, x)$, so that $a(x, y) = 1$ and $a(y, x) = f(x)q(x, y)/f(y)q(y, x)$. Therefore, the transition probability density $p(x, y)$ of the Markov chain $\{X_j : j = 0, 1, \dots\}$ is given by

$$p(x, y) = q(x, y)a(x, y) = q(x, y),$$
$$p(y, x) = q(y, x)a(y, x) = q(y, x)f(x)q(x, y)/(f(y)q(y, x))$$
$$= \frac{f(x)}{f(y)}\, q(x, y).$$

Therefore, $f(x)p(x, y) = f(y)p(y, x)$. The case $f(x)q(x, y) > f(y)q(y, x)$ is treated similarly. In the case of equality, (14.2) holds trivially.

(b) This is a standard result for positive Harris recurrent Markov chains, for which we refer to Robert and Casella (2004, Theorems 6.63, 7.4).

Remark 14.1. To facilitate numerical simulation it is important to choose the *proposal density* $q(x, y)$ well, as well as a good initial state $x_0 = X_0$. One choice is $q(x, y) = g(y)$ where $g(y)$ is a positive density on S. This is referred to as *independent Metropolis–Hastings algorithm*. One assumes here that it is simple to draw observations from the distribution $g(y)\mu(dy)$. This method leads to fast convergence to the limit (14.4) if there is a constant M such that $f(x) \leq Mg(x)$ for all x. Another choice is to have a *symmetric proposal density* $q(x, y) = q(y, x)$, so that it is simple to draw from $q(x, y)\mu(dy)$. In this case the acceptance ratio takes the simpler form $a(x, y) = \min\left\{\frac{f(y)}{f(x)}, 1\right\}$. For example, if $S = \mathbb{R}^d$ one may choose $q(x, y) = \varphi(x - y : c)$ where $\varphi(z : c)$ is the d-dimensional Normal density with mean zero and dispersion matrix cI_d for some properly chosen $c > 0$. This is referred to as the *random walk Metropolis–Hastings algorithm*. If S is an open interval (or an open rectangle) one may make a transformation of it diffeomorphic to $\mathbb{R}^d$ and apply the algorithm. For these we refer to Robert and Casella (2004, Chap. 7).

14.2 Gibbs Sampler

The Metropolis–Hastings algorithm is difficult to apply directly to S of dimension $d > 1$, partly because of the problem with directly generating random vectors and partly because of the slow rate of convergence to stationarity of multidimensional chains. The *Gibbs sampler* alleviates these problems by using several one-dimensional problems to deal with a multidimensional problem. To illustrate this, consider the two-dimensional problem with $f(x, y)$ as the density of (X, Y) on $S \subset \mathbb{R}^2$. Let $f_{Y|X}(y \mid x)$ denote the conditional density of Y (at y), given $X = x$. Similarly define $f_{X|Y}(x \mid y)$.

Suppose it is possible to simulate from the one-dimensional, conditional distributions $f_{X|Y}(\cdot \mid y)$ and $f_{Y|X}(\cdot \mid x)$. The following algorithm for simulating (X_n, Y_n) converging to the distribution $f(x, y)$ is then used in the two-dimensional case, and it can be generalized in an obvious way to higher dimensions.

Two-Stage Gibbs Sampling Algorithm

Step 1.   *Begin with a suitable initial state (x_0, y_0) and generate X_1 with density*
  $f_{X|Y}(\cdot \mid y_0)$.
Step 2.   *Generate Y_1 with density $f_{Y|X}(\cdot \mid X_1)$.*
Step 3.   *Given (X_n, Y_n), generate X_{n+1} with density $f_{X|Y}(\cdot \mid Y_n)$, and generate Y_{n+1} with density $f_{Y|X}(\cdot \mid X_{n+1})$ $(n = 1, 2, \ldots)$.*

Theorem 14.2. *Assume $f(x, y) > 0 \ \forall\, (x, y)$ belonging to an open rectangle $S \subset \mathbb{R}^2$. Then (a) the Markov chain (X_n, Y_n), $n \geq 0$, has the invariant density $f(x, y)$ and (b) the two-dimensional analog of (14.4) holds.*

Proof. (a) The transition probability density of the Markov chain (X_n, Y_n), $n \geq 0$, is given by

$$q(x_1, y_1 \mid x_0, y_0) = f_{X|Y}(x_1 \mid y_0) f_{Y|X}(y_1 \mid x_1) = \frac{f(x_1, y_0)}{f_Y(y_0)} \cdot \frac{f(x_1, y_1)}{f_X(x_1)}$$

so that

$$\int_S q(x_1, y_1 \mid x_0, y_0) f(x_0, y_0) \mu(dx_0) \mu(dy_0)$$

$$= \int_S \frac{f(x_1, y_0) f(x_1, y_1) f(x_0, y_0)}{f_Y(y_0) f_X(x_1)} \mu(dx_0) \mu(dy_0)$$

$$= \int \left(\int f(x_0, y_0) \mu(dx_0) \right) \{ f(x_1, y_0) f(x_1, y_1) / f_Y(y_0) f_X(x_1) \} \, \mu(dy_0)$$

$$= \int \frac{f_Y(y_0) f(x_1, y_0) f(x_1, y_1)}{f_Y(y_0) f_X(x_1)} \mu(dy_0) = f(x_1, y_1),$$

establishing the claim.

(b) The proof of convergence again follows from Theorem 6.63 in Robert and Casella (2004). $\qquad\qquad\square$

An efficient way of implementing the Gibbs sampler is to use the following simulation procedure.

Metropolis–Hastings with Gibbs Sampling

Step 1.   *Choose symmetric proposal distributions q and $\tilde{q}$ for drawing from $f_{X|Y}$ and $f_{Y|X}$, respectively: $q(x, x') = q(x', x)$, $\tilde{q}(y, y') = \tilde{q}(y', y)$.*
Step 2.   *Choose some initial $X_0 = x_0$, $Y_0 = y_0$. Draw W_0 from $q(\cdot, X_0)$, and let*

$$X_1 = \begin{cases} W_0 \text{ with probability } a(X_0, W_0 \mid Y_0) \\ X_0 \text{ with probability } 1 - a(X_0, W_0 \mid Y_0), \end{cases}$$

and draw Z_0 from $\tilde{q}(\cdot, Y_0)$ and let

$$Y_1 = \begin{cases} Z_0 \text{ with probability } \tilde{a}(Y_0, Z_0 \mid X_1) \\ Y_0 \text{ with probability } 1 - \tilde{a}(Y_0, Z_0 \mid X_1), \end{cases}$$

where

$$a(x, x' \mid y) = \min\left\{\frac{f_{X\mid Y}(x' \mid y)}{f_{X\mid Y}(x \mid y)}, 1\right\} = \min\left\{\frac{f(x', y)}{f(x, y)}, 1\right\},$$

$$\tilde{a}(y, y' \mid x) = \min\left\{\frac{f_{Y\mid X}(y' \mid x)}{f_{Y\mid X}(y \mid x)}, 1\right\} = \min\left\{\frac{f(x, y')}{f(x, y)}, 1\right\}.$$

Step 3.   Having drawn (X_n, Y_n), draw (X_{n+1}, Y_{n+1}) following the procedure in Step 2, but with X_0 replaced by X_n and Y_0 by Y_n, and with W_0 replaced by W_n and Z_0 by Z_n. Here W_n is drawn afresh from $q(\cdot, X_n)$ and Z_n from $\tilde{q}(\cdot \mid Y_n)$ $(n \geq 1)$.

The above procedure extends to higher dimensional problems. The following section is devoted to an application.

14.3 Bayes Estimation in the Challenger Disaster Problem: A Project for Students

Consider the space shuttle disaster problem described in Chap. 4, pp. 58–60. Assume the same logistic regression model.

Suggested Model Let Y denote the failure status (response variable), and X the temperature in degrees F at launch time (explanatory variable). Use the *logistic regression model*,

$$P(Y = 1 \mid X = x) = \frac{\exp\{\alpha + \beta x\}}{[1 + \exp\{\alpha + \beta x\}]} = p(x), \text{ say, and}$$

$$P(Y = 0 \mid X = x) = 1 - p(x).$$

Note that one may express the model as

$$\log\left[\frac{p(x)}{(1 - p(x))}\right] = \alpha + \beta x.$$

Hence the name logistic regression.

Assume that the regressor x is *stochastic* and (X_i, Y_i) are i.i.d. random vectors.

For 23 independent Y observations $(y_1, \ldots, y_{23})$ the conditional likelihood function (i.e., the conditional p.d.f. of Y_i, given $X_i = x_i$ $(i = 1, \ldots, 23)$), is

$$\ell(\mathbf{y} \mid \mathbf{x}; \alpha, \beta) = \prod_{i=1,\ldots,23} \left[p(x_i)^{y_i}(1 - p(x_i))^{1-y_i}\right],$$

and the (conditional) log likelihood is

$$\log \ell = \sum_i [y_i(\alpha + \beta x_i)] - \sum_i \log[1 + \exp\{\alpha + \beta x_i\}].$$

Assume that the distribution of X_i does not involve α, β.

(a) Use a suitable prior distribution of (α, β) to compute the posterior distribution of (α, β) on a sufficiently fine grid of points. Do this numerically, by MCMC. For example, you may assume α and β to be independent random variables with α having the Normal distribution $N(\alpha_0, \sigma_0^2)$ and $-\beta$ having a log-normal distribution, i.e., $\xi \equiv \log(-\beta)$ has the distribution $N(\xi_0, \eta_0^2)$. Observe that in the present context $\beta < 0$. For this prior, you may choose the parameters $\alpha_0 = 10$, $\sigma_0^2 = 20$, $\xi_0 = 1$, $\eta_0^2 = 1$, or use the MLEs from Chap. 4, pp. 65, 66.

   [Hint: First compute the posterior distribution of (α, ξ) and then obtain that of (α, β) by simply noting that $\xi = \log(-\beta)$. Denoting by $\pi_1(\alpha)\pi_2(\xi)$ the prior density of (α, ξ), its posterior density is

$$\pi(\alpha, \xi \mid \mathbf{x}, \mathbf{y}) = \frac{\pi_1(\alpha)\pi_2(\xi)/\ell(\mathbf{y} \mid \mathbf{x}; \alpha, \beta)}{\int \ell(\mathbf{y} \mid \mathbf{x}; \alpha, \beta)\pi_1(\alpha)\pi_2(\xi)d\alpha d\xi} = \frac{\ell(\mathbf{y} \mid \mathbf{x}; \alpha, \beta)\pi_1(\alpha)\pi_2(\xi)}{h(\mathbf{x}, \mathbf{y})}$$

[Recall $\ell(\mathbf{y} \mid \mathbf{x}; \alpha, \beta) = \prod_{i=1,\dots,23}[p(x_i)^{y_i}(1 - p(x_i))^{1-y_i}$; $p(x) = \exp\{\alpha + \beta x\}$, $\beta = -e^\xi$.]

   Hence $\pi(\alpha_1, \xi_1 \mid \mathbf{x}, \mathbf{y})/\pi(\alpha_2, \xi_2 \mid \mathbf{x}, \mathbf{y}) = \pi_1(\alpha_1)\pi_2(\xi_1)\ell(\mathbf{y} \mid \mathbf{x}; \alpha_1, \beta_1)/\ell(\mathbf{y} \mid \mathbf{x}; \alpha_2, \beta_2)$ is computable for all $(\alpha_i \xi_i)$ $(i = 1, 2)$.

   Now the conditional (posterior) densities are denoted $f_{\alpha|\xi}$ and $f_{\xi|\alpha}$, and one has

$$\frac{f_{\alpha|\xi}(\alpha_1 \mid \xi)}{f_{\alpha|\xi}(\alpha_2 \mid \xi)} = \frac{\ell(\mathbf{y} \mid \mathbf{x}; \alpha_1, \beta)\pi_1(\alpha_1)}{\ell(\mathbf{y} \mid \mathbf{x}; \alpha_2, \beta)\pi_1(\alpha_2)},$$
$$\frac{f_{\xi|\alpha}(\xi_1 \mid \alpha)}{f_{\xi|\alpha}(\xi_2 \mid \alpha)} = \frac{\ell(\mathbf{y} \mid \mathbf{x}; \alpha, \beta_1)\pi_2(\xi_1)}{\ell(\mathbf{y} \mid \mathbf{x}; \alpha, \beta_2)\pi_2(\xi_2)}. \tag{14.7}$$

Use the *random walk Metropolis–Hastings algorithm* with the proposal density for α, namely $q(\alpha_1, \alpha_2)$, as the Normal density (for α_1) with mean α_2, and some variance, say $\sigma_0^2 = 20$. Similarly, let the proposal density for $\xi = \log(-\beta)$, $q^\sim(\xi_1, \xi_2)$ as the Normal density (for ξ_1) with mean ξ_2 and variance $\eta_0^2 = 1$ (You may, of course, choose different variances for these if you like). Now follow the steps of the *Metropolis–Hastings with Gibbs Sampling* steps.]

(b) (i) Use the posterior distribution of (α, β) in (a) to find the (histograms of the) posterior distributions of the failure probabilities $p(x)$ at launch time temperatures $x = 31\,^\circ\mathrm{F}$, and $x = 65\,^\circ\mathrm{F}$. (ii) Locate the tenth percentile $\delta_{0.10}$, say, of the histogram for the posterior distribution in (i) for the failure probability at $x = 31\,^\circ\mathrm{F}$ (i.e., 10 % of the points of the histogram lie below $\delta_{0.10}$ while 90 % lie above it). Observe that, in the Bayesian paradigm, the statistician has a 90 % faith (or confidence) that the probability of failure at $31\,^\circ\mathrm{F}$ is at least $\delta_{0.10}$.

A Project for Students

Project: Space Shuttle Disaster In 1986, the space shuttle Challenger exploded during take off, killing the seven astronauts aboard. It was determined that the explosion was the result of an O-ring failure, a splitting of a ring of rubber that seals different parts of the ship together. The flight accident was believed to be caused by the unusually cold weather (31 °F) at the time of the launch.

The past O-ring failure data along with temperature at launch time are given below (in increasing order of temperature) for 23 prior flights. The flight numbers denote the (unimportant) time order of launch. The numbers 0 and 1 indicate "no O-ring failure" and "O-ring failure", respectively.

Project Objective Estimate the probability of O-ring failure at temperature $31\,°F$ and at $65\,°F$.

Flight#	14	9	23	10	1	5	13	15	4	3	8	17	2
Failure	1	1	1	1	0	0	0	0	0	0	0	0	1
Temp. in Degrees F	53	57	58	63	66	67	67	67	68	69	70	70	70

Flight#	11	6	7	16	21	19	22	12	20	18
Failure	1	0	0	0	1	0	0	0	0	0
Temp. in Degrees F	70	72	73	75	75	76	76	78	79	81

14.4 Notes and References

Original sources for the material of this chapter may be traced to Metropolis et al. (1953), Hastings (1970), Geman and Geman (1984) and Gelfand and Smith (1990). A standard reference for the subject is Robert and Casella (2004), where one also finds a detailed history of the development of the subject and extensive references. Our presentation is much influenced by Wasserman (2003, Chap. 24), which gives a very readable introduction. For the Challenger disaster problem in Sect. 14.3 we refer to Robert and Casella (2004, pp. 15–19, 281, 282).

For a modern and exciting perspective of the Metropolis–Hastings algorithm, including a dramatic application to cryptography, we refer to Diaconis (2009).

Exercises for Chap. 14

Ex. 14.1. Do the Project.

References

Diaconis, P. (2009). The Markov Chain Monte Carlo revolution. *Bulletin of the American Mathematical Society, 46*(2), 179–20.

Gelfand, A. E., & Smith, A. F. M. (1990). Sampling-based approaches to calculating marginal densities. *Journal of the American Statistical Association, 85*(410), 398–409.

Geman, S., & Geman, D. (1984). Stochastic relaxation, Gibbs distributions, and the Bayesian restoration of images. *IEEE Transactions on Pattern Analysis and Machine Intelligence, 6*(6), 721–741.

Hastings, W. K. (1970). Monte Carlo sampling methods using Markov chains and their applications. *Biometrika, 57*(1), 97–109.

Metropolis, N., Rosenbluth, A. W., Rosenbluth, M. N., Teller, A. H., & Teller, E. (1953). Equations of state calculations by fast computing machines. *Journal of Chemical Physics, 21*(6), 1087–1092.

Robert, C., & Casella, G. (2004). *Monte Carlo statistical methods.* New York: Springer.

Wasserman, L. (2003). *All of statistics: A concise course in statistical inference.* New York: Springer.

Chapter 15
Miscellaneous Topics

Abstract This chapter provides brief introductions to a number of important topics which have not been touched upon in the rest of the book, namely, (1) classification of an observation in one of several groups, (2) principal components analysis which splits the data into orthogonal linear components in the order of the magnitudes of their variances, and (3) sequential analysis in which observations are taken one-by-one until the accumulated evidence becomes decisive.

15.1 Classification/Machine Learning

The problem of classification is that of assigning an observation $X \in \mathscr{X}$ (observation space) to one of a finite number of distributions indexed as $Y \in \mathscr{Y}$ (finite set of *classes*) from which it came. In the computer science literature X is referred to as the *input* and Y the *output* and an algorithm $h : \mathscr{X} \to \mathscr{Y}$ that assigns an observation to a class is sometimes described as *machine learning*. One may think of diagnosing a patient based on her symptoms X [e.g., X may be a vector of measurements of her systolic pressure, cholesterol level (LDL), heart rate, etc.] as having one, say Y, of a set $\mathscr{Y}$ of possible ailments. Assume first that the distributions indexed by $\mathscr{Y} = \{1, \ldots, K\}$ have known densities f_i, $(i \in \mathscr{Y})$, and that in the general population the proportion of the ith class is π_i. An optimal procedure which minimizes the probability of misclassification is the following h^* called the *Bayes classifier*

$$h^*(x) = \operatorname*{argmax}_j \pi_j f_j(x). \tag{15.1}$$

If the set on the right has more than one element, choose any element from it. To prove optimality of h^*, note that the probability of misclassification for any classifier h is

$$\sum_{j=1}^{K} \pi_j \int_{\mathscr{H}} \mathbf{1}_{\{h(x) \neq j\}} f_j(x) dx = 1 - \sum_{j=1}^{K} \int_{\mathscr{H}} \mathbf{1}_{\{h(x)=j\}} \pi_j f_j(x) dx$$

$$= 1 - \sum_{i=1}^{K} \sum_{j=1}^{K} \int_{\mathscr{H}} \mathbf{1}_{\{h(x)=j, h^*(x)=i\}} \pi_j f_j(x) dx$$

© Springer-Verlag New York 2016
R. Bhattacharya et al., *A Course in Mathematical Statistics and Large Sample Theory*, Springer Texts in Statistics, DOI 10.1007/978-1-4939-4032-5_15

$$\leq 1 - \sum_{i=1}^{K}\sum_{j=1}^{K} \int_{\mathscr{H}} \mathbf{1}_{\{h(x)=i\}}\pi_i f_i(x)dx$$

$$= 1 - \sum_{i=1}^{K} \int_{\mathscr{H}} \mathbf{1}_{\{h^*(x)=i\}}\pi_i f_i(x)dx$$

$$= \sum_{i=1}^{K}\pi_i \int_{\mathscr{H}} \mathbf{1}_{\{h^*(x)\neq i\}}\pi_i f_i(x)dx$$

$$= \text{Probability of misclassification for } h^*.$$

Usually one does not know π_j, f_j $(1 \leq j \leq K)$ and estimates them from a random sample, called *training data* in the computer science literature.

As an example, suppose that X is d-dimensional Normal $N(\mu_j, \Sigma_j)$, $1 \leq j \leq K$. Then a straightforward calculation shows (Exercise 15.1)

$$h^*(x) = \operatorname*{argmax}_{j}\left\{ -\frac{1}{2}\log|\Sigma_j| - \frac{1}{2}(x-\mu_j)'\Sigma_j^{-1}(x-\mu_j) + \log\pi_j \right\}, \qquad (15.2)$$

where $|\Sigma_j|$ is the determinant of Σ_j. This *quadratic classifier* is known as *Fisher's quadratic discriminant* originally due to R.A. Fisher. The distance $d(x,\mu) = [(x - \mu)'\Sigma^{-1}(x-\mu)]^{\frac{1}{2}}$ is known as the *Mahalanobis distance*. In the case the dispersion matrices Σ_j, are all the same, namely Σ, it is simple to check that

$$h^*(x) = \operatorname*{argmax}_{j}\left\{ x'\Sigma_j^{-1}\mu_j - \tfrac{1}{2}\mu_j'\Sigma_j^{-1}\mu_j + \log\pi_j \right\}. \qquad (15.3)$$

The function within curly brackets is linear (in x) and is called the *linear discriminant* and the Bayes procedure is referred to as the *linear discriminant analysis, or LDA*.

A more widely used classifier is based on the regression function $\ell(x) = E(Y \mid X = x)$. Although $\ell(x)$ generally does not lie in the finite set $\mathscr{Y}$, it is still used for its simplicity. Indeed, one often assumes $\ell(x)$ to be linear in $x : Y = \beta_0 + \sum_{j=1}^{p}\beta_j X_j + \varepsilon$ with the usual assumptions (See Sect. 6.8, Theorem 6.4, Corollary 6.5). We will consider the case $\mathscr{Y} = \{0,1\}$, i.e., $K = 2$ and write $\ell(x) = P(Y = 1 \mid X = x)$, $x = (x_1, \ldots, x_p)'$. Given a training sample $(Y_i, X_{i1}, \ldots, X_{ip})'$, $1 \leq i \leq n$, the least squares estimate of $\boldsymbol{\beta} = (\beta_0, \beta_1, \ldots, \beta_p)'$ is $\hat{\boldsymbol{\beta}} = (\mathbf{X}'\mathbf{X})^{-1}\mathbf{X}'\mathbf{Y}$ where

$$\mathbf{X} = \begin{bmatrix} 1 & X_{11} & \ldots & X_{1p} \\ 1 & & & \\ \vdots & \vdots & & \\ 1 & X_{n1} & \ldots & X_{np} \end{bmatrix}, \quad \mathbf{Y} = (Y_1, \ldots, Y_n)'. \qquad (15.4)$$

The classification rule for an observation $X = (X_1, \ldots, X_p)'$ is

$$\hat{h}(x) = \begin{cases} 1 & \text{if } \hat{\ell}(x) := X'\hat{\boldsymbol{\beta}} > \frac{1}{2}, \\ 0 & \text{otherwise.} \end{cases} \qquad (15.5)$$

The rationale behind the rule is that $P(Y = 1 \mid X = x) + P(Y = 0 \mid X = x) = 1$. For much more on this issue refer to Hastie et al. (2001, Chap. 4).

A more appropriate model, compared to the classical linear model, is perhaps *logistic regression:* $\ell(x) \equiv P(Y = 1 \mid X = x) = \exp\{\beta_0 + \beta'x\}/[1 + \exp\{\beta_0 + \beta'x\}]$. For estimation of the parameters of the model we refer to Chap. 4 (Project) and Chap. 14, Sect. 14.3. The classifier is then the analog of (15.5), namely:

$$\hat{h}(x) = \begin{cases} 1 & \text{if } \hat{\ell}(x) > \frac{1}{2}, \\ 0 & \text{otherwise.} \end{cases} \tag{15.6}$$

See Hastie et al. (2001, Chap. 4, Sect. 4).

For general parametric models with densities f_j, $1 \leq j \leq K$, governing the K classes one may use the MLE for estimating the parameters from sufficiently large random samples, or training data, in order to compute a classifier $\hat{h}$. If parametric models are not particularly reliable, one may resort to nonparametric density estimation by the *kernel method* described in Chap. 10. If the observation X is high-dimensional then the convergence to the true density is generally very slow. In many data examples and simulations the *nonparametric Bayes* estimation of the density, following Ferguson (1973, 1974), and using appropriate MCMC, seem to yield substantially better approximations. One may also use nonparametric regression for classification (See Ghosh and Ramamoorthi (2002), Chaps. 5, 7, Hjort et al. 2010, and Bhattacharya and Bhattacharya 2012, Chaps. 13, 14).

Finally, we mention the linear classifier known as the *support vector machines* introduced by Vapnik and his co-authors (See Vapnik 1998). This theory provides the precise criterion for separating the training data by a hyperplane, and provides optimal separating hyperplane maximizing its distance from the data. One then uses this to classify an observation into one of two classes ($K = 2$).

15.2 Principal Component Analysis (PCA)

Let $\mathbf{X}$ be a d-dimensional random vector with distribution Q, mean $\boldsymbol{\mu} = (\mu_1, \ldots, \mu_d)'$ and covariance matrix Σ ($d > 1$). Let $u_1, u_2, \ldots, u_d$ be unit length (orthonormal) eigenvectors of Σ with eigenvalues $\lambda_1 \geq \lambda_2 \geq \cdots \geq \lambda_d$. One may express $\mathbf{X}$ as

$$\mathbf{X} = \sum_{i=1}^{d} \langle \mathbf{X}, u_i \rangle u_i = \boldsymbol{\mu} + \sum_{i=1}^{d} \langle \mathbf{X} - \boldsymbol{\mu}, u_i \rangle u_i, \tag{15.7}$$

where $\langle \, , \, \rangle$ denotes Euclidean inner product (and $\|v\| = \langle v, v \rangle$ is the Euclidean norm). The quantities $\langle \mathbf{X} - \boldsymbol{\mu}, u_i \rangle u_i$ are called *principal components* of $\mathbf{X} - \boldsymbol{\mu}$, or of $\mathbf{X}$. Note that $\text{var}\langle \mathbf{X}, u_i \rangle = \text{var}\langle \mathbf{X} - \boldsymbol{\mu}, u_i \rangle = \lambda_i$ ($i = 1, \ldots, d$). The coefficient $\langle \mathbf{X} - \boldsymbol{\mu}, u_1 \rangle$ of the first principal component has the largest variance λ_1. The first principal component is considered to be the most important, followed by the next important component, namely the second principal component, and so on.

The most important use of PCA is in *dimension reduction* especially when d is large. The following result says that the r-dimensional hyperplane passing through $\boldsymbol{\mu}$ and generated by the first r principal components provides the best r-dimensional subspace approximation for $\mathbf{X}$ for the criterion of minimizing expected squared distance.

Theorem 15.1. *Consider the orthogonal projection of* $\mathbf{X}$ *onto an* r-*dimensional hyperplane passing through* $\boldsymbol{\mu}$. *The expected squared distance between* $\mathbf{X}$ *and such a* *projection is minimized over the class of all such hyperplanes when the hyperplane is the translation by $\boldsymbol{\mu}$ of the r-dimensional subspace spanned by $u_1, u_2, \ldots, u_r$. In particular, the optimal projection is $\sum_{i=1}^{r} \langle \mathbf{X} - \boldsymbol{\mu}, u_i \rangle u_i + \boldsymbol{\mu}$.*

Proof. An r-dimensional hyperplane passing through $\boldsymbol{\mu}$ may be expressed as $\boldsymbol{\mu} + H$, where H is spanned by r orthonormal vectors $v_1, v_2, \ldots, v_r$. The projection of $\mathbf{X}$ on this hyperplane is $\boldsymbol{\mu} + \sum_{i=1}^{r} \langle \mathbf{X} - \boldsymbol{\mu}, v_i \rangle v_i$ whose expected squared distance from $\mathbf{X}$ is

$$\mathbb{E} \left\| \mathbf{X} - \boldsymbol{\mu} - \sum_{i=1}^{r} \langle \mathbf{X} - \boldsymbol{\mu}, v_i \rangle v_i \right\|^2$$

$$= E \| \mathbf{X} - \boldsymbol{\mu} \|^2 + \sum_{i=1}^{r} E \langle \mathbf{X} - \boldsymbol{\mu}, v_i \rangle^2 - 2 \sum_{i=1}^{r} E \langle \mathbf{X} - \boldsymbol{\mu}, v_i \rangle^2$$

$$= E \| \mathbf{X} - \boldsymbol{\mu} \|^2 - \sum_{i=1}^{r} E \langle \mathbf{X} - \boldsymbol{\mu}, v_i \rangle^2$$

$$= E \| \mathbf{X} - \boldsymbol{\mu} \|^2 - \sum_{i=1}^{r} v_i' \Sigma v_i. \tag{15.8}$$

The minimizer of this over all orthonormal r-tuples $\{v_1, v_2, \ldots, v_r\}$ is the maximizer

$$\underset{\substack{\{v_1, v_2, \ldots, v_r\} \\ \text{orthonormal}}}{\arg\max} \ \sum_{i=1}^{r} v_i' \Sigma v_i. \tag{15.9}$$

First choose v_1 to maximize $v_1' \Sigma v_1$ over the class of all vectors of norm 1. The maximum value is λ_1 attained by u_1. Next choose v_2 to minimize $v_2' \Sigma v_2$ over the class of all vectors of norm 1 orthogonal to u_1; this maximum is λ_2 attained by $v_2 = u_2$. In this manner we arrive at $\{u_1, u_2, \ldots, u_r\}$ as the maximizer of (15.9), and the minimizer of (15.8). $\qquad\square$

Remark 15.1. One estimates $\boldsymbol{\mu}$ and Σ and, consequently, eigenvalues and eigenvectors from random samples taken from underlying distributions Q. These estimates are then used for constructing principal components and dimension reduction.

Remark 15.2. A word of caution in the use of PCA for purposes of inference such as classification and two-sample tests with high-dimensional data. The idea of dimension reduction via PCA is not that one ignores data pertaining to those principal components with small variances. It simply means that those components may simply be represented by their means. Notice that the optimal r-dimensional hyperplane in Theorem 15.1 passes through $\boldsymbol{\mu}$, which is the mean vector of *all* the coordinates. Indeed, if one wishes to discriminate one distribution Q_1 from another, Q_2, based on a few principal components estimated from random samples from the two distributions, it would be prudent to compare the means of the lowest principal components. For even small differences in means of these components are more likely to be detected, because of small variances, than similar differences in higher level principal components. See Bhattacharya and Bhattacharya (2012, pp. 16, 17).

For applications of PCA in image analysis and pattern recognition we refer to Bishop (2006, Chap. 4) and Hastie et al. (2001, Chap. 4).

15.3 Sequential Probability Ratio Test (SPRT)

The *sequential probability ratio test,* or *SPRT,* was introduced by Wald (1947). We provide here a brief outline of the procedure and its properties.

Usual statistical inference procedures, such as considered in this book so far, are based on fixed sample sizes, which may be either too small or inadequate or too large and wasteful for the specific goals one has. A sequential procedure tries to achieve the desired goals with the smallest possible sample sizes by drawing observations one at a time until the accumulated evidence becomes sufficient for the desired level of accuracy. To illustrate Wald's SPRT in this context we consider the problem of testing a simple null hypothesis H^0 against a simple alternative H^1 with prescribed probabilities α, β of Type 1 and Type 2 errors.

Let $U_1, U_2, \ldots$, be i.i.d. observations with values in some measurable space $(S, \mathscr{S})$. Assume that their common distribution has a probability density $f(u; \theta)$ (with respect to some measure $\nu(du)$). There are two hypotheses concerning the value of θ, namely, $H^0 : \theta = \theta_0$, $H^1 : \theta = \theta_1$, $(\theta_0 \neq \theta_1)$. Assume for simplicity that $f(u; \theta_0)$, $f(u; \theta_1)$ are both strictly positive for all u (outside perhaps a set of zero v-measure). Let X_n be the likelihood ratio: $X_n := \prod_1^n (f(U_j; \theta_1)/f(U_j; \theta_0))$, then, under H^0, $\{X_n\}_{n=1}^\infty$ is a $\{\mathscr{F}_n\}_{n=1}^\infty$-*martingale* with $\mathscr{F}_n := \sigma\{U_1, \cdots, U_n\}$. The sequential probability ratio test (SPRT) of A. Wald may be described as follows. Let $0 < A < 1 < B$ be two positive numbers, and let τ be the first time $\{X_n\}_{n=1}^\infty$ escapes from the interval (A, B)

$$\tau := \inf \{n \geq 1 : X_n \leq A \quad \text{or} \quad X_n \geq B\}. \tag{15.10}$$

Then *accept* H^0 if $X_\tau \leq A$, and *accept* H^1 if $X_\tau \geq B$. Assuming a *parameter identifiability condition* $\nu(\{u : f(u; \theta_0) \neq f(u; \theta_1)\}) > 0$, one may check that $\mathbb{E}^0 \tau < \infty$ and $\mathbb{E}^1 \tau < \infty$, where P^i denotes probability, and $\mathbb{E}^i$ expectation, under H^i $(i = 0, 1)$.[1] Now

$$\mathbb{E}^i X_m \mathbf{1}_{[\tau > m]} \leq BP^i(\tau > m) \longrightarrow 0 \quad \text{as} \quad m \to \infty \quad (i = 0, 1). \tag{15.11}$$

It follows from the *Optional Stopping Theorem*[2] that

$$1 = \mathbb{E}^0 X_1 = \mathbb{E}^0 X_\tau = \mathbb{E}^0 X_\tau \mathbf{1}_{[X_\tau \geq B]} + \mathbb{E}^0 X_\tau \mathbf{1}_{[X_\tau \leq A]}. \tag{15.12}$$

Now, expressing $\mathbf{1}_{[\tau = n]}$ as a function $g_n(U_1, \ldots, U_n)$, one has

$$\mathbb{E}^0 X_\tau \mathbf{1}_{[X_\tau \leq A]} \mathbf{1}_{[\tau = n]} = \mathbb{E}^0 X_n \mathbf{1}_{[X_n \leq A]} \mathbf{1}_{[\tau = n]} = \mathbb{E}^1 \mathbf{1}_{[X_n \leq A]} \mathbf{1}_{[\tau = n]}$$
$$= \mathbb{E}^1 \mathbf{1}_{[X_\tau \leq A]} \mathbf{1}_{[\tau = n]}. \tag{15.13}$$

The second equality in (15.13) holds as a special case of the general equality $\mathbb{E}^0 X_n f(U_1, \cdots, U_n) = \mathbb{E}^1 f(U_1, \cdots, U_n)$ for every nonnegative (or bounded) measurable f on S^n. Summing (15.13) over $n = 1, 2, \ldots$, we get

$$1 = \mathbb{E}^0 X_\tau \geq BP^0(X_\tau \geq B) + P^1(X_\tau \leq A). \tag{15.14}$$

---

[1] See Proposition 3.7 in Bhattacharya and Waymire (2007), noting that $\log X_n$ is a sum of i.i.d. random variables under both H^0 and H^1.

[2] Theorem 3.6 in Bhattacharya and Waymire (2007).

Writing $\alpha := P^0(X_\tau \geq B)$ as the probability of accepting H^1 when H^0 is true, and $\beta := P^1(X_\tau \leq A)$ as the probability of accepting H^0 when H^1 is true, we get

$$B\alpha + \beta \leq 1. \tag{15.15}$$

Similarly, $\{1/X_n\}_{n=1}^\infty$ is a $\{\mathscr{F}_n\}_{n=1}^\infty$-martingale under P^1, and the same argument as above yields

$$1 = \mathbb{E}^1 X_\tau^{-1} = \mathbb{E}^1 X_\tau^{-1} \mathbf{1}_{[X_\tau \leq A]} + \mathbb{E}^1 X_\tau^{-1} \mathbf{1}_{[X_\tau \geq B]}$$
$$\geq A^{-1} P^1(X_\tau \leq A) + P^0(X_\tau \geq B), \tag{15.16}$$

leading to the inequality

$$(A^{-1})\beta + \alpha \leq 1. \tag{15.17}$$

For small values of α and β (i.e., for large B and small A), (15.15) and (15.17) are often treated as (approximate) equalities, and then one has

$$B \simeq \frac{1-\beta}{\alpha}, \quad A \simeq \frac{\beta}{1-\alpha}; \quad \alpha \simeq \frac{1-A}{B-A}, \quad \beta \simeq A\left(\frac{B-1}{B-A}\right). \tag{15.18}$$

This approximation is often applied, but is not always good.

It may be shown that the *SPRT* is *optimal*. This means that in the class of all tests whose error probabilities are no more than the corresponding probabilities of the *SPRT*, the *expected* sample sizes $\mathbb{E}^i \tau$ ($i = 0, 1$) are the smallest for the *SPRT*. See Ferguson (1967, Sect. 7.6, Theorem 2), for a detailed proof of this result. Note that the fixed sample size procedure of finding the smallest n such that the Type 1 and Type 2 errors are no more than α and β, respectively, is also a stopping rule. In order to obtain an approximate value of $\mathbb{E}^i \tau$ ($i = 0, 1$) we consider the $\{\mathscr{F}_n\}_{n=1}^\infty$-martingale $\{S_n - n\mu_i\}$, where

$$S_n := \sum_{j=1}^n \left(\log f(U_j; \theta_1) - \log f(U_j; \theta_0)\right) = \sum_{j=1}^n \log\left(f(U_j; \theta_1)/f(U_j; \theta_0)\right),$$
$$\mu_i := \mathbb{E}^i \log\left(f(U_j; \theta_1)/f(U_j; \theta_0)\right), \quad (i = 0, 1). \tag{15.19}$$

Since $x \to \log x$ is strictly *concave* on $(0, \infty)$, it follows from Jensen's inequality that

$$\mu_i < \log \mathbb{E}^i \left(f(U_j; \theta_1)/f(U_j; \theta_0)\right)$$
$$= \begin{cases} 0 & \text{if } i = 0, \\ \log\left\{\int \frac{f^2(u;\theta_1)}{f(u;\theta_0)}\nu(du)\right\} > 0 & \text{if } i = 1, \end{cases} \tag{15.20}$$

since

$$\int \frac{f^2(u;\theta_1)}{f(u;\theta_0)} \nu(du) = \int \left(\frac{f(u;\theta_1)}{f(u;\theta_0)}\right)^2 f(u;\theta_0)\nu(du)$$
$$> \left(\int \frac{f(u;\theta_1)}{f(u;\theta_0)} f(u;\theta_0)\nu(du)\right)^2 = 1. \tag{15.21}$$

The last inequality follows from the inequality $\mathbb{E}Y^2 \geq (\mathbb{E}Y)^2$, with strict inequality unless Y has all its mass at one point. To rule out the possibility $\mu_0 = -\infty$ and/or $\mu_1 = \infty$, *assume* μ_0, μ_1 to be finite.

The *SPRT* may be expressed as

$$\text{Accept } H_0 \text{ if } S_\tau \leq -a, \quad \text{Accept } H_1 \text{ if } S_\tau \geq b, \tag{15.22}$$
$$a := -\log A, \quad b := \log B.$$

We have already checked that $\mathbb{E}_i \tau < \infty$ $(i = 0, 1)$. Further,

$$\left| \mathbb{E}^i \left(S_m \mathbf{1}_{[\tau > m]} \right) \right| \leq \max\{a, b\} P^i(\tau > m) \longrightarrow 0 \quad \text{as} \quad m \to \infty. \tag{15.23}$$

Therefore, it follows from Wald's identity,[3] applied to the sequence $\{S_n - n\mu_i\}$, that

$$\mathbb{E}^i S_\tau = \mu_i \mathbb{E}^i \tau \quad (i = 0, 1). \tag{15.24}$$

Hence

$$\mathbb{E}^i \tau = \frac{\mathbb{E}^i S_\tau}{\mu_i} \quad (i = 0, 1). \tag{15.25}$$

Again one may 'approximately' calculate $E^i S_\tau$ as follows:

$$\mathbb{E}^i S_\tau \simeq b P^i(S_\tau \geq b) - a P^i(S_\tau \leq -a)$$
$$= b P^i(X_\tau \geq B) - a P^i(X_\tau \leq A),$$
$$\mathbb{E}^0 S_\tau \simeq b\alpha - a(1 - \alpha), \quad \mathbb{E}^1 S_\tau \simeq b(1 - \beta) - a\beta. \tag{15.26}$$

The values of a, b, α, β are then substituted from (15.18), (15.22).

Remark 15.3. Let $A < 1 < B$. The approximations $\alpha' = \frac{1-A}{B-A}$, $\beta' = A\left(\frac{B-1}{B-A}\right)$ of α, β are, in most applications conservative, i.e., $\alpha' < \alpha$, $\beta' < \beta$. In general, one can show that $\alpha' + \beta' \leq \alpha + \beta$ (See Rao, 1965, pp. 401, 402).

Remark 15.4. Sequential procedures for estimation may be derived using the duality between tests and confidence regions (See Sect. 5.9).

Remark 15.5. When the density $f(\,;\theta)$ has a monotone likelihood ratio the SPRT is effective under composite hypotheses $H^0 : \theta \leq \theta_0$, $H^1 : \theta > \theta_1$ $(\theta_0 \leq \theta_1)$ (See Remarks 5.2–5.4 in Chap. 5). See (Siegmund, 1985, pp. 14–19), for a detailed discussion.

Sequential procedures such as the SPRT are especially important in clinical trials with patients for drug testing, and for those cases where sampling is extremely expensive as in the case of car crash safety experiments and in other destructive testing. In such cases the optimality criterion should involve the cost c per item as well as the Type 1, Type 2 errors. An appropriate objective function to minimize in such cases is the *Bayes risk* $\delta\{\text{Prob}(\text{Reject } H_0 \mid \theta_0) + cE(\tau \mid \theta_0)\} + (1 - \delta)\{\text{Prob}(\text{Reject } H_1 \mid \theta_1) + cE(\tau \mid \theta_1)\}$ $(0 < \delta < 1)$. A sequential probability ratio test with appropriate boundary points A, B, minimizes the Bayes risk. See (Lehmann, 1959, pp. 104–110). Both the SPRT minimizing expected sample size for given levels of α, β and the SPRT minimizing appropriate Bayes risks yield substantial reduction in sample sizes and costs.

---

[3] See (Bhattacharya and Waymire, 2007, p. 47).

15.4 Notes and References

Our main sources for Sect. 15.1 on classification are an exposition in Wasserman (2003, Chap. 22), and the book by Hastie et al. (2001). For a theoretical overview of Bayesian nonparametrics we refer to Ghosal (2010). The original basic theory of nonparametric Bayes inference is due to Ferguson (1973, 1974). A useful construction of Ferguson's Dirichlet priors is by the so-called *stick breaking* of Sethuraman (1994). For extensions of nonparametric Bayes theory to manifolds, see Bhattacharya and Dunson (2010, 2012), and Bhattacharya and Bhattacharya (2012, Chaps. 13, 14).

For applications of principal components analysis (Sect. 15.2) to image analysis, see Hastie et al. (2001, Chap. 4), and Bishop (2006, Chap. 4).

Standard texts on sequential analysis (Sect. 15.3) include Wald (1947), Siegmund (1992), and Chernoff (1972). The last two books also provide continuous time versions of the SPRT and Brownian motion approximations of the discrete time SPRT considered here. Nonparametric sequential analysis is considered in the monograph by Sen (1981).

Exercises for Chap. 15

Exercises for Sect. 15.1

Ex. 15.1. (a) Check (15.2) and (15.3).

(b) Refer to the data on skulls in Chap. 8, Exercise 8.13. Assume that the distributions are five-dimensional Normal $N(\mu_A, \Sigma_A)$, $N(\mu_B, \Sigma_B)$. Use the first 15 A skulls and the first 13 B skulls to estimate the unknown parameters (with $\pi_A = 15/28$, $\pi_B = 13/28$). Then apply the quadratic classifier (15.2) to classify the remaining 4 skulls.

(c) Carry out the classification in (b) assuming $\Sigma_A = \Sigma_B$ and using the linear classifier (15.3).

Ex. 15.2. Compute the Bayes classifier (15.1) for K Bernoulli populations with parameters $p_1, p_2, \ldots, p_K$.

Exercises for Sect. 15.2

Ex. 15.3. Refer to the data on skulls in Chap. 8, Exercise 8.13.

(a) Assume $\Sigma_A = \Sigma_B$. From the pooled estimate of the common covariance matrix calculate the eigenvalues and the principal components.

(b) For each of the five principal components in (a) carry out the two-sample t-test for the equality of the means in the population, and list the p-values.

Exercises for Sect. 15.3

Ex. 15.4. Let U_j's be i.i.d. Bernoulli taking values $+1$ and -1 with probabilities p and $q = 1 - p$, respectively.

(a) Compute the sequential probability ratio test for $H_0 : p = \frac{1}{2}$ against $H_1 : p = 2/3$, with the nominal probabilities of error $\alpha = .1$, $\beta = .2$. [i.e., with A and B given by equalities in (15.18)].

(b) Calculate approximate values of $\mathbb{E}^i \tau$ $(i = 0, 1)$ using (15.25), (15.26).

(c) Given successive sample observations $-1, 1, 1, -1, -1, 1, 1, 1, -1, 1, 1, 1, -1, 1,$ $-1, 1, 1, 1, 1$, decide when to stop and what action to take.

References

Bhattacharya, A., & Bhattacharya, R. (2012). *Nonparametric inference on manifolds: With applications to shape spaces.* IMS monograph (Vol. 2). Cambridge: Cambridge University Press.

Bhattacharya, A., & Dunson, D. B. (2010). Nonparametric Bayesian density estimation on manifolds with applications to planar shapes. *Biometrika, 97,* 851–865.

Bhattacharya, A., & Dunson, D. B. (2012). Nonparametric Bayes classification and hypothesis testing on manifolds. *Journal of Multivariate Analysis, 111,* 1–19.

Bhattacharya, A., & Waymire, E. C. (2007). *A Basic course in probability theory.* New York, Springer.

Bishop, C. (2006). Principal component analysis. In *Pattern recognition and machine learning (information science and statistics)* (pp. 561–570). New York: Springer.

Chernoff, H. (1972). Sequential analysis and optimal design. *Regional conferences in applied mathematics* (Vol. 8). Philadelphia: SIAM.

Ferguson, T. S. (1967). *Mathematical statistics: A decision theoretic approach.* New York: Academic Press.

Ferguson, T. S. (1973). A Bayesian analysis of some nonparametric problems. *Annals of Statistics, 1*(2), 209–230.

Ferguson, T. S. (1974). Prior distributions on spaces of probability measures. *Annals of Statistics, 4*(4), 615–629.

Ghosal, S. (2010). Dirichlet process, related priors and posterior asymptotics. In N. L. Hjort et al. (Eds.), *Bayesian nonparametrics* (pp. 36–83). Cambridge: Cambridge University Press.

Ghosh, J. K., & Ramamoorthi, R. V. (2002). *Bayesian nonparametrics.* New York: Springer.

Hastie, T., Tibshirani, R., & Friedman, J. H., (2001). *The elements of statistical learning.* New York: Springer.

Hjont, N. L., Holmes, C., Müller, P., & Walker, S. G., (Eds.). (2010). *Bayesian nonparametrics.* Cambridge: Cambridge University Press.

Lehmann, E. L. (1959). *Testing statistical hypothesis.* New York: Wiley.

Rao, C. R. (1965). *Linear statistical inference and its applications.* New York: Wiley.

Sen, P. K. (1981). *Sequential nonparametrics.* New York: Wiley.

Sethuraman, J. (1994). A constructive definition of Dirichlet priors. *Statistica Sinica, 4,* 639–650.

Siegmund, D. (1985). *Sequential analysis: Tests and confidence intervals.* New York: Springer.

Siegmund, D. (1992). *Sequential analysis: Tests and confidence intervals.* New York: Springer.

Vapnik, V. N. (1998). *Statistical learning theory.* New York: Springer.

Wald, A. (1947). *Sequential analysis.* New York: Wiley.

Wasserman, L. (2003). *All of statistics: A concise course in statistical inference.* New York: Springer.

Appendix A
Standard Distributions

A.1 Standard Univariate Discrete Distributions

I. Binomial Distribution $\mathscr{B}(n,p)$ has the probability mass function

$$f(x) = \binom{n}{x} p^x \qquad (x = 0, 1, \ldots, n).$$

The mean is np and variance $np(1-p)$.

II. The Negative Binomial Distribution $\mathscr{NB}(r,p)$ arises as the distribution of $X \equiv \{$number of failures until the r-th success$\}$ in independent trials each with probability of success p. Thus its probability mass function is

$$f_r(x) = \binom{r+x-1}{x} p^r (1-p)^x \qquad (x = 0, 1, \cdots, \cdots).$$

Let X_i denote the number of failures between the $(i-1)$-th and i-th successes $(i = 2, 3, \ldots, r)$, and let X_1 be the number of failures before the first success. Then $X_1, X_2, \ldots, X_r$ and r independent random variables each having the distribution $\mathscr{NB}(1,p)$ with probability mass function

$$f_1(x) = p(1-p)^x \qquad (x = 0, 1, 2, \ldots).$$

Also,

$$X = X_1 + X_2 + \cdots + X_r.$$

Hence

$$E(X) = rEX_1 = r\left(p\sum_{x=0}^{\infty} x(1-p)^x\right) = rp(1-p)\sum_{x=1}^{\infty} x(1-p)^{x-1}$$

$$= rp(1-p)\sum_{x=1}^{\infty}\left(-\frac{d}{dp}(1-p)^x\right) = -rp(1-p)\frac{d}{dp}\sum_{x=1}^{\infty}(1-p)^x$$

© Springer-Verlag New York 2016
R. Bhattacharya et al., *A Course in Mathematical Statistics and Large Sample Theory*, Springer Texts in Statistics, DOI 10.1007/978-1-4939-4032-5

$$= -rp(1-p)\frac{d}{dp}\sum_{x=0}^{\infty}(1-p)^x = -rp(1-p)\frac{d}{dp}\underbrace{\frac{1}{1-(1-p)}}_{=p}$$

$$= \frac{r(1-p)}{p}. \tag{A.1}$$

Also, one may calculate $\mathrm{var}(X)$ using (Exercise A.1)

$$\mathrm{var}(X) = r\,\mathrm{var}(X_1). \tag{A.2}$$

III. The Poisson Distribution $\mathscr{P}(\lambda)$ has probability mass function

$$f(x) = e^{-\lambda}\frac{\lambda^x}{x!} \qquad (x = 0, 1, 2, \ldots),$$

where $\lambda > 0$ is the mean. To see this let X be a random variable with this distribution. Then

$$E(X) = \sum_{x=0}^{\infty} xe^{-\lambda}\frac{\lambda^x}{x!} = e^{-\lambda}\sum_{x=1}^{\infty}\frac{\lambda^x}{(x-1)!} = \lambda e^{-\lambda}\sum_{y=0}^{\infty}\frac{\lambda^y}{y!}$$

$$= \lambda e^{-\lambda}\cdot e^{\lambda} = \lambda. \qquad (y = x-1)$$

Also,

$$E(X(X-1)) = \sum_{x=0}^{\infty} x(x-1)e^{-\lambda}\frac{\lambda^x}{x!} = e^{-\lambda}\sum_{x=2}^{\infty} x(x-1)\frac{\lambda^x}{x!}$$

$$= e^{-\lambda}\lambda^2\sum_{x=2}^{\infty}\frac{\lambda^{x-2}}{(x-2)!} = e^{-\lambda}\lambda^2\cdot e^{\lambda} = \lambda^2,$$

so that

$$E(X^2) = \lambda^2 + E(X) = \lambda^2 + \lambda,$$
$$\mathrm{var}(X) = E(X^2) - (E(X))^2 = \lambda^2 + \lambda - \lambda^2 = \lambda. \tag{A.3}$$

IV. The Beta-Binomial Distribution $\mathscr{BB}(\alpha, \beta, n)$ is the marginal distribution of X when the conditional distribution of X given (another random variable) $Y = y$ (with values in $[0,1]$) is $\mathscr{B}(n, y)$, where Y has the beta distribution $\mathscr{B}_e(\alpha, \beta)$ (see Sect. A.2). Hence the probability mass function of X is

$$f(x) = \int_0^1 \binom{n}{x} y^x (1-y)^{n-x}\frac{1}{B(\alpha, \beta)} y^{\alpha-1}(1-y)^{\beta-1}dy$$

$$= \frac{\binom{n}{x}}{B(\alpha, \beta)}\int_0^1 y^{x+\alpha-1}(1-y)^{n-x+\beta-1}dy = \frac{\binom{n}{x}B(x+\alpha, n-x+\beta)}{B(\alpha, \beta)}$$

$$= \binom{n}{x}\frac{\Gamma(x+\alpha)\Gamma(n-x+\beta)\Gamma(\alpha+\beta)}{\Gamma(n+\alpha+\beta)\Gamma(\alpha)\Gamma(\beta)} \qquad (x = 0, 1, \ldots, n)$$

[See below for the relation $B(\alpha, \beta) = \Gamma(\alpha)\Gamma(\beta)/\Gamma(\alpha + \beta)$.] Here if $X \sim \mathscr{BB}(\alpha, \beta, n)$,

$$E(X) = EE(X \mid Y) = EnY = nEY = \frac{n\alpha}{\alpha + \beta},$$

$$E(X^2) = EE(X^2 \mid Y) = E[nY(1 - Y) + n^2 Y^2]$$

$$= (n^2 - 1)EY^2 + nEY = (n^2 - 1)\frac{\alpha(\alpha + 1)}{(\alpha + \beta)(\alpha + \beta + 1)} + n\frac{\alpha}{\alpha + \beta},$$

$$\operatorname{var}(X) = E(X^2) - (E(X))^2.$$

(See below for the computation of the moments of the beta distribution.)

A.2 Some Absolutely Continuous Distributions

I. The Uniform Distribution $\mathscr{U}(\alpha, \beta)$ on $[\alpha, \beta]$ has the probability density function (p.d.f.)

$$f(x) = \frac{1}{\beta - \alpha} \quad \text{for } \alpha \leq x \leq \beta,$$

$$= 0 \quad \text{elsewhere.}$$

II. The Beta Distribution $\mathscr{B}_e(\alpha, \beta)$ has p.d.f.

$$f(x) = \frac{1}{B(\alpha, \beta)} x^{\alpha-1}(1 - x)^{\beta-1}, \qquad 0 < x < 1,$$

$$= 0 \quad \text{elsewhere.}$$

Here $\alpha > 0$, $\beta > 0$ and $B(\alpha, \beta)$ is the normalizing constant *(beta function)*

$$B(\alpha, \beta) = \int_0^1 x^{\alpha-1}(1 - x)^{\beta-1}dx.$$

Clearly, if $X \sim \mathscr{B}_e(\alpha, \beta)$, then

$$E(X) = \frac{1}{B(\alpha, \beta)}\int_0^1 x^\alpha(1 - x)^{\beta-1}dx = \frac{B(\alpha + 1, \beta)}{B(\alpha, \beta)},$$

$$E(X^2) = \frac{B(\alpha + 2, \beta)}{B(\alpha, \beta)}, \ldots, E(X^k) = \frac{B(\alpha + k, \beta)}{B(\alpha, \beta)}. \tag{A.4}$$

Recall that the *gamma function* $\Gamma(\alpha)$ is defined by

$$Gamma(\alpha) = \int_0^\infty e^{-x}x^{\alpha-1}dx \qquad (\alpha > 0).$$

On integration by parts one gets

$$\Gamma(\alpha+1) = \int_0^\infty e^{-x}x^\alpha dx = -e^{-x}x^\alpha\Big|_0^\infty + \int_0^\infty \alpha x^{\alpha-1}e^{-x}dx = 0 + \alpha\Gamma(\alpha) = \alpha\Gamma(\alpha).$$

$$\tag{A.5}$$

We now prove

$$B(\alpha, \beta) = \frac{\Gamma(\alpha)\Gamma(\beta)}{\Gamma(\alpha + \beta)}, \qquad \forall\, \alpha > 0,\ \beta > 0. \tag{A.6}$$

Now

$$\Gamma(\alpha)\Gamma(\beta) = \int_0^\infty e^{-x} x^{\alpha - 1} dx \int_0^\infty e^{-y} y^{\beta - 1} dy$$

$$= \int_0^\infty \int_0^\infty e^{-(x+y)} x^{\alpha - 1} y^{\beta - 1} dx dy.$$

Change variables: $z = x + y$, $y = y$, to get $\begin{cases} x = z - y \\ y = y \end{cases}$, with

$$\text{Jacobian} = \left| \det \begin{bmatrix} \frac{\partial x}{\partial t} & \frac{\partial x}{\partial y} \\ \frac{\partial y}{\partial z} & \frac{\partial y}{\partial y} \end{bmatrix} \right| = 1$$

$$\Gamma(\alpha)\Gamma(\beta) = \int_0^\infty e^{-z} \left(\int_0^z (z - y)^{\alpha - 1} y^{\beta - 1} dy \right) dz$$

$$= \int_0^\infty e^{-z} z^{\alpha - 1} z^{\beta - 1} \left(\int_0^z \left(1 - \frac{y}{z} \right)^{\alpha - 1} \left(\frac{y}{z} \right)^{\beta - 1} dy \right) dz$$

$$= \int_0^\infty e^{-z} z^{\alpha + \beta - 2} \left(z \int_0^1 (1 - u)^{\alpha - 1} u^{\beta - 1} du \right) dz \quad \left[u = \tfrac{y}{z},\ du = \tfrac{1}{z}\, dy \right]$$

$$= \int_0^\infty e^{-z} z^{\alpha + \beta - 1} B(\beta, \alpha) dz = \Gamma(\alpha + \beta) B(\beta, \alpha). \tag{A.7}$$

But

$$B(\beta, \alpha) = \int_0^1 u^{\beta - 1} (1 - u)^{\alpha - 1} du = \int_0^1 x^{\alpha - 1} (1 - x)^{\beta - 1} dx$$

$$= B(\alpha, \beta), \qquad (x = 1 - u). \tag{A.8}$$

Hence (A.7) and (A.6). Using (A.4)–(A.6), one gets the k-th moment of a beta $(\mathscr{B}_e(\alpha, \beta))$ random variable X as

$$E(X^k) = \frac{\Gamma(\alpha + k)\Gamma(\beta) \cdot \Gamma(\alpha + \beta)}{\Gamma(\alpha + \beta + k) \cdot \Gamma(\alpha)\Gamma(\beta)} = \frac{(\alpha + k - 1) \cdots (\alpha + 1)\alpha}{(\alpha + \beta + k - 1) \cdots (\alpha + \beta + 1)(\alpha + \beta)}.$$

In particular,

$$E(X) = \frac{\alpha}{\alpha + \beta}, \quad E(X^2) = \frac{(\alpha + 1)\alpha}{(\alpha + \beta + 1)(\alpha + \beta)}, \quad \text{var}(X) = \frac{\alpha\beta}{(\alpha + \beta)^2(\alpha + \beta + 1)}.$$

III. The Gamma Distribution $\mathscr{G}(\alpha, \beta)$ has the p.d.f.

$$f(x) = \frac{1}{\Gamma(\beta)\alpha^\beta}\, e^{-x/\alpha} x^{\beta - 1}, \qquad 0 < x < \infty$$

$$= 0 \quad \text{elsewhere}, \tag{A.9}$$

where $\alpha > 0$, $\beta > 0$. Here α is a *scale parameter*, i.e., if X is $\sim \mathscr{G}(\alpha, \beta)$, then X/α is $\mathscr{G}(1, \beta)$, Note that

$$E\left(\frac{X}{\alpha}\right)^k = \frac{1}{\Gamma(\beta)} \int_0^\infty z^k e^{-z} z^{\beta-1} dz$$

$$= \frac{\Gamma(\beta + k)}{\Gamma(\beta)} = (\beta + k - 1) \cdots (\beta + 1)\beta,$$

so that

$$E(X^k) = \alpha^k (\beta + k - 1) \cdots (\beta + 1)\beta. \tag{A.10}$$

Hence $EX = \alpha\beta$, $\mathrm{var}(X) = \alpha^2 \beta$.

A.2.1 The Normal Distribution $\mathrm{N}(\mu, \sigma^2)$

has p.d.f.

$$f_{\mu,\sigma^2}(x) = \frac{1}{\sqrt{2\pi\sigma^2}} e^{-(x-\mu)^2/2\sigma^2}, \qquad -\infty < x < \infty, \tag{A.11}$$

where $\mu \in (-\infty, \infty)$, $\sigma^2 > 0$. The *standard normal distribution* $\mathrm{N}(0, 1)$ has p.d.f.

$$f(x) = \frac{1}{\sqrt{2\pi}} e^{-x^2/2}, \qquad -\infty < x < \infty. \tag{A.12}$$

To show that (A.12) (and hence (A.11), by transformation $y = \frac{x-\mu}{\sigma}$) is a p.d.f., one needs to show that

$$\int_0^\infty e^{-x^2/2} dx = \sqrt{\frac{\pi}{2}} \tag{A.13}$$

For this use the transformation $z = x^2/2$ to get

$$\int_0^\infty e^{-x^2/2} dx = \int_0^\infty e^{-z} \sqrt{2} \left(\frac{1}{2}\right) z^{-\frac{1}{2}} dz = \frac{1}{\sqrt{2}} \Gamma\left(\frac{1}{2}\right). \tag{A.14}$$

Now, by (A.7) (with $\alpha = \beta = \frac{1}{2}$)

$$\Gamma^2\left(\frac{1}{2}\right) = \Gamma(1) B\left(\frac{1}{2}, \frac{1}{2}\right) = B\left(\frac{1}{2}, \frac{1}{2}\right) \quad \text{(since } \Gamma(1) = \int_0^\infty e^{-x} dx = -e^{-x}|_0^\infty = 1,)$$

$$= \int_0^1 x^{-1/2}(1-x)^{-1/2} dx = \int_0^1 z^{-1}(1-z^2)^{-1/2} 2z\, dz \quad (z = x^{1/2}, \ dx = 2z\, dz)$$

$$= 2\int_0^1 (1-z^2)^{-1/2} dz = 2\int_0^{\pi/2} \frac{\cos\theta\, d\theta}{\cos\theta} \quad (z = \sin\theta, \ dz = \cos\theta\, d\theta)$$

$$= 2\left(\frac{\pi}{2}\right) = \pi.$$

Hence

$$\Gamma\left(\frac{1}{2}\right) = \sqrt{\pi}, \tag{A.15}$$

which when used in (A.14) yields (A.13).

If X is $\mathbf{N}(\mu, \sigma^2)$, then $Z = \frac{X-\mu}{\sigma}$ has the p.d.f. (A.12) (by change of variables). Therefore,

$$E\left(\frac{X-\mu}{\sigma}\right) = E(Z) = \frac{1}{\sqrt{2\pi}} \int_{-\infty}^{\infty} x e^{-x^2/2} dx = 0 \quad (\text{since } x e^{-x^2/2} \text{ is } odd),$$

$$E\left(\frac{X-\mu}{\sigma}\right)^2 = E(Z^2) = \frac{1}{\sqrt{2\pi}} \int_{-\infty}^{\infty} x^2 e^{-x^2/2} dx = \frac{2}{\sqrt{2\pi}} \int_0^{\infty} x^2 e^{-x^2/2} dx$$

$$= \sqrt{\frac{2}{\pi}} \int_0^{\infty} (2z)^{\frac{1}{2}} e^{-z} dz \quad (z = \tfrac{x^2}{2}, \quad dz = x dx)$$

$$= \frac{2}{\sqrt{\pi}} \int_0^{\infty} e^{-z} z^{1/2} dz = \frac{2}{\sqrt{\pi}} \Gamma\left(\frac{3}{2}\right) = \frac{2}{\sqrt{\pi}} \left(\frac{1}{2}\right) \Gamma\left(\frac{1}{2}\right) = \frac{2}{\sqrt{\pi}} \left(\frac{1}{2}\right) \sqrt{\pi}$$

$$= 1.$$

Hence

$$E(X - \mu) = 0, \text{ or } E(X) = \mu,$$
$$E\left(\left(\tfrac{X-\mu}{\sigma}\right)^2\right) = 1, \text{ or, } \text{var}(X) = \sigma^2. \tag{A.16}$$

V. The Chi-Square Distribution χ_k^2 *with k Degrees of Freedom is defined to be the distribution of the sum of squares of k independent standard normal random variables.* To derive its p.d.f. let $X_1, X_2, \ldots, X_k$ be k independent $\mathbf{N}(0,1)$ random variables. Then define the chi-square random variable

$$Z = X_1^2 + X_2^2 + \cdots + X_k^2,$$

and note that, as $\Delta z \downarrow 0$,

$$P(z < Z \le z + \Delta z) = \int \cdots \int \left(\frac{1}{\sqrt{2\pi}}\right)^k e^{-(x_1^2 + \cdots + x_k^2)/2} dx_1 \cdots dx_k$$

$$\{(x_1, \ldots, x_k) : z < \sum_1^k x_i^2 \le z + \Delta z\}$$

$$= \left(\frac{1}{\sqrt{2\pi}}\right)^k \left(e^{-z/2} + O(\Delta z)\right)$$

$$\times \text{volume of the annulus } \{(x_1, \ldots, x_k) : z < \sum x_i^2 < z + \Delta z\}$$

$$= \left(\frac{1}{\sqrt{2\pi}}\right)^k \left(e^{-z/2} + O(\Delta z)\right) \left(c_k \left(\sqrt{z + \Delta z}\right)^k - c_k \left(\sqrt{z}\right)^k\right),$$

writing the volume of a ball of radius r in k-dimension as $c_k r^k$. [Note that $\int_{\{\sum x_i^2 \le r\}} \cdots \int dx_1 \ldots dx_k = r^k \int_{\{\sum x_i^2 \le 1\}} \cdots \int dx_1 \ldots dx_k$, by change of variables $z_i = x_i/r$ $(1 \le i \le k)$.] Since $(\sqrt{z + \Delta z})^k - (\sqrt{z})^k = \frac{d}{dz}(\sqrt{z})^k \cdot \Delta z + O(\Delta z)^2 = \frac{k}{2} z^{\frac{k}{2}-1} \Delta z + (\Delta z)^2$, one has

$$P(z < Z \le z + \Delta z) = c_k \left(\frac{1}{\sqrt{2\pi}}\right)^k \frac{k}{2} e^{-\frac{z}{2}} z^{\frac{k}{2}-1} \Delta z + O(\Delta z)^2. \tag{A.17}$$

Hence the p.d.f. of Z is

$$f(z) = c'_k z^{\frac{k}{2}-1} e^{-\frac{z}{2}}, \quad 0 < z < \infty,$$
$$= 0 \quad \text{elsewhere} \tag{A.18}$$

where c'_k is the normalized constant,

$$c'_k = \left(\int_0^\infty z^{\frac{k}{2}-1} e^{-\frac{z}{2}} dz \right)^{-1} = \left(\int_0^\infty u^{\frac{k}{2}-1} e^{-u} du \left(2^{\frac{k}{2}} \right) \right)^{-1} \quad (u = \tfrac{z}{2})$$
$$= \frac{1}{2^{k/2} \Gamma\left(\frac{k}{2}\right)} \cdot \tag{A.19}$$

Since c'_k may be directly calculated from (A.17), one has

$$c'_k = c_k \left(\frac{1}{\sqrt{2\pi}} \right)^k \frac{k}{2} \cdot \tag{A.20}$$

Comparing (A.19), (A.20) the constant c_k ($=$ volume of the unit ball in $\mathbb{R}^k$) may also be obtained ($c_k = \frac{2\pi^{k/2}}{k\Gamma(k/2)}$). Note also that χ_k^2 is $\mathscr{G}(2, k/2)$ and hence $Z/2 \sim \mathscr{G}(1, \frac{k}{2})$. In particular, if $X \sim \mathbf{N}(0,1)$, $X^2 \sim \chi_1^2$, $\frac{X^2}{2} \sim \mathscr{G}(1, \frac{1}{2})$. This also shows that the sum of independent chi-square random variables $Z_1, Z_2, \ldots, Z_m$ with degrees of freedom $k_1, k_2, \ldots, k_m$, respectively, is a chi-square random variable with degrees of freedom $k = k_1 + \cdots + k_m$ (Exercise A.2). Also, if $Z \sim \chi_k^2$ then using (A.10) [with $\alpha = 1$, $k = 1$, $\beta = k/2$ or the fact that $EX^2 = 1$ if $X \sim \mathbf{N}(0,1)$],

$$EZ = 2E\frac{Z}{2} = 2\frac{k}{2} = k.$$

VI. The Student's t-Distribution t_k with k Degrees of Freedom is defined to be the distribution of $T = X/\sqrt{Z/k}$, where $X \sim \mathbf{N}(0,1)$, $Z \sim \chi_k^2$, and X and Z are *independent.* The (cumulative) distribution function of T is given by

$$P(T \le t) = P\left(X \le t\sqrt{Z/k} \right) = EP\left(X \le t\sqrt{Z/k} \,\Big|\, Z \right)$$
$$= \int_0^\infty P\left(X \le \frac{t}{\sqrt{k}} \sqrt{z} \,\Big|\, Z = z \right) \cdot \frac{1}{2^{k/2} \Gamma(k/2)} z^{k/2-1} e^{-z/2} dz$$
$$= \frac{1}{2^{k/2} \Gamma(k/2)} \int_0^\infty \left[\int^{\sqrt{z/k}\, t} \frac{1}{\sqrt{2\pi}} e^{x^2/2} dx \right] z^{k/2-1} e^{-z/2} dz$$

Differentiating w.r.t. t under the integral sign one gets the p.d.f. of T as

$$f(t) = \frac{1}{2^{k/2} \Gamma\left(\frac{k}{2}\right)} \int_0^\infty \left(\frac{1}{\sqrt{2\pi}} e^{-\frac{zt^2}{wk}} z^{\frac{k}{2}-1} e^{-\frac{z}{2}} \right) \sqrt{\frac{z}{k}} \, dz$$
$$= \frac{1}{\sqrt{k}\, 2^{\frac{k+1}{2}} \sqrt{\pi} \Gamma\left(\frac{k}{2}\right)} \int_0^\infty e^{-\frac{z}{2}\left(1+\frac{t^2}{k}\right)} z^{\frac{k}{2}-\frac{1}{2}} dz$$

$$
= \frac{2^{\frac{k+1}{2}} \left(1 + \frac{t^2}{k}\right)^{-\frac{k+1}{2}}}{\sqrt{k}\, 2^{\frac{k+1}{2}} \sqrt{\pi}\, \Gamma\left(\frac{k}{2}\right)} \, \Gamma\left(\frac{k+1}{2}\right)
$$

$$
= \frac{\Gamma\left(\frac{k+1}{2}\right)}{\sqrt{k\pi}\, \Gamma\left(\frac{k}{2}\right)} \left(1 + \frac{t^2}{k}\right)^{-\frac{k+1}{2}} , \qquad -\infty < t < \infty. \tag{A.21}
$$

VII. The Cauchy Distribution $\mathscr{C}(\alpha, \beta)$ *has p.d.f.*

$$
f(x) = \frac{\beta}{\pi} \cdot \frac{1}{\beta^2 + (x - \alpha)^2} = \frac{1}{\beta} \cdot \frac{1}{\pi \left(1 + \left(\frac{x-\alpha}{\beta}\right)^2\right)} , \qquad -\infty < x < \infty,
$$

where $\beta > 0$, $\alpha \in \mathbb{R}^1$ are parameters. Note that if $X \sim \mathscr{C}(\alpha, \beta)$ then $(X - \alpha)/\beta$ has the standard Cauchy distribution $\mathscr{C}(0, 1)$ with p.d.f.

$$
f(x) = \frac{1}{\pi(1 + x^2)} , \qquad -\infty < x < \infty,
$$

which is the p.d.f. of a Student's t with $k = 1$ d.f. Note that the first moment of the Cauchy distribution does not exist (and, therefore, no higher moment exists).

The final example in this section is important in the theory of testing hypotheses considered in Chap. 5.

VIII. Fisher's F Distribution Let U and V be independent chi-square random variables with degrees of freedom r, s, respectively. (That is, U and V are independent gamma random variables $\mathscr{G}(\frac{r}{2}, 2)$ and $\mathscr{G}(\frac{s}{2}, 2)$.) The *F-statistic* with degrees of freedom (r, s) is then defined by

$$
F = \frac{U/r}{V/s} . \tag{A.22}
$$

Its distribution $\mathscr{F}_{r,s}$, say, is called the *F-distribution* after R.A. Fisher. Its distribution function is computed as follows:

$$
\begin{aligned}
G(x) &:= P(F \leq x) = P\left(U \leq \frac{r}{s} x V\right) \\
&= E\left(\frac{1}{\Gamma\left(\frac{r}{2}\right) 2^{\frac{r}{2}}} \int^{\frac{r}{s} x V} u^{\frac{r}{2}-1} e^{-\frac{u}{2}} \, du\right) \\
&= \frac{1}{\Gamma\left(\frac{r}{2}\right) 2^{\frac{r}{v}}} \int_0^\infty \frac{1}{2^{\frac{s}{2}} \Gamma\left(\frac{s}{2}\right)} v^{\frac{s}{2}-1} e^{-\frac{v}{2}} \left\{\int_0^{\frac{r}{s} x v} u^{\frac{r}{2}-1} e^{-\frac{u}{2}} \, du\right\} dv.
\end{aligned}
$$

Hence the density of $\mathscr{F}_{r,s}$ is given by

$$
\begin{aligned}
f_{r,s}(x) = G'(x) &= \frac{1}{\Gamma\left(\frac{r}{2}\right) \Gamma\left(\frac{s}{2}\right) 2^{\frac{r+s}{2}}} \int_0^\infty v^{\frac{s}{2}-1} e^{-\frac{v}{2}} \left(\frac{rv}{s}\right) \left(\frac{r}{s} x v\right)^{\frac{r}{2}-1} e^{-\frac{r}{s} x \frac{v}{2}} \, dv \\
&= \frac{\left(\frac{r}{s}\right)^{\frac{r}{2}} x^{\frac{r}{2}-1}}{\Gamma\left(\frac{r}{2}\right) \Gamma\left(\frac{s}{2}\right) 2^{\frac{r+s}{2}}} \int_0^\infty v^{\frac{r+s}{2}-1} e^{-\frac{v}{2}\left(1 + \frac{r}{s} x\right)} \, dv \\
&= \frac{\left(\frac{r}{s}\right)^{\frac{r}{2}} \Gamma\left(\frac{r+s}{2}\right)}{\Gamma\left(\frac{r}{2}\right) \Gamma\left(\frac{s}{2}\right) \left(1 + \frac{r}{s} x\right)^{\frac{r+s}{2}}} . \tag{A.23}
\end{aligned}
$$

IX. Logistic Distribution The distribution function of a *standard logistic random variable* X is given by

$$F(x) = \frac{e^x}{1 + e^x}, \quad -\infty < x < \infty,$$

with density

$$f(x) = \frac{e^x}{(1 + e^x)^2}, \quad -\infty < x < \infty.$$

It is easy to check that the density is symmetric about $x = 0$, i.e., $f(x) = f(-x)$ for all x. Hence all odd order moments are zero. The even order moments are given as follows.[1] First note that, for $y > 0$,

$$(1 + y)^{-2} = \sum_{n=0}^{\infty} \binom{-2}{n} y^n = \sum_{n=0}^{\infty} (-1)^n (n + 1) y^n.$$

Using this one obtains

$$
\begin{aligned}
EX^{2m} &= 2 \int_{[0,\infty)} x^{2m} \frac{e^{-x}}{(1 + e^{-x})^2}\, dx \\
&= 2 \sum_{n=0}^{\infty} (-1)^n (n + 1) \int_{[0,\infty)} x^{2m} e^{-(n+1)x} dx \\
&= 2\Gamma(2m + 1) \sum_{n=0}^{\infty} \frac{(-1)^n}{(n + 1)^{2m}} \\
&= 2(2m)! \sum_{n=1}^{\infty} \frac{(-1)^{n-1}}{n^{2m}} = 2(2m)! \left\{ \sum_{r=1}^{\infty} \frac{(-1)^{2r-2}}{(2r - 1)^{2m}} + \sum_{r=1}^{\infty} \frac{(-1)^{2r-1}}{(2r)^{2m}} \right\} \\
&= 2(2m)! \left\{ \sum_{r=1}^{\infty} (2r - 1)^{-2m} - \sum_{r=1}^{\infty} (2\gamma)^{-2m} \right\} \\
&= 2(2m)! \left\{ \sum_{r=1}^{\infty} \gamma^{-2m} - 2 \sum_{r=1}^{\infty} (2\gamma)^{-2m} \right\} = 2(2m)! \left(1 - 2^{-(2m-1)} \right) \sum_{r=1}^{\infty} \gamma^{-2m} \\
&= 2(2m) \left(1 - 2^{-(2m-1)} \right) \zeta(2m),
\end{aligned}
$$

where $\zeta(r) = \sum_{r=1}^{\infty} r^{-2}$ is the *Riemann zeta function*. In particular, $EX^2 = \text{var}(X) = 2\zeta(2) = \pi^2/3$, since[2] $\sum_{r=1}^{\infty} r^{-2} = \pi^2/6$.

Next, if X has the standard logistic distribution, then for any real μ and $h > 0$, the pdf of $Z = hX + \mu$ is

$$f_Z(z) = h f_X \left(\frac{z - \mu}{h} \right) = h^{-1} \frac{\exp\left\{ \frac{z-\mu}{h} \right\}}{\left(1 + \exp\left\{ \frac{z-\mu}{h} \right\} \right)^2}, \quad -\infty < z < \infty.$$

---

[1] See Balakrishnan, N. and Nevzorov, V.B. (2003). *A Primer on Statistical Distributions*, Chap. 22. Wiley, New York.

[2] See Titchmarsh, E.C. (1939). *The Theory of Functions*, 2nd ed., p. 35. Oxford University Press, London.

The mean of Z is μ and its variance is $h^2 \pi^2/3$. The logistic regression in its simplest form used in the student projects in Chaps. 4 and 14, has a binary response variable, say Y, with values 0, 1, and a continuous regressor Z. When $Z = z$, then $Y = 1$ with probability $F_Z(z) = \text{Prob}(Z \le z)$ and $Y = 0$ with probability $1 - F_Z(z)$. One may also apply this in the case of a non-stochastic Z using the same relations formally. For many applications other than logistic regression see, e.g., Balakrishnan and Nevzorov (2003).

A.3 The Multivariate Normal Distribution

(Notation: $\langle x, y \rangle = x'y = \sum_{j=1}^{k} x^{(j)} y^{(j)} \forall x, y \in \mathbb{R}^k$.)

Definition A.1. A random vector $\mathbf{X} = (X^{(1)}, X^{(2)}, \ldots, X^{(k)})'$ with values in $\mathbb{R}^k$ has the *k-dimensional normal distribution* $\mathbf{N}(\mu, \Sigma)$ with *mean* (vector) $\mu = (\mu^{(1)}, \mu^{(2)}, \ldots, \mu^{(k)})'$ and *nonsingular dispersion matrix* Σ if it has the probability density function

$$f(\mathbf{x}) = \frac{1}{(2\pi)^{k/2}(\det \Sigma)^{1/2}} \, e^{-\frac{1}{2}\langle \mathbf{x}-\mu, \, \Sigma^{-1}(\mathbf{x}-\mu)\rangle}$$

$$= \frac{1}{(2\pi)^{k/2}(\det \Sigma)^{1/2}} \, e^{-\frac{1}{2}\Sigma \sum_{i,j=1}^{k} \sigma^{ij}(x^{(i)}-\mu^{(i)})(x^{(j)}-\mu^{(j)})}. \qquad (\text{A.24})$$

Here σ^{ij} is the (i,j) element of Σ^{-1}.

Remark A.1. To show that f is a p.d.f., make the change of variables: $\mathbf{y} = B^{-1}(\mathbf{x} - \mu)$, where B is a symmetric positive definite matrix satisfying $B^2 = \Sigma$. Then $\mathbf{x} - \mu = B\mathbf{y}$, $\Sigma^{-1} = B^{-1}B^{-1}$ (since $(BB)^{-1} = B^{-1}B^{-1}$), and $\langle \mathbf{x} - \mu, \Sigma^{-1}(\mathbf{x} - \mu)\rangle = \langle B\mathbf{y}, B^{-1}B^{-1}B\mathbf{y}\rangle = \langle B\mathbf{y}, B^{-1}\mathbf{y}\rangle = \langle \mathbf{y}, BB^{-1}\mathbf{y}\rangle$ (since $B' = B$) $= \langle \mathbf{y}, \mathbf{y} \rangle = \|\mathbf{y}\|^2 = \sum_{1}^{k}(y^{(j)})^2$. Thus,

$$\int_{\mathbb{R}^k} f(x)dx = \frac{\det B}{(2\pi)^{k/2}(\det \Sigma)^{1/2}} \int_{\mathbb{R}^k} e^{-\frac{1}{2}\sum_{j=1}^{k} y_j^2} dy_1 \ldots dy_k,$$

since the Jacobian matrix is

$$J\left(\frac{\mathbf{x}}{\mathbf{y}}\right) = \begin{bmatrix} \frac{\partial x^{(1)}}{\partial y^{(1)}} & \cdots & \frac{\partial x^{(1)}}{\partial y^{(k)}} \\ \vdots & & \vdots \\ \frac{\partial x^{(k)}}{\partial y^{(1)}} & \cdots & \frac{\partial x^{(k)}}{\partial y^{(k)}} \end{bmatrix} = B,$$

one gets the change of volume elements

$$dx_1 dx_2 \ldots dx_k = \det B \, dy_1 dy_2 \ldots dy_k.$$

Also, $\det \Sigma = \det BB = \det B \cdot \det B$ (since $\det(AB) = \det A \cdot \det B$ for $k \times k$ matrices A, B). Hence $\det B = \sqrt{\det \Sigma}$. One has the positive square root, since B is positive definite. Hence

$$\int_{\mathbb{R}^k} f(x)dx = \frac{1}{(2\pi)^{k/2}} \int_{\mathbb{R}^k} e^{-\frac{1}{2}\sum_{j=1}^{k}(y^{(j)})^2} dy_1 dy_2 \ldots dy_k$$

$$= \prod_{j=1}^{k} \left(\frac{1}{\sqrt{2\pi}} \int_{\mathbb{R}^k} e^{-\frac{y_j^2}{2}} dy_j \right) = \prod_{j=1}^{k} 1 = 1.$$

This proof also shows that if $\mathbf{X}$ is distributed as $\mathbf{N}(\boldsymbol{\mu}, \Sigma)$, then $\mathbf{Y} \equiv B^{-1}(\mathbf{X} - \boldsymbol{\mu})$ is distributed as $\mathbf{N}(\mathbf{0}, I)$ where I is the $k \times k$ *identity matrix* (having 1's on the diagonal, an zero's off the diagonal). The distribution $\mathbf{N}(\mathbf{0}, I)$ is called the k-*dimensional standard normal distribution*. Notice that $\mathbf{Y} = (Y^{(1)}, Y^{(2)}, \ldots, Y^{(k)})'$ is distributed as $\mathbf{N}(\mathbf{0}, I)$, if and only if $Y^{(1)}, Y^{(2)}, \ldots, Y^{(k)}$ are k independent 1-dimensional standard normal random variables.

Conversely, given any positive definite matrix Σ ($k \times k$), and any vector $\boldsymbol{\mu} \in \mathbb{R}^k$, the random vector $\mathbf{X} = (X^{(1)}, X^{(2)}, \ldots, X^{(k)})'$ defined by

$$\mathbf{X} = B\mathbf{Y} + \boldsymbol{\mu} \tag{A.25}$$

is distributed as $\mathbf{N}(\boldsymbol{\mu}, \Sigma)$, if $\mathbf{Y}$ is k-dimensional standard normal. Here B is a symmetric positive definite matrix satisfying $B^2 = \Sigma$. If Σ is merely *nonnegative definite*, then also the definition (A.25) makes sense and defines a random vector $\mathbf{X}$ whose distribution is denoted by $\mathbf{N}(\boldsymbol{\mu}, \Sigma)$. If Σ is nonnegative definite and of rank less than k (i.e., at least one eigenvalue of Σ is zero), then $\mathbf{N}(\boldsymbol{\mu}, \Sigma)$ defined above via (A.25) is called a *singular k-dimensional normal distribution*. The representation (A.25) yields

$$E(\mathbf{X}) = BE\mathbf{Y} + \boldsymbol{\mu} = B\mathbf{0} + \boldsymbol{\mu} = \boldsymbol{\mu},$$

$$\text{cov}(X^{(i)}, X^{(j)}) = \text{cov}\left(\sum_{r=1}^{k} b_{ir} Y^{(r)} + \mu^{(i)}, \sum_{r'=1}^{k} b_{jr'} Y^{(r')} + \mu^{(j)} \right)$$

$$= \text{cov}\left(\sum_{r=1}^{k} b_{ir} Y^{(r)}, \sum_{r'=1}^{k} b_{jr'} Y^{(r')} \right)$$

$$= \sum_{r,r'=1}^{k} b_{ir} b_{jr'} \text{cov}(Y^{(r)}, Y(r')) = \sum_{r=1}^{k} b_{ir} b_{jr}$$

$$= (i,j) \text{ element of } BB = (i,j) \text{ element of } \Sigma = \sigma_{ij}. \tag{A.26}$$

This justifies the name *mean vector* for $\boldsymbol{\mu}$ and *dispersion* (i.e., covariance *matrix*) for Σ.

In general, if $\mathbf{X}$ is a k-variate Normal random vector (singular, or non-singular), distributed as $\mathbf{N}(\boldsymbol{\mu}, \Sigma)$, and if $\mathbf{Y} = \boldsymbol{\nu} + C\mathbf{X}$, where $\boldsymbol{\nu} \in \mathbb{R}^d$ and C is a $d \times k$ matrix, then $\mathbf{Y}$ has the Normal distribution $\mathbf{N}(\boldsymbol{\beta}, \Gamma)$, whose mean $\boldsymbol{\beta}$ and covariance (or, dispersion) matrix are given by

$$\boldsymbol{\beta} = E\mathbf{Y} = \boldsymbol{\nu} + CE\mathbf{X} = \boldsymbol{\nu} + C\boldsymbol{\mu},$$

$$\Gamma = \text{cov}\mathbf{Y} = \text{cov}(C\mathbf{X}) = C\Sigma C'. \tag{A.27}$$

Note that if the $d \times d$ matrix $C\Sigma C'$ is of rank k_0 ($\leq \min\{k, d\}$), then $\mathbf{Y}$ may be represented as $\mathbf{Y} = B\mathbf{Z} + \boldsymbol{\beta}$, where B is the non-negative definite symmetric matrix satisfying $B^2 = C\Sigma C'$, and Z is a d-dimensional standard Normal $N(\mathbf{O}, I)$.

One may prove this directly using (A.25), or using moment generating functions that we discuss in Appendix B.

Let now $\mathbf{X}$ be $\mathbf{N}(\boldsymbol{\mu}, \Sigma)$, a k-dimensional Normal distribution with Σ nonsingular. Then, for any given m, $1 < m \le k$, $\mathbf{Y} = (X_m, \ldots, X_k)'$ has the Normal distribution $\mathbf{N}(\boldsymbol{\mu}_m^k, \Sigma_m^k)$, where

$$\boldsymbol{\mu}_m^k = (\mu_m, \ldots, \mu_k)', \quad \Sigma_m^k = ((\sigma_{ij}))_{m \le i, j \le k}. \tag{A.28}$$

This follows as a special case of the transformation in the preceding paragraph, with $d = k - m + 1$, $\boldsymbol{\nu} = \mathbf{0}$, $C = \begin{bmatrix} 0 & 0 \\ 0 & I_d \end{bmatrix}$ (I_d is the $d \times d$ identity matrix). Indeed, since $\mathbf{Y}$ is a linear transformation of $\mathbf{X}$ and, hence, Normal, its mean and covariance are easily seen to be given by (A.28). *The (marginal) density of* $\mathbf{Y} = (X_m, \ldots, X_k)'$ *is given by*

$$f_{\mathbf{Y}}(x_m, \ldots, x_k) = 2\pi^{-d/2} \left(\det \Sigma_m^k \right)^{-1/2} \exp \left\{ -\frac{1}{2} \sum_{i,j=m}^{k} \gamma^{ij} (x_i - \mu_i)(x_j - \mu_j) \right\} \tag{A.29}$$

where $((\gamma^{ij}))_{m \le i, j \le k} = \Gamma^{-1}$. Hence the conditional density of $W = (X_1, \ldots, X_{m-1})'$, given $X_m = x_m, \ldots, X_k = x_k$, is

$$f_{W \mid x_m, \ldots, x_k}(x_1, \ldots, x_{m-1}) = \frac{f_{\mathbf{X}}(x_1, \ldots, x_k)}{f_{\mathbf{Y}}(x_m, \ldots, x_k)}$$

$$= c \exp \left\{ -\frac{1}{2} \sum_{1 \le i, j \le m-1} \sigma^{ij}(x_i - \mu_i)(x_j - \mu_j) \right.$$

$$\left. - \sum_{i=1}^{m-1} \left(\sum_{j=m}^{k} \sigma^{ij}((x_j - \mu_j))(x_i - \mu_i) \right) + \frac{1}{2} \sum_{m \le i, j \le k} \gamma^{ij}(x_i - \mu_i)(x_j - \mu_j) \right\} \tag{A.30}$$

$$= c_1(x_m, \ldots, x_k) \exp \left\{ -\frac{1}{2} \sum_{1 \le i, j \le m-1} \sigma^{ij}(x_i - \ell_i(x_m, \ldots, x_k))(x_j - \ell_j(x_m, \ldots, x_k)) \right\},$$

where $c_1(x_m, \ldots, x_k)$ depends (possibly) on $x_m, \ldots, x_k$ and $\boldsymbol{\mu}$ and Σ, and $\ell_i(x_m, \ldots, x_k)$ is the affine linear function of $x_m, \ldots, x_k$, given by

$$\ell_i(x_m, \ldots, x_k) = \mu_i - A_i \left(\Sigma^{-1} \right)_{m-1}^{k-m+1} \begin{pmatrix} x_m - \mu_m \\ \vdots \\ x_k - \mu_k \end{pmatrix}. \tag{A.31}$$

Here $\left(\Sigma^{-1} \right)_{m-1}^{k-m+1}$ is the $(m-1) \times (k-m+1)$ matrix comprising the first $m-1$ rows and the last $k-m+1$ columns of Σ^{-1}, and A_i is the i-th row of the $(m-1) \times (m-1)$ matrix

$$A = \left(((\sigma^{ij}))_{1 \le i, j \le m-1} \right)^{-1}. \tag{A.32}$$

Thus *the conditional distribution of* $W = (X_1, \ldots, X_{m-1})'$, *given* $X_m = x_m, \ldots, X_k = x_k$, *is an* $(m-1)$-*dimensional Normal distribution* $\mathbf{N}(\boldsymbol{\ell}(x_m, \ldots, x_k), \mathbf{A})$, *with the mean vector* $\boldsymbol{\ell}$, *and covariance (or, dispersion) matrix* $\mathbf{A}$.

As a special case, with $m = 2$, the conditional distribution of X_1, given $X_2 = x_2, \ldots, X_k = x_k$, is Normal $\mathbf{N}(\ell_1, 1/\sigma^{11})$, where (Exercise A.3)

$$\ell_1 = \ell_1(x_2, \ldots, x_k) = \mu_1 - \frac{1}{\sigma^{11}} \sum_{j=2}^{k} \sigma^{1j}(x_j - \mu_j). \tag{A.33}$$

A Useful Property of the Multivariate Normal Distribution $N(\boldsymbol{\mu}, \Sigma)$

Consider all vectors $\mathbf{x}$ in the following as column vectors, with $\mathbf{x}'$ as the transpose of $\mathbf{x}$, a row vector.

Proposition A.1. *Let* $\mathbf{X}$ *be a* k-*dimensional Normal random vector* $N(\boldsymbol{\mu}, \Sigma)$, *where* Σ *is positive definite. Then* $(\mathbf{X} - \boldsymbol{\mu})'\Sigma^{-1}(\mathbf{X} - \boldsymbol{\mu})$ *has the chi-square distribution with* k *degrees of freedom.*

Proof. Let $\mathbf{Y} = \mathbf{X} - \boldsymbol{\mu}$. Then Y is $N(0, \Sigma)$. Let B be a $k \times k$ matrix such that $BB' = \Sigma^{-1}$. (See the Lemma below.) Then $\Sigma = (B')^{-1}B^{-1}$. Now the random vector $\mathbf{Z} = \mathbf{B}'\mathbf{Y}$ is $N(0, B'\Sigma B)$. But $B'\Sigma B = B'(B')^{-1}B^{-1}B = I_k$, where I_k is the $k \times k$ identity matrix. Thus $\mathbf{Z}$ is a standard k-dimensional Normal random vector whose coordinates $Z_1, \ldots, Z_k$ are one-dimensional independent standard Normal random variables. Hence $(\mathbf{X} - \boldsymbol{\mu})'\Sigma^{-1}(\mathbf{X} - \boldsymbol{\mu}) = \mathbf{Y}'\Sigma^{-1}\mathbf{Y} = \mathbf{Y}'BB'\mathbf{Y} = \mathbf{Z}'\mathbf{Z} = Z_1^2 + \cdots + Z_k^2$ has the chi-square distribution with k degrees of freedom. Q.E.D. $\qquad\square$

Note If $\mathbf{Y}$ is a k-dimensional random vector with covariance matrix $\Sigma = ((\sigma_{ij}))$, and $\mathbf{Z} = c\mathbf{Y}$ for some $m \times k$ matrix C, then the $m \times m$ covariance matrix of $\mathbf{Z}$ is $C\Sigma C'$. For, the covariance between the i-th and j-th components of Z is $\operatorname{cov}(Z_i, Z_j) = \operatorname{cov}(\sum_{1 \leq r \leq k} C_{ir}Y_r, \sum_{1 \leq s \leq k} C_{js}Y_s) = \sum_{r,s} C_{ir}C_{js}\sigma_{rs} = \sum_s(\sum_r C_{ir}\sigma_{rs})C_{js} = \sum_s(C\Sigma)_{is}C_{js}$, which is the (i,j) element of the matrix $C\Sigma C'$.

Lemma A.1. *Let* $\boldsymbol{\Gamma}$ *be a* $k \times k$ *symmetric and positive definite matrix. Then there exists a* $k \times k$ *nonsingular matrix* B *such that* $\boldsymbol{\Gamma} = BB'$.

Proof. Let $\lambda_1, \ldots, \lambda_k$ be the (positive) eigenvalues of Γ, counting multiplicities, and let $a_1, \ldots, a_k$ be corresponding eigenvectors of unit length each. Then $\boldsymbol{\Gamma}a_i = \lambda_i a_i$ $(i = 1, \ldots, k)$, and the matrix A comprising the k (column) vectors $a_1, \ldots, a_k$ is orthonormal, i.e., $AA' = \mathbf{I}_k$, and satisfies $\boldsymbol{\Gamma}A = A(\operatorname{Diag}(\lambda_1, \ldots \lambda_k))$. Define $B = A(\operatorname{Diag}(\sqrt{\lambda_1}, \ldots, \sqrt{\lambda_k}))$. Then $BB' = A(\operatorname{Diag}(\lambda_1, \ldots, \lambda_k))A' = \boldsymbol{\Gamma}AA' = \boldsymbol{\Gamma}$. Q.E.D $\qquad\square$

Exercises for Appendix A

Ex. A.1. Calculate $\mathrm{var}(X_1)$ and $\mathrm{var}(X)$ in (A.2).

Ex. A.2. Show that the sum of independent chi-square random variables $Z_1, Z_2, \ldots, Z_m$ with degrees of freedom $k_1, k_2, \ldots, k_m$, respectively, is a chi-square random variable with degrees of freedom $k_1 + k_2 + \cdots + k_m$.

Ex. A.3. For the case of the Normal distribution (A.24), with $k = 2$, show that
$$\sigma^{11} = \frac{\sigma_{22}}{\sigma_{11}\sigma_{22} - \sigma_{12}^2} = \frac{1}{\sigma_{11}(1-\rho^2)}, \ \sigma^{12} = \sigma^{21} = -\frac{\sigma_{12}}{\sigma_{11}\sigma_{22} - \sigma_{12}^2} = -\frac{1}{\sqrt{\sigma_{11}\sigma_{22}}}\left(\frac{\rho}{1-\rho^2}\right), \ \text{so}$$
that the conditional distribution of X_1, given $X_2 = x_2$, is Normal with mean $\ell_1(x_2) = \mu_1 + \sqrt{\frac{\sigma_{11}}{\sigma_{22}}}\,\rho(x_2 - \mu_2)$ and variance $\sigma_{11}(1 - \rho^2)$.

Appendix B
Moment Generating Functions (M.G.F.)

Definition B.1. The *m.g.f. of a random variable* X (or of its *distribution*) is defined as

$$\varphi(z) = Ee^{zX} \qquad z \in \mathbb{R}^1.$$

Theorem B.1. *Suppose the m.g.f.* $\varphi(\xi)$ *(or,* $\varphi(\boldsymbol{\xi})$*) of a random variable (vector)* X *is finite in a neighborhood of zero (origin =* $\mathbf{0}$*). Then* φ *determines the distribution of* X.

Proof. See Proposition 4.2 in Chap. 4.

Theorem B.2. *Suppose the m.g.f.* φ *of a random variable* X *is finite in an interval nondegenerate) around zero. (a) Then all moments of* X *are finite and*

$$EX^k = \left.\frac{d^k \varphi(z)}{dz^k}\right|_{z=0}.$$

(b) $\varphi(z) = \sum_0^\infty \mu_k \left(\frac{z^k}{k!}\right)$, *where* $\mu_k = EX^k$.

Proof. Follows from the proof of Proposition 4.2 in Chap. 4. $\qquad\square$

Theorem B.3. *If* $X_1, X_2, \ldots, X_n$ *are independent with finite m.g.f.'s* $\varphi_1, \varphi_2, \ldots, \varphi_n$ *in a neighborhood of zero, then the m.g.f.* φ *of* $X_1 + X_2 + \cdots + X_n$ *is*

$$\varphi(z) = \varphi_1(z)\varphi_2(z)\ldots\varphi_n(z),$$

in a neighborhood of zero.

Proof.

$$\varphi(z) = Ee^{z(X_1+X_2+\cdots+X_n)} = (Ee^{zX_1})(Ee^{zX_2})\cdots(Ee^{zX_n})$$
$$= \varphi_1(z)\varphi_2(z)\cdots\varphi_n(z).$$

$\qquad\square$

© Springer-Verlag New York 2016

R. Bhattacharya et al., *A Course in Mathematical Statistics and Large Sample Theory*, Springer Texts in Statistics, DOI 10.1007/978-1-4939-4032-5

Example B.1. X is $\mathscr{P}(\lambda)$. Then

$$\varphi(z) = \sum_{x=0}^{\infty} e^{zx} e^{-\lambda} \lambda^x / x! = e^{-\lambda} \sum_{x=0}^{\infty} (\lambda e^z)^x / x!$$
$$= e^{-\lambda} e^{\lambda e^z} = e^{\lambda(e^z - 1)} < \infty, \quad -\infty < z < \infty.$$

Example B.2. X is $\mathscr{B}(n, p)$:

$$\varphi(z) = \sum_{x=0}^{n} e^{zx} \binom{n}{x} p^x q^{n-x} = \sum_{x=0}^{n} \binom{n}{x} (pe^z)^x q^{n-x}$$
$$= (pe^z + q)^n < \infty, \quad -\infty < z < \infty.$$

Example B.3. X is $\mathscr{NB}(r, p)$:

$$\varphi(z) = \sum_{x=0}^{\infty} e^{zx} \binom{r + x - 1}{r - 1} p^r q^x.$$

Alternatively, $\varphi(z) = E e^{z(X_1 + X_2 + \cdots + X_r)}$, where X_i is the number of failures between the $(i - 1)$-th and i-th successes. Because $X_1, X_2, \ldots, X_r$ are i.i.d., one has

$$\varphi(z) = E e^{zX_1} e^{zX_2} \ldots e^{zX_r} = (E e^{zX_1})(E e^{zX_2}) \cdots (E e^{zX_r})$$
$$= \varphi_1(z)\varphi_1(z) \cdots \varphi_1(z) = (\varphi_1(z))^r,$$

where

$$\varphi_1(z) = E e^{zX_1} = \sum_{x=0}^{\infty} e^{zx} pq^x = p \sum_{x=0}^{\infty} (qe^z)^x = \frac{p}{1 - qe^z} < \infty$$
$$\text{if } e^z < \frac{1}{q}, \text{ i.e., } z < \ln \frac{1}{q}.$$
$$(\ln \frac{1}{q} > 0).$$

Therefore,

$$\varphi(z) = \left(\frac{p}{1 - qe^z} \right)^r.$$

Example B.4. Let X be gamma $\mathscr{G}(\alpha, \beta)$ with p.d.f.

$$f(x) = \frac{1}{\Gamma(\beta)\alpha^\beta} e^{-x/\alpha} x^{\beta - 1}, \quad 0 < x < \infty$$
$$= 0, \quad \text{elsewhere.}$$

Its m.g.f. is

$$\varphi(z) = E e^{zX} = \int_0^{\infty} e^{zx} f(x) dx = \frac{1}{\Gamma(\beta)\alpha^\beta} e^{-\left(\frac{1}{\alpha} - z \right)x} x^{\beta - 1} dx$$
$$= \infty \quad \text{if } z \geq \frac{1}{\alpha}.$$

For $z < \frac{1}{\alpha}$, change variables $y = \left(\frac{1}{\alpha} - z\right) x$, to get

$$\varphi(z) = \frac{1}{\Gamma(\beta)\alpha^\beta} \int_0^\infty e^{-y} y^{\beta-1} dy \Big/ \left(\frac{1}{\alpha} - z\right)^\beta$$

$$= \frac{\alpha^\beta}{\Gamma(\beta)\alpha^\beta (1 - \alpha z)^\beta} \cdot \Gamma(\beta) = \frac{1}{(1 - \alpha z)^\beta}, \quad -\infty < z < \frac{1}{\alpha}.$$

Example B.5. Let $X_1, X_2, \ldots, X_n$ be i.i.d. gamma $\mathscr{G}(\alpha, \beta)$. Then $Y = X_1 + X_2 + \cdots + X_n$ *is gamma* $\mathscr{G}(\alpha, n\beta)$. For the m.g.f. of Y is, by Theorem B.3,

$$\varphi_Y(z) = \frac{1}{(1 - \alpha z)^{n\beta}}$$

which is the m.g.f. of a gamma $\mathscr{G}(\alpha, n\beta)$.

Example B.6. $X \sim N(\mu, \sigma^2)$. First let $\mu = 0$, $\sigma^2 = 1$. Then

$$\varphi(z) = Ee^{zX} = \frac{1}{\sqrt{2\pi}} \int_{-\infty}^\infty e^{zx} e^{-x^2/2} dx = \frac{1}{\sqrt{2\pi}} \int_{-\infty}^\infty e^{-\frac{1}{2}(x-z)^2 + \frac{z^2}{2}} dx$$

$$= \frac{1}{\sqrt{2\pi}} e^{z^2/2} \int_{-\infty}^\infty e^{-\frac{1}{2}y^2} dy \qquad (y = x - z)$$

$$= e^{z^2/2}.$$

Now consider the general case $X \sim \mathbf{N}(\mu, \sigma^2)$, and let $Y = \frac{X-\mu}{\sigma}$. Then $Y \sim \mathbf{N}(0, 1)$ and

$$Ee^{zX} = Ee^{z(\sigma Y + \mu)} = Ee^{z\mu + \sigma z Y} = e^{\mu z} Ee^{\sigma z Y}$$

$$= e^{\mu Z} e^{\sigma^2 z^2/2} = e^{\mu z + \sigma^2 z^2/2}.$$

Example B.7. Consider $X \sim N(\mu, \sigma^2)$. By Example B.6 above, for the case $\mu = 0$, $\sigma^2 = 1$, expanding the m.g.f. φ of X in a power series,

$$\varphi(z) = e^{z^2/2} = 1 + \frac{z^2}{2} + \frac{1}{2!}\left(\frac{z^2}{2}\right)^2 + \frac{1}{3!}\left(\frac{z^2}{2}\right)^3 + \frac{1}{4!}\left(\frac{z^2}{2}\right)^4 + \cdots ,$$

one has $EX^k = 0$ for all odd integers k, and

$$EX^2 = \varphi''(0) = 1, \quad EX^4 = \varphi^{(iv)}(0) = \frac{1}{8}\left(\frac{d}{dz^4} z^4\right)_{z=0} = \frac{4!}{8} = 3.$$

In general, the term-by-term differentiation of the power series for $\varphi(z)$, evaluated at $z = 0$, yields

$$EX^{2k} = \left[\frac{d}{dz^{zk}} \cdot \frac{\left(\frac{z^2}{2}\right)^k}{k!}\right]_{z=0} = \frac{(2k)!}{k!2^k} = (2k - 1)(2k - 3)\cdots 3.1. \tag{B.1}$$

A Normal random variable $\mathbf{N}(\mu, \sigma^2)$, say X, may be expressed as $X = \mu + \sigma Y$ where Y is $\mathbf{N}(0, 1)$. Hence, the *raw moments* of X are

$$\mu_1' = EX = \mu, \quad \mu_2' = EX^2 = \mu^2 + \sigma^2, \quad \mu_3' = EX^3 = \mu^3 + 3\mu\sigma^2,$$
$$\mu_4' = EX^4 = \mu^4 + 6\mu^2\sigma^2 + 3\sigma^4,$$

and the *centered moments* of X, $\mu_k \equiv E(X - \mu)^k$, are given by

$$\mu_k = 0 \qquad \forall \text{ odd integers } k,$$
$$\mu_{2k} = E(X - \mu)^{2k} = \sigma^{2k} EY^{2k} = (2k - 1)(2k - 3)\cdots 3.1.\sigma^{2k}, \quad (k = 1, 2, \dots). \tag{B.2}$$

II. Multivariate M.G.F. Let $\mathbf{X} = (X^{(1)}, X^{(2)}, \dots, X^{(k)})'$ be a k-dimensional random vector. Then the m.g.f. of $\mathbf{X}$ is

$$\varphi(\boldsymbol{\xi}) = Ee^{\langle \boldsymbol{\xi}, \mathbf{X} \rangle} = Ee^{\boldsymbol{\xi}, \mathbf{X}} = Ee^{\sum_{j=1}^k -\xi^{(j)} X^{(j)}}$$

for $\xi = (\xi^{(1)}, \dots, \xi^{(k)})' \in \mathbb{R}^k$.

Theorem B.4. *Suppose that the m.g.f. $\varphi(\boldsymbol{\xi})$ of a random vector $\mathbf{X} = (X^{(1)}, \dots, X^{(k)})$ is finite in a neighborhood of $\boldsymbol{\xi} = 0$ (i.e., for $|\xi| < \delta$, for some $\delta > 0$). (a) Then all moments of X are finite and, using multi-indices $\boldsymbol{\nu}$,*

$$\mu_\nu \equiv EX^\nu \equiv E\left[(X^{(1)})^{\nu^{(1)}} (X^{(2)})^{\nu^{(2)}} \cdots (X^{(k)})^{\nu^{(k)}}\right]$$

$$= [D^\nu \varphi(\boldsymbol{\xi})]_{\xi=0} \equiv \left[\left(\frac{\partial}{\partial \xi^{(1)}}\right)^{\nu^{(1)}} \left(\frac{\partial}{\partial \xi^{(2)}}\right)^{\nu^{(2)}} \cdots \left(\frac{\partial}{\partial \xi^{(k)}}\right)^{\nu^{(k)}} \varphi(\boldsymbol{\xi})\right]_{\xi=0}$$

for all $\nu = (\nu^{(1)}, \dots, \nu^{(k)}) \in (\mathbb{Z}^+)^k$ (= the set of all k-tuples of nonnegative integers).

$$(b) \quad \varphi(\boldsymbol{\xi}) = \sum_\nu \frac{\mu_\nu}{\nu!} \boldsymbol{\xi}^\nu \qquad |\xi| < \delta,$$

      where $\boldsymbol{\xi}^\nu = (\xi^{(1)})^{\nu^{(1)}} (\xi^{(2)})^{\nu^{(2)}} \cdots (\xi^{(k)})^{\nu^{(k)}}$, $\nu! = \nu^{(1)}!\nu^{(2)}!\cdots\nu^{(k)}!$.

(c) *If $\mathbf{X}_1, \mathbf{X}_2, \dots, \mathbf{X}_n$ are independent k-dimensional random vectors whose m.g.f.'s $\varphi_1(\xi), \varphi_2(\xi), \dots, \varphi_k(\xi)$ are finite for $|\xi| < \delta$, for some $\delta > 0$, then the m.g.f. $\varphi(\boldsymbol{\xi})$ of $\mathbf{X} = \mathbf{X}_1 + \mathbf{X}_2 + \cdots + \mathbf{X}_n$ is finite for $|\xi| < \delta$ and is given by*

$$\varphi(\boldsymbol{\xi}) = \varphi_1(\boldsymbol{\xi})\varphi_2(\boldsymbol{\xi}) \cdots \varphi_n(\boldsymbol{\xi}).$$

Proof. The term-by-term differentiation below is justified by the proof of Proposition 4.2 in Chap. 4. (a) Note that, if Q is the distribution of $\mathbf{X}$,

$$\frac{\partial}{\partial \xi^{(1)}} \varphi(\boldsymbol{\xi}) = \frac{\partial}{\partial \xi^{(1)}} \int_{\mathbb{R}^k} e^{\boldsymbol{\xi}' \mathbf{x}} dQ(\mathbf{x})$$

$$= \int_{\mathbb{R}^k} \frac{\partial}{\partial \xi^{(1)}} e^{\xi^{(1)} x^{(1)} + \cdots + \xi^{(i)} x^{(i)} + \cdots + \xi^{(k)} x^{(k)}} dQ(\mathbf{x})$$

$$= \int_{\mathbb{R}^k} x^{(1)} e^{\xi^{(1)} x^{(1)} + \cdots + \xi^{(i)} x^{(i)} + \cdots + \xi^{(k)} x^{(k)}} dQ(\mathbf{x}).$$

Continuing in this manner

$$\left(\frac{\partial}{\partial \xi^{(1)}}\right)^{\nu^{(1)}} \varphi(\boldsymbol{\xi}) = \int^{\mathbb{R}^k} (x^{(1)})^{\nu^{(1)}} e^{\xi^{(1)} x^{(1)} + \cdots + \xi^{(i)} x^{(i)} + \cdots + \xi^{(k)} x^{(k)}} dQ(\mathbf{x}).$$

Differentiating w.r.t. $\xi^{(2)}$, $\nu^{(2)}$-times, one gets

$$\left(\frac{\partial}{\partial \xi^{(1)}}\right)^{\nu^{(2)}} \left(\frac{\partial}{\partial \xi^{(1)}}\right)^{\nu^{(1)}} \varphi(\boldsymbol{\xi})$$

$$= \int_{\mathbb{R}^k} \left(\frac{\partial}{\partial \xi^{(2)}}\right)^{\nu^{(2)}} \left[\left(x^{(1)}\right)^{\nu^{(1)}} e^{\boldsymbol{\xi}'\mathbf{x}}\right] dQ(\mathbf{x})$$

$$= \int_{\mathbb{R}^k} \left(x^{(1)}\right)^{\nu^{(1)}} \left(x^{(2)}\right)^{\nu^{(2)}} e^{\boldsymbol{\xi}'\mathbf{x}} dQ(\mathbf{x}).$$

Continuing in this manner one has

$$D^{\nu}\varphi(\boldsymbol{\xi})\bigg| \equiv \left(\frac{\partial}{\partial \xi^{(1)}}\right)^{\nu^{(1)}} \left(\frac{\partial}{\partial \xi^{(2)}}\right)^{\nu^{(2)}} \cdots \left(\frac{\partial}{\partial \xi^{(k)}}\right)^{\nu^{(k)}} \varphi(\boldsymbol{\xi})$$

$$= \int_{\mathbb{R}^k} (x^{(1)})^{\nu^{(1)}} (x^{(2)})^{\nu^{(2)}} \cdots (x^{(k)})^{\nu^{(k)}} \cdot e^{\boldsymbol{\xi}'\mathbf{x}} dQ(\mathbf{x}).$$

Hence

$$D^{\nu}\varphi(\boldsymbol{\xi})\bigg|_{\xi=0} = \int_{\mathbb{R}^k} (x^{(1)})^{\nu^{(1)}} \cdots (x^{(k)})^{\nu^{(k)}} dQ(\mathbf{x}) = \mu_{\nu} = E\mathbf{X}^{(\nu)}.$$

(b)

$$\varphi(\boldsymbol{\xi}) = Ee^{\boldsymbol{\xi}'\mathbf{X}} = Ee^{\sum_{i=1}^{k} \xi^{(i)} X^{(i)}} = \sum_{r=0}^{\infty} \frac{1}{r!} E\left(\sum_{i=1}^{k} \xi^{(i)} X^{(i)}\right)^{r}$$

$$= \sum_{r=0}^{\infty} \frac{1}{r!} E\left(\sum_{|\nu|=r} \binom{r}{\nu^{(1)}}\binom{r-\nu^{(1)}}{\nu^{(2)}} \cdots \binom{\nu^{(k)}}{\nu^{(k)}}\right. \times$$

$$\times \left. \left(\xi^{(1)} X^{(1)}\right)^{\nu^{(1)}} \left(\xi^{(2)} X^{(2)}\right)^{\nu^{(2)}} \cdots \left(\xi^{(k)} X^{(k)}\right)^{\nu^{(k)}}\right),$$

where $|\nu| = \nu^{(1)} + \cdots + \nu^{(k)}$ and the second sum is over all $\nu \in (\mathbb{Z}^+)^k$ such that $|\nu| = r$.

$$\binom{r}{\nu^{(1)}}\binom{r-\nu^{(1)}}{\nu^{(2)}} \cdots \binom{\nu^{(k)}}{\nu^{(k)}} = \frac{r!}{\nu^{(1)}!\nu^{(2)}!\cdots \nu^{(k)}!} .$$

Hence, writing $\nu! = \nu^{(1)}!\cdots \nu^{(k)}!$, one has

$$\varphi(\boldsymbol{\xi}) = \sum_{r=0}^{\infty} \sum_{|\nu|=r} \frac{\boldsymbol{\xi}^{\nu}}{\nu!} E\left[(X^{(1)})^{\nu^{(1)}} (X^{(2)})^{\nu^{(2)}} \cdots (X^{(k)})^{\nu^{(k)}}\right]$$

$$= \sum_{r=0}^{\infty} \sum_{|\nu|=r} \frac{\mu_{\nu}}{\nu!} \boldsymbol{\xi}^{\nu} = \sum_{\nu \in (\mathbb{Z}^+)^k} \frac{\mu_{\nu}}{\nu!} \boldsymbol{\xi}^{\nu}.$$

(c) Essentially the same proof as that of Theorem B.3. $\square$

Example B.8. $\mathbf{X} \sim (\mathbf{N}(\boldsymbol{\mu}, \Sigma))$. Assume Σ is $k \times k$ symmetric positive definite. Let B be a symmetric positive definite matrix such that $BB = \Sigma$. Then define $Y = B^{-1}(X - \mu)$. We have seen that $Y \sim \mathbf{N}(\mathbf{0}, I)$, i.e., the coordinates of Y are k independent standard normal random variables. The m.g.f. of $\mathbf{Y}$ is

$$\psi(\boldsymbol{\xi}) \equiv Ee^{\boldsymbol{\xi}'\mathbf{Y}} = Ee^{\sum_{i=1}^{k} \xi^{(i)} Y^{(i)}} = \prod_{i=1}^{k} Ee^{\xi^{(i)} Y^{(i)}}$$

$$= \prod_{i=1}^{k} e^{(\xi^{(i)})^2/2} = e^{|\boldsymbol{\xi}|^2/2}.$$

Thus the m.g.f. of $\mathbf{X}$ is

$$\varphi(\boldsymbol{\xi}) = Ee^{\boldsymbol{\xi}'\mathbf{X}} = Ee^{\boldsymbol{\xi}'(B\mathbf{Y}+\boldsymbol{\mu})} = e^{\boldsymbol{\xi}'\boldsymbol{\mu}} Ee^{\boldsymbol{\xi}'B\mathbf{Y}}$$

$$= e^{\boldsymbol{\xi}'\boldsymbol{\mu}} Ee^{(B\boldsymbol{\xi})'\mathbf{Y}} = e^{\boldsymbol{\xi}'\boldsymbol{\mu}} \psi(B\boldsymbol{\xi}) = e^{\boldsymbol{\xi}'\boldsymbol{\mu}} e^{|B\boldsymbol{\xi}|^2/2}$$

$$= e^{\boldsymbol{\xi}'\boldsymbol{\mu}+(B\boldsymbol{\xi})'(B\boldsymbol{\xi})/2} = e^{\boldsymbol{\xi}'\boldsymbol{\mu}+\boldsymbol{\xi}'B'B\boldsymbol{\xi}/2} = e^{\boldsymbol{\xi}'\boldsymbol{\mu}+\boldsymbol{\xi}'BB\boldsymbol{\xi}/2}$$

$$= e^{\boldsymbol{\xi}'\boldsymbol{\mu}+\frac{1}{2}\boldsymbol{\xi}'\Sigma\boldsymbol{\xi}}.$$

Appendix C
Computation of Power of Some Optimal Tests: Non-central t, χ^2 and F

Example C.1. Consider the UMP test φ^* of size $\alpha \in (0,1)$ for $H_0 : \mu \leq \mu_0$ against $H_1 : \mu > \mu_0$ based on $\mathbf{X} = (X_1, \ldots, X_n)$ where X_i's are i.i.d. $\mathbf{N}(\mu, \sigma_0^2)$, with $\sigma_0^2 > 0$ known and $\mu \in \mathbb{R} = \Theta : \varphi^*(\mathbf{x}) = 1$ if $\sqrt{n}(\overline{x} - \mu_0)/\sigma_0 > z_\alpha$ and $\varphi^*(\mathbf{x}) = 0$ otherwise. The power function is

$$
\begin{aligned}
\gamma(u) \equiv \gamma_{\varphi^*}(\mu) &= P_\mu(\sqrt{n}(\overline{X} - \mu_0)/\sigma_0 > z_\alpha) \\
&= P_\mu\left(\sqrt{n}(\overline{X} - \mu)/\sigma_0 > z_\alpha - \frac{\sqrt{n}(\mu - \mu_0)}{\sigma_0}\right) \\
&= 1 - \Phi\left(z_\alpha - \frac{\sqrt{n}(\mu - \mu_0)}{\sigma_0}\right) \qquad (\mu \in \Theta_1 = (\mu_0, \infty)), \quad \text{(C.1)}
\end{aligned}
$$

where Φ is the (cumulative) distribution function of a standard normal random variable.

Similarly, the power function of the UMP unbiased test for $H_0 : \mu = \mu_0$ against $H_1 : \mu \neq \mu_0$ is

$$
\begin{aligned}
\gamma(u) &= P_\mu\left(\left|\frac{\sqrt{n}(\overline{X} - \mu_0)}{\sigma_0}\right| > \frac{z_\alpha}{2}\right) \\
&= 1 - P_\mu\left(-\frac{z_\alpha}{2} \leq \frac{\sqrt{n}(\overline{X} - \mu_0)}{\sigma_0} \leq \frac{z_\alpha}{2}\right) \\
&= 1 - P_\mu\left(-\frac{z_\alpha}{2} - \frac{\sqrt{n}(\mu - \mu_0)}{\sigma_0} \leq \frac{\sqrt{n}(\overline{X} - \mu)}{\sigma_0} \leq z_{\alpha/2} - \frac{\sqrt{n}(\mu - \mu_0)}{\sigma_0}\right) \\
&= 1 - \left\{\Phi\left(z_{\alpha/2} - \frac{\sqrt{n}(\mu - \mu_0)}{\sigma_0}\right) - \Phi\left(-z_{\alpha/2} - \frac{\sqrt{n}(\mu - \mu_0)}{\sigma_0}\right)\right\} \quad (\mu \neq \mu_0).
\end{aligned}
$$

$$\text{(C.2)}$$

Example C.2. The UMP test for $H_0 : \sigma^2 \leq \sigma_0^2$ against $H_1 : \sigma^2 > \sigma_0^2$ based on i.i.d. $\mathbf{N}(\mu_0, \sigma^2)$ random variables X_i $(1 \leq i \leq n)$ is to reject H_0 iff $\sum_{i=1}^{n}(X_i - \mu_0)^2/\sigma_0^2 > \chi^2_{1-\alpha}(n)$ [upper α-point of chi-square distribution with n d.f.]. Its power is

© Springer-Verlag New York 2016
R. Bhattacharya et al., *A Course in Mathematical Statistics and Large Sample Theory*, Springer Texts in Statistics, DOI 10.1007/978-1-4939-4032-5

$$\gamma(\sigma^2) = P_{\sigma^2}\left(\sum_{i=1}^{n}(X_i - \mu_0)^2/\sigma_0^2 > \chi_{1-\alpha}^2(n)\right)$$

$$\equiv P_{\sigma^2}\left(\sum_{i=1}^{n}(X_i - \mu_0)^2/\sigma^2 > \frac{\sigma_0^2}{\sigma^2}\chi_{1-\alpha}^2(n)\right) = \int_{\frac{\sigma_0^2}{\sigma^2}\chi_{1-\alpha}^2(n)}^{\infty} k_n(u)\,du$$

$$= 1 - K_n\left(\frac{\sigma_0^2}{\sigma^2}\chi_{1-\alpha}^2(n)\right) \qquad (\sigma^2 > \sigma_0^2), \tag{C.3}$$

where k_n is the p.d.f., and K_n the cumulative distribution function of the chi-square distribution with n d.f.

Example C.3. Consider now the test for $H_0 : \mu \le \mu_0$ against $H_1 : \mu > \mu_0$ based on i.i.d. X_i $(1 \le i \le n)$ which are $\mathbf{N}(\mu, \sigma^2)$ with σ^2 unknown. The UMP unbiased test of size α rejects H_0 iff $\sqrt{n}(\overline{X} - \mu_0)/s > t_{1-\alpha}(n-1)$ [the upper α-point of Student's t with $n-1$ d.f.]. Here $s^2 = \sum(X_i - \overline{X})^2/(n-1)$; and $\frac{(n-1)s^2}{\sigma^2}$ has a chi-square distribution with $n-1$ d.f. and is independent of $\sqrt{n}(\overline{X} - \mu)/\sigma$ which is $\mathbf{N}(0,1)$ under P_{μ,σ^2}. Thus the power of the test is

$$\gamma(\mu; \sigma^2) = P_{\mu,\sigma^2}\left(\frac{\sqrt{n}(\overline{X} - \mu_0)}{s} > t_{1-\alpha}(n-1)\right)$$

$$= E_{\mu,\sigma^2}\left[P_{\mu,\sigma^2}\left(\frac{\sqrt{n}(\overline{X} - \mu_0)}{\sigma} > \frac{s}{\sigma}t_{1-\alpha}(n-1)\Big| s\right)\right]$$

$$= E_{\mu,\sigma^2}\left[P_{\mu,\sigma^2}\left(\frac{\sqrt{n}(\overline{X} - \mu)}{\sigma} > \frac{\sqrt{n}(\mu - \mu_0)}{\sigma} + \frac{s}{\sigma}t_{1-\alpha}(n-1)\Big| s\right)\right]$$

$$= 1 - \int_0^{\infty} \Phi\left(-\frac{\sqrt{n}(\mu - \mu_0)}{\sigma} + \frac{(t_{1-\alpha}(n-1))}{\sqrt{n-1}}\sqrt{u}\right) k_{n-1}(u)\,du. \tag{C.4}$$

In obtaining (C.4) we used (i) the independence of $\overline{X}$ and s and (ii) the fact that $U \equiv (n-1)\left(\frac{s}{\sigma}\right)^2$ is a chi-square random variable with $n-1$ d.f. (As before, k_ν is the p.d.f. of a chi-square with ν d.f.)

Replacing $t_{1-\alpha}(n-1)$ by a general argument $t > 0$, and differentiating (C.4) w.r.t. t (and changing the sign of the derivative) one arrives at the p.d.f. of the so-called *non-central t-distribution with the noncentrality parameter* $\Delta = \frac{\sqrt{n}(\mu - \mu_0)}{\sigma}$ *and d.f.* $\nu = n-1$, as

$$f_{\nu,\Delta}(t) \equiv f(t) = \int_0^{\infty} \varphi\left(-\Delta + \frac{t}{\sqrt{\nu}}\sqrt{u}\right) \cdot \frac{\sqrt{u}}{\sqrt{\nu}} k_\nu(u)\,du \tag{C.5}$$

where φ is the standard normal density. Simplifying a little, this density is (Recall: $k_\nu(u) = \frac{1}{2^{\nu/2}\Gamma\left(\frac{\nu}{2}\right)} e^{-\frac{u}{2}}u^{\frac{\nu}{2}-1}$)

$$f(t) = \frac{1}{\sqrt{2\pi}\,2^{\nu/2}\Gamma\left(\frac{\nu}{2}\right)\sqrt{\nu}} \int_0^{\infty} u^{\frac{\nu-1}{2}} \exp\left\{-\frac{u\left(1+\frac{t^2}{\nu}\right) + \Delta^2 - \frac{2\Delta t}{\sqrt{\nu}}\sqrt{u}}{2}\right\} du$$

$$= \frac{e^{-\frac{\nu\Delta^2}{2(\nu+t^2)}}}{\sqrt{\nu}\,\sqrt{2\pi}\,2^{\nu/2}\Gamma\left(\frac{\nu}{2}\right)} \int_0^{\infty} u^{\frac{\nu-1}{2}} \exp\left\{-\frac{1}{2}\left(\sqrt{u}\sqrt{\frac{\nu+t^2}{\nu}} - \frac{\Delta t}{\sqrt{\nu+t^2}}\right)^2\right\} du.$$

So, one has the *non-central t p.d.f.*

$$f(t) = c_\nu (t^2 + \nu)^{-(\nu+1)/2} \exp\left\{ -\frac{\nu\Delta^2}{2(\nu+t^2)} \right\} \int_0^\infty \exp\left\{ -\frac{1}{2}\left(x - \frac{\Delta t}{\sqrt{\nu+t^2}} \right) \right\}^2 dx,$$

$$(C.6)$$

with a change of variables $u \to x = \sqrt{\frac{\nu+t^2}{\nu}}\, x$. Here

$$c_\nu = \frac{\nu^{\nu/2}}{\sqrt{\pi}\,\Gamma\left(\frac{\nu}{2}\right) 2^{(\nu-1)/2}} \cdot$$

The power function (C.4) may now be simply expressed as

$$\gamma(\mu; \sigma^2) = \int_{t_{1-\alpha}(n-1)}^\infty f_{\nu,\Delta}(t)dt \quad \text{with } \nu = n-1, \ \Delta = \frac{\sqrt{n}(\mu - \mu_0)}{\sigma}, \qquad (C.7)$$

$(\mu > \mu_0,\ \sigma^2 > 0)$.

Example C.4. To test $H_0 : \sigma^2 \leq \sigma_0^2$ against $H_1 : \sigma^2 > \sigma_0^2$ based on i.i.d. $\mathbf{N}(\mu, \sigma^2)$ random variables X_i (μ not known), the UMP unbiased test is to reject H_0 iff $\frac{\sum(X_i - \overline{X})^2}{\sigma_0^2} \equiv \frac{(n-1)s^2}{\sigma_0^2} > \chi_{1-\alpha}^2(n-1)$. Its power is

$$P_{\mu,\sigma^2}\left(\frac{(n-1)s^2}{\sigma_0^2} > \chi_{1-\alpha}^2(n-1) \right) \equiv P_{\mu,\sigma^2}\left(\frac{(n-1)s^2}{\sigma^2} > \frac{\sigma_0^2}{\sigma^2}\chi_{1-\alpha}^2(n-1) \right)$$

$$= \int_{\frac{\sigma_0^2}{\sigma^2}\chi_{1-\alpha}^2(n-1)} k_{n-1}(u)du \qquad (\sigma^2 > \sigma_0^2).$$

Example C.5 (The Non-Central Chi-Square Distribution). Finally, we consider the noncentral chi-square distribution. If Y is $\mathbf{N}(\mu, 1)$ then the distribution of Y^2 is said to be a noncentral chi-square distribution with 1 d.f. The sum of squares $\sum Y_i^2$ of n independent $N(\mu_i, 1)$ normal random variables Y_i s said to have the *noncentral chi-square distribution with* d.f. n and *noncentrality parameter* $\sum_1^n \mu_i^2$.

Proposition C.1. *Let Y_i be $\mathbf{N}(\mu_i, 1)$, $1 \leq i \leq n$, and $Y_1, Y_2, \ldots, Y_n$ independent. Then the p.d.f. of $V := \sum_1^n Y_i^2$ is given by the noncentral chi-square p.d.f. with d.f. n and noncentrality parameter $\Delta = \sum_1^n \mu_i^2$, namely,*

$$f(v) \equiv f(v; n, \Delta) = \sum_{j=0}^\infty p\left(j; \frac{\Delta}{2} \right) k_{n+2j}(v) \qquad (C.8)$$

where

$$p\left(j; \frac{\Delta}{2} \right) = e^{-\frac{\Delta}{2}} \left(\frac{\Delta}{2} \right)^j \frac{1}{j!} \qquad (j = 0, 1, 2, \ldots) \qquad (C.9)$$

is the probability (mass) function of a Poisson random variable with mean (parameter) $\frac{\Delta}{2}$, and k_{n+2j} is the chi-square p.d.f. with d.f. $n + 2j$.

Proof. The p.d.f. of $V_i := Y_i^2$ is (use a two-to-one change of variables $\{y, -y\} \to v = y^2$, or differentiate (w.r.t. v) the cumulative distribution function of V_i, namely, $\int_{-\sqrt{v}}^{\sqrt{v}} \frac{1}{\sqrt{2\pi}} e^{-\frac{(y-\mu_i)^2}{2}} dy$)

$$f_i(v) = \frac{1}{2\sqrt{2\pi}}\, v^{-\frac{1}{2}}\left(e^{-\frac{1}{2}v-\frac{1}{2}\mu_i^2}\right)\left\{e^{-\mu_i\sqrt{v}} + e^{\mu_i\sqrt{v}}\right\}$$

$$= \frac{1}{2\sqrt{2\pi}}\, v^{-\frac{1}{2}}\left(e^{-\frac{1}{2}v}\cdot e^{-\frac{1}{2}\mu_i^2}\right)\left\{\sum_{j=0}^{\infty}\frac{2(\mu_i^2 v)^j}{(2j)!}\right\}$$

$$= \sum_{j=0}^{\infty}\frac{1}{j!}\,e^{-\frac{\mu_i^2}{2}}\left(\frac{\mu_i^2}{2}\right)^{j}\cdot\frac{1}{\sqrt{2\pi}}\frac{j!}{(2j)!}\,2^j\cdot 2^{j-\frac{1}{2}}\left(\frac{v}{2}\right)^{j-\frac{1}{2}}e^{-\frac{v}{2}}$$

$$= \sum_{j=0}^{\infty}p\left(j;\frac{\mu_i^2}{2}\right)\cdot\frac{1}{\sqrt{2\pi}}\frac{2^{j-\frac{1}{2}}}{(2j-1)(2j-3)\cdots 3.1}\,e^{-\frac{v}{2}}\left(\frac{v}{2}\right)^{j-\frac{1}{2}}$$

$$= \sum_{j=0}^{\infty}p\left(j;\frac{\mu_i^2}{2}\right)\frac{1}{2\Gamma\left(j+\frac{1}{2}\right)}\left(\frac{v}{2}\right)^{\frac{2j-1}{2}}e^{-\frac{v}{2}}$$

$$= \sum_{j=0}^{\infty}p\left(j;\frac{\mu_i^2}{2}\right)k_{2j+1}(v).$$

The p.d.f. of $Y_1^2 + Y_2^2$ is the convolution of $f_1(v) := f(v;1,\mu_1^2)$ and $f_2(v) := f(v;1,\mu_2^2)$, namely,

$$(f_1 * f_2)(v) = \sum_{j=0}^{\infty}\sum_{j'=0}^{\infty}p\left(j;\frac{\mu_1^2}{2}\right)p\left(j';\frac{\mu_2^2}{2}\right)(k_{1+2j} * k_{1+2j'})(v)$$

$$= \sum_{j=0}^{\infty}\sum_{j'=0}^{\infty}\frac{e^{-(\mu_1^2+\mu_2^2)/2}}{j!\,j'!}\left(\frac{\mu_1^2}{2}\right)^{j}\left(\frac{\mu_2^2}{2}\right)^{j'}k_{2+2(j+j')}(v).$$

For the convolution of two (central) chi-square p.d.f. s is a chi-square p.d.f. whose d.f. is the sum of the d.f.'s of its two components. This equals

$$\sum_{s=0}^{\infty}\left\{\sum_{(j,j'):j+j'=s}e^{-(\mu_1^2+\mu_2^2)/2}\cdot\frac{1}{j!\,j'!}\left(\frac{\mu_1^2}{2}\right)\left(\frac{\mu_2^2}{2}\right)^{j'}\right\}k_{2+2s}(v)$$

$$= \sum_{s=0}^{\infty}e^{-(\mu_1^2+\mu_2^2)/2}\frac{(\mu_1^2+\mu_2^2)^s}{2^s\,s!}\,k_{2+2s}(v) = f(v;2,\mu_1^2+\mu_2^2). \tag{C.10}$$

For the last step in (C.10), one uses the combinatorial identity:
$\sum_{\{(j,j')\in(\mathbb{Z}^+)^2:j+j'=s\}}a^j\,b^{j'}/j!\,j'! = \frac{(a+b)^s}{s!}$. *(Binomial expansion of $(a+b)^s$).*
The general result follows by induction. $\qquad\qquad\qquad\qquad\qquad\qquad\square$

Remark C.1. The non-central chi-square appears in the *power function of F-tests* in linear models.

The final example in this section enables one to compute the power of F tests in linear models.

Example C.6 (The Non-Central F Distribution $\mathscr{F}_{r,s}(\Delta)$). Recall the F-distribution $\mathscr{F}_{r,s}$ in A.2, VIII. This is the distribution of the F-statistic

$$F = \frac{U/r}{V/s}, \tag{C.11}$$

where U and V are independent chi-square random variables with degrees of freedom r, s, respectively. If we let U be a *noncentral chi-square distribution* with d.f. r and noncentrality parameter Δ, then the Proposition above (Proposition C.1) says that U may be thought of as a chi-square random variable with a *random degree of freedom* $r + 2\gamma$, γ having the Poisson distribution $\mathscr{P}(\Delta/2)$ (with mean $\Delta/2$). Therefore, conditionally given $\gamma = j$, the distribution of $(U/(r + 2j))/(V/s)$ is the (central) F-distribution $\mathscr{F}_{r+2j,s}$, with p.d.f. $f_{r+2j,s}$ (see Sect. A.2). Hence the p.d.f. of $(U/r)/(V/s) = \left(\frac{r+2j}{r}\right)[(U/(r+2j))/(V/s)]$, conditionally given $\gamma = j$, is

$$
f_{r+2j,s}\left(\frac{r}{r+2j}\,x\right)\left(\frac{r}{r+2j}\right) \qquad (j = 0, 1, \dots). \tag{C.12}
$$

Therefore, the density of the non-central F-statistic (C.11) is

$$
f_{r,s,\Delta}(x) = \sum_{j=0}^{\infty} e^{-\Delta/2}\left(\frac{\Delta}{2}\right)^{j}\frac{1}{j!}\left(\frac{r}{r+2j}\right)f_{r+2j,s}\left(\frac{r}{r+2j}\,x\right), \qquad 0 < x < \infty. \tag{C.13}
$$

By using the formula for $f_{r+2j,s}$ one gets

$$
f_{r,s,\Delta}(x) = e^{-\Delta/2}\sum_{j=0}^{\infty}\left(\frac{\Delta}{2}\right)^{j}\frac{1}{j!}\,\frac{r^{\frac{r}{2}+j}s^{\frac{s}{2}}\,\Gamma\left(\frac{r+s}{2}+j\right)}{\Gamma\left(\frac{r}{2}+j\right)\Gamma\left(\frac{s}{2}\right)}\cdot\frac{x^{\frac{r}{2}+j-1}}{(s+rx)^{\frac{r+s}{2}+j}}\cdot \tag{C.14}
$$

For the use of the N–P Lemma to an observed random variable X given by (C.11), one considers the ratio

$$
\frac{f_{r,s,\Delta}(x)}{f_{r,s}(x)} = e^{-\frac{\Delta}{2}}\sum_{j=0}^{\infty}\left(\frac{\Delta}{2}\right)^{j}\frac{1}{j!}\,\frac{\Gamma\left(\frac{r+s}{2}+j\right)\Gamma\left(\frac{r}{2}\right)}{\Gamma\left(\frac{r}{2}+j\right)\Gamma\left(\frac{r+s}{2}\right)}\left(\frac{rx}{s+rx}\right)^{j}. \tag{C.15}
$$

Each summand is an increasing function of x and, hence, so is the sum.

Appendix D
Liapounov's, Lindeberg's and Polya's Theorems

Liapounov's Central Limit Theorem *Let $X_{j,n}$ $(1 \leq j \leq k_n; \, n = 1, 2, \dots)$ be a triangular array of independent random variables each with zero mean. Write $s_n^2 = \Sigma_{j=1}^{k_n} \mathrm{var}(X_{j,n})$, $\rho_{3,n} = \Sigma_{j=1}^{k_n} E|X_{j,n}|^3$. If $\rho_{3,n}/s_n^3 \to 0$ as $n \to \infty$, then*

$$\sum_{j=1}^{k_n} \frac{X_{j,n}}{s_n} \xrightarrow{\mathscr{L}} N(0,1). \tag{D.1}$$

Liapounov's Theorem follows from a more general theorem due to Lindeberg:

Lindeberg's Central Limit Theorem [1] Let the triangular array $X_{j,n}$ $(1 \leq j \leq k_n; \, n = 1, 2, \dots)$ have mean zero and finite variance. Write $s_n^2 = \sum_{j=1}^{k_n} \mathrm{var}\, X_{j,n}$, $\varepsilon_{j,n} = X_{j,n}/s_n$. If the *Lindeberg condition*

$$\sum_{j=1}^{k_n} E\varepsilon_{j,n}^2 \mathbf{1}_{[|\varepsilon_{j,n}|>\eta]} \longrightarrow 0 \quad \text{as } n \to \infty \tag{D.2}$$

holds for every constant $\eta > 0$, then

$$\sum_{j=1}^{k_n} \varepsilon_{j,n} \xrightarrow{\mathscr{L}} N(0,1). \tag{D.3}$$

Polya's Theorem *Suppose $F_n \xrightarrow{\mathscr{L}} F$, and $F(x)$ is continuous on $(-\infty, \infty)$. Then $\sup_x |F_n(x) - F(x)| \to 0$ as $n \to \infty$.*

Proof. We need to show that, under the given hypotheses, given any $\varepsilon > 0$ there exists an integer $n(\varepsilon)$ such that

$$\sup_x |F_n(x) - F(x)| < \varepsilon \qquad \text{for all } n > n(\varepsilon).$$

---

[1] For proof of Lindeberg's and Liapounov's CLTs, see Bhattacharya and Waymire (2007), pp. 99–103.

Find $A_\varepsilon > 0$ such that $F(-A_\varepsilon) < \varepsilon/4, 1 - F(A_\varepsilon) < \varepsilon/4$. Since $F_n(-A_\varepsilon) \to F(-A_\varepsilon)$ and $F_n(A_\varepsilon) \to F(A_\varepsilon)$, there exists a positive integer N_1 such that $\forall\, n > N_1$, $|F_n(-A_\varepsilon) - F(-A_\varepsilon)| < \varepsilon/4, |1 - F_n(A_\varepsilon) - \{1 - F(A_\varepsilon)\}| = |F_n(A_\varepsilon) - F(A_\varepsilon)| < \varepsilon/4$. This implies,

$$F_n(-A_\varepsilon) < \frac{\varepsilon}{4} + \frac{\varepsilon}{4} = \frac{\varepsilon}{2}, \quad 1 - F_n(+A_\varepsilon) < \frac{\varepsilon}{4} + \frac{\varepsilon}{4} = \frac{\varepsilon}{2} \quad \forall\, n > N_1. \tag{D.4}$$

Since F is continuous, it follows that it is *uniformly* continuous on $[-A_\varepsilon, A_\varepsilon]$. Hence there exists $\delta > 0$ such that $|F(x) - F(y)| < \varepsilon/2$, whenever $x, y \in [-A_\varepsilon, A_\varepsilon]$ and $|x - y| < \delta$. Choose points $x_0 = -A_\varepsilon < x_1 < x_2 < \cdots < x_k = A_\varepsilon$ such that $|x_{j+1} - x_j| < \delta$. There exists, for each j, an integer M_j such that $|F_n(x_j) - F(x_j)| < \varepsilon/2$ for all $n > M_j$ $(j = 0, 1, \ldots, k)$. Now let

$$n(\varepsilon) = \max\{N_1, M_0, M_1, \ldots, M_k\}. \tag{D.5}$$

Then if $x \in [x_j, x_{j+1}]$, one has $\forall\, n > M_{j+1}$

$$F_n(x) - F(x) \leq F_n(x_{j+1}) - F(x_j)$$
$$= F_n(x_{j+1}) - F(x_{j+1}) + F(x_{j+1}) - F(x_j) < \frac{\varepsilon}{2} + \frac{\varepsilon}{2} = \varepsilon,$$

and $\forall\, n > M_j$ one has

$$F(x) - F_n(x) \leq F(x_{j+1}) - F_n(x_j) = F(x_{j+1}) - F(x_j) + F(x_j) - F_n(x_j) < \frac{\varepsilon}{2} + \frac{\varepsilon}{2} = \varepsilon,$$

i.e.

$$|F_n(x) - F(x)| < \varepsilon, \quad \forall\, n > n(\varepsilon), \text{ if } x \in [-A_\varepsilon, A_\varepsilon]. \tag{D.6}$$

On the other hand, if $x < -A_\varepsilon$, then $\forall\, n > N_1$

$$|F_n(x) - F(x)| \leq F_n(x) + F(x) \leq F_n(-A_\varepsilon) + F(-A_\varepsilon) < \frac{\varepsilon}{2} + \frac{\varepsilon}{4} < \varepsilon, \tag{D.7}$$

while if $x > A_\varepsilon$, then for all $n > N_1$

$$|F_n(x) - F(x)| = |1 - F_n(x) - \{1 - F(x)\}| \leq 1 - F_n(x) + 1 - F(x)$$
$$\leq 1 - F_n(A_\varepsilon) + 1 - F(A_\varepsilon) < \frac{\varepsilon}{2} + \frac{\varepsilon}{4} < \varepsilon. \tag{D.8}$$

Combining (D.6)–(D.8) one has

$$|F_n(x) - F(x)| < \varepsilon \quad \forall\, x \in (-\infty, \infty) \text{ if } n > n(\varepsilon).$$

$$\square$$

Solutions of Selected Exercises in Part I

Chapter 1

1.2 Let us index the N_i members of the ith stratum as $\{x_{ij} : 1 \leq j \leq N_i\}$, $i = 1, 2, \ldots, k$. Also, let X_{ij} $(j = 1, \ldots, n_i)$ denote a random sample of size n_i (with replacement) from the ith stratum $(i = 1, \ldots, k)$. Then, $E(\overline{X}_i) = m_i \equiv \frac{1}{N_i} \sum_{j=1}^{N_i} x_{ij} =$ mean of the ith stratum, $\text{var}(\overline{X}_i) = \frac{v_i}{n_i}$ where $v_i = \frac{1}{N_i} \sum_{j=1}^{N_i} (x_{ij} - m_i)^2 =$ variance of the ith stratum.

(a) Let $\overline{Y} = \sum_{i=1}^{k} w_i \overline{X}_i$ $(w_i = \frac{N_i}{N})$. Then (i) $E\overline{Y} = \sum_{i=1}^{k} w_i E(\overline{X}_i) = \sum_{i=1}^{k} w_i m_i = \sum_{i=1}^{k} \frac{N_i}{N} \left(\frac{1}{N_i} \sum_{j=1}^{N_i} x_{ij} \right) = \frac{1}{N} \sum_{i=1}^{k} \sum_{j=1}^{N_i} x_{ij} = \frac{x_1 + x_2 + \cdots + x_N}{N} = m$ and
(ii)

$$\text{var}(\overline{Y}) = \sum_{i=1}^{k} w_i^2 \, \text{var}(\overline{X}_i) = \sum_{i=1}^{k} w_i^2 \frac{v_i}{n_i} \, . \tag{S.1}$$

(b) One may express the population variance as

$$v = \frac{1}{N} \sum_{i=1}^{N} (x_i - m)^2 = \frac{1}{N} \sum_{i=1}^{k} \sum_{j=1}^{N_i} (x_{ij} - m)^2$$

$$= \frac{1}{N} \sum_{i=1}^{k} \sum_{j=1}^{N_i} (x_{ij} - m_i + m_i - m)^2$$

$$= \frac{1}{N} \sum_{i=1}^{k} \xi \sum_{j=1}^{N_i} (x_{ij} - m_i)^2 + \sum_{j=1}^{N_i} (m_i - m)^2 + 2(m_i - m) \sum_{j=1}^{N_i} (x_{ij} - m_i)$$

$$= \frac{1}{N} \sum_{i=1}^{k} \{N_i, v_i + N_i (m_i - m)^2 + 0\} = \sum_{i=1}^{k} \frac{N_i}{N} v_i + \sum_{i=1}^{k} \frac{N_i}{N} (m_i - m)^2$$

$$= \sum_{i=1}^{k} w_i v_i + \sum_{i=1}^{k} w_i (m_i - m)^2 . \tag{S.2}$$

© Springer-Verlag New York 2016
R. Bhattacharya et al., *A Course in Mathematical Statistics and Large Sample Theory*, Springer Texts in Statistics, DOI 10.1007/978-1-4939-4032-5

(c) Let $n_i = nw_i \; \forall \; i = 1, \ldots, k$. Then, using (S.1) and (S.2).

$$E(\overline{Y} - m)^2 = \sum_{i=1}^{k} w_i^2 \frac{v_i}{nw_i} = \frac{1}{n} \sum_{i=1}^{k} w_i v_i,$$

$$E(\overline{X} - m)^2 = \frac{v}{n} = \frac{1}{m} \sum_{i=1}^{k} w_i v_i + \frac{1}{n} \sum_{i=1}^{k} w_i (m_i - m)^2$$

$$> E(\overline{Y} - m)^2,$$

unless $m_i = m \; \forall \; i = 1, \ldots, k$. Only in the latter case $E(\overline{Y} - m)^2 = E(\overline{X} - m)^2$.

(d) Suppose v_i's are known. Then $E(\overline{Y} - m)^2$ is minimized (with respect to n_i's) by solving the equations [see (S.1)]

$$\frac{\partial}{\partial n_i} \left\{ \sum_{i=1}^{k} w_i^2 \frac{v_i}{n_i} + \lambda \left(\sum_{i'=1}^{k} n_{i'} \right) \right\} = 0 \quad (i = 1, \ldots, k),$$

where λ is the so-called Lagrange multiplier. That is, $-w_i^2 \frac{v_i}{n_i^2} + \lambda = 0$, or $w_i^2 \frac{v_i}{n_i^2} = \lambda$, or, $n_i^2 = \frac{w_i^2 v_i}{\lambda}$, or, $n_i = \frac{w_i \sqrt{v_i}}{\sqrt{\lambda}}$ $(i = 1, 2, \ldots, k)$.

Summing over i, one gets $n = \frac{\sum_{i=1}^{k} w_i \sqrt{v_i}}{\sqrt{\lambda}}$, or, $\sqrt{\lambda} = \frac{\sum_{i=1}^{k} w_i \sqrt{v_i}}{n}$. Hence the optimal choice for n_i is

$$n_i = \frac{n w_i \sqrt{v_i}}{\sum_{i=1}^{k} w_i \sqrt{v_i}} \quad (i = 1, 2, \ldots, k).$$

Thus, n_i is proportional to the size of the stratum and to the standard deviation of the stratum. $\qquad\square$

Chapter 2

2.1 In Example 2.1, p. 12, prove (2.4) and (2.6).

Proof of (2.4). Recall that one can express $\frac{(n-1)s^2}{\sigma^2}$ as

$$\frac{(n-1)s^2}{\sigma^2} = \sum_{i=2}^{n} Y_i^2 \quad (Y_i \text{ i.i.d. } N(0,1)).$$

Then

$$E\left(\frac{(n-1)s^2}{\sigma^2}\right)^2 = E\frac{(n-1)^2 s^4}{\sigma^4} = \frac{(n-1)^2}{\sigma^4} E s^4 = E\left(\sum_{i=2}^{n} Y_i^2\right)^2. \qquad (S.3)$$

But the best expectation equals (using $EY_1^4 = 3$),

$$E\left(\sum_{i=2}^n Y_i^2\right)^2 = E\left(\sum_{i=2}^n Y_i^4\right) + E\left(\sum_{2\le i\ne j\le n} Y_i^2 Y_j^2\right)$$

$$= \sum_{i=2}^n EY_i^4 + \sum_{2\le i\ne j\le n} (EY_i^2)(EY_j^2) = 3(n-1) + (n-1)(n-2)$$

$$= (n-1)(3+n-2) = (n-1)(n+1).$$

Hence the last equality in (S.3) yields

$$Es^4 = \frac{\sigma^4}{(n-1)^2}(n-1)(n+1) = \frac{n+1}{n-1}\sigma^4,$$

so that

$$E_\theta(s^2 - \sigma^2) = E_\theta s^4 = \sigma^4 - 2\sigma^2 E_\theta s^2 = E_\theta s^4 - \sigma^4 = \frac{2\sigma^4}{n-1}.$$

Proof of (2.6). Here $\theta = (\mu, \sigma^2)$, $d(\mathbf{X}) = (\overline{X}, s^2)$. Therefore, $R(\theta, d) = E_\theta|\theta - d(\mathbf{X})|^2 = E_\theta[(\overline{X} - \mu)^2 + (s^2 - \sigma^2)^2] = \frac{\sigma^2}{n} + \frac{2\sigma^4}{n-1}$.

Chapter 3

3.6 $X_1, \ldots, X_n$ i.i.d. Bernoulli, $P_\theta(X_i = 1) = \theta$, $P_\theta(X_i = 0) = 1-\theta$, $\theta \in [0,1] = \Theta$. the (joint) distribution of $\mathbf{X} = (X_1, \ldots, X_n)$ is

$$P_\theta(\mathbf{X} = \mathbf{x}) = \theta^r(1-\theta)^{n-r} \quad (r = \textstyle\sum_{j=1}^n x_j, \ \mathbf{x} = (x_1, \ldots, x_n) \in \{0,1\}^n).$$

By Example 3.5, the Bayes estimator for the beta prior $\mathscr{B}_e(\alpha, \beta)$ under loss $L(\theta, a) = C(\theta - a)^2$ ($c > 0$ does not depend on θ) is given by

$$d_o(\mathbf{x}) = \frac{r + \alpha}{n + \alpha + \beta}.$$

Its risk function is

$$R(\theta, d_0) = cE_\theta\left(\varepsilon - \frac{\sum_1^n X_j + \alpha}{n + \alpha + \beta}\right)^2$$

$$= c\left[\operatorname{var}_\theta(d_0(\mathbf{X})) + (E_\theta d_0(\mathbf{X}) - \theta)^2\right]$$

$$= c\left[\frac{n\theta(1-\theta)}{n + \alpha + \beta^2} + \left(\frac{n\theta + \alpha}{n + \alpha + \beta} - \theta\right)^2\right]$$

$$= c\left[\frac{n\theta(1-\theta)}{(n + \alpha + \beta)^2} + \frac{\alpha + \theta(\alpha + \beta)^2}{(n + \alpha + \beta)^2}\right]$$

$$= \frac{c}{(n + \alpha + \beta)^2}[n\theta(1-\theta) + (\alpha - \theta(\alpha + \beta))^2]$$

For $\alpha = \beta = \frac{\sqrt{n}}{2}$, this simplifies to

$$R(\theta, d_0) = \frac{c\frac{n}{4}}{(n + \sqrt{n})^2} = \frac{c}{4(\sqrt{n} + 1)^2},$$

a Bayes rule which is an equalizer rule. Hence it is **minimax** (Theorem 3.7). Also, the property P1 (p. 29) holds with $\widetilde{\Theta} = (0, 1)$. Hence d_0 is admissible,

$$d_0(\mathbf{X}) = \left(\sum_1 X_{jo} + \frac{\sqrt{n}}{2} / (n + \sqrt{n}) \right) = \frac{\overline{X} + \frac{1}{2\sqrt{n}}}{1 + \frac{1}{\sqrt{n}}}.$$

3.7 $X_1, \ldots, X_n$ i.i.d. Poisson, with common pmf $P_\theta(X_i = x) = e^{-\theta}\theta^x / x!$, $x \in \{0, 1, 2, \ldots\} = \mathbb{Z}_+$, $\theta \in (0, \infty) = \Theta$. The (joint) pmf of $\mathbf{X} = (X_1, \ldots, X_n)$ is

$$f(\mathbf{x} \mid \theta) = e^{-n\theta}\theta^{\sum_1^n x_j} / \prod_{j=1}^n x_j! \quad (\mathbf{x} = (x_1, \ldots, x_n) \in \mathbb{Z}_+^n = \mathscr{H}).$$

(a) Let $L(\theta, a) = \frac{e^\theta}{\theta}(\theta - a)^2$. For the gamma prior $\tau = \mathscr{G}(\alpha, \beta)$, the Bayes risk of an estimator d is

$$r(\tau, d) = \sum_{\mathbf{x} \in \mathscr{X}} \frac{1}{\prod x_j!} \int_0^\infty \frac{e^\theta}{\theta}(\theta - d(\mathbf{x}))^2 e^{-n\theta}\theta^{\sum x_j}$$

$$= \sum_{\mathbf{x} \in \mathscr{X}} \frac{1}{\prod x_j!} \int_0^\infty \frac{1}{\Gamma(\beta)\alpha^\beta}(\theta^2 - 2\theta d(\mathbf{x}) + d^2(\mathbf{x}))e^{-\theta/\frac{\alpha}{(n-1)\alpha+1}} \cdot \theta^{\sum x_j + \beta - 2} d\theta.$$

If $\mathbf{x} = (0, \ldots, \theta)$ then the integral diverges (at 0) if $\beta \le 1$ ($\beta > 0$), unless one sets $d((0, \ldots, 0)) = 0$. For all other $\mathbf{x}$ (and for $\mathbf{x} = (0, \ldots, 0)$ if $\beta > 1$), one has the integral equal to

$$\frac{1}{\Gamma(\beta)\alpha^\beta} \cdot \Gamma(\Sigma x_j + \beta - 1)\alpha'^{\beta'} \cdot \int_0^\infty (\theta - d(\mathbf{x}))^2 g(\theta \mid \alpha', \beta') d\theta,$$

where $g(\theta \mid \alpha', \beta')$ is the pdf of the gamma distribution $\mathscr{G}(\alpha', \beta')$, with $\alpha' = \frac{\alpha}{(n-1)\alpha+1}$, $\beta' = \sum_1^n x_j + \beta - 1$. This is minimized by the mean of this gamma distribution (for each $\mathbf{x}$, if $\beta > 1$), i.e., the Bayes estimator is given by

$$d_0(\mathbf{x}) = \alpha'\beta' = \frac{\alpha}{(n-1)\alpha + 1}\left(\sum_1^n x_j + \beta - 1 \right).$$

(b) For the case $\beta = 1$, this becomes

$$d_0(\mathbf{x}) = \frac{\alpha}{(n-1)\alpha + 1} \sum_1^n x_j, \quad \mathbf{x} \in \mathscr{X} = \mathbb{Z}_+^n, \tag{S.4}$$

which automatically satisfies the restriction $d_0((0, \ldots, 0)) = 0$ imposed earlier. If one takes $\alpha = 1$, $\beta = 1$, then (S.4) becomes

$$d_0(\mathbf{x}) = \overline{x}.$$

Hence $\bar{x}$ is admissible under the loss function $L(\theta, a) = \frac{e^\theta}{\theta}(\theta - a)^2$. [Note that P_1 holds with $\widetilde{\Theta} = \Theta$, since $P_\theta(\mathbf{x}) > 0 \ \forall \ \mathbf{x}$, and $P_{\theta_0}(A) = 0$ implies A is empty, whatever be $\theta_0 \in \Theta$.]

(c) We now show that $d_0 = \overline{X}$ is admissible under squared error loss $L(\theta, a) = (\theta - a)^2$. Let $\widetilde{L}(\theta, a) = \frac{e^\theta}{\theta}(\theta - a)^2 \equiv \frac{e^\theta}{\theta} L(\theta, a)$. Suppose $d_0 = \overline{X}$ is not admissible under L. Then there exists an estimator d_1 such that

$$R(\theta, d_1) \le R(\theta, d_0) \ \forall \ \theta, \quad R(\theta_0, d_1) < R(\theta_0, d_0)$$

for some $\theta_0 > 0$. Multiplying both sides of the inequalities by $\frac{e^\theta}{\theta}$, one obtains the risk function $\widetilde{R}(\theta, d_1)$, $\widetilde{R}(\theta, d_0)$ under $\widetilde{L}$ satisfying the inequalities

$$\widetilde{R}(\theta, d_1) \le \widetilde{R}(\theta, d_0) \ \forall \ \theta, \quad \widetilde{R}(\theta_0, d_1) < \widetilde{R}(\theta_0, d_0).$$

But this implies d_0 is inadmissible under loss $\widetilde{L}$—a contradiction.

(d) We now show that $\overline{X}$ is minimax under the loss function $L(\theta, a) = (\theta - a)^2/\theta$. The risk function of $\overline{X}$ is

$$R(\theta, \overline{X}) = E_\theta(\theta - \overline{X})^2/\theta = \frac{1}{\theta} \operatorname{var}_\theta(\overline{X}) = \frac{1}{\theta} n\theta = \frac{1}{\theta} \ \forall \ \theta \in \Theta = (0, \infty).$$

[Note that the variance of X_1 is θ.] Thus $\overline{X}$ is an equalizer rule. We will now apply Theorem 3.6 to find a sequence of priors τ_N such that the Bayes risks of the corresponding Bayes estimators d_N, say, satisfy

$$\lim_{N \to \infty} r(\tau_N, d_N) = \frac{1}{n}. \tag{S.5}$$

Now for the gamma prior $\mathscr{G}(\alpha, \beta)$, the Bayes estimator is obtained by minimizing the Bayes risk $r(\tau, d)$ over the class of all estimators d:

$$r(\tau, d) = \sum_{\mathbf{x} \in \mathscr{X}} \frac{1}{\prod_1^n x_j!} \int_0^\infty \frac{(\theta - d(\mathbf{x}))^2}{\theta} e^{-n\theta} \theta^{\sum_i^n x_j} \frac{\theta^{\beta-1} e^{-\theta/d}}{\Gamma(\beta)\alpha^\beta} d\theta$$

$$= \frac{1}{\Gamma(\beta)\alpha^\beta} \sum_{\mathbf{x} \in \mathscr{X}} \frac{1}{\prod_1^n x_j!} \int_0^\infty (\theta - d(\mathbf{x}))^2 e^{-\theta/\frac{\alpha}{n\alpha+1}} \theta^{\sum x_j + \beta - 2} d\theta$$

$$= \frac{1}{\Gamma(\beta)\alpha^\beta} \cdot \Gamma(\beta')\alpha'^{\beta'} \sum_{\mathbf{x} \in \mathscr{X}} \frac{1}{\prod_1^n x_j!} \int_0^\infty (\theta - d(\mathbf{x}))^2 g(\theta \mid \alpha', \beta') d\theta,$$

where $g(\theta \mid \alpha', \beta')$ is the pdf of the gamma distribution $\mathscr{G}(\alpha', \beta')$, and $\alpha' = \frac{\alpha}{n\alpha+1}$, $\beta' = \sum_1^n x_j + \beta - 1$.

Let us choose $\beta > 1$. Then, as in (b), the last integral is minimized (for each $\mathbf{x} \in \mathscr{X}$) by $\alpha'\beta' = \frac{\alpha(\sum_1^n x_j + \beta - 1)}{n\alpha+1}$. Hence the Bayes estimator for the prior $\mathscr{G}(\alpha, \beta)$, $\beta > 1$, is

$$d_0(\mathbf{x}) = \frac{\alpha(\sum_{j=1}^n x_j + \beta - 1)}{n\alpha + 1}.$$

We now show that for $\alpha = \alpha_N$, $\beta = \beta_N$ with $\alpha_N \uparrow \infty$ and $\beta_N \downarrow 1$ as $N \uparrow \infty$, the Bayes estimator d_N for the prior $\tau_N = \mathscr{G}(\alpha_N, \beta_N)$ has Bayes risks $r(\tau_N, d_N)$ satisfying (S.5). Now for the prior $\mathscr{G}(\alpha, \beta)$, $\beta > 1$, the Bayes risk is

$$r(\tau, d_0) = E\left[\left(\frac{\alpha(\sum_1^n X_j + \beta - 1)}{n\alpha + 1} - \mathcal{O}\right)^2 / \mathcal{O}\right] = E\left[\mathcal{O} - 2d_0(\mathbf{X}) + \frac{d_0^2(\mathbf{X})}{\mathcal{O}}\right]$$

$$= \alpha\beta - 2E\left[\frac{\alpha(n\theta + \beta - 1)}{n\alpha + 1}\right] + E\left[\frac{\mathrm{var}_\mathcal{O}(d_0(\mathbf{X})) + (E_\mathcal{O} d_0(\mathbf{X}))^2}{\mathcal{O}}\right]$$

$$= \alpha\beta - 2\left[\frac{\alpha(n\alpha + 1\beta - 1)}{n\alpha + 1}\right]$$

$$+ E\left[\left(\frac{\alpha}{n\alpha + 1}\right)^2 \frac{n\mathcal{O}}{\mathcal{O}} + \left(\frac{\alpha}{n\alpha + 1}\right)^2 \frac{(n\mathcal{O} + \beta - 1)^2}{\mathcal{O}}\right]$$

$$= \alpha\beta - \frac{2\alpha[(n\alpha + 1)\beta - 1]}{n\alpha + 1}$$

$$+ n\left(\frac{\alpha}{n\alpha + 1}\right)^2 + \left(\frac{\alpha}{n\alpha + 1}\right)^2 \left[n^2 E(\mathcal{O}) + 2(\beta - 1) + \frac{(\beta - 1)^2}{\mathcal{O}}\right]$$

$$= \alpha\beta - 2\alpha\beta + \frac{2\alpha}{n\alpha + 1} + n\left(\frac{\alpha}{n\alpha + 1}\right)^2$$

$$+ \left(\frac{\alpha}{n\alpha + 1}\right)^2 \left[n^2 \alpha\beta + 2(\beta - 1) + \frac{\beta - 1}{\alpha}\right].$$

Note that, for $\beta > 1$, $E(1/\mathcal{O}) = \frac{1}{\Gamma(\beta)\alpha^\beta} \int_0^\infty \frac{1}{\theta} e^{-\theta/\alpha} \theta^{\beta-1} d\theta = \frac{\Gamma(\beta-1)\alpha^{\beta-1}}{\Gamma(\beta)\alpha^\beta} = \frac{1}{\alpha(\beta-1)}$.
Also use $(1 + \frac{1}{n\alpha})^{-2} = 1 - \frac{2}{n-\alpha} + \theta(\frac{1}{\alpha^2})$ as $\alpha \to \infty$ to get

$$r(\tau, d_0) = \alpha\beta - 2\alpha\beta + \frac{2\alpha}{n\alpha+1} + n\left(\frac{1}{n+\frac{1}{\alpha}}\right)^2 + \left(1 + \frac{1}{n\alpha}\right)^2 \left[\alpha\beta + \frac{2(\beta-1)}{n^2} + \frac{\beta-1}{n^2\alpha}\right]$$

$$= \alpha\beta - 2\alpha\beta + \frac{2}{n}(1 + o(1)) + n\left(\frac{1}{n^2} + o(1)\right)$$

$$+ \alpha\beta - \frac{2\beta}{n} + \frac{2(\beta - 1)}{n^2} + o(1) \ (\text{as } \alpha \to \infty)$$

$$= \frac{2}{n} + \frac{1}{n} - \frac{2\beta}{n} + \frac{2(\beta - 1)}{n^2} + o(1) \text{ as } \alpha \to \infty.$$

$$\longrightarrow \frac{1}{n} \text{ as } \alpha \to \infty, \beta \downarrow 1.$$

□

Chapter 4

4.3

(a) Find the UMVU estimator of θ^i in Example 4.5 ($k = 1, 2, \ldots, n$).

Solution. The complete sufficient statistic for θ is $T = \sum_{1 \leq j \leq n} X_j$. A simple unbiased estimator of θ^k is $d(\mathbf{X}) = X_1 X_2, \ldots, X_k$, which takes the value 1 with probability θ^k (when $= X_1 = 1$, $X_2 = 1, \ldots, X_k = 1$), and 0 with probability $1 - \theta^k$. The UMVU estimate is then

$$d^*(\mathbf{X}) = E(X_1, X_2, \ldots, X_k \mid T) = P(X_1, X_2, \ldots, X_k = 1 \mid T).$$

The last probability is zero if $T < k$, and, for $T = r \geq k$, it equals

$$\frac{P(X_1, X_2, \ldots, X_k = 1, T = r)}{P(T=r)} = \theta^k\,{}^{n-k}C_{r-k}\theta^{r-k}(1-\theta)^{n-k-(r-k)} \Big/ \left[{}^nC_r\theta^r(1-\theta)^{n-r}\right]$$

$$= \frac{{}^{n-k}C_{r-k}}{{}^nC_r}.$$

Hence, $d^*(\mathbf{X}) = 0$ for $T = \sum_{1 \leq j \leq n} X_j = 0, 1, \ldots, k-1$, and it equals $r(r-1)\ldots(r-k+1)/[n(n-1)\ldots(n-k+1)]$, for $T = r$ where $r = k, k+1, \ldots, n$. This can be succinctly expressed as

$$d^*(\mathbf{X}) = \frac{T(T-1)\ldots(T-k+1)}{[n(n-1)\ldots(n-k+1)]}.$$

(b) (i) For $X_1, \ldots, X_n$ i.i.d. uniform on $(0, \theta)$, it is shown in Example 4.6 that $M = \max\{X_1, \ldots, X_n\}$ is a complete sufficient statistic. Hence to find the UMVU estimator $g(M)$ of $\sin\theta$, we may solve for g satisfying the following. (Note that the pdf of M is $(n/\theta^n)t^{n-1}.1_{\{0<t<\theta\}}.$)

$$\sin\theta = E_\theta g(M) = \frac{n}{\theta^n}\int_{(0,\theta)} g(t)t^{n-1}dt, \quad \text{or,} \quad \left(\frac{\theta^n}{n}\right)\sin\theta = \int_{(0,\theta)} g(t)t^{n-1}dt.$$

Differentiation with respect to θ leads to the equation $\theta^{n-1}\sin\theta + (\theta^n/n)\cos\theta = g(\theta)\theta^{n-1}$. Hence $g(\theta) = \sin\theta + (\theta/n)\cos\theta$. That is, the UMVU estimator of $\sin\theta$ is $\sin M + (M/n)\cos M$.

(ii) In the same manner the UMVU estimator $g(M)$ of e^θ satisfies the equation $\theta^n e^\theta/n = \int_{(0,\theta)} g(t)t^{n-1}dt$, and differentiation leads to $\theta^{n-1}e^\theta + \theta^n e^\theta/n = g(\theta)\theta^{n-1}$, i.e., $g(\theta) = e^\theta + \theta e^\theta/n$. Therefore, the UMVU estimator of e^θ is $e^M(1 + M/n)$.

4.7 Let $\mathbf{X} = (X)_{ij} : 1 \leq i \leq m, 1 \leq j \leq n) = (\mathbf{X}_j, 1 \leq j \leq n)$ be $n \geq 2$ i.i.d. random vectors $\mathbf{X}_j = (X_{1j}, X_{2j}, \ldots, X_{mj})$, $1 \leq j \leq n$, from the m-dimensional Normal distribution $N(\boldsymbol{\mu}, \Sigma)$ where $\boldsymbol{\mu} = (\mu_1, \ldots, \mu_m)$ is the mean vector, and Σ is the $m \times m$ symmetric positive definite covariance matrix of $\mathbf{X}_j$ $(j = 1, \ldots, n)$. From Example 4.12 it follows that $T \equiv [(\overline{X}_{i.} = \frac{1}{n}\sum_{j=1}^n X_{ij}, 1 \leq i \leq m), (m_{ii} = \frac{1}{n}\sum_{j=1}^n X_{ij}^2, 1 \leq i \leq m), (m_{ii'} = \frac{1}{n}\sum_{j=1}^n X_{ij}X_{i'j}, 1 \leq i < i' \leq m)]$ is a complete sufficient statistic for $(\boldsymbol{\mu}, \Sigma)$.

(a) Since $E_\theta \overline{X}_{i.} = \mu_i$, $1 \leq i \leq m$, the UMVU estimator of μ_i is $\overline{X}_{i.}$ (a function of T). Hence the UMVU estimator of $\boldsymbol{\mu}$ is $\mathbf{m} = (\overline{X}_{1.}, \ldots, \overline{X}_{m.})$.

(b) One has $E_\theta(\sum_{j=1}^n X_{ij}^2 - n\overline{X}_{i.}^2) = (n-1)\sigma_{ii.}$ For X_{ij}, $1 \leq j \leq n)$, are i.i.d. Normal $N(\mu_i, \sigma_{ii})$, so that it follows from Example 1.1 in Chap. 1 that the *sample variance* $\mathscr{S}_{ii} = \sum_{j=1}^n(X_{ij} - \overline{X}_{i.})^2/(n-1)$ is an unbiased estimator of the population variance σ_{ii}. Hence $\frac{\sum_{j=1}^n(\sum_{j=1}^n X_{ij}^2 - n\overline{X}_{ij}^2)}{n-1} = \mathscr{S}_{ii}$, a function $g_i(T)$, say, is the UMVU estimator of $\sigma_{ii.}$ Similarly, for $i < i'$, writing $\mu_{ii'} = EX_{ij}X_{i'j}$,

$$E_\theta\left[\sum_{j=1}^{n} X_{ij}X_{i'j} - n\overline{X}_{i.}\overline{X}_{i'.}\right] = E_\theta\left[n\mu_{ii'} - \sum_{j=1}^{n} X_{ij}\sum_{j'=1}^{\lambda} X_{i'j'}/n\right]$$

$$= n\mu_{ii'} - \frac{1}{n}E_\theta\sum_{j=1}^{n} X_{ij}X_{i'j} - \sum_{\substack{j=1 \\ j'\neq j}}^{n} (E_\theta X_{ij})(E_\theta X_{i'j'})/n$$

$$= n\mu_{ii'} - \mu_{ii'} - (n-1)\mu_i\mu_{i'}$$

$$= (n-1)(\mu_{ii'} - \mu_i\mu_{i'}) = (n-1)E(X_{ij} - \mu_i)(X_{i'j} - \mu_{i'})$$

$$= (n-1)\sigma_{ii'}.$$

[Note: $E(X_{ij} - \mu_i)(X_{i'j} - \mu_{i'}) = EX_{ij}X_{i'j} - \mu_i\mu_{i'} - \mu_i\mu_{i'} - \mu_i\mu_{i'} + \mu_i\mu_{i'} = \mu_{ii'} - \mu_i\mu_{i'}$.] Hence the *sample covariance*

$$s_{ii'} \equiv \frac{1}{n-1}\sum_{j=1}^{n}(X_{ij} - \overline{X}_{i.})(X_{i'j} - \overline{X}_{i'.})$$

$$= \frac{1}{n-1}\left[\sum_{j=1}^{n} X_{ij}X_{ij'} - n\overline{X}_{i.}\overline{X}_{i'.}\right]$$

$$= \frac{n}{n-1}\left[m_{ii'} - \overline{X}_{i.}\overline{X}_{i'.}\right] = g(T), \text{ say,}$$

is the UMVU estimator of the population covariance $\sigma_{ii'}$. It follows that

$$S = ((r_{ii'}))_{1\leq i,i'\leq m} \text{ is the UMVU estimator of } \Sigma.$$

[Here $\text{var}(S) \equiv \sum_{1\leq i,i'\leq m} E(s_{ii'} - \sigma_{ii'})^2$.]

4.8 Let $X_1,\ldots,X_n$ be i.i.d. $N(\mu,\sigma^2)$, $\theta = (\mu,\sigma^2) \in \mathbb{R}\times(0,\infty)$. Find the UMVU estimator of μ/σ.

Solution. Since $\overline{X}$ and s^2 are independent, one has

$$E_\theta\frac{\overline{X}}{s} = (E_\theta\overline{X})(E_\theta(1/s)) = \mu/E_\theta(1/s).$$

Now one may write

$$\frac{1}{s} = \left[\frac{(n-1)/\sigma^2}{(n-1)s^2/\sigma^2}\right]^{\frac{1}{2}} = \frac{(n-1)^{\frac{1}{2}}}{\sigma}\cdot\frac{1}{U^{1/2}},$$

where U has the chisquare distribution with $n-1$ degrees of freedom, i.e., a gamma a distribution $\mathcal{G}(2,\frac{n-1}{2})$. Hence

$$E_\theta\frac{1}{s} = \frac{(n-1)^{\frac{1}{2}}}{\sigma}E_\theta\frac{1}{U^{\frac{1}{2}}} = \frac{(n-1)^{\frac{1}{2}}}{\sigma}\int_0^\infty u^{-\frac{1}{2}}\frac{1}{\Gamma(\frac{n-1}{2})2^{(n-1)/2}}e^{-u/2}u^{\frac{n-1}{2}-1}du$$

$$= \frac{(n-1)^{\frac{1}{2}}}{\sigma\Gamma(\frac{n-1}{2})2^{\frac{n-1}{2}}}\int_0^\infty e^{-u/2}u^{\frac{n-2}{2}-1}du$$

$$= \frac{(n-1)^{\frac{1}{2}} \Gamma(\frac{n-2}{2}) 2^{\frac{n-2}{2}}}{\sigma \Gamma(\frac{n-1}{2}) 2^{\frac{(n-1)}{2}}}$$

$$= \frac{(n-1)^{\frac{1}{2}} \Gamma(\frac{n-2}{2})}{\sigma \sqrt{2} \Gamma(\frac{n-1}{2})} = \frac{d_n}{\sigma}, \quad d_n = \frac{(n-1)^{\frac{1}{2}} \Gamma(\frac{n-2}{2})}{\sqrt{2} \Gamma(\frac{n-1}{2})}.$$

Hence $E_\theta \frac{1}{d_n s} = \frac{1}{\sigma}$, and $\frac{1}{d_n}(\overline{X}/s)$ is the UMVU estimator of μ/σ. *This requires $n > 2$.*]

4.11 Let $X_1, \ldots, X_n$ be i.i.d. beta random variables $Be(\theta, 1)$, with common p.d.f.

$$f(x \mid \theta) = \theta x^{\theta - 1}, \ 0 < x < 1, \ \theta \in (0, \infty).$$

Find the UMVU estimators of (a) θ, (b) $\frac{1}{\theta}$.

Solution. (a) $f(x \mid \theta) = \theta \cdot \frac{1}{x} \cdot e^{\theta \ln x}$, belongs to the 1-parameter exponential family with natural parameter $\theta \in (0, \infty)$. The complete sufficient statistic for θ is $T = \sum_{j=1}^n \ln X_j$.

(b) We need to find a function $g(T)$ such that $E_\theta g(T) = \theta$. Now consider the random variable $Y_j = -\ln X_j$, then the p.d.f. of Y_j is (Note: $x = e^{-y}$.)

$$f_Y(y \mid \theta) = \theta e^y e^{-\theta y} e^{-y} = \theta e^{-\theta y}, \quad y \in (0, \infty).$$

That is, Y_j is gamma $\Gamma(\frac{1}{\theta})$, and (with

$$EY_j = \frac{1}{\theta}.$$

Hence $\sum_{j=1}^n Y_j / n \equiv -\sum_{j=1}^n \ln X_j / n$ is the UMVU estimator of $\frac{1}{\theta}$. $\qquad \square$

Chapter 5

5.6 Let X_j, $1 \le j \le n$, be i.i.d. with density

$$f(x \mid \theta) = \theta x^{\theta - 1}, \quad 0 < x < 1 \ (\theta \in \Theta = (0, \infty)).$$

Find the UMPU test of size $\alpha \in (0, 1)$ for $H_0 : \theta = 1$ against $H_1 : \theta \neq 1$.

Solution. The joint density is

$$f_n(\mathbf{x} \mid \theta) = \theta^n \prod_{j=1}^n x_j^{\theta - 1} = \theta^n \left(\prod_{j=1}^n x_j \right)^{-1} e^{\theta \sum_{j=1}^n \ln x_j}$$

which is a one-parameter exponential family with natural parameter $\pi = \theta$ and a complete sufficient statistic $T = -\sum_{j=1}^n \ln x_j$. From HW Set 4 (Problem #4), we know that T has the gamma distribution $\mathscr{G}(\frac{1}{\theta}, n)$. By Example 5.4, the UMPU test for $N_0 : \theta = 1$ against $H_1 : \theta \neq 1$, (or, $H_0 : \pi = -1$, $H_1 : \pi \neq -1$) is given by

$$\varphi^*(\mathbf{x}) = \begin{cases} 1 \text{ if } T < t_1 \text{ or } T > t_2 \\ 0 \text{ otherwise,} \end{cases}$$

where $t_1 < t_2$ are determined by

$$\int_{t_1}^{t_2} \frac{1}{\Gamma(n)} e^{-t} t^{n-1} dt = 1 - \alpha, \qquad \left(\frac{t_2}{t_1}\right)^n = e^{t_2 - t_1}.$$

5.9 Let $X_1, \ldots, X_m$ and $Y_1, \ldots, Y_n$ be independent random samples from exponential distributions with means θ_1, θ_2, respectively [i.e., from $\mathscr{G}(\theta_1, 1)$ and $\mathscr{G}(\theta_2, 1)$]. Find the UMPU test of size α for $H_0 : \theta_1 \leq \theta_2$ against $H_1 : \theta_1 > \theta_2$.

Solutions. The (joint) pdf of the observation vector $(X_1, \ldots, X_m, Y_1, \ldots, Y_n)$ is

$$f_n(\mathbf{x}, \mathbf{y} \mid \theta_1, \theta_2) = \frac{1}{\theta_1^m \theta_2^n} e^{-\sum_1^m x_i/\theta_1 - \sum_1^1 y_j/\theta_2}$$

$$= \frac{1}{\theta_1^m \theta_2^n} e^{-\left(\frac{1}{\theta_1} - \frac{1}{\theta_2}\right) \sum_1^m x_i - \frac{1}{\theta_2} \left(\sum_1^m x_i + \sum_1^n y_j\right)}$$

which is a 2-parameter exponential family with natural parameter $\pi_1 = \frac{1}{\theta_2} - \frac{1}{\theta_1}$, $\pi_2 = -\frac{1}{\theta_2}$, and complete sufficient statistic $T = (T_1, T_2)$, where $T_1 = \sum_{i=1}^m X_i$, $T_2 = \sum_1^m x_i + \sum_1^n y_j$. The null hypothesis is $H_0 : \pi_1 \leq 0$ and the alternative is $H_1 : \pi_1 > 0$. Here the natural parameter space is $\Pi = (-\infty, \infty) \times (-\infty, 0)$, and the boundary is $\Pi_B \equiv \{(0, \pi_2), \pi_2 \in (-\infty, 0)\} = \{0\} \times (-\infty, 0)$. The UMPU test of size α is given by

$$\varphi^*(t) = \begin{cases} 1 \text{ if } T_1 > t_1(t_2) \\ 0 \text{ if } T_1 \leq t_1(t_2), \end{cases}$$

where t_2 is determined by

$$P_{(0, \pi_2)}(T_1 > t_1(t_2) \mid T_2 = t_2) = \alpha \ \forall \ \pi_2 \in (-\infty, 0),$$

i.e.,

$$P_{(0, \pi_2)}\left(\frac{T_1}{T_2} > \eta(t_2) \mid T_2 = t_2\right) = \alpha \ \forall \ \pi_2 \in (-\infty, 0),$$

where $\eta(t_2) = t_1(t_2)/t_2$. Now the distribution of $U \equiv \frac{T_1}{T_2}$ under Π_B, i.e., under $\pi_1 = 0$ (or, $\theta_1 = \theta_2$) is that of a ratio $U = \frac{V_1}{V_1 + V_2}$ where $V_1 = \sum_1^m X_i/\theta_1$ is $\mathscr{G}(1, m)$, $V_2 = \sum_1^n Y_j/\theta_1$ is $\mathscr{G}(1, n)$, with V_1 and V_2 independent. This *distribution of U does not depend on π_2* (i.e., it is the same for all $\boldsymbol{\pi} = (0, \pi_2) \in \Pi_B$). Hence, by Basu's Theorem, T_1/T_2 is independent of T_2 under Π_B. Therefore, the UMPU test is to *reject H_0 iff*

$$\frac{T_1}{T_2} > \eta,$$

where $T_1/T_2 = V_1/(V_1 + V_2)$ *has the beta distribution function* Beta (m, n). (See the next Exercise.) Hence η is the $(1 - \alpha)^{th}$ quantile of this Beta (m, n) distribution.

5.10 Let U_1 and U_2 be independent gamma random variables $\mathscr{G}(\theta, m)$ and $\mathscr{G}(\theta, n)$. Prove that $Z_1 \equiv U_1/(U_1 + U_2)$ and $Z_2 \equiv U_1 + U_2$ are independent random variables with Z_1 having the beta distribution Beta(m, n) and Z_2 have the gamma distribution $\mathscr{G}(\theta, m + n)$.

Solution. The joint density of (U_1, U_2) is

$$f_{(U_1,U_2)}(u_1, u_2) = \frac{1}{\Gamma(m)\Gamma(n)\theta^{m+n}} \, e^{-(u_1+u_2)/\theta} u_1^{m-1} u_2^{n-1} \ (0 < u_1 < \infty,\ 0 < u_2 < \infty).$$

The Jacobian of the transformation $(u_1, u_2) \rightarrow (\frac{u_1}{u_1+u_2}) = z_1,\ u_1 + u_2 = z_2$, is

$$J\left(\frac{z_1, z_2}{u_1, u_2}\right) = \begin{bmatrix} \frac{\partial z_1}{\partial u_1} & \frac{\partial t_1}{\partial u_2} \\[2mm] \frac{\partial z_2}{\partial u_1} & \frac{\partial z_2}{\partial u_2} \end{bmatrix} = \begin{bmatrix} \frac{u_2}{(u_1+u_2)^2} & \frac{-u}{(u_1+u_2)^2} \\[2mm] 1 & 1 \end{bmatrix},$$

whose determinant is $1/(u_1 + u_2) = 1/z_2$. Hence the joint density of Z_1 and Z_2 is given by [Note: $u_1 = z_1 z_2$, $u_2 = z_2(1 - z_1)$.]

$$g_{(Z_1,Z_2)}(z_1, z_2) = f_{(U_1,U_2)}(u_1, u_2) \cdot z_2 \,|_{z_1, z_2} = \frac{e^{-z_2/\theta}(z_1 z_2)^{m-1}}{\Gamma(m)\Gamma(n)\theta^{m+1}} [z_2(1 - z_1)]^{n-1} z_2$$

$$= \frac{1}{\Gamma(m)\Gamma(n)} z_1^{m-1}(1 - z_1)^{n-1} \cdot \frac{e^{-z_2/\theta}}{\theta^{m+n}} z_2^{m+n-1}$$

$$= \frac{\Gamma(m_n)}{\Gamma(m)\Gamma(n)} z_1^{m-1}(1 - z_1)^{n-1} \cdot \frac{1}{\Gamma(m+n)\theta^{m+n}} e^{-z_2/\theta} z_2^{m+n-1},$$

$$(0 < z_1 < 1,\ 0 < z_2 < \infty).$$

5.18 In this *two-way layout with one observation per cell*, the model is given by (5.147) and (5.148), but with $\eta_{ij} = 0 \ \forall \ i, j$, i.e.,

$$X_{ij} = \mu + \delta_i + \gamma_j + \varepsilon_{ij} \ (\varepsilon_{ij}\text{'s are i.i.d. } N(0, \sigma^2)), \ 1 \le i \le I, 1 \le j \le J, \quad \text{(S.6)}$$

where $\mu \in \mathbb{R}$, $\sum_{i=1}^{I} \delta_i = 0$, $\sum_{j=1}^{J} \gamma_j = 0$. Hence

$$\boldsymbol{EX} = \mu + \delta_i + \gamma_j \text{ lies in (and spans) an Euclidean space}$$
$$\text{of } \mathbf{dimension}\ 1 + I - 1 + J - 1 = I + J - 1 = k. \quad \text{(S.7)}$$

Note that in this model, if one writes $\theta_{ij} = EX_{ij}$ then $\mu = \overline{\theta}_{..} = \frac{1}{IJ}\sum_{i,j}\theta_{ij}$, $\delta_i = \overline{\theta}_{i.} - \overline{\theta}_{..}$, $\gamma_j = \overline{\theta}_{.j} - \overline{\theta}_{..}$, where a dot $(\cdot)$ indicates averaging over the corresponding (missing) index. To find the minimum of $\|\mathbf{X} - \boldsymbol{EX}\|^2$ in this model, it is algebraically convenient to write

$$X_{ij} - EX_{ij} = X_{ij} - \mu - \delta_i - \gamma_j = (\overline{X}_{..} - \mu) + (\overline{X}_{i.} - \overline{X}_{..} - \delta_i)$$
$$+ (\overline{X}_{.j} - \overline{X}_{..} - \gamma_j) + (X_{ij} - \overline{X}_{i.} - \overline{X}_{.j} + \overline{X}_{..}), \quad \text{(S.8)}$$

and check, as in Example 5.19, that the sum over i, j of the products of the $\binom{4}{2} = 6$ pairs of the four terms on the right are all zero. Thus

$$\|\mathbf{X} - \boldsymbol{EX}\|^2 = IJ(\overline{X}_{..} - \mu)^2 + J\sum_{i=1}^{I}(\overline{X}_{i.} - \overline{X}_{..} - \delta_i)^2 + I\sum_{j=1}^{J}(\overline{X}_{.j} - \overline{X}_{..} - \gamma_j)$$

$$+ \sum_{i=1}^{I}\sum_{j=1}^{J}(X_{ij} - \overline{X}_{i.} - \overline{X}_{.j} + \overline{X}_{..})^2, \quad \text{(S.9)}$$

so that the minimum of (S.9) over $E\mathbf{X}$ is attained by taking $\mu = \overline{X}_{..}$, $\delta_i = \overline{X}_{i.} - \overline{X}_{..}$, $\gamma_j = \overline{X}_{.j} - \overline{X}_{..}$, which are the least squares estimators of the corresponding parameters. One then has

$$\min_{E\mathbf{X}} \|\mathbf{X} - E\mathbf{X}\|^2 = \sum_{i=1}^{I}\sum_{j=1}^{J}(X_{ij} - \overline{X}_{i.} - \overline{X}_{.j} + \overline{X}_{..})^2, \tag{S.10}$$

and the $IJ = n$ coordinates $\hat{m}_{ij}$ of $\hat{\mathbf{m}}$ are given by

$$\hat{m}_{ij} = \overline{X}_{..} + (\overline{X}_{i.} - \overline{X}_{..}) + (\overline{X}_{.j} - \overline{X}_{..}) = \overline{X}_{i.} + \overline{X}_{.j} - \overline{X}_{..} \ (1 \le i \le I, 1 \le j \le J). \tag{S.11}$$

(a) To test H_0 : *All the IJ means $\theta_{ij} \equiv EX_{ij}$ are equal, that is, $\boldsymbol{\delta_i = 0} \ \forall \ \boldsymbol{i}$ and $\boldsymbol{\eta_j = 0} \ \forall \ \boldsymbol{j}$* one has [see (S.8)]

$$X_{ij} - EX_{ij} = X_{ij} - \mu = (\overline{X}_{..} - \mu) + (X_{ij} - \overline{X}_{..}). \tag{S.12}$$

The sum over i, j of the product of the two terms on the right vanishes, so that the minimum of

$$\|\mathbf{X} - E\mathbf{X}\|^2 = IJ(\overline{X}_{..} - \mu)^2 + \sum_{i=1}^{I}\sum_{j=1}^{J}(X_{ij} - \overline{X}_{..})^2, \tag{S.13}$$

is attained by setting $\mu = \overline{X}_{..}$, and then

$$\min_{E\mathbf{X} \text{ under } H_0} \|\mathbf{X} - E\mathbf{X}\|^2 = \sum_{i=1}^{I}\sum_{j=1}^{J}(X_{ij} - \overline{X}_{..})^2. \tag{S.14}$$

The least squares estimator of the mean vector $\mathbf{m} = E\mathbf{X}$ is given by

$$\hat{m}_{ij} = \hat{\mu} = \overline{X}_{..} \quad \forall \ i, j. \tag{S.15}$$

Hence, by (S.11) and (S.15),

$$\|\hat{\mathbf{m}} - \hat{\mathbf{m}}\|^2 = \sum_{i,j}\left\{(\overline{X}_{i.} - \overline{X}_{..}) + (\overline{X}_{.j} - \overline{X}_{..})\right\}^2 = J\sum_{i=1}^{I}(\overline{X}_{i.} - \overline{X}_{..})^2 + I\sum_{j=1}^{J}(\overline{X}_{.j} - \overline{X}_{..})^2.$$

From (S.10), (S.15), the UMP unbiased and invariant test is to

$$reject \ H_0 \ iff \ \frac{\{J\sum_{i=1}^{I}(\overline{X}_{i.} - \overline{X}_{..})^2 + I\sum_{j=1}^{J}(\overline{X}_{.j} - \overline{X}_{..})^2\}/(I + J - 2)}{\sum_{i=1}^{I}\sum_{j=1}^{J}(X_{ij} - \overline{X}_{i.} - \overline{X}_{.j} + \overline{X}_{..})^2/(I - 1)(J - 1)}$$
$$> F_{1-\alpha}(I + J - 1, (I - 1)(J - 1)), \tag{S.16}$$

for $k = 1 + I - 1 + J - 1 = I + J - 1$, and $n = IJ$, so that $n - k = (I - 1)(J - 1)$, and the number of linearly independent constraints under H_0 is $I - 1 + J - 1 = I + J - 2 = 4$.

(b) To test $\boldsymbol{H_0 : \delta_i = 0} \ \forall \boldsymbol{i}$, [i.e., the $\overline{\theta}_{i.} = \cdots = \overline{\theta}_{I.}$ under the model (S.6)] note that, under H_0, $EX_{ij} = \mu + \gamma_j$, so that [see (S.8)]

$$X_{ij} - EX_{ij} = (\overline{X}_{..} - \mu) + (\overline{X}_{i.1} - \overline{X}_{..}) + (\overline{X}_{.j} - \overline{X}_{..} - \gamma_j) + (X_{ij} - \overline{X}_{i.} - \overline{X}_{.j} + \overline{X}_{..}),$$
(S.17)

and

$$\|\mathbf{X} - E\mathbf{X}\|^2 = IJ(\overline{X}_{..} - \mu)^2 + J\sum_{i=1}^{I}(\overline{X}_{i.} - \overline{X}_{..})^2 + I\sum_{j=1}^{J}(\overline{X}_{.j} - \overline{X}_{..} - \gamma_j)^2$$
$$+ \sum_{i=1}^{I}\sum_{j=1}^{J}(X_{ij} - \overline{X}_{i.} - \overline{X}_{.j} + \overline{X}_{..})^2,$$
(S.18)

which is minimized by taking $\mu = \overline{X}_{..}$, $\gamma_j = \overline{X}_{.j} - \overline{X}_{..}$, to yield

$$\min_{E\mathbf{X} \text{ under } H_0} \|\mathbf{X} - E\mathbf{X}\|^2 - J\sum_{i=1}^{I}(\overline{X}_{i.} - \overline{X}_{..})^2 + \sum_{i=1}^{I}\sum_{j=1}^{J}(X_{ij} - \overline{X}_{.i} - \overline{X}_{.j} + \overline{X}_{..}).$$
(S.19)

Hence the numerator sum of squares of the F-statistic is the difference between (S.19) and (S.18), namely,

$$J\sum_{i=1}^{I}(\overline{X}_{i.} - \overline{X}_{..})^2.$$
(S.20)

One may obtain (S.20) also as $\|\hat{\hat{m}} - \hat{m}\|^2$, where $\hat{m}_{ij}$'s are given by (S.11), and $\hat{\hat{m}}_{ij} = \overline{X}_{..} + \overline{X}_{.j} - \overline{X}_{..} = \overline{X}_{.j}$, $(1 \le i \le I, 1 \le j \le J)$. Hence the UMPU invariant test for $H_0 : \delta_i = 0 \; \forall i$ is

$$\text{Reject } H_0 \text{ iff } \frac{J\sum_{i=1}^{I}(\overline{X}_{i.} - \overline{X}_{..})^2/(I-1)}{\sum_{i=1}^{I}\sum_{j=1}^{J}(X_{ij} - \overline{X}_{i.} - \overline{X}_{.j} + \overline{X}_{..})^2/(I-1)(J-1)}$$
$$> F_{1-\alpha}(I-1, (I-1)(J-1)).$$
(S.21)

For there are $r = I - 1$ linearly independent constraints under H_0.

(c) The UMPU invariant test for $\mathbf{H_0 : \gamma_j = 0 \; \forall j}$ is entirely analogous to the case (b):

$$\text{Reject } H_0 \text{ iff } \frac{I\sum_{j=1}^{J}(\overline{X}_{.j} - \overline{X}_{..})^2/(J-1)}{\sum_{i=1}^{I}\sum_{j=1}^{J}(X_{ij} - \overline{X}_{i.} - \overline{X}_{.j} + \overline{X}_{..})^2/(I-1)(J-1)}$$
$$> F_{1-\alpha}(J-1, (I-1)(J-1)).$$
(S.22)

Index

© Springer-Verlag New York 2016

385

R. Bhattacharya et al., *A Course in Mathematical Statistics and Large Sample Theory*, Springer Texts in Statistics, DOI 10.1007/978-1-4939-4032-5

CLT, 128
   for U-statistics, 215
   for the MLE, 168
composite alternative hypothesis, 69
composite null hypothesis, 69
conditional density, 23, 26
conditional distribution, 189, 355
conditional p.d.f., 189
confidence interval, 153
confidence level, 89
confidence region, 89
   Bonferroni, 136
   regression lines, 142
   UMA, 91
confidence regions, 153
consistency, 117
   of a test, 204
   of sample mean, 119
   of sample moments, 120
consistent, 172, 216
contaminated normal, 152
Control group, 88
converge in distribution, 128
convergence
   almost sure, 119
convergence in law, 128
convex set, 176
Cornish-Fisher expansions, 293
correlation model, 143, 151
coverage error, 64, 257
Cramér-Rao bound, 165
Cramér-Rao Inequality, 55
Cramér-Rao Information Inequality, 56, 166
   multiparameter, 174
critical region, 68
cross-validation, 275
   generalized, 276
cumulant generating function, 281
curse of dimensionality, 275

D
decision rule, 11, 12
deductive inference, 3
delta method, 131
design matrix, 146, 261
differentiable manifold, 305
dimension reduction, 335
dispersion matrix, 352
distribution
   F, 350
   beta, 25
   Cauchy, 350
   empirical, 257
   logistic, 351
   multivariate normal, 352
   negative binomial, 61
   normal, 347
   shifted empirical, 261
   standard normal, 347

distribution function, 126
distributions
   absolutely continuous, 345
duality
   between confidence regions, 90

E
Edgeworth expansion, 290
Edgeworth expansions, 281
efficient
   asymptotically, 168
error
   coverage, 257
   Type I, 14
   Type II, 14
Errors in variables models, 154
estimable, 45, 98
estimate
   pooled, 86
estimation, 6
estimator
   asymptotically efficient, 176
   best linear unbiased, 98
   consistent, 118
   least squares, 99
   Maximum Likelihood, 19
   unbiased, 45
   uniformly minimum variance unbiased, 45
expectation, 118
explanatory variable, 121
exponential families, 47
exponential family
   one-parameter, 50, 51
   two-parameter, 50, 51
Exponential family: one-parameter, 47

F
false discovery rate, 318
Fieller's method, 160
Fisher information, 58
Fisher Linkage Model, 59
Fisher-Yates test, 97, 209, 214
Fréchet function, 304
Fréchet mean, 304
frequency chi-square test, 220
function
   estimable, 98
   maximal invariant, 100
   moment generating, 357
functional model, 155

G
gamma
   density, 21
   distribution, 346
   function, 345
Gauss-Markov Theorem, 98